93.50
70I

*Developments in Geotechnical Engineering 15*

SEISMIC RISK AND ENGINEERING DECISIONS

Further titles in this series:

1. G. SANGLERAT
THE PENETROMETER AND SOIL EXPLORATION

2. Q. ZARUBA AND V. MENCL
LANDSLIDES AND THEIR CONTROL

3. E.E. WAHLSTROM
TUNNELING IN ROCK

4. R. SILVESTER
COASTAL ENGINEERING, I and II

5. R.N. YOUNG AND B.P. WARKENTIN
SOIL PROPERTIES AND BEHAVIOUR

6. E.E. WAHLSTROM
DAMS, DAM FOUNDATIONS, AND RESERVOIR SITES

7. W.F. CHEN
LIMIT ANALYSIS AND SOIL PLASTICITY

8. L.N. PERSEN
ROCK DYNAMICS AND GEOPHYSICAL EXPLORATION

9. M.D. GIDIGASU
LATERITE SOIL ENGINEERING

10. Q. ZARUBA AND V. MENCL
ENGINEERING GEOLOGY

11. H.K. GUPTA AND B.K. RASTOGI
DAMS AND EARTHQUAKES

12. F.H. CHEN
FOUNDATIONS ON EXPANSIVE SOILS

13. L. HOBST AND J. ZAJIC
ANCHORING IN ROCK FORMATIONS

14. B. VOIGHT (Editor)
ROCKSLIDES AND AVALANCHES, I and II

*Developments in Geotechnical Engineering 15*

# SEISMIC RISK AND ENGINEERING DECISIONS

by

C. LOMNITZ and E. ROSENBLUETH (Editors)

*Instituto de Geofísica, Universidad Nacional Autónoma de México*
*Instituto de Ingeniería, Universidad Nacional Autónoma de México*

ELSEVIER SCIENTIFIC PUBLISHING COMPANY

Amsterdam — Oxford — New York 1976

ELSEVIER SCIENCE PUBLISHERS B.V.
Molenwerf 1
P.O. Box 211, 1000 AE Amsterdam, The Netherlands

*Distributors for the United States and Canada:*

ELSEVIER SCIENCE PUBLISHING COMPANY INC.
52, Vanderbilt Avenue
New York, N.Y. 10017

First edition 1976
Second impression 1983

**Library of Congress Cataloging in Publication Data**
Main entry under title:

Seismic risk and engineering decisions.

   (Developments in geotechnical engineering ; 15)
   Includes bibliographies and index.
   1. Earthquakes and building.
I. Rosenblueth, Emilio, 1926-    II. Lomnitz, Cinna.
III. Series.
TA654.6.S45    624'.176    76-41743
ISBN 0-444-41494-0

ISBN: 0-444-41494-0

© Elsevier Science Publishers B.V., 1976
All rights reserved. No part of this publication may be reproduced, stored in a retrieval system or transmitted in any form or by any means, electronic, mechanical, photocopying, recording or otherwise, without the prior written permission of the publisher, Elsevier Science Publishers B.V., P.O. Box 330, 1000 AH Amsterdam, The Netherlands

Printed in The Netherlands

# CONTENTS

CHAPTER 1  INTRODUCTION ..... 1
(E. Rosenblueth and C. Lomnitz)

CHAPTER 2  EARTHQUAKES AND EARTHQUAKE PREDICTION ..... 3
(C. Lomnitz and S.K. Singh)

2.1  The earthquake process ..... 3
2.2  Models of the earthquake process ..... 3
    2.2.1  An earthquake mechanism: Reid's theory ..... 4
    2.2.2  Plate tectonics ..... 7
    2.2.3  Stochastic models for large earthquakes ..... 11
2.3  The search for earthquake predictors ..... 14
    2.3.1  Earthquakes at plate boundaries: seismic gaps ..... 14
    2.3.2  Earthquake precursors: foreshocks ..... 16
    2.3.3  Monitoring crustal movements ..... 18
    2.3.4  Premonitory changes in seismic velocities ..... 19
    2.3.5  Other premonitory effects: the dilatancy model ..... 21
    2.3.6  Elastic and viscoelastic modeling of tectonic plates ..... 25
References ..... 27

CHAPTER 3  GEOLOGICAL CRITERIA FOR EVALUATING SEISMICITY ..... 31
(C.R. Allen)

3.1  Introduction ..... 31
3.2  California ..... 34
3.3  Turkey ..... 40
3.4  Japan ..... 49
3.5  Philippines ..... 56
3.6  China ..... 59
3.7  Thrust faults ..... 63
3.8  Conclusions ..... 64
Acknowledgments ..... 66
References ..... 67

CHAPTER 4  SOIL DYNAMICS: BEHAVIOR INCLUDING LIQUEFACTION ..... 71
(E. Faccioli and D. Reséndiz)

4.1  Introduction ..... 71
    4.1.1  Nature of soils: Phases and stresses ..... 71
    4.1.2  Drainage conditions in earthquake problems ..... 71
    4.1.3  Independent variables and test conditions in soil dynamics ..... 72
    4.1.4  Stable and unstable soil conditions ..... 73
4.2  Stress—strain relationship under stable conditions ..... 76
    4.2.1  Introduction ..... 76

|   |       | 4.2.2 | Shear modulus for small-amplitude vibration | 77 |
|---|-------|-------|---|---|
|   |       | 4.2.3 | Internal damping | 81 |
|   |       | 4.2.4 | Stress—strain relationships in large-amplitude cyclic deformation | 82 |
|   |       | 4.2.5 | Strength under cyclic loading | 83 |
|   | 4.3   | Local amplification | | 87 |
|   |       | 4.3.1 | Nature of the phenomenon. Problems of interpretation | 87 |
|   |       | 4.3.2 | Analytical models | 91 |
|   |       | 4.3.3 | Effects of weak interbedded layers | 102 |
|   |       | 4.3.4 | Modification of design spectra according to soil profile | 103 |
|   | 4.4   | Compaction and loss of strength | | 106 |
|   |       | 4.4.1 | Introductory remarks | 106 |
|   |       | 4.4.2 | Volume change under vibration | 107 |
|   |       | 4.4.3 | Loss of strength of loose saturated sands and soft cohesive soils | 112 |
|   |       | 4.4.4 | Cyclic mobility of cohesionless soils in laboratory tests | 115 |
|   |       | 4.4.5 | Effects of permeability, drainage path an boundary conditions | 118 |
|   |       | 4.4.6 | Evaluation of liquefaction and cyclic mobility potentials | 119 |
|   |       | 4.4.7 | Loss of strength and fatigue effects in cohesive soils | 123 |
|   | 4.5   | Soil exploration | | 124 |
|   |       | 4.5.1 | Considerations on field exploration programs | 124 |
|   |       | 4.5.2 | Exploration methods | 125 |
|   | References | | | 132 |

CHAPTER 5   THE PYSICS OF EARTHQUAKE STRONG MOTION   141
(J.N. Brune)

| 5.1 | Introduction | | 141 |
|---|---|---|---|
| 5.2 | Physical parameters of the earthquake source | | 141 |
|   | 5.2.1 | The eleastic rebound mechanism | 141 |
|   | 5.2.2 | Point source theory | 142 |
|   | 5.2.3 | Far-field approximation | 144 |
|   | 5.2.4 | Far-field radiation from a double-couple point source | 144 |
|   | 5.2.5 | Seismic moment and fault slip | 145 |
|   | 5.2.6 | Energy and stress | 147 |
|   | 5.2.7 | Frictional heat generation and seismic efficiency | 149 |
|   | 5.2.8 | Stress drop, fault displacement, and source dimension | 151 |
|   | 5.2.9 | Effective accelerating stress | 152 |
| 5.3 | Earthquake modeling | | 152 |
|   | 5.3.1 | Instantaneous stress pulse model | 152 |
|   | 5.3.2 | Crack tip stress singularity | 159 |
|   | 5.3.3 | Focussing of energy by rupture propagation | 160 |
|   | 5.3.4 | Approximate solutions to the dynamic problem of propagating ruptures | 161 |
|   | 5.3.5 | Foam-rubber model | 162 |
|   | 5.3.6 | Geometrical and boundary-condition effects | 168 |
|   | 5.3.7 | Complexity, scattering and attenuation effects | 171 |
| 5.4 | Conclusion — Estimates of maximum probable near-source ground motion | | 173 |
| Acknowledgements | | | 174 |
| References | | | 174 |

CHAPTER 6   SEISMICITY   179
(L. Esteva)

| 6.1 | On seismicity models | | 179 |
|---|---|---|---|
| 6.2 | Intensity attenuation | | 182 |
|   | 6.2.1 | Intensity attenuation on firm ground | 182 |

| | | | |
|---|---|---|---|
| 6.3 | Local seismicity | | 189 |
| | 6.3.1 | Magnitude-recurrence expressions | 190 |
| | 6.3.2 | Variation with depth | 195 |
| | 6.3.3 | Stochastic models of earthquake occurrence | 195 |
| | 6.3.4 | Influence of the seismicity models on seismic risk | 206 |
| 6.4 | Assessment of local seismicity | | 208 |
| | 6.4.1 | Bayesian estimation of seismicity | 208 |
| 6.5 | Regional seismicity | | 218 |
| | 6.5.1 | Intensity-recurrence curves | 218 |
| | 6.5.2 | Seismic probability maps | 220 |
| | 6.5.3 | Microzoning | 220 |
| References | | | 222 |

## CHAPTER 7  TSUNAMIS   225
(R.L. Wiegel)

| | | | |
|---|---|---|---|
| 7.1 | Introduction | | 225 |
| | 7.1.1 | Some data | 225 |
| | 7.1.2 | Relationships among earthquake magnitudes, aftershock areas, tectonic displacements and tsunami damage | 226 |
| | 7.1.3 | Landslide and subaqueous slide-generated tsunamis | 232 |
| 7.2 | Theory of the generation of tsunamis | | 235 |
| | 7.2.1 | Initial elevation or depression of the water surface | 235 |
| | 7.2.2 | Vertical displacement of bottom: linear theory | 244 |
| | 7.2.3 | Moving boundary: linear theory and hydraulic-model studies | 244 |
| | 7.2.4 | Large high-speed horizontal motion of vertical plane boundary | 248 |
| | 7.2.5 | Exact two-dimensional numerical solution for waves generated by a moving boundary | 249 |
| 7.3 | Tsunami sources and travel across the ocean | | 254 |
| | 7.3.1 | Sources | 254 |
| | 7.3.2 | Directional characteristics | 257 |
| | 7.3.3 | Travel across the ocean | 260 |
| 7.4 | Effects along the coast | | 263 |
| | 7.4.1 | Refraction | 264 |
| | 7.4.2 | Wave trapping | 264 |
| | 7.4.3 | Mach reflection | 264 |
| | 7.4.4 | Resonance | 270 |
| | 7.4.5 | Run-up and draw-down | 271 |
| 7.5 | Distribution functions | | 275 |
| | 7.5.1 | Entrance to San Francisco Bay, California, and other locations | 278 |
| | 7.5.2 | Risk | 279 |
| 7.6 | Combined tide and tsunami probabilities | | 280 |
| References | | | 283 |

## CHAPTER 8  STRUCTURAL RESPONSE TO EARTHQUAKES   287
(E.H. Vanmarcke)

| | | | |
|---|---|---|---|
| 8.1 | Introduction | | 287 |
| 8.2 | Common ground-motion representations | | 289 |
| | 8.2.1 | Response spectra | 289 |
| | 8.2.2 | Simulated earthquakes | 292 |
| | 8.2.3 | Spectral-density functions | 295 |

| | | |
|---|---|---|
| 8.3 | Random vibration-based prediction of response spectra | 298 |
| | 8.3.1 Stationary response variance | 299 |
| | 8.3.2 Transient response variance | 302 |
| | 8.3.3 Other pertinent response statistics | 305 |
| | 8.3.4 Prediction of maximum response | 308 |
| | 8.3.5 Compatibility of ground-motion representations | 315 |
| 8.4 | Multi-degree-of-freedom systems | 316 |
| | 8.4.1 Response-spectrum approach | 317 |
| | 8.4.2 Random vibration approach | 319 |
| | 8.4.3 Time-integration analysis | 322 |
| 8.5 | Light secondary systems | 323 |
| | 8.5.1 A response spectrum-based method | 324 |
| | 8.5.2 Random vibration approach | 325 |
| | 8.5.3 Time-integration method | 327 |
| 8.6 | Inelastic systems | 327 |
| | 8.6.1 Inelastic reponse spectra | 329 |
| | 8.6.2 A probabilistic model | 330 |
| | 8.6.3 Time-integration analysis | 334 |
| Acknowledgements | | 334 |
| References | | 334 |

## CHAPTER 9  DESIGN
(R.V. Whitman and C.A. Cornell)                                              339

| | | |
|---|---|---|
| 9.1 | Analysis of total risk | 340 |
| 9.2 | Analysis of total risk: two-state systems | 342 |
| 9.3 | Analysis of total risk: multiple damage states | 349 |
| 9.4 | Analysis of total risk: distributed targets | 354 |
| 9.5 | Design for seismic risk: optimization of an individual project | 358 |
| 9.6 | Desing for seismic risk: structural building codes | 369 |
| 9.7 | Design for seismic hazard: lifelines | 374 |
| | 9.7.1 Possible performance criteria | 374 |
| | 9.7.2 Modeling of lifeline systems | 376 |
| References | | 379 |

## CHAPTER 10  SEISMOLOGICAL INSTRUMENTATION
(T.V. McEvilly)                                                             381

| | | |
|---|---|---|
| 10.1 | Introduction | 381 |
| 10.2 | Applications | 381 |
| 10.3 | Requirements: General | 382 |
| 10.4 | Peak-reading instruments | 382 |
| | 10.4.1 Peak ground motion | 382 |
| | 10.4.2 Peak structural motion | 384 |
| | 10.4.3 Peak structural deformations | 385 |
| 10.5 | Conventional seismographic systems — Design considerations | 386 |
| | 10.5.1 Basic design parameters | 386 |
| | 10.5.2 Bandwidth | 387 |
| | 10.5.3 Sensitivity | 388 |
| | 10.5.4 Response curve | 389 |
| | 10.5.5 Setting specifications | 391 |
| | 10.5.6 Peripheral considerations | 397 |

| | | |
|---|---|---:|
| 10.6 | Conventional seismographic systems — Component elements | 398 |
| | 10.6.1 General constraints | 398 |
| | 10.6.2 The complete seismograph | 399 |
| | 10.6.3 Seismometers | 401 |
| | 10.6.4 Signal conditioning | 406 |
| | 10.6.5 Recording | 409 |
| | 10.6.6 Timing | 411 |
| | 10.6.7 Telemetry | 412 |
| 10.7 | Calibration | 413 |
| References | | 414 |
| INDEX | | 415 |

*Chapter 1*

# INTRODUCTION

EMILIO ROSENBLUETH and CINNA LOMNITZ

*Instituto de Ingeniería, Universidad Nacional Autónoma de México, Mexico*
*Instituto de Geofísica, Universidad Nacional Autónoma de México, Mexico*

Papers written in the last few years have enriched greatly the literature in earthquake engineering. The discipline is in rapid evolution and revision, more so than most other branches of civil engineering. Several factors incide in thus coloring design to resist earthquakes: the birth and consolidation of the theory of plate tectonics in the 1960s, causing an upheaval in ideas about the origin of earthquakes and opening the possibility of predicting these phenomena; a trend toward explicit optimization in all of engineering, which inevitably gives rise to revision of established approaches, especially in matters so plagued with huge uncertainties; expansion of human settlements and their accompanying structures onto very active seismic zones, raising questions of high economic repercussions in land use and in the effects of earthquake features so far regarded as not significant; need to design structures of ever increasing importance and whose failure would be catastrophic, especially nuclear power plants, with the ensuing public pressure for extreme safety; the advent of efficient methods of analyzing problems in continuum mechanics; and the commercial availability of instruments capable of recording strong motions with unprecedented precision and accuracy, which have made earthquake engineers reexamine their estimates of ground-motion parameters.

The avalanche of technical papers has crystallized in some excellent text and reference books covering the areas of geotectonics, seismology, and structural analysis and design. Not all questions have been answered about these subjects, of course, but the state of the art is covered to an extent such as to awaken some confidence in the reader. He feels that he has a reasonable idea of how and why earthquakes originate and evolve as they traverse various geologic formations and of how structures respond to given specific disturbances. He is still much amiss with respect to the disturbances for which he should analyze and design. This is the purpose of the present work: to help bridge the crevasse of decision making between earthquake characteristics and structural behavior.

When we say *should* analyze and design we imply an ethical question. This we shall dodge. Somehow, every engineer will presumably fix his scale of values and choose the path of rational behavior: he will aim at producing optimum designs within the framework of that scale of values. Such scales differ among individuals and are biased by the roles that individual engineers assume. This is forcefully brought out in the preliminary results of a survey, reported in Chapter 9. The survey does not tell what an engineer's utility scale should be, but it gives much food for thought, which may help the reader reach a consensus with himself.

The book contains four parts. Chapters 2—5 give the background on earthquake generation and characteristics. Chapter 6 is in a way the kern: it reviews the present state of matters in seismicity assessment and treats uncertainties explicitly. Chapters 7 and 8 tell of what earthquakes do to large bodies of water and structures. These three parts set the stage for rational decision making, which is covered in Chapter 9. Chapter 10 delves into instrumentation, with emphasis on choice of equipment to fit specific requirements.

Short-range earthquake prediction, fascinating and promising as it is, receives little more than a fleeting glimpse in the chapter on seismology. In the realm of engineering decisions short-range earthquake prediction is interesting mainly for its value in helping us better to understand seismic mechanisms; the decisions it immediately leads to are more in the area of social organization: evacuation of suspect auditoriums, schools, and temples; preparedness of rescue squads, emergency vehicles, and hospitals and clinics. Deterministic long-range prediction would have a decisive influence on design but long-range prediction is still quite uncertain to say the least, and very diffuse prediction in a statistical sense is tantamount to seismicity assessment. Emphasis is laid throughout on bases for achieving better design rather than for knowing when our structures might fail.

Efforts were made to give the book unity of treatment and of style. This was accomplished to some extent. Inevitably, though, differences in professional traditions and in personal preferences among authors of different chapters are very noticeable. Some chapters lean more to the qualitative and descriptive and others to the quantitative and mathematical; some to deterministic dynamics while others to stochastic processes. It is hoped that this diversity will lend the work renewed freshness and that the reader will profit from the multiplicity of vantage points.

Some chapters (2, 3, and 10) require little background beyond superficial familiarity with geology and dynamics. Others (4, 5, and 7) assume that the reader has a good preparation in dynamics. And yet others (6, 8, and 9) presuppose some knowledge of probability and statistics.

The authors will be satisfied if this book aids some experienced consulting engineers in assessing seismic risk and making rational decisions when locating and designing important engineering works and when drafting building codes and land use regulations while it supplies the advanced student of engineering with bases for benefiting from his future experience.

*Chapter 2*

EARTHQUAKES AND EARTHQUAKE PREDICTION

C. LOMNITZ and S.K. SINGH

*Instituto de Geofísica, Universidad Nacional Autónoma de México, Mexico*


2.1. THE EARTHQUAKE PROCESS

With the development of the hypothesis of sea-floor spreading, seismic activity began to be regarded as a global process. Before 1960 the dominant tendency in seismology had been to divide the earth's lithosphere into seismic regions, and to analyze the seismic activity in each region in terms of its assumed specific characteristics. The set of these characteristics, including the locations, magnitudes, and focal depths of historical earthquakes, was called the "seismicity" of the region. It was assumed that brevity of the historical record was the only real obstacle which prevented a full understanding of the seismicity of a given region to a level that would be adequate for all practical purposes.

In this chapter we discuss some recent developments which have implications for earthquake prediction. Historically earthquake prediction is perhaps the oldest problem of seismology; but no substantial progress was achieved in this field for over a century. Today, even though several successful predictions of individual earthquakes have been reported, these results are still not entirely consistent. Systematic prediction of earthquakes is now an objective that seems attainable within a few years, though the precise methods and their scopes of application are still in the stage of development and discussion.


2.2 MODELS OF THE EARTHQUAKE PROCESS

A physical process is a sequence of events governed by underlying regularities in time and space. In the case of earthquakes (as in other physical processes), we consider two kinds of representations: dynamic models, and stochastic models. These two categories of theoretical constructs are complementary: the more we know about the dynamics of a process, the less we need to know about its statistics.

Accurate earthquake observations by instrumental methods date back to the beginnings of the century: an exceedingly short period of record for statistical purposes, considering the time scale of geological events. The accurate detection and location of small earthquakes, such as may occur in a seismic region in an average year, has no appreciable statistical relevance since about 96% of the seismic energy in the global system is liberated by large earthquakes, of magnitude 6.5 and above. In other words, the structure of the process is governed almost entirely by the occurrence of events which are rare in terms of the human lifespan. In this case, the development of a workable model of earthquake prediction depends to a great extent on advances in research on the dynamics of the process, including the general mechanism of earthquake generation, and the physical changes which occur in the source region prior to an earthquake.

## 2.2.1. An earthquake mechanism: Reid's theory

An earthquake is an energy transient in the earth's lithosphere. It is detected as a localized burst of mechanical energy, which propagates outward from the focal area in the form of seismic waves. The focal area itself is generally inaccessible to direct observation, as it is located at depths of up to 700 km in the interior of the earth. Most of our information on the earthquake mechanism depends on the analysis of seismic waves recorded at stations on the earth's surface.

One of the earliest dynamical models of the earthquake mechanism is due to the collaboration between a geologist, Andrew C. Lawson, and a civil engineer, Harry F. Reid. Their model of the San Andreas Fault, developed in 1908, has the following basic features:

(1) An earthquake occurs when the lithosphere breaks along a discrete fracture surface, which may be pre-identified as a geological fault.

(2) The earthquake is preceded by a gradual buildup of elastic strains on both sides of the fault.

(3) At the time of the earthquake the two sides of the fault are mutually displaced in an amount that corresponds exactly to the full release of the elastic strains about the fault (Fig. 2.1).

The origin of the strain accumulation on the fault was not explicitly discussed in Reid's work; hence the charges of "mysticism" brought against him by some geophysicists, such as Matuzawa. However, since about 1960 there has been little doubt that the basic model of plate tectonics was already implicit in Reid's theory: "The only way in which the indicated strains could have been set up is by a relative displacement of the land on opposite sides of the fault and at some distance from it. We conclude that the strains were set up by a slow relative displacement of the land on opposite sides of the fault and practically parallel with it; and that these displacements extended to a considerable distance from the fault." (Reid, 1910)

The evidence used by Reid for his model was almost entirely derived from geodetic measurements. Fig. 2.2 summarizes the observations of geodetic displacements Reid used to infer the prior distribution of strains about the

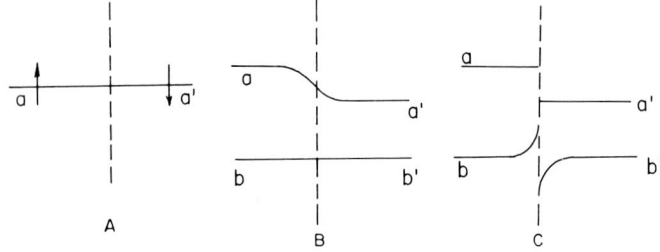

PLAN VIEW

Fig. 2.1. Reid's (1910) elastic rebound model for a strike-slip earthquake. Dotted line shows the fault trace. A. Unstrained condition, $a-a'$ normal to the fault trace. B. Strained condition, $a-a'$ deformed antisymmetrically, $b-b'$ a line normal to the fault trace. C. Relief of strain by faulting and elastic rebound, $b-b'$ shows the seismic movement.

San Andreas Fault. These observations were based on the differences in positions of bench-marks between the triangulations of 1874—1892 and 1906—1907. The bench-marks are not located on a single traverse; rather, they are scattered over a broad region both north and south of the latitude of San Francisco. Reid assumed that scatter of the observations could be fitted by a smooth curve, as in Fig. 2.3. The possibility of secondary faulting was not explored at that time. Today it would seem more natural to account for the observations by a combination of seismic fault displacement and fault creep, both on the San Andreas Fault and on its system of secondary faults throughout the region. Some fault slices have apparently stayed behind the general plate motion; e.g. the Mt. Tamalpais triangulation point appears to have moved relatively *backwards* (to the south) between the surveys of 1851—1866 and 1874—1892. Such observations indicate that the strain distribution in the region was more complex than Reid and Lawson had anticipated.

Brune (1974) has recently suggested that the main modifications of Reid's theory in recent times amount to: (1) a better understanding of the origins of the regional strains, i.e. the development of plate tectonics; (2) the discovery of fault creep; and (3) extension of the range of possibilities for the states of stress that could have been responsible for the earthquake.

One might add that, even making allowance for these extensions of knowledge, the mechanism suggested by Reid is still greatly oversimplified. The oversimplification is made clear by comparing Fig. 2.3, in which the relative displacements of an unspecified number of fault blocks and fault slices have been fitted to a smooth curve, with Fig. 2.2 which contains the geodetic measurements themselves. The displacements are referred to the baseline Mt. Mocho—Mt. Diablo. The inconsistencies (such as the southward motion of Monterey Bay, for example) are systematic and not random in geographical distribution, as Reid had assumed.

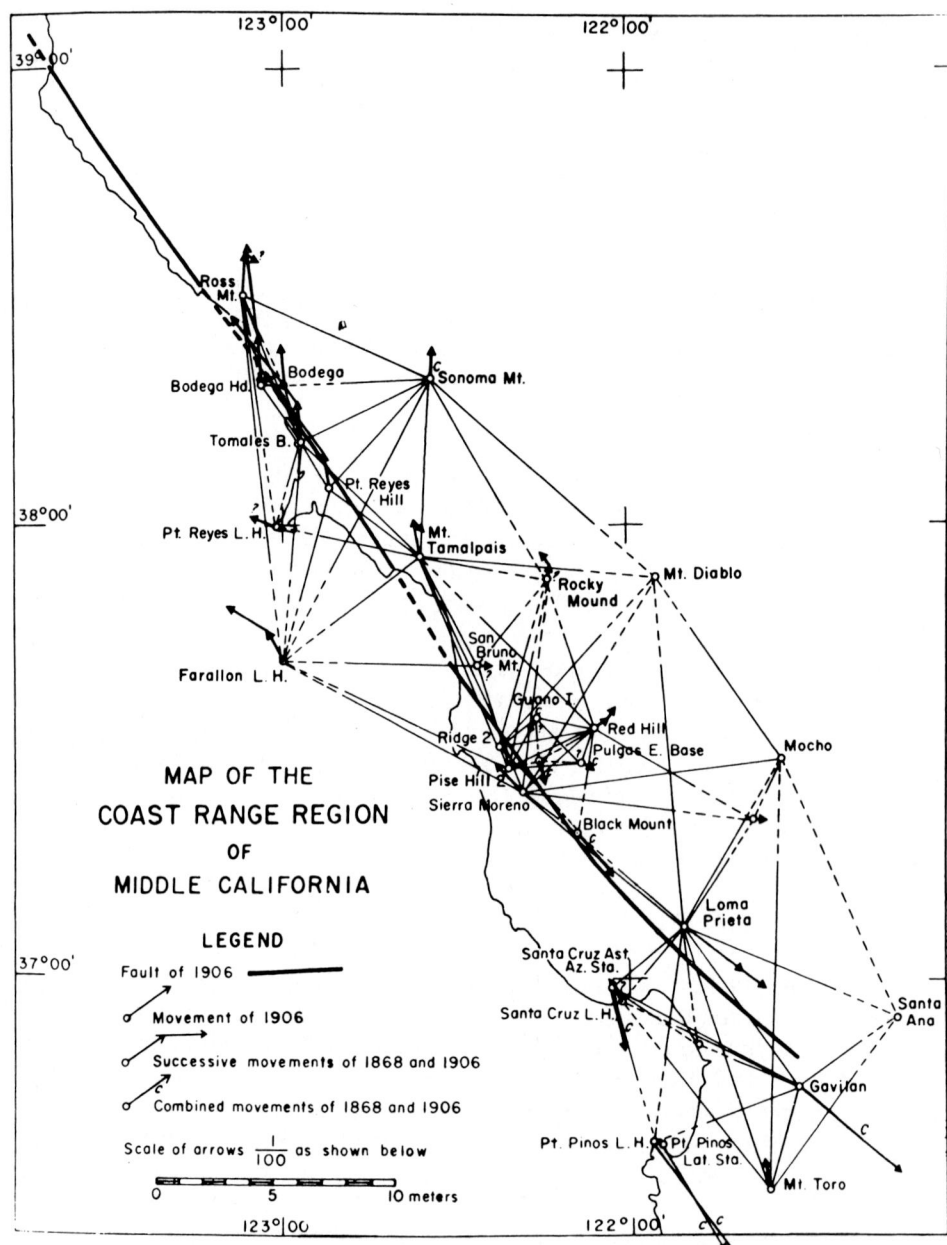

Fig. 2.2. Geodetic displacements referred to the Mocho—Mt. Diablo baseline (reproduced from the Report of the San Francisco Earthquake, 1910).

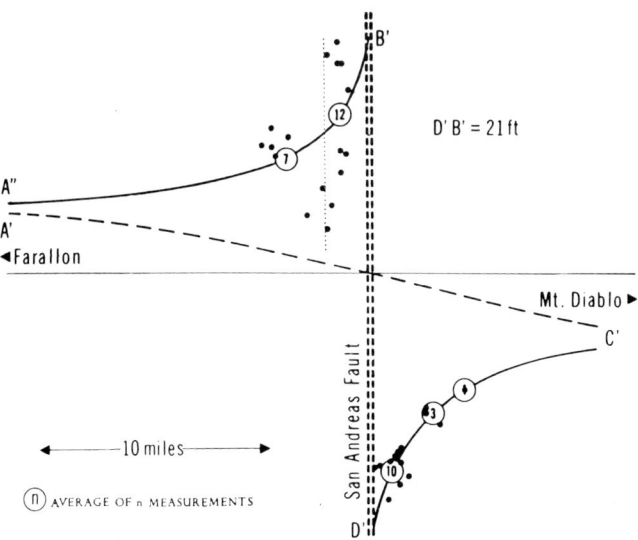

Fig. 2.3. Geodetic displacements from the 1906 earthquake projected onto a plane normal to the San Andreas fault. (Redrawn after Reid, 1910).

The simplified pattern of strain accumulation proposed by Reid may be allowable as long as it remains understood that a generalized, long-term deformation of a plate boundary is being described. In terms of local seismic risk a more detailed analysis is required, as witnessed by the seismic activity on secondary faults of the San Andreas System, and the geographical distribution of minor seismicity and fault creep in northern and central California (Bolt et al., 1968; Nason et al., 1974).

*2.2.2 Plate tectonics*

According to Le Pichon et al. (1973), "plate tectonics is a unifying working hypothesis which provides a kinematic model of the upper layer of the earth". It assumes that the outer layer of the earth, called the lithosphere, is much more rigid than the underlying asthenosphere. The lithosphere is considered to be divided into a small number of plates endowed with motion, both with reference to the earth's mantle, and relative to each other. As a first approximation intra-plate deformations can be neglected in comparison with differential motions at plate boundaries: this is where most earthquakes do occur.

Because of the rigidity of the plates, we may utilize the kinematics of rigid bodies on the surface of a sphere to describe the plate motions. There are three types of plate boundaries.

(1) *plate divergence* at mid-oceanic ridges, where new surface is created symmetrically from an axis of upwelling;

(2) *plate convergence* at oceanic trenches, where old surface is destroyed asymmetrically by subduction of one plate under another; and

(3) *plate transcursion* along transform faults. A transform fault is a section of plate boundary parallel to the vector of relative motion, i.e. where the surface of both plates is conserved.

Notwithstanding the disclaimers of some of its early proponents, plate tectonics cannot stand as a purely kinematic hypothesis. Plate motion must be driven by a global system of forces, and the creation and consumption of lithosphere at the ridge crests and ocean trenches must correspond to a return flow of material at depth. Hence plate tectonics necessarily implies a dynamic model of circulation in the outer shell of the earth. The lithosphere is the external rigid portion of mantle circulation; its thickness is usually taken to be 80—100 km but it is subject to rheological definition.

Whenever a plate is subducted under another plate it becomes seismically active. At shallow depths ($h < 70$ km) the earthquakes occur along the interface between the two plates and throughout the overriding lithospheric plate boundary, particularly if it is continental. At greater depths, sometimes down to 700 km, the earthquake foci occur in a zone of 20—50 km width which delineates the center of the subducted plate. The study of focal mechanism of these earthquakes shows that they are generated by compressional or extensional stresses which tend to be oriented along the dip of the plate (Isacks and Molnar, 1971). The most consistent explanation of these observations appears to be provided by gravitational sinking of the lithospheric plate, possibly as part of the convective circulation pattern of the earth's mantle.

The major plates of the world are six: Pacific, Americas, India, Africa, Eurasia, Antarctica. There are several smaller plates which have been accepted by many investigators: Cocos, Caribbean, Nazca, Arabian, Philippine, Somalian (Fig. 2.4). A number of other small plates have been proposed on the basis of tectonics, in various complex regions of the world: in Iran, in the Red Sea, in the Gulf of California, in the Bismarck Sea, in the Mediterranean Sea, and so on. Figure 2.4 shows the relative velocities as well as their boundaries. Thus, the global pattern of plate motions represents the visible portion of the general circulation in the earth's mantle. The peculiar zigzag pattern of orthogonal ridge segments and transform faults can be reproduced experimentally in hot wax (Oldenburg and Brune, 1972).

Only a few subduction zones reach down to depths of the order of 700 km: Tonga, Japan, Chile, Indonesia, the Philippines, New Hebrides, and the Solomon Islands. Other subduction zones may reach to shallower depths, presumably because they are younger, or because the plate boundary has been shifting in geological time. The oldest subduction zones that are now seismically active go back to Triassic times (about 200 million years ago). This was the time when the early supercontinent, Pangaea, began breaking up into the major continental masses known today (see Fig. 2.5). However,

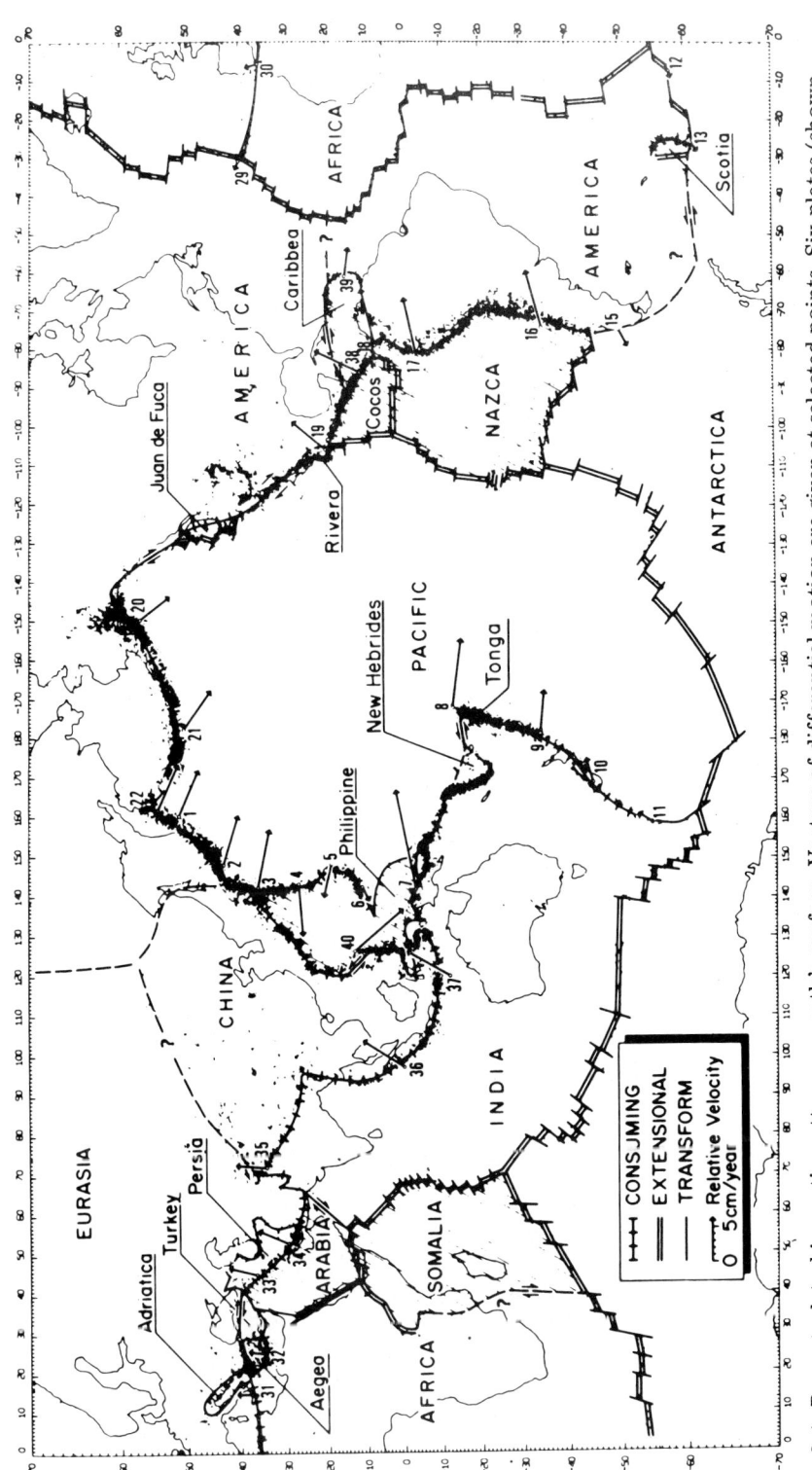

Fig. 2.4. Present plate kinematic pattern on earth's surface. Vectors of differential motion are given at selected points. Six plates (shown hachured) have been added to the original six large-plate model of Le Pichon. Seismicity from 1961 to 1967 is also shown. (After Le Pichon et al., 1973.)

*fossil* subduction zones of both earlier and later dates are known to exist under the continents.

About 75% of all earthquakes correspond to the subduction boundaries of the Pacific Plate and of its adjoining minor plates. The remaining seismic activity occurs along a complex sequence of plate boundaries which extends from the Himalayas into Central Asia and China, and westward across Afghanistan, Iran, Turkey, and the Mediterranean to the Azores. Less than 3% of the earth's seismic-energy release occurs on the mid-oceanic rises, and in the interiors of the plates. Practically 99% of all earthquakes occur along plate boundaries (Lomnitz, 1974).

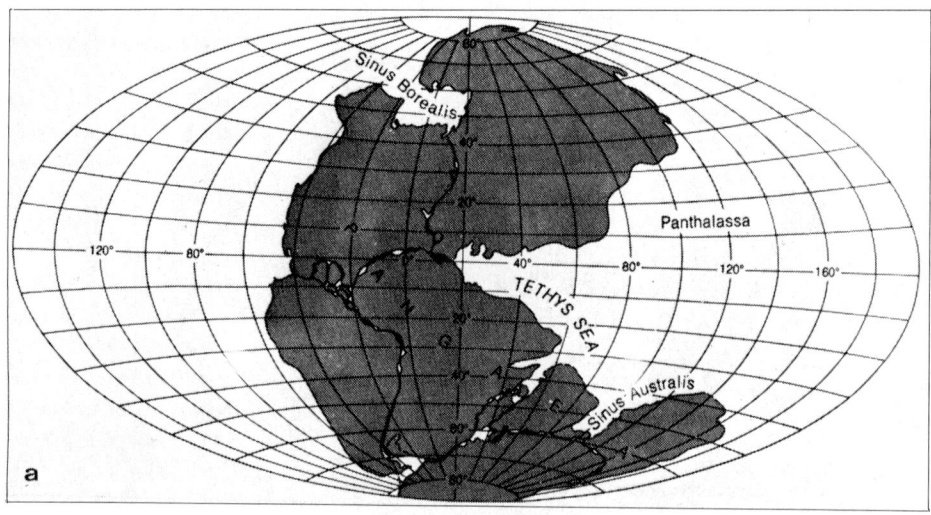

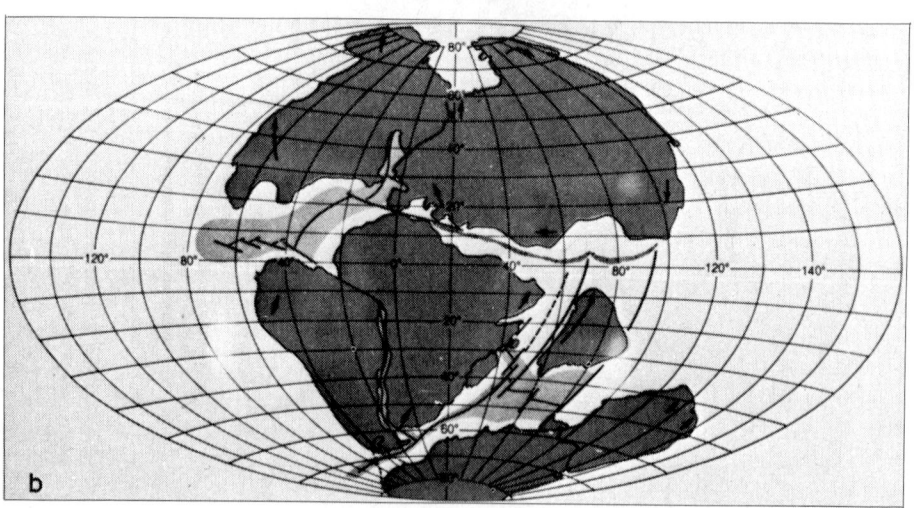

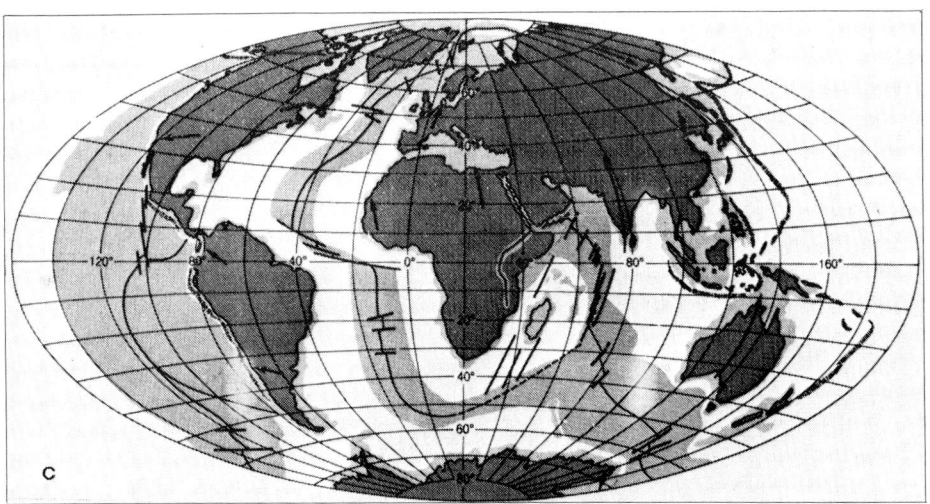

Fig. 2.5. Breaking up of the ancient landmass Pangaea. (a) 200 million years ago. (b) 135 million years ago after 65 million years of drift. Arrows depict motion of the continents and light shaded areas indicate new ocean floor generated in the preceding 40 million years. (c) Present world geography. Motion of the continents and new ocean floor generated in the past 65 million years are shown (After Dietz and Holden, 1970.)

### 2.2.3 Stochastic models for large earthquakes

Earthquake data are very inhomogeneous, because of the uneven geographical distribution of stations. Most earthquakes occur along the boundaries of the Pacific Ocean, which itself contains very few seismographic stations. Moreover, the worldwide density of stations has increased markedly from decade to decade. At the turn of the century there were only a score; at present there are thousands of stations. It seems likely that the three-letter code which is being used to designate seismographic stations will be exhausted within the next few years. Thus, the completeness of a sample of earthquake data is strongly dependent on the geographical area and on the particular time interval covered.

Nevertheless, certain broad empirical distributions of earthquake parameters have been derived. Figure 2.6 is a compilation of commonly used relationships between magnitude, epicentral distance, intensity, ground velocity, and ground acceleration for shallow earthquakes. The figure is based on mean values proposed by Rosenblueth and Esteva (1964). The Richter magnitude $M$ is a parameter that describes the total energy $E$ (in ergs) of seismic waves radiated by the focus. The following approximate relationship holds (Richter, 1958, p. 366):

$$\log_{10}E = 11.4 + 1.5M \tag{2.1}$$

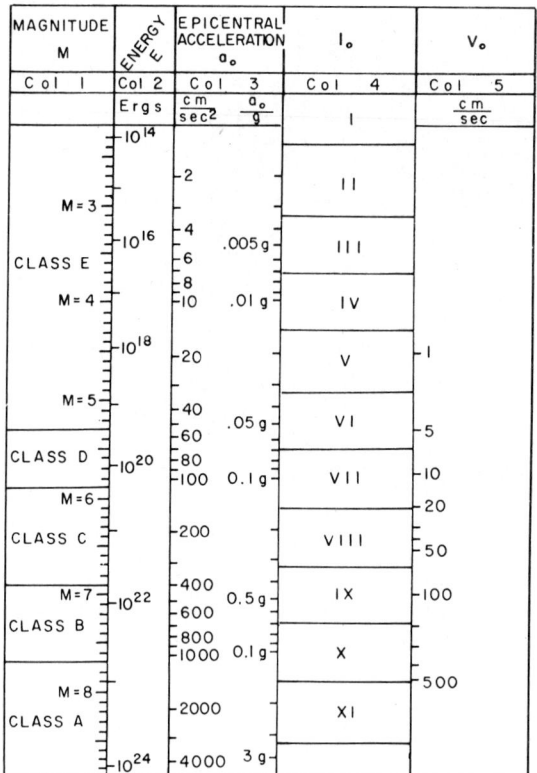

Fig. 2.6. A summary of rough relationships between magnitude—energy—epicentral acceleration, and between acceleration, intensity and ground velocity (order of magnitude approximations).

Another version uses a value of the constant of 11.8 instead of 11.4, which yields energies more than twice as high; however, this difference is within the range of probable error of many magnitude determinations.

The body-wave magnitude $m$ is based on the amplitudes of body waves (P and S waves) rather than on the surface-wave amplitudes, which depend on the focal depth of the earthquake. A commonly-used relationship between the two magnitudes is (Richter, 1958, p. 348):

$$M = 1.59m - 3.97 \tag{2.2}$$

Since the two magnitude scales are not directly comparable. eq. 2.1 should be used only for rough energy calculations.

Seismic effects of explosions may be compared to earthquakes if the yield, or TNT equivalence, of the explosion is known. The body-wave magnitude of an underground explosion is approximately:

$$m = 0.67 \log_{10} Y + K \pm 0.3 \tag{2.3}$$

where $Y$ is the yield in kilotons. The constant $K$ is 4.25 for close coupling in granite, and 3.25 for close coupling in dry alluvium.

The best available means of describing the ground effects of an earthquake in the absence of instruments is the *modified Mercalli* (MM) intensity scale. This scale describes a range of observations and bodily sensations characterizing twelve different levels of ground shaking. The latitude of personal interpretation allowed by the MM scale may lead to discrepancies of the order of one Mercalli degree between equally competent investigators. Some utilize the highest-ranking single effect observed in a locality, while others make a mental average of effects observed in a variety of structures or features. In either case, the intensity estimate depends on the amount of time spent in the locality and on the type of previous experience of the investigator.

The structure of the MM scale is not linear. The intensity range from I to V is not particularly relevant in terms of earthquake risk. About 90% of all seismic damage corresponds to MM intensities VI, VII, and VIII; these intensities span the range of horizontal velocities from 5 to 50 cm/sec, which is of greatest interest to engineers. Accelerations of the order of gravity occur spottily in the immediate epicentral area of destructive earthquakes. The large intensity variations observed in sedimentary basins subjected to earthquakes may be due to variability in soil conditions, and to patterns of constructive and destructive interference related to the normal modes of vibration of the basin. Other intensity scales in current use include the Japanese scale of five degrees, and the MSK scale which is analogous to the MM scale.

The frequency distribution of earthquake magnitudes, especially in the middle range ($0 < M < 7$), is reasonably well approximated by the exponential distribution:

$$f(M) = \beta e^{-\beta M}, \qquad M \geqslant 0 \tag{2.4}$$

where $f(M)$ is the probability density function of $M$ in a given volume of the earth's crust. Parameter $\beta$, which equals the reciprocal of the mean of $M$, is a regional constant. For historical reasons the $b$-value, $b = \beta \log_{10} e$, is more generally used than $\beta$. Depending on the region, the focal depth, and the stress level, the $b$-value may fluctuate between 0.3 and 1.5.

A low $b$-value is normally associated with a high stress drop, and vice versa. Aftershocks tend to have high $b$-values, because of the fact that a significant proportion of the available tectonic stress is released in the main shock. Earthquakes in oceanic ridge areas tend to have higher $b$-values than earthquakes in subduction zones, presumably because of a smaller capacity for stress accumulation in ridge areas.

Large earthquakes are rare events in statistical terms. Since the distribution of independent rare events in time tends toward the Poisson distribution we may use a Poisson model as a first approximation for the occurrence of large earthquakes (Epstein and Lomnitz, 1966):

$$p(n) = \frac{\lambda^n}{n!} e^{-\lambda} \tag{2.5}$$

where $p(n)$ = probability that $n$ large earthquakes occur during a given unit, and $\lambda$ = mean number of large earthquakes per time unit. Parameter $\lambda$ depends on the region and on the magnitude or intensity level selected for the particular application. Thus, if an average of three destructive earthquakes occur per century in a given region, the Poisson probability for one destructive earthquake to occur in a decade is $p(1) = 0.3 \exp(-0.3) = 0.22$. Of course, the probability for *at least* one destructive earthquake to occur is larger, namely $p(1) + p(2) + \ldots = 1 - p(0) = 0.26$. Such estimates are useful because they provide approximate predictions of earthquake risk, which may be relevant to engineering decisions.

If the distribution of earthquake magnitudes is known, the distribution of *largest* magnitudes may be derived. For the exponential-Poisson model represented by eqs. 2.4 and 2.5 the probability density function $G(y)$ of the maximum-magnitude earthquake $y$ in $D$ years is obtained from:

$$\ln[-\ln G(y)] = \ln \alpha - \beta y \tag{2.6}$$

where $\alpha$ is the mean estimated number of earthquakes with $M \geqslant 0$ during a design period $D$, and $\beta$ is obtained from eq. 2.4. Then the "earthquake risk", i.e. the probability of occurrence of an earthquake of magnitude $y$ or greater in a $D$-year period, is (Lomnitz, 1974):

$$R_D(y) = 1 - \exp(-\alpha D \, e^{-\beta y}) \tag{2.7}$$

The timespan covered by the available data is a serious limitation, as the earthquake-data coverage is almost always short compared to the design period $D$ for important structures. Scarcity of statistical information must be compensated for by resorting to Bayesian statistics. In this approach we begin with a conceptual (stochastic) model of the process, which provides us with prior probabilities, and through application of Bayes' theorem, arrive at posterior, improved probabilities (Newmark and Rosenblueth, 1971, chapter 8). When the amount of data is extensive, the frequentist (classical statistical) and the Bayesian treatments give results in close agreement.

At present there are no tested methods of estimation for predicting the earthquake risk $R_D$ at a point locality, though several methods have been proposed (cf. Chapter 6). Other types of information used to infer earthquake hazard from large earthquakes include geological and tectonic studies (Chapter 3) and studies related to source physics (Chapter 5).

## 2.3 THE SEARCH FOR EARTHQUAKE PREDICTORS

### 2.3.1 Earthquakes at plate boundaries: seismic gaps

If the hypothesis of plate tectonics is correct, the probability of occurrence of earthquakes should be roughly uniformly distributed along a given plate boundary. However, continental margins can frequently be divided into major geotectonic units, which may have different seismic regimes. Continental margins behind subduction zones are built up of fragments shored up against the primitive continental core, or *shield*, by successive subduction episodes in geologic history. Plate boundaries have shifted or changed their locations, and the relative motion between plates has undergone variations in rate and direction.

An oceanic plate may accrete to a continent in two different ways: (a) by wedging of sediments scraped off the top of the oceanic plate, thus building up the continental shelf; (b) by intrusion of lithospheric materials rising from the subducted plate in the form of molten magma, which invades the continental crust from below. Where the continental margin is in a state of tension the lighter magmatic differentiates may rise in the gravity field and may be extruded through volcanic vents.

The range of processes we have outlined is studied by geologists under broad descriptive terms such as *orogeny*, *magmatism*, *volcanism*, *metamorphism*, *plutonism*, and so on. Gravitational sliding from the uplifted areas complicates the picture. Most of these tectonic processes are accompanied by shallow earthquakes, while the major shocks at intermediate and large depths tend to take place in the subducted oceanic plate. Since slippage along the margin of a plate boundary should be continuous, the quiescent gaps between active segments of the boundary represent potential danger areas. Fedotov (1965), Tobin and Sykes (1968), Mogi (1969), and Sykes (1971) have identified such gaps in the Japan—Kurile—Kamchatka arc, the northeast Pacific, eastern Japan, and the Aleutian arc. Since 1965 three major shocks have occurred in the gaps which had previously been identified by Fedotov.

The idea of seismic gaps was developed further by Kelleher (1970; 1972). It appears that plate boundaries tend to rupture repeatedly along discrete zones, which are activated all at once; the earliest identification of such regions of "sympathetic activity" is due to Tsuboi (1958). Gajardo and Lomnitz (1960) isolated four such regions along the Chilean subduction zone. Kelleher et al. (1973; 1974) have given a set of criteria which they believe may assist in forecasting large shallow earthquakes in seismic gaps. Essentially, these criteria are of two kinds: historical evidence of past major earthquakes, and quiescence for periods of the order of 30 years or more. There is also some preliminary evidence that earthquakes along rectilinear plate boundaries (Chile, Anatolia) may occur as an orderly progression in time and space. These criteria have been used in producing charts of seismic gaps (Fig. 2.7).

The idea of seismic gaps also appears to be valid for transcurrent plate boundaries. The two most recent major earthquakes on the San Andreas Fault have occurred in seismic gaps (Fig. 2.8). On the other hand, areas of

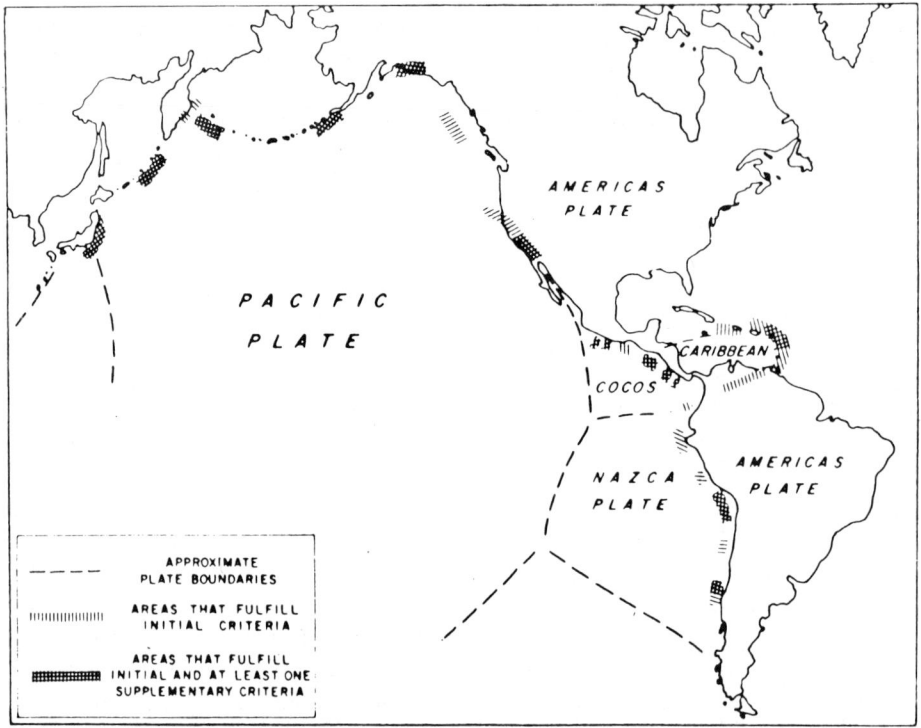

Fig. 2.7. Areas (shaded and double shaded) of special seismic potential for large shallow earthquakes along major plate boundaries of the Pacific and the Caribbean obtained by applying a set of criteria (see text). (From Kelleher et al., 1973.)

continuous creep and abundant low-level seismicity (such as Hollister area) have not produced any large shocks in historic time. One of the major objectives of seismicity studies may be the determination of the boundaries and special characteristics of such major units which make up a plate boundary.

The predictive value of seismic gaps is subject to a number of unresolved questions. Intervals of less than 30 years between large earthquakes in a given region are not uncommon; it would be dangerous to assume that the occurrence of a large shock guarantees a period of quiet of several decades in a given sector. Other seismic gaps have enjoyed periods of quiet of the order of centuries; hence, assignment of relative risks to different sectors of a plate boundary is not a simple function of the seismic activity of the gaps in the immediate past.

*2.3.2 Earthquake precursors: foreshocks*

The following parameters have been reported to present some potential as earthquake precursors: ground tilt, ground elevation, tectonic strain, seismic velocities, electromagnetic field strength, electric resistivity, and changes in

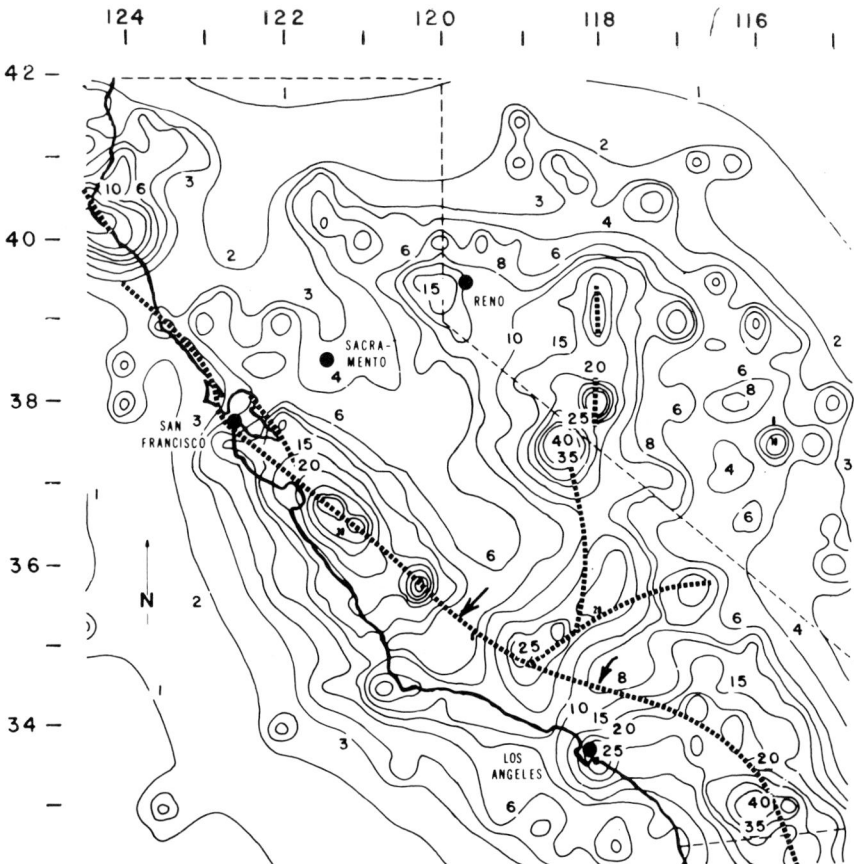

Fig. 2.8. 1932—1962 contours of estimated probability of incidence of an event of acceleration 0.1 g in 30 years in California (from Lomnitz, 1974). Notice that two minima, near Parkfield and near San Fernando, were locations for subsequent strong earthquakes (shown by arrows).

ground-water flow. Observations on alleged precursory changes in animal behavior and effects of sky luminescence have also been reported (Derr, 1973).

The single precursory phenomenon which is most reliably established is the occurrence of foreshocks. The vast majority of large earthquakes is preceded by smaller foreshocks in the same region; there is considerable evidence that the main shock itself may often be triggered by a smaller event.

Foreshocks as warnings of an impending major shock have been traditionally utilized by the population in seismic regions such as Chile and Japan; thousands of lives have been saved by foreshocks. As yet there are no means of determining whether a major earthquake will be triggered after a given event, or whether the activity will subside without further consequences.

A foreshock occurs at a time of high strain energy accumulation in the region. At such a time, any small shock causes stress changes which may accelerate or precipitate a major event. Because of the abnormally high state of stress in the region, the mean magnitude of foreshocks tends to be higher than average, i.e. the $b$-value is abnormally low. However, the number of foreshocks is usually too small for a reliable statistical estimate of the $b$-value to be made in each case.

Foreshocks can have their own aftershock sequences: the destructive South Chile earthquake of 21 May 1960 (magnitude 7.5) was regarded as a main shock, until a much larger earthquake occurred 33 hours later. This foreshock sequence included another six earthquakes of magnitude 7—7.5 during the eleven hours which immediately preceded the main shock.

*Earthquake swarms* are sequences that occur in regions of active tensional tectonics or in geological environments which have been associated with magmatic intrusion, volcanism, and hydrothermal activity. Swarms are usually quite shallow. The mean magnitude of the earthquakes tends to increase during the swarm; a large shock may occur at any time in the sequence. The destructive earthquake of 3 May 1965 in San Salvador occurred three months after the initiation of swarm activity. Major earthquakes associated with swarms are not known to exceed magnitude 6.5: apparently the strain accumulation in this type of shallow activity is not large enough for a major shock.

*2.3.3 Monitoring crustal movements*

The motion of large plates of lithosphere implies the presence of tangential stresses in the earth. Existence of gravity anomalies is indirect evidence of such deviations from hydrostatic equilibrium. Soviet workers (e.g., Artemjev et al., 1972) have explored the predictive value of gravity measurements. They have produced some correlations between the distribution of earthquakes and isostatic gravity anomaly as well as its horizontal gradient. Such correlations are potentially useful in seismic-risk estimation, particularly in areas where other data are lacking.

Reid's elastic rebound model (Reid, 1910) had postulated a slow elastic buildup over the years preceding a major earthquake. More recent data have indicated that the actual process of crustal deformation is more complex (Tsuboi, 1933; Mescherikov, 1968). Present writers tend to distinguish a four-stage process connected with the earthquake cycle (Scholz, 1972). These stages are: (1) secular deformation; (2) an accelerated pre-earthquake deformation; (3) earthquake; and (4) post-earthquake deformation.

Secular deformation caused by slow and steady accumulation of strain is well documented for Japan, certain areas in the USSR, and the San Andreas Fault in California. Near plate boundaries, this strain accumulation is caused by relative motion of the adjacent plates whose value is fairly well known from plate tectonics. This relative motion is partly accomodated by aseismic slip and partly by seismic slip. It is of some importance in seismic-risk estimation to know what part of the energy is being accumulated for an earthquake. Knowing the long-term steady-state relative motion between two adjacent plates and thus the long-term slip rate along a given fault zone, we can write the potential seismic slip at present by (Brune and Lomnitz, 1974):

$$\text{potential seismic slip at present} = (\text{plate motion} - \text{seismic slip} - \text{aseismic slip in the last } t \text{ years} \quad (2.8)$$

The seismic slip can be obtained from seismic moments of the earthquakes (Brune, 1968; Davies and Brune, 1971) occurring in the past $t$ years, whereas aseismic slip can only be determined by direct observation. Unfortunately it seems that the sample length $t$ of the seismicity data required for the determination of potential slip for many areas may be 1000 years or more, which is not available.

Rikitake (1974) has determined from geodetic measurements that the ultimate crustal strain near the epicenter at which an earthquake occurs has a mean value $\epsilon$ of $5.3 \cdot 10^{-5}$ with a standard deviation $\sigma$ of $3.3 \cdot 10^{-5}$. Assuming that the strain is completely relieved by a large earthquake and that it increases linearly with time, the probability of a large event can be calculated from observed strain rate and $\epsilon$- and $\sigma$-values given above.

Pre-earthquake movements, characterized by an acceleration in the rate of crustal deformation over the secular rate, have been reported before many earthquakes. The detection of such movements is important for eventual short-term prediction. The possible cause of these movements is discussed later in this chapter.

Movement at the time of an earthquake is usually caused by slip on a fault plane which can be inferred by fitting a dislocation model (static and dynamic) to the observed data.

Post-earthquake movements often decay logarithmically. Aseismic slip on the fault occurs in the same direction as the movement during the earthquake if the earthquake slip was overdamped, and in the opposite direction if it was underdamped (Fitch and Scholz, 1971; Scholz, 1972).

*2.3.4 Premonitory changes in seismic velocities*

Beginning in 1962, a group of Soviet seismologists have found changes in the ratio of compressional-wave velocity $v_p$ to shear-wave velocity $v_s$ before moderate local earthquakes in the Garm region of Tadjikistan (Kondratenko and Nersesov, 1962; Nersesov et al., 1969; Semenov, 1969). The earthquakes occurred sometime after the $v_p/v_s$ anomaly had returned to its normal value. The duration of the anomaly depended on the magnitude of the impending earthquake, but the amplitude of the anomaly itself appeared to be independent of the magnitude (Fig. 2.9). Similar observations have been reported for small-magnitude earthquakes in New York State (Aggarwal et al., 1973). Whitcomb et al. (1973) showed that the $v_p/v_s$ anomaly existed for nearly 3.5 years before the San Fernando, California earthquake (9 Feb. 1971) which had a magnitude of 6.4. Ohtake (1973) has found similar phenomena before shallow Japanese earthquakes.

Most of these results have been obtained using small earthquakes located near the source region of a later main event. The seismic stations were also close to the source region. Temporal changes in the velocity under a station have also been obtained by systematically observing the mean teleseismic P wave travel-time residual at a station for a large number of earthquakes. Using this technique a decrease in $v_p$ was found to have occurred at the Matsushiro seismic station before the Matsushiro swarm of 1965 in Japan (Wyss and Holcomb, 1973), in agreement with Ohtake's (1973) result. Wyss and Johnston (1974) have studied similar residuals at stations in New Zealand and have again found a decrease in $v_p$ before large earthquakes.

The results on the $v_p/v_s$ anomaly may be summarized as follows:

(1) The duration of the precursory anomaly increases with the magnitude of the earthquake.

(2) The amplitude of the anomaly is independent of the magnitude of the earthquake.

(3) The decrease in $v_p/v_s$ is of the order of 10% or more.

(4) The effect is due to a decrease in $v_p$; the value of $v_s$ remains nearly constant.

(5) The ratio $v_p/v_s$ returns to normal (and may even overshoot) just before the earthquake.

(6) The premonitory decrease in $v_p/v_s$ occurs probably more slowly than the return to the normal value.

(7) The $v_p/v_s$ anomaly is found more frequently in thrust earthquakes than in strike-slip earthquakes.

The first confirmation of Soviet $v_p/v_s$ observations was made by Aggarwal et al. (1973) in the area of Blue Mountain Lake, in upstate New York. They detected premonitory changes of up to 13% in $v_p/v_s$. Later, the travel times were closely monitored in the same region, and another similar decrease in $v_p/v_s$ was detected after 30 July 1973, leading to a prediction for an earthquake of magnitude 2.5—3.0 to occur within a few days following 1 August. The magnitude was estimated from the spatial extent of the $v_p/v_s$ anomaly and the time from the magnitude itself.

An earthquake of magnitude 2.6 actually occurred in the Blue Mountain Lake region on 3 August 1973, thus confirming the prediction (Aggarwal et al., 1975). The careful measurements of seismic velocities in this region showed that:

(1) Decrease in $v_p$ is significantly larger than the decrease in $v_p/v_s$.

(2) The velocity anomaly depends on azimuth.

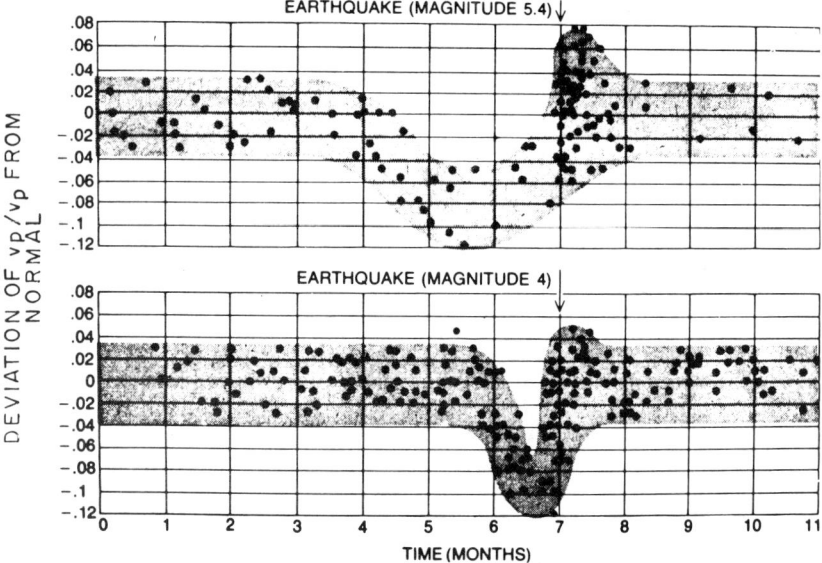

Fig. 2.9. Soviet data on premonitory change in the normal $v_p/v_s$ value before two earthquakes. Points represent small events in the region. Shaded band indicates statistical scatter of the observations. Duration of the $v_p/v_s$ anomaly appears to increase with the magnitude of the impending earthquake. (After Sadovsky et al., 1972.)

(3) The velocity anomaly depends on depth, since at shallow depths no anomaly is detected.

The report by Aggarwal et al. represents the first fully documented case of prediction of location, time, and magnitude of an earthquake. It is worth noting that the predicted earthquake involved shallow thrust faulting and occurred in a small region whose activity had been previously monitored so that correlations between velocity anomalies and earthquake occurrence could be made.

*2.3.5 Other premonitory effects: the dilatancy model*

Other premonitory phenomena, reported before earthquakes, include changes in electrical conductivity (Sadovsky et al., 1972; Barsukov, 1972; Mazella and Morrison, 1973), increases in the rate of flow of water in the rocks near the epicentral area (Ulomov and Mavashev, 1967; Tsuneishi and Nakamura, 1970), and anomalous rates of change of the geomagnetic field and the telluric currents. Some of these changes, coupled with the recognition that pore fluids may play a vital role in tectonic processes and that the diffusion of pore fluid may be a rate-controlling process in the aftershock sequence (Nur and Booker, 1972), have led to the postulation of a dilatancy—fluid diffusion model of the earthquake mechanism.

This model, first proposed by Nur (1972) to explain Soviet data on the $v_p/v_s$ anomaly, is based on laboratory observations of dilatancy in rocks under compression (Brace et al., 1966), and of velocity changes as a function of water saturation in rocks (Nur and Simmons, 1969). Nur (1972) suggested that the decrease in $v_p/v_s$ could be a result of dilatancy of rocks in the focal zone. The subsequent increase in $v_p/v_s$ could be caused by the flow of water into the dilated region.ND in rocks refers to an increase of volume relative to elastic changes (Brace et al., 1966).

The model has also been used and further elaborated to explain the observations by Aggarwal et al. (1973), Whitcomb et al. (1973), Scholz et al. (1973), Nur et al. (1973), Anderson and Whitcomb (1973), and Scholz and Kranz (1974). The dilatancy—fluid diffusion hypothesis may be summarized as follows.

Accumulation of strain in the focal region causes dilatancy of rocks. Dilatancy, which is produced by the formation and propagation of cracks, begins to occur at stresses close to half the strength of the rock (Brace et al., 1966). If pore fluid is present and if the dilatancy occurs at a rate faster than the flow in the dilatant region, the increase in porosity causes undersaturation. Expansion during dilatancy must be sufficiently large so that the effect on the pressure differential causing the back flow, and the permeability of the dilatant region will all be sufficiently large during the anomalous period. Data on crustal uplift before earthquakes put some constraints on the dilatant volumetric strain (Hanks, 1974; Singh, 1973), indicating that $\Delta V/V$ in the field was at least one order of magnitude smaller than the volumetric strains observed in the laboratory (Brace et al., 1966). This implies a rather small increase in pore space at depth. In order to explain the large observed drop in $v_p/v_s$, Hanks (1974) postulates that the incipient dilatant volume in the crust is saturated with water close to a pressure such that liquid and vapor phases coexist for the temperature at that depth. As the pore pressure drops below this critical value, due to dilatancy, vaporization causes the bulk modulus of the rock-fluid medium to drop suddenly, causing the observed decrease in $v_p$ whereas $v_s$ remains essentially unchanged.

Increase in porosity is assumed to stop eventually because of strengthening of the rock due to the increase of effective stress, a phenomenon known as dilatancy-hardening (Frank, 1965). The fluid flow into the dilatant region should continue, however, bringing the rock back to the saturated state and the $v_p/v_s$ anomaly back to the normal value. The time elapsed in returning to the saturated state would depend on the size of the dilatant volume, the permeability, and the pressure differential causing the back flow. As the pore pressure rises the rock weakens and the earthquake is triggered. The time elapsed between the end of the $v_p/v_s$ anomaly and the triggering of the earthquake is about 15—25% of the total duration $t$ of the anomaly, where $t$ is measured from the onset of the $v_p/v_s$ decrease. If the flow rate is large as compared to the dilatancy rate, no undersaturation occurs and no premonitory $v_p/v_s$ anomaly results. Note also that the magnitude of the $v_p/v_s$ anomaly is a function of saturation but is independent of the dilatant volume which affects the size of the earthquake. The magnitude $M$ of the earthquake appears instead to be related to the duration of the anomalous period $t$ (in days):

$$\log_{10} t = 0.67 M - 1.35 \tag{2.9}$$

The longest dimension $L$ (km) of the aftershock zone and $t$ are claimed to be related by $t = L^2/c$ (with $c = 5.8 \cdot 10^4$ cm$^2$/sec), thus implying a causal mechanism involving diffusion of cracks and fluid near the source region (Whitcomb et al., 1973; Scholz et al., 1973).

Dilatancy means deformation of the epicentral area and transient fluid flow into the dilatant region: these are the first-order effects implied by the model. Nur (1974), analyzing the Matsushira, Japan earthquake swarm as to gravity, deformation and water-flow data, finds a confirmation of the model; this has lately been questioned. The deformation data contain information regarding the size, shape, location, and the volumetric strain of the dilatant region. The importance of collecting such data and a method for their interpretation has been discussed by Singh and Sabina (1975).

The decrease in apparent electrical resistivity $\rho_a$ prior to earthquakes has similarly been explained by the creation of new conducting paths opened by the increase in pore space (Brace and Orange, 1968). However, an initial increase in $\rho_a$ could occur due to undersaturation of the crack volume. Anomalies in the geomagnetic field could be caused by the change in electrical resistivity or by the piezomagnetic effect resulting from an increase in effective stress (Stacey, 1964, Shamsi and Stacey, 1969; Nagata, 1972). Observed $b$-values in the pre-earthquake period have also been cited in support of the dilatancy—fluid diffusion model (Scholz et al., 1973). Figure 2.10 shows, schematically, the predicted changes in various physical parameters based on the dilatancy—fluid diffusion model, assuming that the rate of dilatancy is faster than the flow of fluid. If, instead, the flow of fluid is faster than the rate of dilatancy the curves in Fig. 2.10 would have to be altered (see Nur, 1974).

Scientists from the Institute of Physics of the Earth in Moscow have proposed another model, called the dilatancy—instability model, in which dilatancy is supposed to lead to an avalanche-like growth of cracks, resulting in strain instability near the main fault. This would cause the stress to decrease, cracks partially to close, and the rock to recover some of its original properties. In this model, water diffusion does not play any role and the predicted changes in physical parameters are somewhat different than those from the dilatancy—fluid diffusion model. Due to insufficiency of data it is difficult to choose between the two models.

The claim that the dilatancy model provides a common physical basis for nearly all premonitory phenomena observed prior to shallow earthquakes

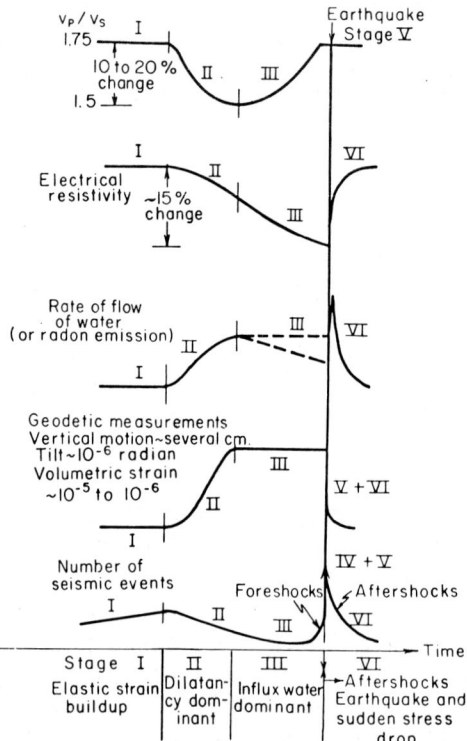

Fig. 2.10. Premonitory changes in physical parameters during an earthquake cycle as predicted by dilatancy—fluid diffusion model (rate of dilatancy faster than rate of fluid flow). Roman numerals indicate various stages of the cycle. Short-term fluctuations (stage IV), which are observed before some large earthquakes, are omitted. (After Scholz et al., 1973.)

(Scholz et al., 1973) is based on the qualitative nature of the model. The role of water is not well known. Thus, for example, an increase or a decrease in the electrical resistivity near the focal region is explained by the model simply by changing the relative role of water. Many important questions remain unanswered. How universal is the model? Although both thrust type and strike-slip type earthquakes have been reported to be preceded by $v_p/v_s$ anomaly, there are several cases of strike-slip type earthquakes where, apparently, no such anomaly occurred (e.g., McEvilly and Johnson, 1973; Cramer and Kovach, 1974). The observation on $v_p/v_s$ anomalies from controlled explosions are essential to decide whether or not the anomalies observed from seismic sources within epicentral regions may be related to a source effect. Such data, so far, are generally inconclusive.

The available instances of prediction have generally utilized a single premonitory effect; the combination of effects observed prior to the Matsushiro seismic swarm was probably caused by a magmatic intrusion, rather than by dilatancy. There are also some inconsistencies in the model, such as the validity of eq. 2.8 for all geological environments when the dilatancy—fluid diffusion model predicts a considerable variation in the duration $t$ of the $v_p/v_s$ anomaly depending on the geological setting. The question of the maximum depth at which dilatancy can occur is still unclear.

In the recent paper by Aggarwal et al. (1975) it is shown that prior to the predicted earthquake, at least, the $v_p/v_s$ anomaly was not due to the source parameters of individual small earthquakes but was caused by temporal changes in the rock properties in the focal region, and that $v_s$ decreased by 12% below normal. This is the first time that such a large decrease in $v_s$ has been reported and, if common, must be accounted for in the theories of physical basis for premonitory phenomena.

Non-hydrostatic stresses would also cause anisotropy of S waves. In an anisotropic medium the S wave splits into two distinct components which travel with different velocities. Gupta (1973) suggests that such S-wave splitting provides a premonitory change and a potential method for predicting earthquakes.

In conclusion, while the dilatancy model holds promise as a potential aid in shallow-earthquake prediction, further laboratory and field data are required in order to assess the merits of the model and to determine the general range of its application.

*2.3.6 Elastic and viscoelastic modeling of tectonic plates*

If most earthquakes are rupture phenomena caused by the interaction of a few rigid plates, it seems at least feasible to predict major events by monitoring the motion of the plates and the stress concentrations along their mutual boundaries. Each earthquake tells us something about the state of stress in its focal region. Seismic slip along the plate boundary can be estimated from the seismic moments (Davies and Brune, 1971). These quantities may be used in computer models to determine the probable changes in the stress pattern of the lithosphere.

Artyushkov (1973) has shown on general grounds that shallow-seated lateral inhomogeneities in the thickness and density may cause important stress concentrations in the earth's lithosphere. These stress concentrations may explain the seismicity tied to certain features of the relief (foothills, marginal areas of plateaus, and uplifts), which had also been noticed and classified by Montessus de Ballore (1906). The seismic regime of intra-plate regions and similar areas of low seismicity may therefore be of considerable interest in defining the dependence between earthquake risk and crustal inhomogeneities. While most models of the lithosphere used in computation neglect lateral thickness inhomogeneities altogether, it would be relatively simple to introduce more realistic models of the lithosphere in regions where

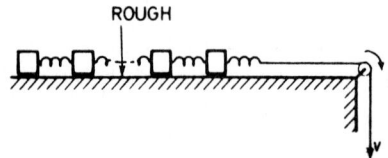

Fig. 2.11. Masses and springs arranged in a linear array over a horizontal surface as a laboratory model for earthquake series. The leading spring is connected by a long thread to a motor which moves at a constant velocity and causes impulsive motion of the masses. By introducing viscosity into a corresponding numerical model, aftershocks may be generated following a major shock. The results show that the statistical properties of the events are governed by the friction on the fault surface. (From Burridge and Knopoff, 1967.)

the crustal and subcrustal structure is relatively well known.

Tectonic stress changes may also be monitored by observing changes in $b$-values (Lomnitz, 1973; Wyss, 1973). Decreases in the $b$-value are assumed to be correlated with increases in the mean magnitude of earthquake, and hence with increases in tectonic stress. Such a correlation is borne out by the experimental work of Mogi (1962a,b; 1963) and Scholz (1968a,b,c). The decreases in $b$-values were mentioned in connection with foreshocks; however, Gibowicz (1973) has also observed them in aftershock sequences, where they precede large aftershocks.

Burridge and Knopoff (1967) have proposed a simplified mechanical model which may be used in simulating a plate boundary (Fig. 2.11). Each block represents a seismic "province", "region", or "gap" (depending on the terminology). The variations in seismicity can be simulated by varying the weights and frictional resistances of the blocks. Such features might eventually be incorporated into geometrical models of subduction zones, or other types of plate boundaries.

In general, the field of modelling of tectonic processes by means of mechanical, statistical, or computational models is still in the qualitative stage. Comparatively little effort has been invested in these methods so far.

An extensive bibliographical research on earthquake prediction, with particular attention to Soviet and Eastern European sources, is being carried out at the Freiberg School of Mining (Schmidt, 1971; 1973).

REFERENCES

Aggarwal, Y.P., Sykes, L.R., Armbruster, J. and Sbar, M.L., 1973. Premonitory changes in seismic velocities and prediction of earthquakes. *Nature*, 241: 101—104.
Aggarwal, Y.P., Sykes, L.R., Simpson, D.W. and Richards, P.G., 1975. Spatial and temporal variations in $t_s/t_p$ and in P wave residuals at Blue Mountain Lake, New York: application to earthquake prediction. *J. Geophys. Res.*, 80: 718—732.

Anderson, D.L. and Whitcomb, J.H., 1973. The dilatancy—diffusion model of earthquake prediction. In: R.L. Kovach and A. Nur (Editors), *Proceedings of the Conference on Tectonic Problems of the San Andreas Fault Zone*. Stanford University Press, Palo Alto, Calif., pp. 417—426.

Artemjev, M.E., Bune, V.I., Dubrovsky, V.A. and Kambarov, N. Sh., 1972. Seismicity and isostasy. *Phys. Earth Planet. Inter.*, 6: 256—262.

Artyushkov, E.V., 1973. Stresses in the lithosphere caused by crustal thickness inhomogeneities. *J. Geophys. Res.*, 78: 7675—7708.

Barsukov, O.M., 1972. Variations of electrical resistivity of mountain rocks with tectonic changes. In: E.F. Savarensky and T. Rikitake (Editors), *Forerunners of Strong Earthquakes. Tectonophysics*, 14: 273—277.

Bolt, B.A., Lomnitz, C. and McEvilly, T.V., 1968. Seismological evidence on the tectonics of central and northern California and the Mendocino escarpment. *Bull. Seismol. Soc. Am.*, 58: 1725—1767.

Brace, W.F. and Orange, A.S., 1968. Electrical resistivity changes in saturated rocks during fracture and frictional sliding. *J. Geophys. Res.*, 73: 1433—1445.

Brace, W.F., Paulding Jr., B.W. and Scholz, C.H., 1966. Dilatancy in the fracture of crystalline rocks. *J. Geophys. Res.*, 71: 3939—3953.

Brune, J.N., 1968. Seismic moment, seismicity and rate of slip along major fault zones. *J. Geophys. Res.*, 73: 777—784.

Brune, J.N., 1974. Current status of understanding quasi-permanent fields associated with earthquakes. *Trans. Am. Geophys. Union*, 55: 820—827.

Brune, J.N. and Lomnitz, C., 1974. Recent seismological development relating to earthquake hazard. *Geofís. Int. México*, 13: 49—63.

Burridge, R. and Knopoff, L., 1967. Model and theoretical seismicity. *Bull. Seismol. Soc. Am.*, 57: 341—371.

Cramer, C.H. and Kovach, R.L., 1974. A search for teleseismic travel-time anomalies along the San Andreas fault zone. *Geophys. Res. Lett.*, 1: 90—92.

Davies, G. and Brune, J.N., 1971. Regional and global fault slip rates from seismicity. *Nature Phys. Sci.*, 229: 101—107.

Derr, J.S., 1973. Earthquake lights: a review of observations and present theories. *Bull. Seismol. Soc. Am.*, 63: 2177—2187.

Dietz, R.S. and Holden, J.C., 1970. Reconstruction of Pangaea: Breakup and dispersion of continents, Permian to present. *J. Geophys. Res.*, 75: 4939—4956.

Epstein, B. and Lomnitz, C., 1966. A model for the occurrence of large earthquakes. *Nature*, 211: 954—956.

Fedotov, S.A., 1965. Regularities of the distribution of strong earthquakes of Kamchatka, the Kurile Islands, and northeastern Japan. *Tr. Inst. Fiz. Zemli Akad. Nauk SSSR*, 36 (203): 66—93.

Fitch, T.J. and Scholz, C.H., 1971. Mechanism of underthrusting in southwest Japan: a model of convergent plate interactions. *J. Geophys. Res.*, 76: 7260—7292.

Frank, F.C., 1965. On dilatancy in relation to seismic sources. *Rev. Geophys.*, 3: 484—503.

Gajardo, E. and Lomnitz, C., 1960. Seismic provinces of Chile. *Proc. 2nd Conf. Earthquake Eng. Tokyo*, pp. 1529—1540.

Gibowicz, S.J., 1973. Variation of frequency—magnitude relationship during Taupo earthquake Swarm of 1964—65. *N.Z. J. Geol. Geophys.*, 16: 18—51.

Gupta, I.N., 1973. Premonitory variations in S-wave velocity anisotropy before earthquakes in Nevada. *Science*, 182: 1129—1132.

Hanks, T.C., 1974. Constraints on the dilatancy—diffusion model of the earthquake mechanism. *J. Geophys. Res.*, 79: 3023—3025.

Isacks, B. and Molnar, P., 1971. Distribution of stresses in the descending lithosphere from global survey of focal-mechanism solutions of mantle earthquakes. *Rev. Geophys. Space Phys.*, 9: 103—174.

Kelleher, J., 1970. Space—time seismicity of the Alaska—Aleutian seismic zone. *J. Geophys. Res.*, 75: 5745—5756.

Kelleher, J., 1972. Rupture zones of large South American earthquakes and some predictions. *J. Geophys. Res.*, 77: 2087—2103.

Kelleher, J., Sykes, L. and Oliver, J., 1973. Possible criteria for predicting earthquake locations and their application to major plate boundaries of the Pacific and the Caribbean. *J. Geophys. Res.*, 78: 2547—2585.

Kelleher, J., Savino, J., Rowlett, H. and McCann, W., 1974. Why and where great thrust earthquakes occur along island arcs. *J. Geophys. Res.*, 79: 4889—4899.

Kondratenko, A.M. and Nersesov, I.L., 1962. Some results of the study on change in the velocity of longitudinal wave and relation between the velocities of longitudinal and transverse waves in a focal zone. *Tr. Inst. Fiz. Zemli, Akad. Nauk SSSR*, 25: 130—150.

Le Pichon, X., Francheteau, J. and Bonnin, J., 1973. *Plate Tectonics*. Elsevier, Amsterdam, 330 pp.

Lomnitz, C., 1973. A statistical argument for the existence of a discontinuity in some subduction zones. *J. Geophys. Res.*, 78: 2612—2615.

Lomnitz, C., 1974. *Global Tectonics and Earthquake Risk*. Elsevier, Amsterdam, 320 pp.

Mazella, A. and Morrison, H.P., 1973. Deep resistivity associated with earthquakes on the San Andreas fault. *Trans. Am. Geophys. Union*, 54: 1136 (abstract).

McEvilly, T.V. and Johnson, L.R., 1973. Earthquake of strike-slip type in central California: evidence on the question of dilatancy. *Science*, 182: 581—583.

Mescherikov, J.A., 1968. Recent crustal movements in seismic regions; geodetic and geomorphic data. *Tectonophysics*, 6: 29—39.

Mogi, K., 1962a. Study of the elastic shocks caused by the fracture of heterogeneous materials and its relation to the earthquake phenomena. *Bull. Earthquake Res. Inst. Tokyo Univ.*, 40: 125—173.

Mogi, K., 1962b. Magnitude—frequency relation for elastic shocks accompanying fractures of various materials and some related problems in earthquakes. *Bull. Earthquake Res. Inst. Tokyo Univ.*, 40: 831—853.

Mogi, K., 1963. The fracture of a semi-infinite body caused by an inner stress origin and its relation to the earthquake phenomena, 2. *Bull. Earthquake Res. Inst. Tokyo Univ.*, 41: 595—614.

Mogi, K., 1969. Some features of recent seismic activity in and near Japan, 2. Activity before and after great earthquakes. *Bull. Earthquake Res. Inst. Tokyo Univ.*, 47: 395—417.

Montessus de Ballore, F., 1906. *Les Tremblements de Terre*. Colín, Paris.

Nagata, T., 1972. Application of tectonomagnetism to earthquake phenomena. In: E.F. Savarensky and T. Rikitake (Editors). *Forerunners of Strong Earthquakes. Tectonophysics*, 14: 263—271.

Nason, R.D., Philippsborn, F.R. and Yamashita, P.A., 1974. Catalog of creepmeter measurements in central California from 1968 to 1972. *U.S. Geol. Surv. Open-File Rep.*, 74-31, 287 pp.

Nersesov, I.L., Semenov, A.N. and Simbireva, I.G., 1969. Space—time distribution of the travel-time ratios of the transverse and longitudinal waves in the Garm area. In: *The Physical Basis of Foreshocks*. Nauka, Moscow.

Newmark, N. and Rosenblueth, E., 1971. *Fundamentals of Earthquake Engineering*. Prentice-Hall, Englewood Cliffs, N.J., 640 pp.

Nur, A., 1972. Dilatancy, pore fluids, and premonitory variations of $t_s/t_p$ travel times. *Bull. Seismol. Soc. Am.*, 62: 1217—1222.

Nur, A., 1974. Matsushiro, Japan, earthquake swarm: confirmation of the dilatancy—fluid diffusion model. *Geology*, 2: 217—221.

Nur, A. and Booker, J.R., 1972. Aftershocks caused by pore fluid flow. *Science*, 175: 885—887.

Nur, A. and Simmons, G., 1969. The effect of saturation on velocity of low-porosity rocks. *Earth Planet. Sci. Lett.*, 7: 183—193.

Nur, A., Bell, M.L. and Talwani, P., 1973. Fluid flow and faulting, 1. A detailed study of the dilatancy mechanism and premonitory velocity changes. In: R.L. Kovach and A. Nur (Editors), *Proceedings of the Conference on Tectonic Problems of the San Andreas Fault*. Stanford University Press, Palo Alto, Calif., pp. 391—404.

Ohtake, M., 1973. Change in the $v_p/v_s$ ratio related with occurrence of some shallow earthquakes in Japan. *J. Phys. Earth*, 21: 173—184.

Oldenburg, D.W. and Brune, J.N., 1972. Ridge transform fault spreading pattern in freezing wax. *Science*, 178: 301—304.

Reid, H.F., 1910. The mechanics of the earthquake. In: *The California Earthquake of April 18, 1906. Report of the State Earthquake Investigation Commission*, 2. Carnegie Institution of Washington, D.C., 192 pp.

Richter, C.F., 1958. *Elementary Seismology*. Freeman and Co., San Francisco, 768 pp.

Rikitake, T., 1974. Probability of earthquake occurrence as estimated from crustal strain. *Tectonophysics*, 23: 299—312.

Rosenblueth, E. and Esteva, L., 1964. Espectros de temblores a distancias moderadas y grandes. *Soc. Mex. Ing. Sísmica*, March

Sadovsky, M.A., Nersesov, I.L., Nigmatullaev, S.K., Latynina, L.A., Lukk, A.A., Semenov, N.A., Simbireva, I.G. and Ulomov, V.I., 1972. The processes preceding strong earthquakes in some regions of Middle Asia. In: E.F. Savarensky and T. Rikitake (Editors), *Forerunners of Strong Earthquakes. Tectonophysics*, 14: 295—307.

Schmidt, P., 1971. Zu Fragen der Erdbebenprognose. *Acta Geodaet. Geophys. Montanist. Acad. Sci. Hung.*, 6: 449—457.

Schmidt, P., 1973. Zu Fragen der Erdbebenprognose (continuation). *Acta Geodaet. Geophys. Montanist. Acad. Sci. Hung.*, 8: 451—460.

Scholz, C.H., 1968a. An experimental study of the fracturing process in brittle rock. *J. Geophys. Res.*, 74: 1447—1454.

Scholz, C.H., 1968b. Microfracturing and the inelastic deformation of rock. *J. Geophys. Res.*, 73: 1417—1432.

Scholz, C.H., 1968c. The frequency—magnitude relation of microfracturing in rock and its relation to earthquakes. *Bull. Seismol. Soc. Am.*, 58: 399—415.

Scholz, C.H., 1972. Crustal movements in tectonic areas. In: E.F. Savarensky and T. Rikitake (Editors), *Forerunners of Strong Earthquakes. Tectonophysics*, 14: 201—217.

Scholz, C.H. and Cranz, R., 1974. Notes on dilatancy recovery. *J. Geophys. Res.*, 79: 2132—2135.

Scholz, C.H., Sykes, L.R. and Aggarwal, Y.P., 1973. Earthquake prediction: a physical basis. *Science*, 181: 803—810.

Semenov, A.N., 1969. Variations in the travel-time of transverse and longitudinal waves before violent earthquakes. *Izv. Akad. Sci. USSR, Ser. Earth Phys.*, 4: 245—248 (English transl.).

Shamsi, S. and Stacey, F.D., 1969. Dislocation models and seismomagnetic calculations for California 1906 and Alaska 1964 earthquakes. *Bull. Seismol. Soc. Am.*, 59: 1435—1448.

Singh, S.K., 1973. Premonitory elevation change before an earthquake based on dilatancy—diffusion model. *Geofís. Int., México*, 13: 279—289.

Singh, S.K. and Sabina, F., 1975. Epicentral deformation based on dilatancy—fluid diffusion model. *Bull. Seismol. Soc. Am.*, 65: 845—854.

Stacey, F.D., 1964. The seismomagnetic effect. *Pure Appl. Geophys.*, 58: 5—22.

Sykes, L.R., 1971. Aftershock zones of great earthquakes, seismicity gaps, and earthquake prediction for Alaska and the Aleutians. *J. Geophys. Res.*, 76: 8021—8041.

Tobin, D. and Sykes, L.R., 1968. Seismicity and tectonics of the northeast Pacific ocean. *J. Geophys. Res.*, 73: 3821—3845.

Tsuboi, C., 1933. Investigation of the deformation of the earth's crust found by precise geodetic means. *Jap. J. Astron. Geophys.*, 10: 93—248.

Tsuboi, C., 1958. Earthquake province — domain of sympathetic seismic activities, *J. Phys. Earth*, 6: 35.

Tsuneishi, Y. and Nakamura, K., 1970. Faulting associated with the Matsushiro swarm earthquake. *Bull. Earthquake Res. Inst. Tokyo Univ.*, 48: 29—51.

Ulomov, V.I. and Mavashev, B.Z., 1967. Forerunners of the strong tectonic earthquake. *Dokl. Akad. Nauk Moscow SSSR*, 179: 319—321.

Whitcomb, J.H., Garmany J.D. and Anderson, D.L., 1973. Earthquake prediction: variation of seismic velocities before the San Fernando earthquake. *Science*, 180: 632—635.

Wyss. M., 1973. Towards a physical understanding of the earthquake frequency distribution. *Geophys. J.R. Astron. Soc.*, 31: 341—359.

Wyss, M. and Holcomb, D.J., 1973. Earthquake prediction based on station residuals. *Nature*, 245: 139—140.

Wyss, M. and Johnston, A.C., 1974. A search for teleseismic P residual changes before large earthquakes in New Zealand. *J. Geophys. Res.*, 79: 3283—3290.

*Chapter 3*

GEOLOGICAL CRITERIA FOR EVALUATING SEISMICITY *

CLARENCE R. ALLEN

*Seismological Laboratory, California Institute of Technology, Pasadena, Calif., U.S.A.*

3.1 INTRODUCTION

It has become increasingly clear in recent years that, for most parts of the world, neither the catalogs of instrumentally recorded earthquakes nor the histories of felt earthquakes cover a sufficiently long time to allow valid extrapolations of future earthquake activity, except on very broad regional scales. The physical basis of tectonic processes implies that large earthquakes should not occur randomly in time, at least during periods of only a few hundred years and within single tectonic provinces. And no amount of sophisticated statistics or extreme value theory can throw much light on the nature and frequency of large events based on a time sample that is too short to include any such events, unless a specific physical model is also assumed. Even in regions with continuing seismic activity in the magnitude-8 range, such events have seldom recurred in the exact same localities within the historic record, so statistical extrapolations from these events must be limited to broad regions and not to specific areas or sites. Such generalized extrapolations can, of course, be exceedingly valuable, but the current need is for a better understanding of the likelihood and recurrence rates of large earthquakes on a more local scale — the type of question that is being asked repeatedly in the design of critical structures such as dams and nuclear facilities, as well as in the formulation of building codes that are both adequately conservative and economically realistic.

The purpose of this chapter is to argue that the geological record, and the late Quaternary history in particular, is a far more valuable tool in estimating seismicity and associated seismic hazard than has generally been appreciated in most parts of the world. By looking back many thousands of years, one is able to overcome many of the statistical shortcomings of the instrumental and historic records. More specifically, it is argued that surface faulting during large, shallow earthquakes is more universal than has been recognized, and that the geomorphic effects of these movements and radiometric age

---

* Adapted from the Address as Retiring President of the Geological Society of America, Miami Beach, Florida, November 1974. Published with the permission of the Geological Society of America.

dating of earlier events provide tools for understanding seismicity whose promise has not been fully realized.

One might argue that the future possible success of earthquake predictions will obviate the need for seismicity studies based upon records of the past. The writer is optimistic that the earthquake-prediction effort will indeed succeed and that predictions will eventually become sufficiently reliable to permit meaningful short-term public response. A recent visit to China reinforces his view that successful predictions, when achieved, can and will be tremendously beneficial in saving lives. Clearly tens and perhaps even hundreds of thousands of lives could be saved by an accurate short-term earthquake prediction in parts of China, owing to population density, housing materials, and the ability of the social system to respond quickly to such a warning. It is little wonder that earthquake prediction has become a major national effort in that country.

On the other hand, no one currently visualizes that accurate earthquake predictions will be made fifty or one hundred years before an event, which is the time scale of concern in designing major structures and in formulating building codes. Thus earthquake prediction will, at best, solve only a part of the hazard-reduction problem, and it will not replace the need for estimates of long-term seismicity and recurrence rates of large events. The two efforts involve different types of public responses and should be considered partners rather than competitors in the total hazard-reduction program — together, of course, with engineering studies.

It is significant that the earthquake catalogs of those parts of the world with the longest historical records are the very ones which give us the greatest pause in extrapolating these records into the future. This should be a lesson in terms of the temptation to draw far-reaching conclusions from a relatively short seismic history such as characterizes North America, and from such single events as the Charleston and New Madrid earthquakes. As an example of a long-term history, consider the Chinese record based on valid chronicles of seismic activity for large parts of China extending back almost 3000 years — more than a quarter of the Holocene epoch. The startling thing about this record, as pointed out by our Chinese colleagues, is the lack of uniformity in both space and time. Most large earthquakes have indeed occurred in relatively well-defined zones, and many of these zones have been the loci of continuing moderate activity, but there are so many conspicuous exceptions within the 3000-year record as to make one cautious in drawing generalizations. For example, Mei (1960a,b) has plotted cumulative strain release from 466 BC to the present for the Kansu and North China region (Fig. 3.1), an area four times larger than that of California and Nevada combined. This is a period during which she feels that the record of large shocks is relatively complete. The seismic activity during the first and last parts of the period is high, but during an 800-year period from 200 to 1000 AD large shocks are almost lacking. Yet the seismic hazard in this region cannot be

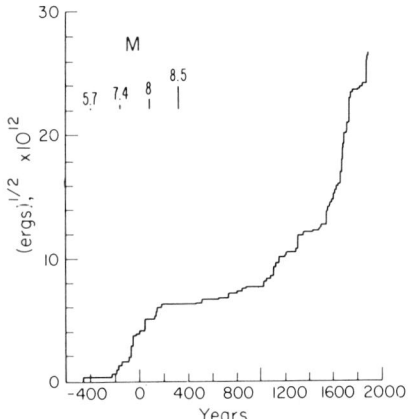

Fig. 3.1. Cumulative seismic strain release, 466 BC to the present, for Kansu and North China region, China. (From Mei, 1960a.)

considered low; the historic record includes at least two shocks of magnitude near 8.5, one of which — in 1556 — was the most disastrous earthquake in history, causing more than 820,000 deaths. The other great event — that of 1668 — occurred in a part of the region which neither before nor since has been characterized by high continuing activity. Significantly, however, both of these disastrous events occurred in areas of major throughgoing Quaternary faults, as will be emphasized later in this chapter.

Ambraseys (1970) has described a somewhat similar situation in the eastern Mediterranean region, noting that over the past 2000 years, "earthquake frequency is subject to marked fluctuations rather than to any general monotonic variation". In Turkey, for example, a remarkable series of earthquakes, sequential in both time and space, commenced with the great Erzincan shock of 1939 in eastern Turkey. Figure 3.2 shows those shocks of greater than magnitude 7 that have occurred since that time, along with the associated fault displacements (Ketin, 1948; Ketin and Roesli, 1953; Allen, 1969; Ambraseys, 1970). However, in talking to many farmers along virtually the entire length of the faulted zone, all of whom distinctly remember the surface fracturing and other effects of the 1939 and subsequent events, the writer was unable to find any instances of handed-down family stories or reports of earlier similar occurrences. This suggests that this sequence of events had not been repeated for at least 100 or 200 years prior to 1939. It is interesting to speculate as to the kind of seismic zoning map of Turkey which might have been constructed in 1938. Ambraseys (1970) points out that had the much earlier history been known, the fault could also have been identified as dangerous, for a somewhat similar sequence of earthquakes occurred between 967 and 1050 AD and still earlier activity is known from the 3rd, 5th, and 7th centuries. Ambraseys concludes, on the basis of the histor-

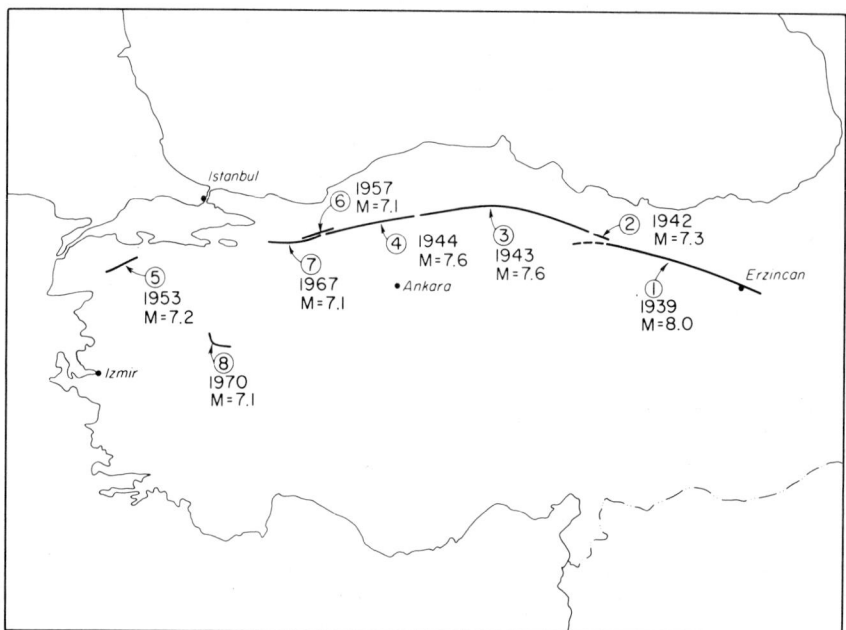

Fig. 3.2. Faulting (heavy lines) associated with earthquakes of magnitude 7.0 and greater in Turkey since 1939.

ic record between 10 AD and 1100 AD, that quiescent periods averaged about 150 years in length, and this is consistent with the observation suggesting the absence of faulting for at least 100—200 years prior to 1939. Many other examples of long-term temporal variations in seismicity could be cited, such as the high activity in Korea that ended abruptly about three centuries ago (Richter, 1958).

The purpose in mentioning these examples is to emphasize the difficulties and dangers in attempting to estimate future seismic activity on the basis of the historic record alone, particularly if that record is short. In the sections that follow, illustrations will be drawn from several parts of the world to indicate how the geologic record can be used to supplement or even supersede the historic record in the evaluation of seismicity. Attention will be focused first on seismotectonic relationships in California — the most thoroughly studied seismic area in North America — and other parts of the world will then be looked at for comparison.

3.2 CALIFORNIA

The seismicity of California is related to motion along the plate boundary between the North American and Pacific plates (Atwater, 1970; Minster et al., 1974), dominated by the San Andreas fault system but showing many manifestations of a plate boundary more complicated than that of a single, ideal transform fault. Five seismotectonic generalizations can be applied to California:

(1) Virtually all large earthquakes — those exceeding magnitude 6.0 — have occurred because of ruptures on faults that *had* been recognized, *could* have been recognized, or *should* have been recognized by field geologists prior to the events (Allen et al., 1965).

(2) All of these faults have been characterized by a history of earlier Quaternary (and probably Holocene) displacements.

(3) All earthquakes are shallow — not exceeding about 20 km in depth — and most of those larger than magnitude 6.0 have been accompanied by surface faulting, as have many of even lesser magnitudes.

(4) The larger earthquakes have generally occurred on the longer faults, although there has been sufficiently wide variation to indicate caution in blindly applying any single formula for this relationship.

(5) Generally, only a small segment of the entire length of a fault zone has broken during any single earthquake, although there are some conspicuous and significant exceptions (Albee and Smith, 1966).

The one-to-one relationship between active faults and earthquakes in California is so clear that, if one were faced with the task of making a seismic zoning map of the State based either solely on the geologic data or solely on the geophysical and historical data, the more realistic map could be made with the geologic data. Combining the two approaches will give a still more realistic map, and in recent years there has been a satisfying trend in this direction as field geologists have become more and more involved in seismic-hazard problems. This trend has also resulted in more controversies, but most of them have helped bring into the open many of the very difficult problems that previously were not being squarely faced.

Figure 3.3, a generalized map of faults with Quaternary displacements in southern California, illustrates the importance of geologic data. An immediate reaction to this map might be that so much of the area is characterized by Quaternary faulting that seismic hazard must be considered high throughout most of the region, and this is exactly the conclusion that should be drawn — regardless of what the very recent seismic history indicates. Admittedly, the map could be much improved in its relevance to present-day seismicity if Holocene displacements could be separated from earlier Quaternary movements, although the gross relationships would probably be little changed. A satisfactory seismic zoning map, neglecting differential soils effects, could be constructed from a map of faults with Holocene displacements, giving greater weights in terms of large earthquakes and long-period, long-duration ground-motion effects to those fault zones with the greatest continuity and length, such as the San Andreas and Garlock. Still greater improvement could be attained by identifying those faults that have had

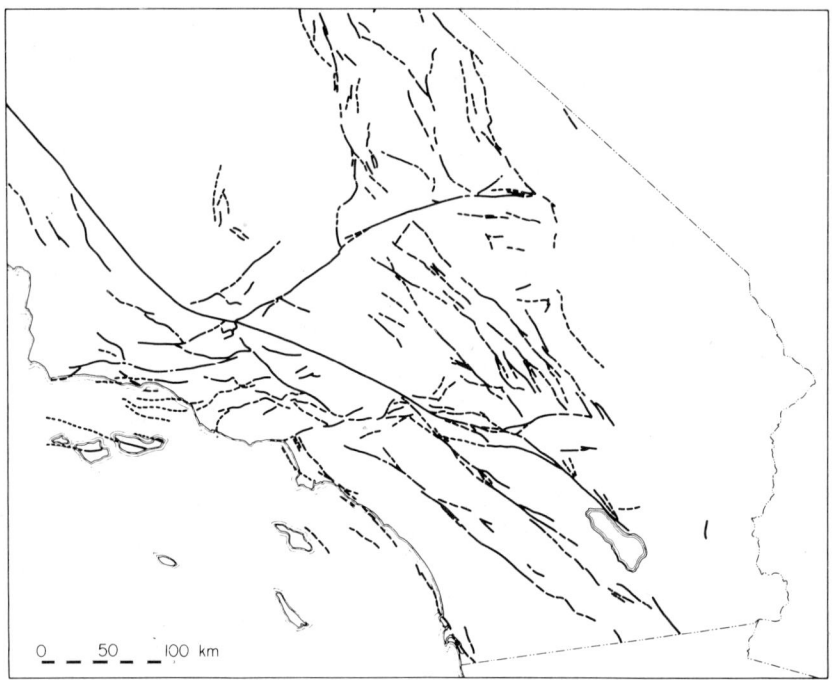

Fig. 3.3. Faults with Quaternary displacements in southern California. (Generalized and simplified from Jennings, 1973.)

*recurrent* Holocene displacements, a currently challenging task.

A comparison of this generalized map of Quaternary fault displacements with the actual distribution of large earthquakes in this region since 1912 — an interval of 63 years — is instructive (Fig. 3.4). With few exceptions, large earthquakes have indeed occurred in areas of Quaternary faulting, and even the smaller earthquakes (not shown in Fig. 3.4) have generally followed the same trends. On the other hand, there are significant areas of Quaternary (and Holocene) faulting that have not experienced major earthquakes during this particular 63-year period, and these areas should be considered equally hazardous with many of those areas that have. Particularly noteworthy in terms of inactivity are the Garlock fault and the Death Valley region and, of course, the central segment of the San Andreas fault, which last broke in 1857 and has been almost completely inert seismically since. In a sense, this is nothing more than a statement of the phenomenon of the temporal "seismic gap", which has long been recognized in California (Omori, 1906; Allen et al., 1965) and has recently been applied more systematically to worldwide earthquake occurrences (Kelleher et al., 1973).

What is clearly needed to refine our seismic probability map is a better understanding of the Holocene movements on the various faults. Which ones

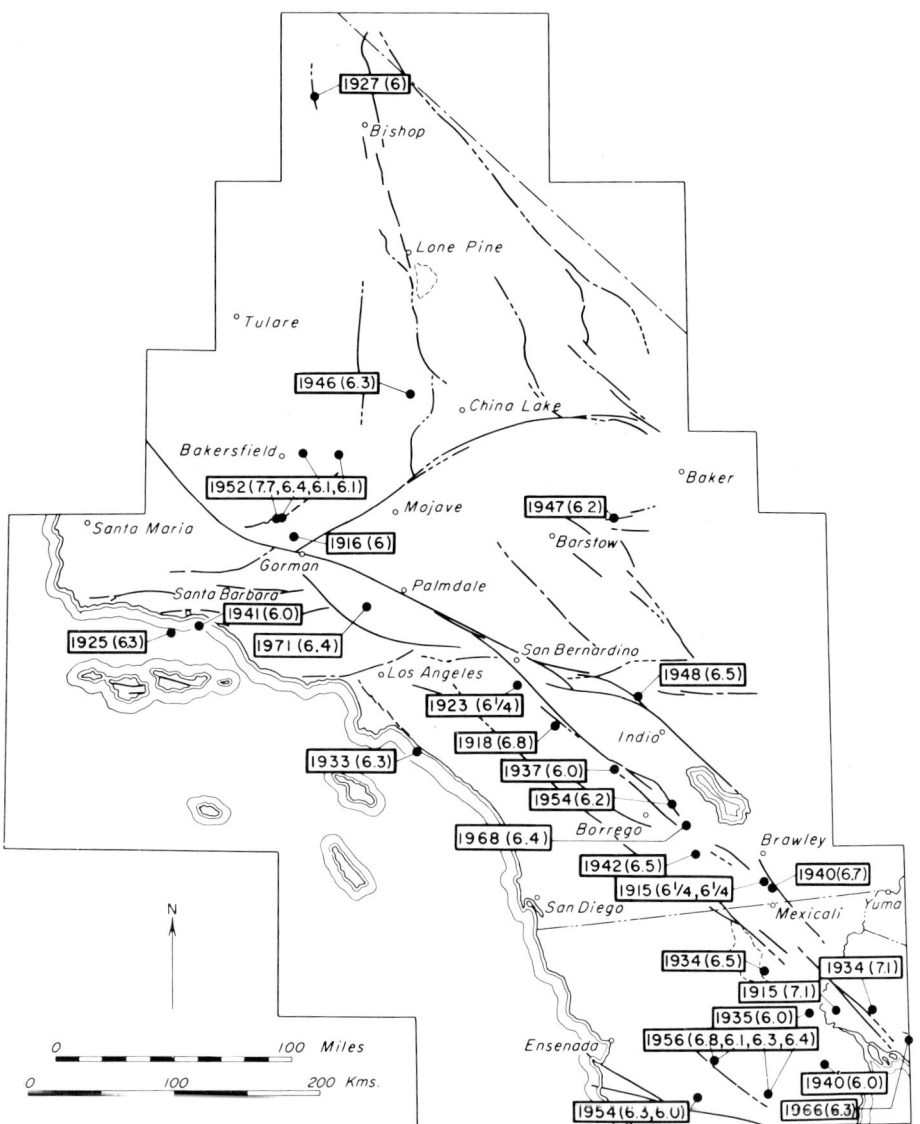

Fig. 3.4. Earthquakes of magnitude 6.0 and greater in the southern California region, 1912–1974.

have moved most recently, and which ones have moved most often? An example of the kind of information that can be obtained with detailed investigation is provided by study of the 1968 Borrego Mountain earthquake in southeastern California ($M$ = 6.4). Following the earthquake, Clark et al. (1972) analyzed displacements of Holocene strata that were exposed in a trench excavated across one of the fault breaks (Fig. 3.5). Using carbon-14 dating, they found a remarkably linear relationship between the ages of strata and total offsets, implying a uniform rate of tectonic movement over the past 3000 years (Fig. 3.6). Furthermore, if one assumes that all past displacements have occurred during earthquakes comparable in size to the 1968 event, then such earthquakes have occurred about once every 200

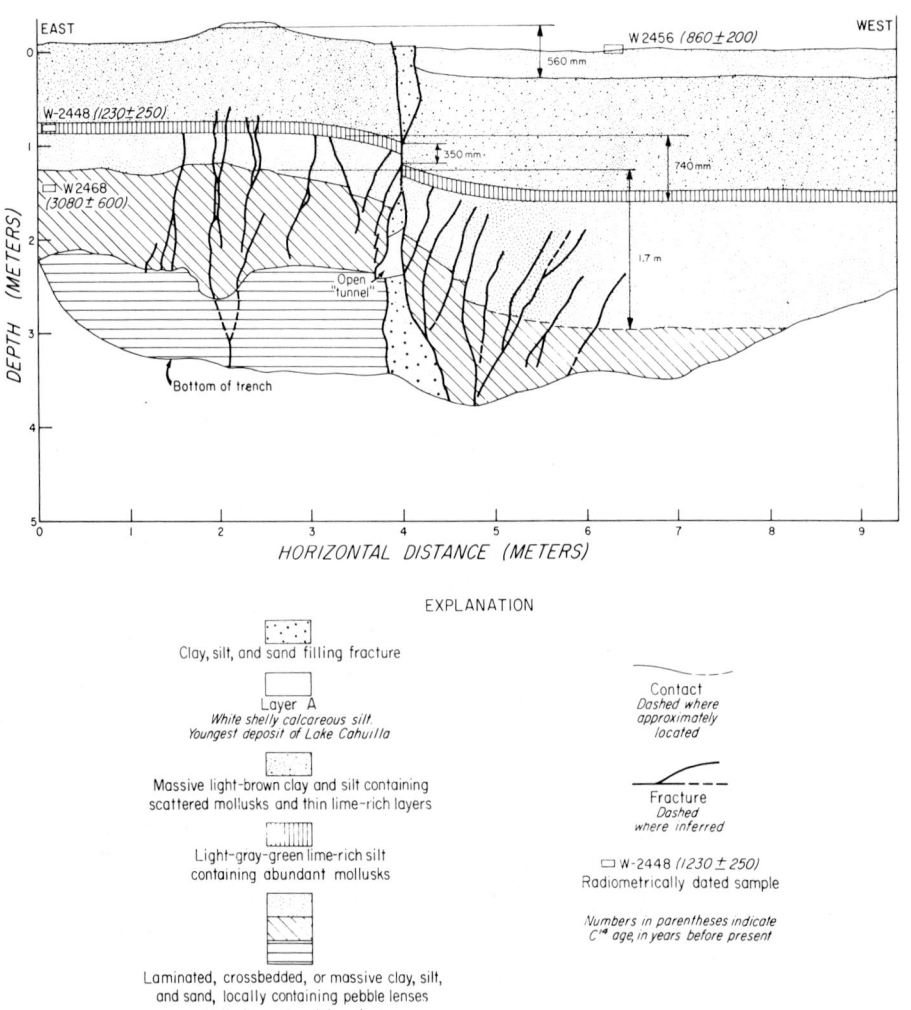

Fig. 3.5. Diagram from Clark et al. (1972) showing wall of trench excavated across one branch of the fault causing the 1968 Borrego Mountain earthquake in California ($M$ = 6.4). The 1968 earthquake is represented by the small displacement of the ground surface, and deeper and older strata (note carbon-14 ages) are offset by progressively greater amounts.

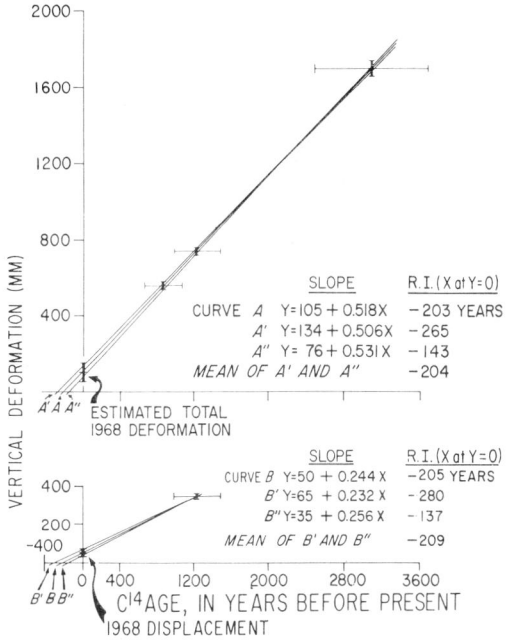

Fig. 3.6. Graph from Clark et al. (1972) showing relation of vertical offset to $^{14}$C age of strata in trench of Fig. 3.5. Upper graph shows total vertical deformation (drag plus slip), whereas bottom graph shows fault slip only. Various curves represent alternate deformation rates based on extreme and estimated values for 1968 deformation. R.I. is recurrence interval for breaks of same size as that of 1968.

years. This is the first reasonably good control we have had on earthquake recurrence rates on southern California faults, and despite the many assumptions involved, this technique represents a powerful tool for seismic-hazard evaluation. The need is to gather such data from more localities and to extend them farther back in time.

Not pointed out by Clark et al. (1972) is the fact that these data also put limits on the maximum sizes of earthquakes that have occurred during these intervals. Had one of the displacements taken place during a magnitude-8 earthquake, for example, the data of Bonilla and Buchanan (1970) suggest that the displacement should have been nine times that of the magnitude-6.4 event of 1968. Such a large displacement is not compatible with at least the two most recent intervals plotted in Fig. 3.6. While many debatable assumptions are also involved in this interpretation, it demonstrates that geological data have the possibility of giving information not only on recurrence rates of large earthquakes, but also on their maximum credible magnitudes, both of which are critical to engineers in the design of earthquake-resistant structures.

The claim has sometimes been made that the seismotectonic relationships of California are so unique that they have little relevance to the determination of seismicity and seismic hazard in most other parts of the world. This is said to be because: (1) California earthquakes are all very shallow and accompanied by an inordinate amount of surface faulting that is not typical of most other areas; (2) the desert environment and brief agricultural history of parts of California allow for preservation of many geomorphic forms of Quaternary deformation that are unidentifiable in most parts of the world; and (3) the dominantly strike-slip nature of California faults is clearly atypical and leads to a lack of appreciation of other types of faulting that characterize many regions.

While it is true that California earthquakes may be atypically shallow, it is nevertheless the very shallow earthquakes that are the principal cause of seismic disasters throughout the world, and it is these earthquakes on which we must focus our attention.

As stated before, surface faulting is far more common world-wide during large shallow earthquakes than has generally been appreciated; too many earthquakes have not been studied adequately by field geologists, and critical tools of the geologist such as aerial photographs have too often been lacking. While it is true that California has long been renowned for its strike-slip faults, we now know that the plate boundary represented by the San Andreas and associated faults is by no means a unique feature within the continents. Nor is the experience in California limited to strike-slip events; perhaps the largest earthquake in California's recorded history was a dominantly normal-faulting event, and the most recent disaster — that of 1971 in San Fernando — was caused by predominant thrusting.

## 3.3 TURKEY

Inasmuch as both California and Turkey have experienced great earthquakes with large strike-slip displacements, it is instructive to compare the two regions in terms of seismotectonic relationships. One principal question is: Could the North Anatolian fault, which is the locus of most of the activity shown in Fig. 3.2, have been identified in the field as an active fault *prior* to the sequence of large earthquakes that began in 1939? The answer to this question is resoundingly "yes". Physiographic features of Quaternary displacement along the North Anatolian fault are so spectacular that it would surely be illustrated in textbooks along with the San Andreas and Alpine faults if aerial photographs were generally available. Typical features of active faulting are shown in Figs. 3.7—3.9, which are based on field work made possible by cooperation of the Mineral Research and Exploration Institute of Turkey and Istanbul Technical University. Such fault topography characterizes the fault along the entire length that has broken since 1939,

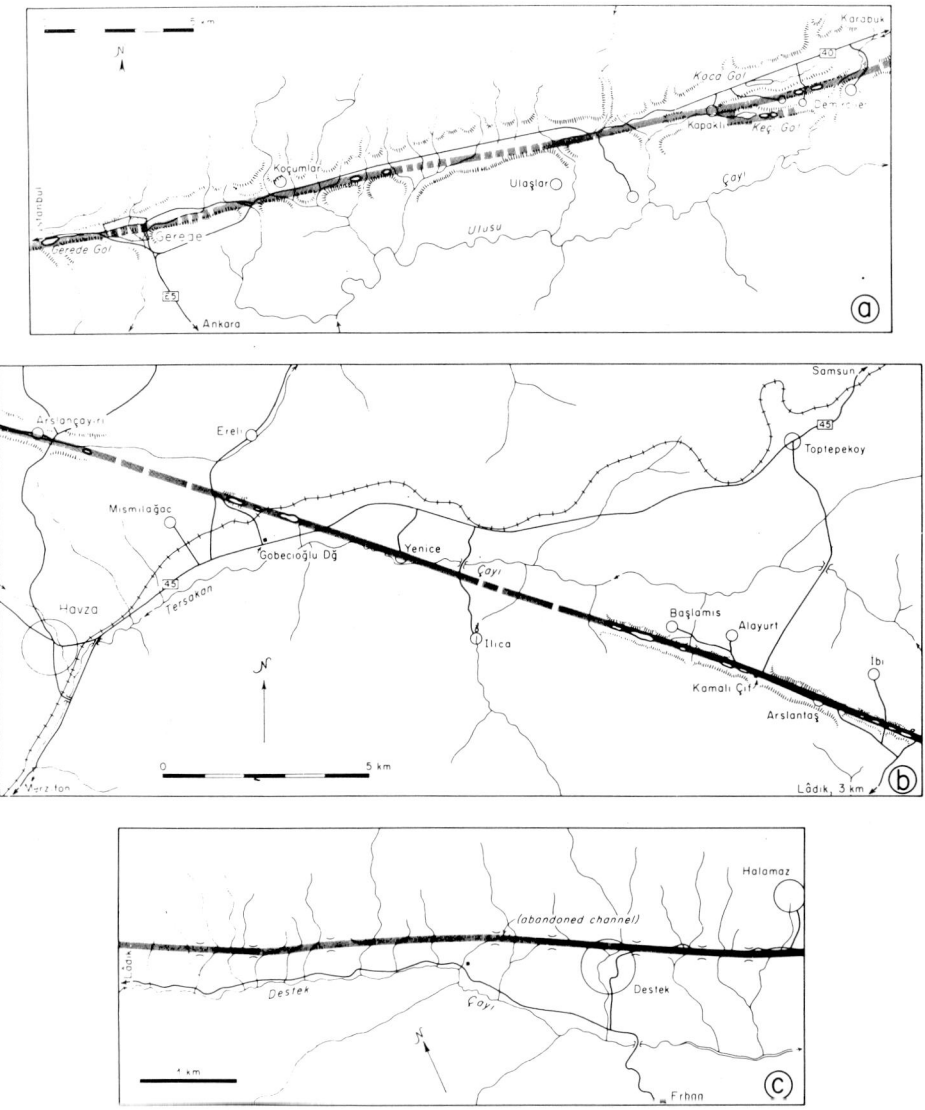

Fig. 3.7. Sketch maps of three widely separated areas along the North Anatolian fault, Turkey, the most recent zone of movement of which is indicated by the shaded line. (a) Fault zone near Gerede, along main Istanbul—Ankara highway about 110 km northwest of Ankara; note sag ponds. (b) Area near Havza where fault crosses main Amasya—Samsun railroad and highway, about 270 km northeast of Ankara. (c) Details of right-lateral stream offsets along fault near Destek, 50 km southeast of Havza.

although the extent of the fault still farther east and west is a matter of some controversy.

Fig. 3.8. View west along rift valley of North Anatolian fault from a point about 1 km west of Gerede (Fig. 3.7a). Note sag pond.

Ambraseys (1971) has plotted the epicenters of all earthquakes exceeding magnitude 5 that occurred between 1900 and 1965 in the eastern Mediterranean region (Fig. 3.10) and compares their distribution with that of several thousand earlier shocks obtained from historical records going back about 2000 years. Ambraseys points out a general correspondence between the very recent record and the longer historical record but also notes some conspicuous anomalies that deserve attention. Had not the recent events starting in 1939 been included, the broad band of current seismicity along the North Anatolian fault would have been poorly defined. More significant is the wide band of early historic activity extending from eastern Turkey southwest into

Fig. 3.9. Recent scarp (arrow) about 5 km east of Gerede (Fig. 3.7a), resulting at least in part from horizontal displacement during 1944 earthquake. Stream in foreground is right-laterally offset.

Syria, Lebanon, and Israel — a zone that experienced many disastrous earthquakes up until about 1200 AD but has been relatively quiescent since.

One may go one step beyond Ambraseys' historical approach and emphasize that virtually all of these areas of historic seismicity could have been identified — and perhaps with less research effort — by a study of Quaternary faulting. Indeed, some other hazardous areas that do not happen to have been active within the past 2000 years might show up in addition. As examples, we shall look at three specific areas — one in eastern Turkey with-

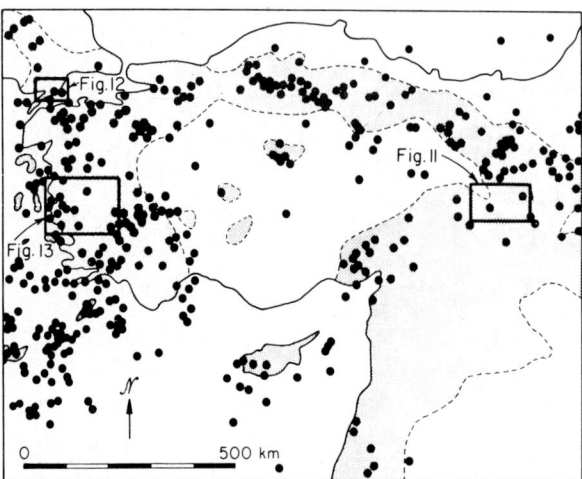

Fig. 3.10. Map of Turkey and adjacent areas, based on Ambraseys (1971). Black dots are earthquakes exceeding magnitude 5, 1900—1965. Stippled area represents area of historical seismicity, I—XVII centuries. Inserts show locations of Figs. 3.11—3.13.

in the southwestward-trending, currently quiescent zone, a second one north of the Dardanelles along the general projected trend of the North Anatolian fault zone, and a third in the block-faulted area near Izmir along the Aegean coast.

Figure 3.11 is part of an ERTS * image showing a segment of the southwestward trending fault zone that is at least semi-continuous between eastern Turkey and the Dead Sea rift (Allen, 1969). Despite the current seismic quiescence of this fault system (Fig. 3.10), the clear lines of Quaternary displacement indicate in themselves that its seismic potential is by no means nil. The fault system is conjugate to the North Anatolian fault — almost a mirror image of the Garlock—San Andreas relationship — and is left-lateral in its sense of motion. Its length and continuity suggest that it is capable of generating large earthquakes, and a major earthquake did occur in this region in 995 AD diverting rivers and changing stream courses near Palu in the area of the northeast corner of Fig. 3.11 (Ambraseys, 1970). Farther northeast along this same fault zone, the 1971 Bingöl earthquake ($M = 6.7$) was associated with a 20-cm left-lateral displacement (Seymen and Aydın, 1972; Arpat and Şaroğlu, 1972).

An area just north of the Dardanelles was shaken by a major earthquake of magnitude 7.75 in 1912 — the so-called Şarköy-Mürefte earthquake. Although no clear-cut fault displacement has heretofore been documented in association with this event, the obvious fault trend shown in Fig. 3.12 (and

---

* Earth Resources Technology Satellite.

Fig. 3.11. ERTS image (No. 1485-07323) of northeastward trending fault zone (black arrows) in eastern Turkey.

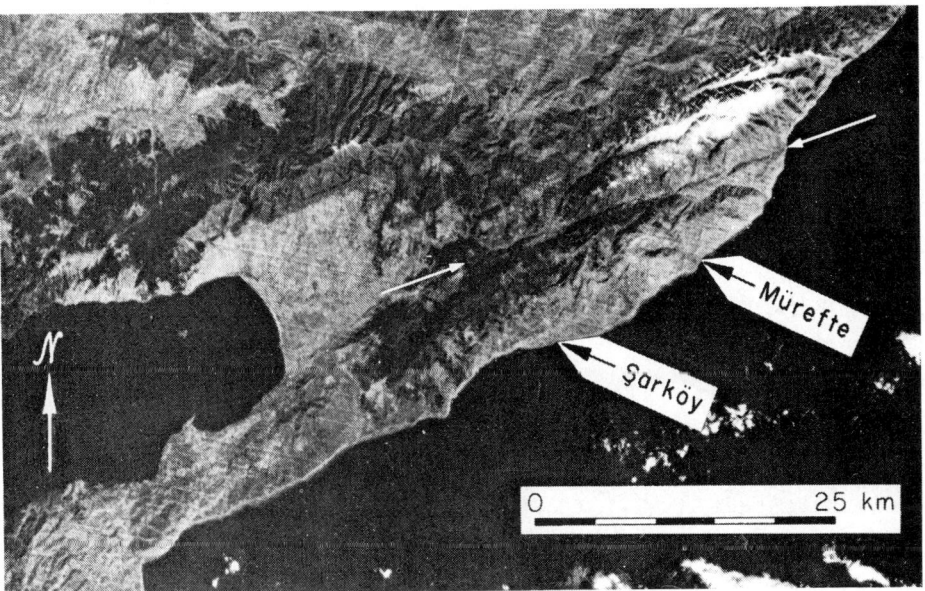

Fig. 3.12. ERTS image (No. 1152-08262) of area of 1912 Şarköy—Mürefte earthquake north of the Dardanelles, Turkey. Surface faulting extended along segment of fault between white arrowheads, and perhaps still farther west.

much better in larger-scale photographs) is so suggestive that the writer visited this area in 1974 together with S. Sipahioğlu of Kandilli Observatory, with the objective of talking to old farmers who might still remember the 1912 event. It quickly became clear that significant surface faulting, probably of a strike-slip nature, had indeed occurred in 1912 along this exact line, for at least as great a distance as that between the two arrowheads in Fig. 3.12.

The Menderes massif east of Izmir (Fig. 3.13) has long been recognized as a major fault block bounded by normal faults, and focal mechanisms confirm this style of current faulting in southwestern Turkey (McKenzie, 1972). Aerial photographs and field observations showing spectacular scarps in alluvium and numerous tilted and deformed terraces leave no doubt that the faults bordering the Menderes massif are active. To further document this activity, we talked to numerous old farmers living between Aydın and Nazilli, on the south flank of the massif, concerning a major earthquake that heavily damaged these two cities in 1899. We were able to determine un-

Fig. 3.13. ERTS Image (Nos. 1115-08210 and 1115-08213) of Menderes massif area, Turkey. Black dashed lines indicate extent of surface faulting in 1899 and 1969 earthquakes. Black dots indicate Roman cities destroyed by earthquake of 17 AD. (After Ambraseys, 1971.)

equivocally that vertical displacements up to 2 m had in fact taken place in 1899 along the extent of the line shown in Fig. 3.13, at least from Yılmazköy on the west to Arslanlı on the east. Furthermore, it was clear in field localities that this displacement took place exactly along the line of earlier Holocene displacements. A somewhat smaller earthquake in 1969 ($M = 6.1$) was associated with minor vertical displacements, also along pre-existing Holocene breaks, near Alaşehir on the north side of the massif. The total extent of faulting (Fig. 3.13) was somewhat longer than that reported by Arpat and Bingöl (1969), extending discontinuously at least 30 km from Hacialiler on the east to Yeniköy on the west. Both the 1899 and 1969 breaks constitute only segments of major active fault zones that have much greater total extent, particularly to the west, and it is obvious that this entire area could easily be delineated as one of significant seismic hazard, even in the absence of any historic record.

Ambraseys (1971) has carried out an investigation of the earthquake of 17 AD, which was one of the great disasters of the eastern Aegean region. Several major Roman cities were destroyed (Fig. 3.13), and it seems reasonable that this event may have been associated with a major fault displacement along the north side of the Menderes massif centered farther west than that of 1969. The city of Manisa, then known as Magnesia, was one of those de-

Fig. 3.14. Normal fault exposed in terrace gravels, 8 km east of Manisa, Turkey. Limestone (left) is faulted against alluvium.

stroyed, and features of Holocene faulting in this vicinity (Figs. 3.14 and 3.15) are among the most spectacular in the region.

Thus it would appear that seismotectonic relationships in Turkey are remarkably similar to those in California, and many of the same generalizations hold. In particular, all earthquakes greater than about magnitude 6.5 on which we have been able to obtain field data appear to have had associated surface faulting, of both strike-slip and dip-slip variety. Furthermore the distribution of faults with Quaternary displacements seems to be a valid general guide to modern seismicity. Are there not other geological lines of evidence pertinent to seismicity besides geomorphic evidence of past fault displacements? Much of western Turkey is blanketed by widespread Neogene continental deposits, most of which are essentially flat-lying in the virtually aseismic parts of central Anatolia but are highly deformed approaching western Turkey and the Aegean Coast, where seismicity is high. Perhaps some measure of deformation, such as the maximum dip in every 25-km square, might also constitute a technique of constructing a seismicity map. A somewhat similar method has been tried by Clark et al. (1965) in New Zealand. Nevertheless, the most relevant information for current seismicity lies in the Holocene history, and the farther back one goes into geological history, the less pertinent will be the record of older deformation to present tectonic processes.

Fig. 3.15. Fault scarp (arrow) in terrace gravels, 8 km east of Manisa, Turkey.

Turkey appears to be no more unique in the eastern Mediterranean area than is California in western North America. Brief field studies carried out in connection with UNESCO Balkan Seismicity Project indicate that at least Greece and Bulgaria share many of the same seismotectonic relationships, particularly the close association between major earthquakes and surface faulting, much of which has never been adequately documented. It is significant that Ambraseys (in press) similarly concludes — but primarily from historical studies — that surface faulting throughout the entire Near and Middle East "is not a phenomenon as rare as it was thought to be", even in association with relatively small earthquakes.

3.4 JAPAN

Let us now turn to a different part of the world, in which broad seismotectonic relationships are more worrisome than in California and Turkey. Japan is likewise an area with both a high seismicity and a lengthy recorded earthquake history.

Most of the greatest Japanese earthquakes have been centered offshore, and large parts of Japan are adjacent to active offshore plate boundaries capable of generating major shocks fairly frequently. But numerous disastrous earthquakes have also occurred within the Japanese islands proper, and the question is: Do these earthquakes follow any patterns that might be decipherable from geologic studies alone? The answer is clearly "yes", but in a somewhat qualified way. Most large, damaging earthquakes originating on land have been associated with surficial displacements on faults with an earlier history of Quaternary activity. On the other hand, this generalization has limited value in terms of seismic zoning because there are so many faults throughout Japan that have had Quaternary displacements. Furthermore, some of the largest earthquakes have occurred on seemingly innocuous faults, and some of the most spectacular and throughgoing faults have had no significant earthquakes on them during the entire 2000-year historic record.

As an example, consider the fault pattern in the part of central Honshu between Tokyo and Kyoto (Fig. 3.16). Some of these faults are probably more continuous than indicated because continuity is obscured by massive landsides in this mountainous and typhoon-prone region, but nevertheless the basic pattern is one of a mosaic of individual blocks bounded by relatively short faults of dominantly northwest and northeast trends. The senses of historic and Quaternary displacements on these faults are consistent with east—west compression, and one gets the impression of almost random chattering as the mosaic of blocks is gradually squeezed and crunched. Some of the chattering has, however, been far from inconsequential; several of the most disastrous earthquakes of Japanese history have occurred within the

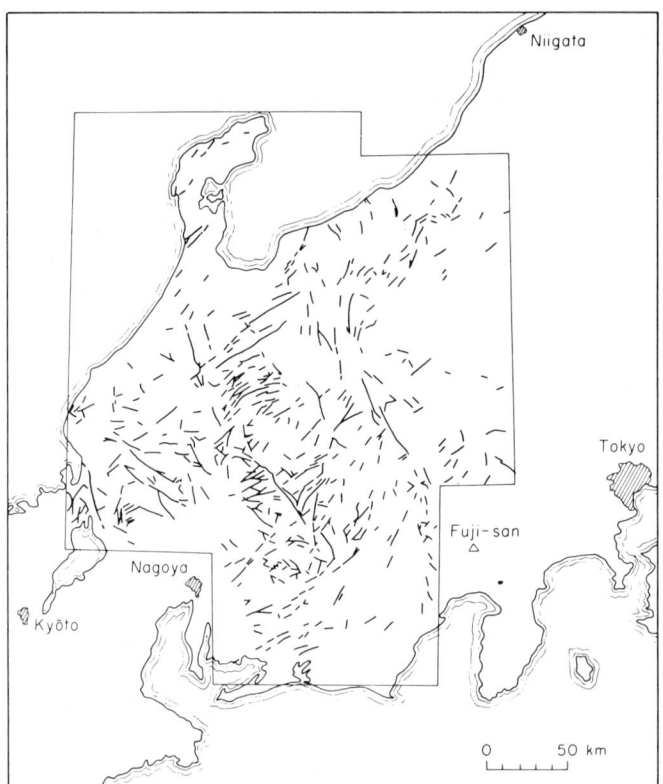

Fig. 3.16. Quaternary faults in area of central Japan between Tokyo and Kyoto. (After Matsuda and Okada, 1968.)

region, including the 1891 Nobi (or Mino-Owari) shock, which was associated with displacement on one of the fault systems shown in Fig. 3.16 north of Nagoya. A somewhat similar event may have occurred here in 745 AD. From the geologic point of view, the disconcerting aspect of the 1891 earthquake was that, although the breaking was indeed along a fault with a prior history of Quaternary displacement, virtually the entire known length of the individual fault zone broke during this one shock (Fig. 3.17) (Matsuda, 1974), and the break jumped back and forth between several members of the zone. At the present time, there is no geologic reasoning for concluding that this particular zone was a more likely candidate for an earthquake than hundreds of others, and it is a relatively short, discontinuous fault to generate an event as great as that of 1891, which was certainly in the magnitude-8 range. If this fault is capable of generating a magnitude-8 shock, then many others in central Japan must have a similar capability.

A somewhat analogous event was the 1930 Izu earthquake of magnitude

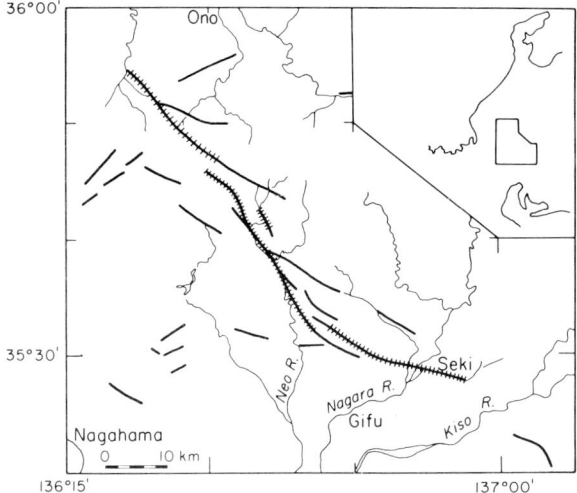

Fig. 3.17. Map of Nobi (Mino-Owari) earthquake area, central Japan, after Matsuda (1974). Heavy black lines are Quaternary faults; those with cross-hatching broke in association with 1891 earthquake.

7.1, caused by displacement on the Tanna fault (Fig. 3.18) (Matsuda and Okada, 1968). A similar displacement took place in 841 AD. Kuno (1936) demonstrated that the Tanna fault had a post-Middle Pleistocene displacement of 1000 m, so again the earthquake clearly occurred on an active fault. But virtually the entire known length of this Quaternary fault broke during the 1930 event, again unlike typical events in California, where only fractional breakage is the rule (Albee and Smith, 1966).

The major Tango earthquake of 1927, of magnitude 7.75, also was associated with remarkably short fault segments which do not appear to represent important geological structures (Yamasaki and Tada, 1928). Yet it is clear from the focal mechanism, depth of focus, and geodetic observations that movement on these faults was in fact the prime cause of the 1927 event, and not merely some sort of surficial geological damage related to a more profound seismic event at greater depth (Kanamori, 1973). The same interpretation applies to the 1891 and 1930 events. We must conclude that the surficial fault structure of Japan, complex as it is, truly reflects the tectonic processes taking place there today in terms of large earthquakes.

By way of contrast, there are at least two major faults of considerable length and of large structural discordance in Japan that would appear to be likely candidates for great earthquakes, but which have experienced no significant seismic activity within the 2000-year history. The most obvious is the Median Tectonic Line, a profound geological discontinuity extending more than 600 km across central and western Japan (Figs. 3.19 and 3.20),

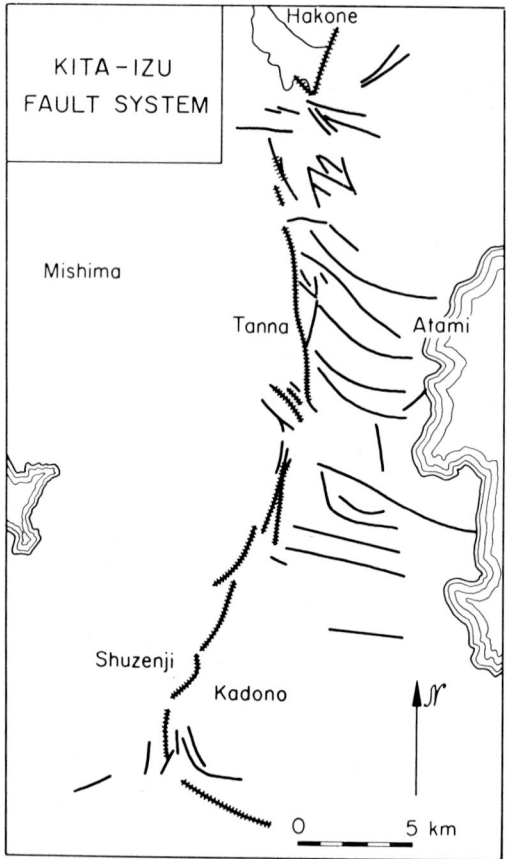

Fig. 3.18. Map of Kita-Izu fault system, central Japan. (After Matsuda and Okada, 1968.) Heavy black lines are Quaternary faults; those with cross-hatching broke in association with 1930 earthquake.

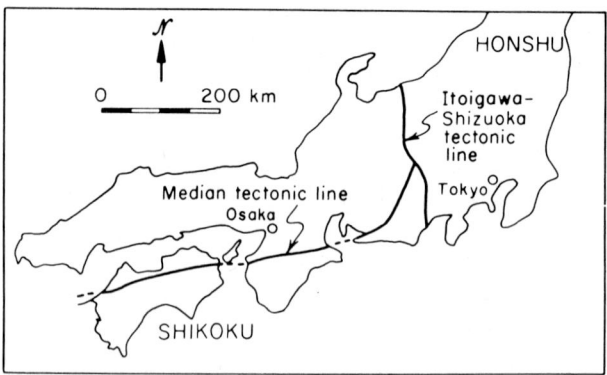

Fig. 3.19. Map of central Japan showing locations of Median Tectonic Line and Itoigawa—Shizuoka Tectonic Line.

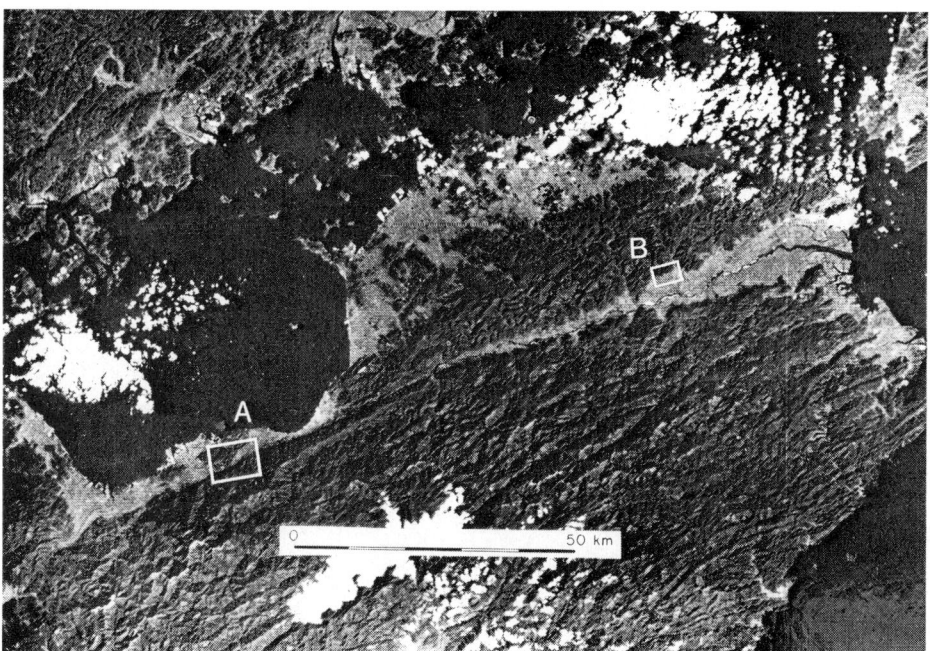

Fig. 3.20. ERTS image (No. 1112-01120) of northern Shikoku, Japan showing trace of Median Tectonic Line. Insert A shows location of Fig. 3.21; insert B, Fig. 3.22.

long segments of which have been shown by Kaneko (1966) and by Okada (1968, 1970, 1973) to have a history of significant Quaternary displacement. Several of these displaced features are portrayed in Fig. 3.21, and there is no question that at least through most of Shikoku, the fault has had both vertical and right-lateral slip in Quaternary time. Numerous closed depressions preserved along its trace suggest that some of this movement has been Holocene in age. From a thorough field study and radiocarbon dating of numerous displaced features, Okada (1973) concludes that the rate of late Quaternary displacement is some 5—10 m per thousand years — a significant rate in comparison with other regional active faults of the world. Thus, from geologic evidence alone, the fault should be recognized as an active feature over a length capable of generating large earthquakes, and this conclusion coupled with the absence of large earthquakes over the past 2000 years makes the Median Tectonic Line in this region a particularly likely candidate for a truly great earthquake in the not-too-far-distant future. This conclusion of Okada's (1970) has been the subject of considerable debate among Japanese geologists and seismologists.

To demonstrate that physiographic features of active faulting can be easily identified on aerial photographs even in areas of heavy rainfall and extensive

Fig. 3.21. Vertical aerial photo of area along Median Tectonic Line (arrows), 10 km east of Niihama City, Shikoku. Note that fault cuts terrace surface at A, with scarp 4—6 m high in gravels of 20,000—30,000 year age and younger (Okada, 1973).

cultural modification, a part of the Median Tectonic Line in the Yoshino River Valley of eastern Shikoku is shown on Fig. 3.22. The linear scar cutting across the photograph near the base of the hills is one of the principal members of the fault system. The difficulty in recognizing this feature from ground studies alone illustrates the value of high-quality, large-scale aerial photographs as a most important geologic tool in seismotectonic field studies. Under many circumstances, further insight is provided by specially flown low-altitude, low-sun-angle photos, as described by Slemmons (1969) and Sherard et al. (1974).

A second major geological discontinuity in Japan, characterized by significant late Quaternary movement but few if any major historic earthquakes, is the western boundary of the Fossa Magna, the so-called Itoigawa—Shizuoka Tectonic Line (Fig. 3.19). Like the Median Tectonic Line, the Itoigawa—Shizuoka Line is also a profound structural discontinuity, with large vertical displacement since Miocene time (Matsuda et al., 1967). It has not commonly been considered a throughgoing active fault system, although isolated segments have been recognized as displacing Quaternary deposits (Research Group for Quaternary Tectonic Map, 1973). However, long seg-

Fig. 3.22. Vertical aerial photo of Chichio fault (arrows), a member of the Median Tectonic Line system 3 km northwest of Ichiba-machi, Tokushima Prefecture. At right edge of photo, scarp is 15 m high in gravels (Okada, 1970).

ments of the zone are concealed by massive landslides, particularly near and northwest of Lake Suwa, and in my opinion it is a more continuous and significant active fault zone that has generally been realized. In many areas, Holocene alluvium is displaced along several fault strands (Fig. 3.23). Despite the fact that few major historic earthquakes have occurred in this region, and despite the fact that the seismic potential for large earthquakes is considered relatively low on the basis of the 1700—1967 record (Mogi, 1967), one may argue that the Itoigawa—Shizuoka Tectonic Line, like the Median Line, should be considered dangerous on the basis solely of geologic evidence.

What conclusions can be drawn from the rather disconcerting seismotectonic relationships in Japan? Even the 2000-year historic record is inadequate in itself for predicting seismic hazard, and the abundance of active faults throughout many parts of Japan indicates that large earthquakes can and will occur over large parts of the country. The great offshore earthquakes add further to this unhappy situation, but it now seems to be increasingly recognized that Japanese building codes must be strict, and that they must be relatively uniform throughout almost the entire country.

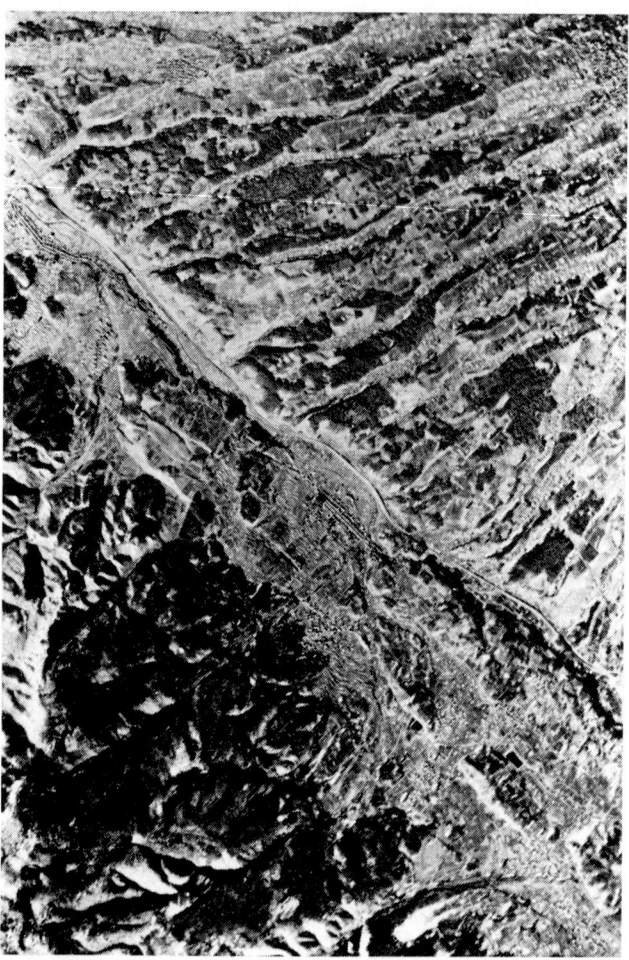

Fig. 3.23. Vertical aerial photo of Itoigawa—Shikuoka Tectonic Line about 75 km northwest of Fuji-San (Fig. 3.16). Fault trends from upper left to lower right. Note numerous displaced terraces and multiple strands within zone. Village in center is Misayamakobe, Nagano Prefecture.

3.5 PHILIPPINES

To demonstrate that geomorphic features of late Quaternary displacement are identifiable even in areas of tropical vegetation, let us turn briefly to the Philippines. On the basis of field work and a study of aerial photographs (Allen, 1962), the trace of the Philippine fault was delineated roughly along the trend that had earlier been suggested by Willis (1937) (Fig. 3.24). The writer argued from primarily geomorphic evidence that the fault was a major active regional structure of predominantly left-handed displacement (Fig.

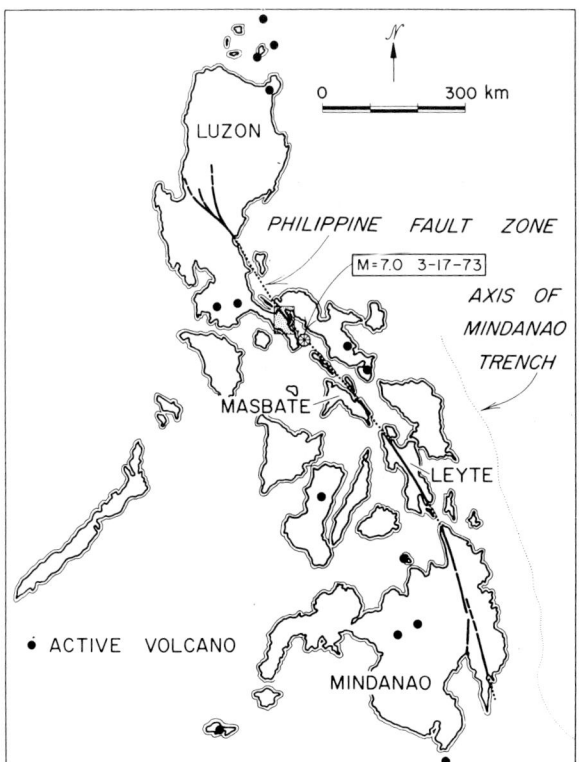

Fig. 3.24. Trace of Philippine fault zone and epicenter of 1973 Ragay Gulf earthquake. Stippled area is location of Fig. 3.26.

3.25). It was, however, somewhat disconcerting that no major earthquakes were known to have occurred along the fault during the historic record, albeit a short one, and both the continuity and sense of movement of the fault have recently been the subject of some controversy (Rutland, 1968; Gervasio, 1971). Thus a magnitude-7.3 earthquake along the projected trace of the fault in 1973 in the Ragay Gulf of southeastern Luzon (Figs. 3.26 and 3.27) was of particular interest. Subsequent investigation in the field, together with Edgar Morante of the Philippine Weather Bureau and E.V. Tamesis of the University of the Philippines, showed that the Philippine fault had indeed broken entirely across the Tayabas Isthmus of Luzon, with a maximum left-lateral displacement of 3.2 m at the point where the fault offset the beach line on the south coast at Cabong Norte (Figs. 3.26 and 3.27) (Morante and Allen, 1974). Two points should be emphasized:

(1) The fault broke essentially along the line earlier identified from aerial photographs, despite the jungle environment.

(2) The surface displacement probably would have gone unnoticed had

Fig. 3.25. Aerial view of scarps of Philippine fault zone east from near Bitulok, Luzon, toward Dingalan Bay (background). Photograph taken with help of W.R. Merrill.

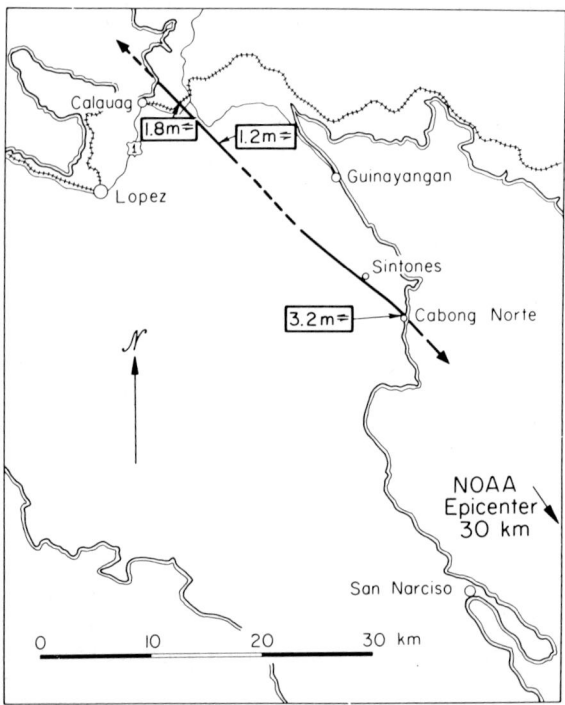

Fig. 3.26. Trace of segment of Philippine fault (heavy line) that broke in association with 1973 Ragay Gulf earthquake. See Fig. 3.24 for location. Figures in boxes represent left-lateral displacements measured in field. The break occurred along the trace earlier identified by Allen (1962) and Kimura et al. (1968).

Fig. 3.27. 3.2-meter left-lateral offset of beach line at time of 1973 Ragay Gulf earthquake, 200 m north of village of Cabong Norte (Fig. 3.26). Offset was confirmed by numerous other displaced features nearby, such as rows of Coconut trees. Break followed low pre-existing scarp.

not a railroad track been deformed attracting attention of authorities.

It is interesting to contemplate how many similar fault ruptures have gone unnoticed around the world in past years simply because no one looked for them, or because they were not considered worthy of detailed description and mapping. Within California, for example, some ten individual surface fault displacements have been described in association with earthquakes occurring in the past thirty years, as compared to only three in the preceding thirty years. It is difficult to believe that this is a result of secular changes in seismic behavior.

3.6 CHINA

China is particularly intriguing because of its many great, disastrous earthquakes and because earthquake-related deformation appears to be taking place well within a single tectonic plate — or at least along plate or miniplate boundaries that are not understood in the overall context of worldwide plate tectonics.

Mentioned earlier was the great 1556 earthquake near Sian, which caused more than 820,000 deaths. Could this region have been recognized as earthquake-prone on the basis of the geology alone? The ERTS image of Fig. 3.28 provides a quick affirmative answer; perhaps no area in the world displays more dramatic evidence of Quarternary normal faulting than the Shansi graben area of northern China, which includes Sian and the Wei Ho Valley at its southwestern end. Several good candidates for the fault causing the 1556 earthquake are visible in Fig. 3.28, and little doubt exists but that the bounding faults of the graben structure are very active features — a conclusion confirmed in the field during our recent visit there. Another of the great earthquakes in Chinese history occurred in 1303 farther northeast in the Shansi graben, near Lin-fen, and the ERTS image of that area (Fig. 3.29) shows similar relationships. Detailed seismotectonic work in the Shansi graben (Deng et al., 1973) indicates that the most intense earthquake activity has been along those bounding faults that have had the greatest Cenozoic dis-

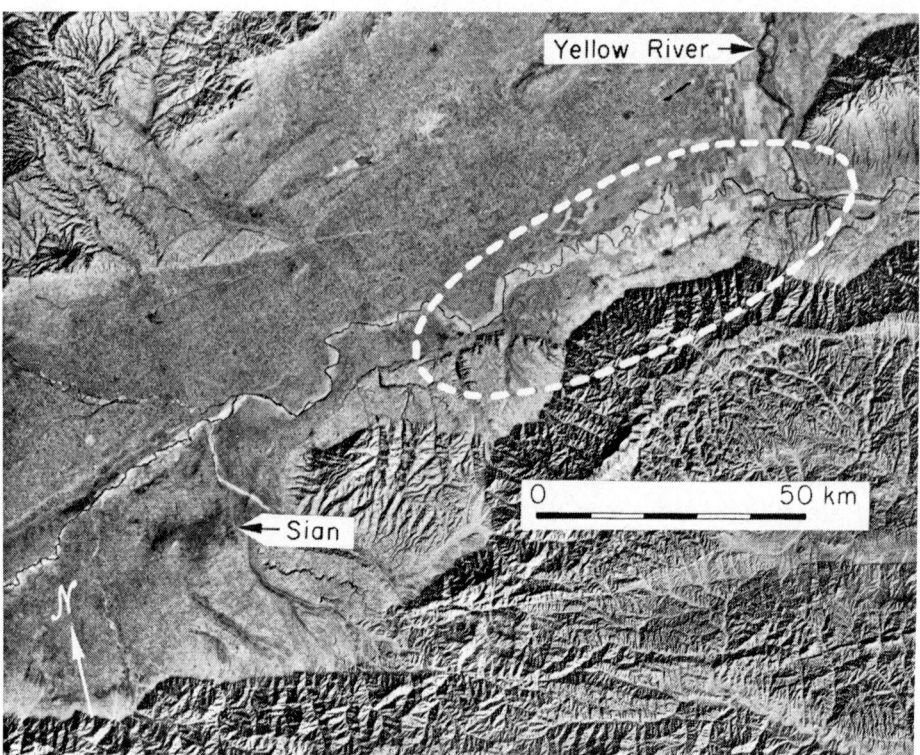

Fig. 3.28. ERTS image (No. 1575-02461) of Sian region, Shansi Province, China. White dashed line is outline of area of shaking greater than intensity XI during great 1556 earthquake. (After Kuo, 1957.)

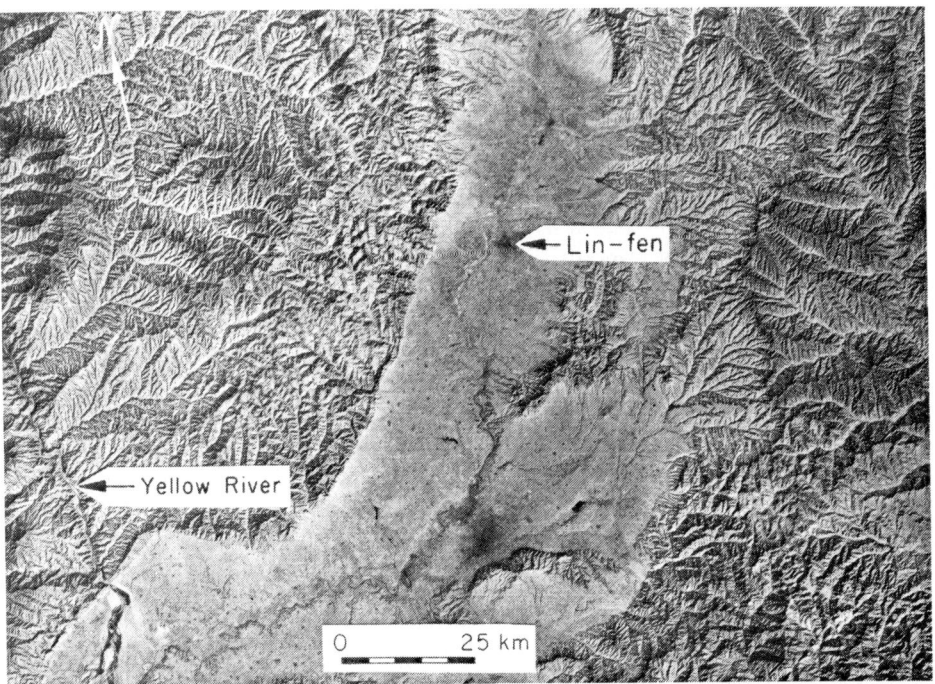

Fig. 3.29. ERTS image (No. 1524-02400) of Lin-fen region, Shansi Province, China. Epicenter of great 1303 earthquake is thought to be just north of Lin-fen. Note numerous fault scarps. Intricately dissected area is underlain mainly by Pleistocene loess.

placements, a conclusion based on subsurface stratigraphy as well as the more visible geomorphology.

Also mentioned earlier was the great earthquake of 1668, centered roughly half way between Shanghai and Peking in a region that before and since has been one of low seismicity. Furthermore, the epicentral area of the earthquake was almost *totally* quiescent for more than 150 years prior to 1668. Nevertheless (Sun Jie, unpublished data), the epicentral area of the 1668 event is athwart a major regional fault zone with abundant evidence from drilling records and geophysical surveys of significant Quaternary vertical displacement. This kind of geologic information is considered particularly important in China, because the historic record is long enough to document the fact that many of the largest earthquakes, in addition to that of 1668, have been preceded by long periods of almost complete seismic quiescence that might otherwise lead to a false sense of security. The largest earthquake of contemporary China occurred in Haiyuan, Ningsia Province, in 1920, and the epicentral area of the earthquake and its aftershocks is reported to have been a significant "seismic gap" for 280 years prior to 1920 (Li Meng-yuan, unpublished data).

The close relationship between active faults and major shallow earthquakes seems to be well accepted by Chinese geologists and geophysicists. For example, it was reported by Li Zhi-Yi, from a study of recent earthquakes in Yunnan and western Szechwan Provinces, that 61% of all earthquakes of intensity VII and greater had occurred along or near geologically mapped "big, active faults"; 80% of those equal to or greater than intensity IX were so located, and 100% of those equal to or greater than intensity X. One of the reasons for recognition of this close association between faults and earthquakes may be that there is probably no country in the world where active faults are so beautifully displayed as in China. In addition to the spectacular normal faults of the Shansi graben, numerous great linear faults of probably strike-slip origin are visible in western China. One of these (Fig. 3.30) along the axis of the Altyn Tagh (A-erh-chin) Range may be the longest continental strike-slip fault in the world, with active strands extending some 1200 km from near 80°E to 95°E. No great earthquakes are known along this and parallel structures, but the exceedingly sparse population probably explains this. Mei (1960a,b) notes that epicentral maps of China

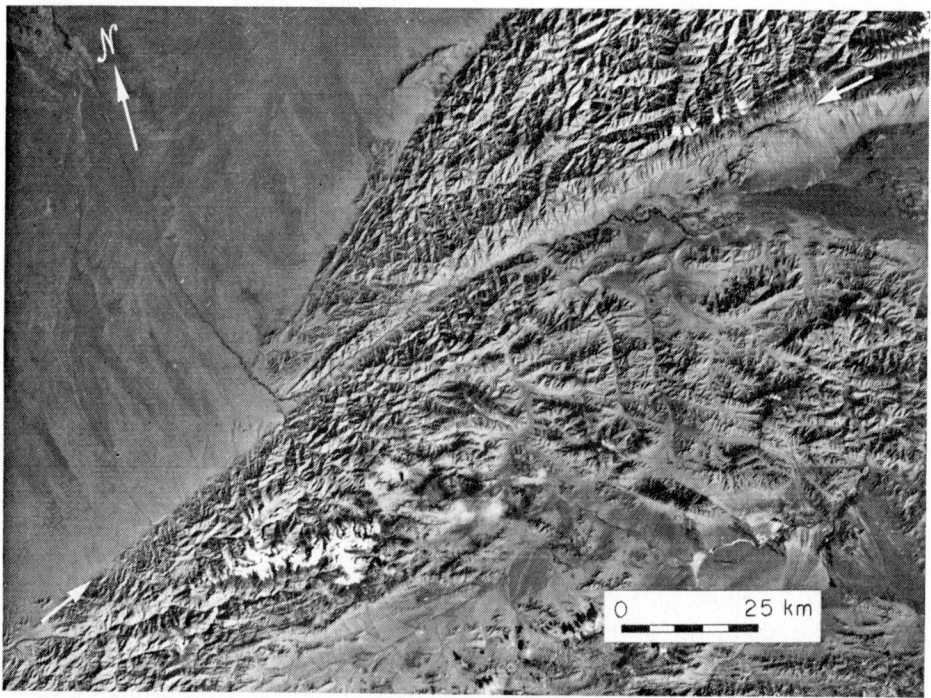

Fig. 3.30. ERTS image (No. 1074-04253) of active fault (arrows) along axis of Altyn Tagh Range, China. Linearity of fault for almost 1200 km suggests strike-slip origin. Center of image is about 37°30′N, 86°15′E.

based on historical data prior to 1900 suggest that almost all activity is limited to the highly populated eastern half of the country, whereas maps produced from instrumental data since 1900 show that in reality the great bulk of continuing activity for both large and small shocks is in sparsely populated western China.

3.7 THRUST FAULTS

Thus far little has been said about thrust faults, which have caused many of the world's most disastrous earthquakes, although probably not in as high a proportion as was once thought. They represent a difficult problem for the geologist, because thrust faults are difficult to map, and determination of the degree of activity may be even more difficult. The problem is that the traces of active thrust faults tend to be very irregular, and they often lie within areas of high relief because of the very nature of their movements. They are often concealed by massive landslides, an indirect result of the thrust displacements, and the upper plates tend to be imbricated into a complex series of individual breaks (Sherard et al., 1974). Some of these complications are illustrated in Fig. 3.31.

Fig. 3.31. Slumping along scarp of thrust fault formed in association with 1957 Gobi—Altai earthquake in Mongolia ($M$ = 8.3) (Florensov and Solonenko, 1963). Vertical movement here is 5—6 m. Note man for scale (arrow). Photo by N. Florensov.

Perhaps no part of the world better dramatizes the problem of thrust faults than does the Himalayan front, which has been the locus of four earthquakes exceeding magnitude 8.3 in the past 75 years, two of which are among the greatest ever recorded. Effects of these four events have been limited to specific and widely separated local areas, and the degree to which the rest of the Himalayan front shares the same degree of seismic potential is difficult to determine because too little is known about the Quaternary history of the frontal fault system. Until more is known on this subject, we must probably assume that the entire Himalayan front — at least the 2500 km from one syntaxial bend to the other — is capable of generating earthquakes as large as those of 1889, 1905, 1934, and 1950. This conclusion can have severe impact on the construction of badly needed facilities such as major dams. Still it seems that only on the basis of extensive regional geologic studies emphasizing the Holocene history will we have any justification for modifying it.

The San Fernando earthquake of 1971 in southern California also points up some problems and challenges in understanding active thrust faults. Where active strike-slip and normal faults are more obvious to the geologist, as in California, thrust faults may not receive the attention that they deserve in a balanced evaluation of over-all seismic hazard. Despite some reports to the contrary, significant parts of the San Fernando fault zone that slipped during the earthquake *had* been recognized and mapped prior to the event, and it *had* been recognized as an active feature in that it clearly displaced Quaternary strata (Proctor et al., 1972). Nevertheless, it is true that it had not generally been acknowledged as of high seismic potentiality, and the event came as a geologic surprise to most. It is also significant that on the day after the earthquake, much of the fault trace was difficult to recognize in the field because of slumping and landsliding that tended to conceal or eradicate the evidence of surficial displacement. To anyone who studied this fault trace in the field, it is no surprise that active thrust faults have not been more adequately identified and delineated. One lesson is certainly clear: trenching and core-drilling techniques are essential if such faults are to be adequately studied or even recognized. Particularly for thrust faults, aerial photo interpretation is no substitute for detailed geological mapping in the field. One of the challenges that geologists face in California and other seismic areas is to make sure that active thrust faults receive the attention they deserve in balanced seismotectonic studies.

3.8 CONCLUSIONS

What lessons can be drawn from this brief look at a number of seismic areas around the world?

(1) Surface faulting during large shallow earthquakes has been more com-

mon and universal than generally appreciated in many parts of the world. All too often in the past, geological field work searching for and documenting such displacements has been insufficient. Particular care should be exercised in distinguishing between effects of faulting and massive landsliding, which turns out to be a more difficult field problem than generally recognized when one tries it in a recently shaken area.

(2) Because of widespread faulting during large shallow earthquakes, perhaps the most significant geologic criterion for identifying areas of high seismicity is the late Quaternary record of similar events in the recent past. Most pertinent of all is the Holocene record. Some schools of seismotectonics have held that the *older* geologic history is especially important in establishing present-day seismic habits, particularly in terms of the boundaries of tectonic provinces that have had differing geologic histories. While present-day processes are to a large degree inherited from the past, understanding the Quaternary period is much more important than understanding earlier periods and this is where attention should first be concentrated. The farther back in geologic time that one goes, the more irrelevant are relationships to present-day tectonic processes and current seismicity. This is not an argument against general geological mapping of major faults. In fact, it is sometimes only on the basis of the distribution of older rocks that we are able to unravel where the recent breaks go and how they are mechanically related to other members of the system. But when all is said and done, it is the most recent movements on the fault that are the most relevant to earthquakes in the near future.

(3) Those parts of the world that have the longest historical records of earthquakes are the areas that should give us the greatest pause in extrapolating that history into the future, because it is clear that even a 2000- or 3000-year history is not a sufficiently valid statistical sample to use as a firm guide to over-all activity. In areas such as California and Nevada, where our historical record barely exceeds one century, we must be exceedingly cautious in extrapolating from this very short history. The problem gets even more difficult as we get farther and farther away from active plate boundaries and into areas of low long-term seismicity. What conclusions, for example, can be drawn from the single great earthquake at Charleston, South Carolina, in 1886? Is Charleston really any more dangerous in terms of another similar earthquake tomorrow than is Washington, D.C. or New York City? The single historical event tells us essentially nothing in itself except that earthquakes of the same magnitude must therefore be considered credible events, however unlikely, throughout the same entire tectonic province, at least until we understand from geological and geophysical studies why the Charleston earthquake occurred where it did and how other areas are truly different. While it is true that the one event at Charleston demonstrates that there is a structure capable of producing a large earthquake there, the fact that we have not identified that structure gives us little confidence that similar structures do

not indeed exist elsewhere. The Charleston area should be the subject of a considerably more intensive seismotectonic research effort, in view of the tremendous stakes involved in the construction of new and critical facilities such as nuclear power plants throughout the east coast area.

(4) In view of the difficulties of interpreting the historic record, and in view of the large variation of geological environments in which major earthquakes have occurred, geologists and geophysicists must continue to be exceedingly conservative in their estimates of the likelihoods of major damaging earthquakes in specific areas. We have been surprised too often in the past, and we cannot afford to be surprised too many times in the future. Every year, more and more is at stake in terms of the effects on humans of major earthquakes. At the same time, if we are to retain credibility as scientists, we must make sure that our scientific judgements are divorced from other considerations and that our conservatism is a scientific conservatism alone. Particularly in the cases of critical structures such as dams and nuclear facilities, there is an acceptable level of risk to society that cannot and should not be zero; but this determination must be made by the society as a whole. It is neither fair to ask geologists and geophysicists unilaterally to assign an acceptable level of risk at the same time they are attempting to evaluate the probability of occurrence of an unlikely disastrous event, nor should geologists and geophysicists assume that it is their right to make this judgement any more than other members of society who must both take the risk and accept the cost. The judgement of whether a given structure is or is not "safe" is a dual question that involves both the scientific probability of a disastrous event and an assignment of a level of risk that is considered *sufficiently* safe. The two questions, while both important, must be answered separately before the final conclusion is reached.

Lastly, it is worth reiterating that the most important single contribution to gaining a better understanding of long-term seismicity, which is critical to the siting and design of safe structures and to the establishment of realistic building codes, is to learn more — region by region — of the late Quaternary history of deformation, and particularly that of the Holocene epoch. More specifically, there is a special need for geomorphological studies in these regions, better and more radiometric dates, and accurate detailed geological field mapping utilizing trenches and boreholes as well as surface exposures. Studies of these kinds offer the best hope of inferring what has happened during earthquakes within the very recent geologic past, and therefore what is likely to happen again in the near future.

ACKNOWLEDGMENTS

In addition to those specifically cited in the text, the writer is deeply indebted to numerous groups and individuals in many parts of the world for assistance and participation in field studies. The advice of T. Matsuda of the University of Tokyo and of A. Okada of Aichi Prefectural University was especially useful in field work carried out in Central Japan. Particularly appreciated is the assistance of various governmental organizations and educational institutions in Turkey, Greece, Bulgaria, Japan, Philippines, China, and Pakistan. Foreign travel was supported by UNESCO, National Science Foundation (U.S.—Japan Cooperative Science Program), National Academy of Sciences (Committee on Scholarly Communication with the People's Republic of China), Carnegie Institution of Washington (G.K. Gilbert Award), and the California Institute of Technology (John Barber Fund). The writer appreciates critical review of the manuscript by A.L. Albee, H. Kanamori, and R.P. Sharp.

REFERENCES

Albee, A.L. and Smith, J.L., 1966. Earthquake characteristics and fault activity in southern California. In: R. Lung and R. Proctor (Editors), *Engineering Geology in Southern California*. Association of Engineering Geologists, Glendale, pp. 9—33.

Allen, C.R., 1962. Circum-Pacific faulting in the Philippines—Taiwan region. *J. Geophys. Res*, 67: 4795—4812.

Allen, C.R., 1969. Active faulting in northern Turkey. *Calif. Inst. Technology Contrib.* No. 1577, 32 pp.

Allen, C.R., St. Amand, P., Richter, C.F. and Nordquist, J.M., 1965. Relationship between seismicity and geologic structure in the southern California region. *Bull. Seismol. Soc. Am.*, 55: 753—797.

Ambraseys, N.N., 1970. Some characteristic features of the Anatolian fault zone. *Tectonophysics*, 9: 143—165.

Ambraseys, N.N., 1971. Value of historical records of earthquakes. *Nature*, 232: 375—379.

Ambraseys, N.N., in press. Studies in historical seismicity and tectonics: Near and Middle East. In: W. Brice (Editor), *Historical Geography of the Middle East*. Academic Press, London, Ch. 2.

Arpat, E. and Bingöl, E., 1969. The rift system of western Turkey; thoughts on its development. *Turkey Miner. Res. Explor. Inst. Bull.*, No. 73: 1—9.

Arpat, E. and Şaroğlu, F., 1972. The east Anatolian fault system; thoughts on its development. *Turkey Miner. Res. Explor. Inst. Bull.*, No. 78: 33—39.

Atwater, T., 1970. Implications of plate tectonics for the Cenozoic tectonic evolution of western North America. *Geol. Soc. Am. Bull.*, 81: 3513—3592.

Bonilla, M.G. and Buchanan, J.M., 1970. Interim report on worldwide historic surface faulting. *U.S. Geol. Surv., Open File Rep.*

Clark, M.M., Grantz, A. and Rubin, M., 1972. Holocene activity on the Coyote Creek fault as recorded in sediments of Lake Cahilla. *U.S. Geol. Surv. Prof. Pap.*, 787: 112—130.

Clark, R.M., Dibble, R.R., Fyfe, H.E., Lensen, G.J. and Suggate, R.P., 1965. Tectonic and earthquake risk zoning. *R. Soc. N.Z. Trans*, 1: 113—126.

Deng, Chi-tung, Wang, Ke-lu, Wang, Yi-lu, Tang, Han-jun, Wu, Yu-wen and Ding, Menglin, 1973. On the tendency of the seismicity and the geological framework of the seismic belt of the Shansi graben. *Sci. Geol. Sin.*, 1973 (1): 37—47 (in Chinese).

Florensov, N.A. and Solonenko, V.P. (Editors), 1963. *The Gobi—Altai Earthquake.* USSR Academy of Sciences, Moscow, 392 pp. (in Russian).
Gervasio, F.C., 1971. Geotectonic development of the Philippines. *Philipp. Geol.*, 25: 18—38.
Jennings, C.W., 1973. Preliminary fault and geologic map, State of California. *Calif. Div. Mines Geol., Prelim. Rep.*, 13.
Kanamori, H., 1973. Mode of strain release associated with major earthquakes in Japan. *Ann. Rev. Earth. Planet. Sci.*, 1: 213—239.
Kaneko, S., 1966. Transcurrent displacement along the median line, southwestern Japan. *N.Z. J. Geol. Geophys.*, 9: 45—59.
Kelleher, J., Sykes, L. and Oliver, J., 1973. Possible criteria for predicting earthquake locations and their applications to major plate boundaries of the Pacific and the Caribbean. *J. Geophys. Res.*, 78: 2547—2585.
Ketin, I., 1948. Über die tektonisch-mechanischen Folgerungen aus den grossen anatolischen Erdbeben des letzten Dezenniums. *Geol. Rundsch.*, 36: 77—84.
Ketin, I. and Roesli, F., 1953. Makroseismische Untersuchungen über das nordwestanatolische Beben vom 18 Marz 1953. *Eclogae Geol. Helv.*, 46: 187—208.
Kimura, T., Tokiyama, A., Gonzalez, B.A. and Andal, D.F., 1968. Geologic structure in the Tayabas Isthmus district, Philippines. *Geol. Paleontol. Southeast Asia*, 4: 156—178.
Kuno, H., 1936. On the displacement of the Tanna fault since Pleistocene. *Bull. Earthquake Res. Inst., Tokyo Univ.*, 14: 619—631.
Kuo, Tseng-chien, 1957. On the Shensi earthquake of January 23, 1556. *Acta Geophys. Sin.*, 6: 59—68 (in Chinese).
Matsuda, T., 1974. Surface faults associated with the Nobi (Mino-Owari) earthquake of 1891, Japan. *Tokyo Univ. Earthquake Res. Inst. Spec. Bull.*, 13: 85—126 (in Japanese).
Matsuda, T. and Okada, A., 1968. Studies of active faults in Japan. *Quat. Res. Tokyo*, 7: 188—199 (in Japanese).
Matsuda, T., Nakamura, K. and Sugimura, A., 1967. Late Cenozoic orogeny in Japan. *Tectonophysics*, 4: 349—366.
McKenzie, D., 1972. Active tectonics of the Mediterranean region. *Geophys. J. R. Astron. Soc.*, 30: 109—185.
Mei, Shi-yun, 1960a. Characteristics of earthquake activity in China. *Acta Geophys. Sin.*, 9: 1—19 (in Chinese).
Mei, Shi-yun, 1960b. The seismic activity of China. *Izv. Geophys. Ser.*, 1960: 381—395 (in Russian).
Minster, J.B., Jordan, T.H., Molnar, P. and Haines, E., 1974. Numerical modelling of instantaneous plate tectonics. *Geophys. J. R. Astron. Soc.*, 36: 541—576.
Mogi, K., 1967. Regional variations in magnitude—frequency relation of earthquakes. *Bull. Earthquake Res. Inst., Tokyo Univ.*, 45: 313—325.
Morante, E.M. and Allen, C.R., 1974. Displacement on the Philippine fault during the Ragay Gulf earthquake of 17 March 1973. *Geol. Soc. Am. Abstr. Progr.*, 5: 744—745 (abstract).
Okada, A., 1968. Strike-slip faulting of late Quaternary along the median dislocation line in the surroundings of Awa-Ikeda, northwestern Shikoku. *Quat. Res. Tokyo*, 7: 15—26 (in Japanese).
Okada, A., 1970. Fault topography and rate of faulting along the median tectonic line in the drainage basin of the River Yoshino, northeastern Shikoku, Japan. *Geogr. Rev. Japan*, 43: 1—21 (in Japanese).
Okada, A., 1973. Quaternary faulting along the median tectonic line in the central part of Shikoku. *Geogr. Rev. Japan*, 46: 295—322 (in Japanese).
Omori, F., 1906. Preliminary note on the cause of the California earthquake of 1906.

In: D.S. Jordon (Editor), *The California Earthquake of 1906*. Robertson, San Francisco, pp. 281—318.
Proctor, R.J., Crook, R., Jr., McKeown, M.H. and Moresco, R.L., 1972. Relation of known faults to surface ruptures, 1971 San Fernando earthquake, southern California. *Geol. Soc. Am. Bull*, 83: 1601—1618.
Research Group for Quaternary Tectonic Map, 1973. *Quaternary Tectonic Map of Japan.* Natl. Research Center for Disaster Prevention, Tokyo.
Richter, C.F., 1958. *Elementary Seismology*. Freeman and Cooper, San Francisco, 768 pp.
Rutland, R.W.R., 1968. A tectonic study of part of the Philippine fault zone. *Geol. Soc. London Q. J.*, 123 (for 1967): 293—325.
Seymen, I. and Aydın, A., 1972. The Bingöl earthquake fault and its relation to the North Anatolian fault zone. *Turkey Miner. Res. Explor. Inst. Bull.*, 79: 1—8.
Sherard, J.L., Cluff, L.S. and Allen, C.R., 1974. Potentially active faults in dam foundations. *Geotechnique*, 24: 367—428.
Slemmons, D.B., 1969. New methods of studying regional seismicity and surface faulting. *Am. Geophys. Union Trans.*, 50: 397—398.
Willis, B., 1944. Geologic observations in the Philippine archipelago. Natl. Res. Counc. Philipp. Bull., 13.
Yamasaki, N. and Tada, F., 1928. The Oku-Tango earthquake of 1927. *Bull. Earthquake Res. Inst. Tokyo Univ.*, 4: 159—177.

*Chapter 4*

SOIL DYNAMICS: BEHAVIOR INCLUDING LIQUEFACTION

EZIO FACCIOLI and DANIEL RESÉNDIZ

*U.N.E.S.C.O., Proyecto Dinámica de Suelos, Universidad Nacional Autónoma de México, Mexico*
*Instituto de Ingeniería, Universidad Nacional Autónoma de México, Mexico*

4.1 INTRODUCTION

*4.1.1 Nature of soils: Phases and stresses*

Soil is an aggregate of discrete particles whose voids are filled with air and/or water. Hence, soil is a two-or-three phase material whose state of stress is fully described only if the stresses corresponding to each phase are given. If, for practical reasons and simplicity, only water-saturated and dry soils are considered, it is still necessary to deal in the first case with the stress in the solid skeleton (effective stress) and that in water (pore pressure); only in coarse cohesionless dry soils will the total stress equal the effective stress.

Pore water pressure may result from steady-state flow of water through the pores of soil (steady-state pore pressure) or from transient flow induced by a tendency of the pore space to change in volume (induced pore pressure).

Steady-state pore pressure does not depend on soil properties but on hydraulic conditions which are independent of the soil response to external loads. This component of pore pressure is therefore an independent variable from the standpoint of soil mechanics. On the other hand, induced pore pressure does depend on the mechanical properties of soils, i.e. on the permeability and susceptibility of soil to volume change under stress.

*4.1.2 Drainage conditions in earthquake problems*

In general, the drainage condition, i.e. the degree of pore pressure dissipation during the loading process, depends on the rate of stress, the coefficient of permeability of the soil, and the geometry and boundary conditions of the prototype.

Given the permeability of natural soils to water, in soil dynamics the rate of deformation is high enough to make water migration during the loading process negligible under field conditions. Therefore, most fully saturated or partially saturated soils deform under conditions of nearly constant water content; in fully saturated soils this means that deformation occurs at constant volume.

Conversely, given the high soil permeability to air and the very high compressibility of air, most dry coarse-grained soils undergo free-drainage deformation or nearly so. Free-drainage conditions may also prevail in extremely coarse (particle size in the order of one inch or larger), saturated soils.

Intermediate cases obviously exist. They are those of dry soils with particle size in the order of 0.1 mm (fine sands), and saturated soils with permeability in the order of $10^{-1}$ cm/sec. For these intermediate cases it is safe and probably acceptable to proceed under the assumption of constant water content deformation. Cases where this may be unacceptably conservative would be handled by estimating upper and lower bounds of soil response from the alternate assumptions of fully-drained and constant-volume deformation, or else, in saturated soils, by solving the equations describing the process of volume change of the soil under load simultaneously with the equations of transient flow of water in a porous medium. Details of this last approach are outside the scope of this chapter.

*4.1.3 Independent variables and test conditions in soil dynamics*

In general, the mechanical behavior of a soil element depends on its initial state (void ratio, degree of saturation, structure, and state of stress) as well as on the way stress increments are applied (stress path, stress rate, and drainage conditions). Therefore, the mechanical properties of soils should ideally be determined on samples for which the preceding variables are the same as in the prototype.

It has been repeatedly found in the laboratory that the effects of those seven variables can be taken into account with good approximation by means of three independent factors: strain rate, stress path, and state of effective stress, the latter being the dominant factor. This is an expression of the so-called principle of effective stress.

According to this principle, every soil property would have to be determined and expressed in terms of effective stress. Practical use of the principle depends on the possibility of predicting pore pressure in field conditions. This is straightforward in free-draining soils as indicated in Section 4.1.2, since in them pore pressure is independent of deformation, and duplication of field conditions to determine soil properties in the laboratory simply requires running drained tests at the appropriate stress level. However, prediction of pore pressure in less pervious soils is far more complex, except when the acting stress changes slowly enough to prevent development of induced pore pressure, which is not the case in soil dynamics.

However, under a given load process at constant water content the induced pore pressure is essentially a function of the initial state of the soil element (Bishop and Eldin, 1950) or, in other words, it is a function of the prior effective-stress history. If we consider this fact together with the principle of effective stress, we conclude that the induced pore pressure, and hence the effective-stress path and the whole behavior of two soil elements strained at constant water content, are the same if and only if three factors are reproduced in both elements, namely, the initial state, the total stress path *, and the loading rate (though most experimental evidence indicates that, within the range of loading rates of interest in earthquake engineering, the latter factor generally has a second-order effect only). This conclusion constitutes the basis for selecting the relevant variables and test conditions in engineering problems where soil is involved.

As explained in Section 4.1.2, soil dynamics problems may be grossly classified in two categories: one where free-drainage conditions prevail and another where soil deformation occurs at constant water content. In the first, pore pressure is an independent variable and drained tests on dry or saturated specimens have to be used to determine the relevant soil properties. Problems in the second category are the most common in earthquake engineering; in them, pore pressure is a dependent variable and hence it is simpler, more reliable, and more direct not to include it explicitly in soil response analysis. † Dealing with problems in this category requires tests on representative samples where at least two of the three relevant factors pointed out (initial state and stress path) are carefully controlled to reproduce the conditions of the prototype.

*4.1.4 Stable and unstable soil conditions*

Consider two specimens of a cohesionless soil at initial void ratios $e_1$ and $e_2$, subject to consolidated-drained triaxial tests such that shear in both specimens starts at the same effective consolidation pressure $\bar{\sigma}_c$. If $e_1$ is sufficiently high and $e_2$ low enough, the behavior of the specimens will exhibit the qualitative difference illustrated in Fig. 4.1. In specimen 1 the shear stress increases continuously with strain up to a maximum at point $s_1$ and

---

* In a saturated soil only the deviatoric stress path has to be reproduced, since any increment of hydrostatic stress leaves the effective stress unchanged.
† Dealing explicitly with pore pressure in this type of problem is tantamount to a circular manipulation of variables. Suppose one decides to deal explicitly with pore pressure (i.e. work in terms of effective stress) in a soil dynamics problem where the soil deforms at constant water content. To do so, one has to predict the induced pore pressure in the field, which requires performing undrained tests on representative samples where the initial state and the field stress-path are reproduced. These are the same tests necessary to make a total-stress type of analysis except that, if an effective stress analysis is preferred, induced pore pressures have to be measured. These measurements are then used to predict the induced pore pressure in the field and, from this and the total stress, to predict effective stress. Therefore, if pore-pressure measurements are perfect, the effective-stress method of analysis is exactly equivalent to the total-stress method, except that the latter is simpler and more direct. Since pore-pressure measurements are far from perfect, when induced pore pressure exists the total-stress method of analysis is not only simpler, but better.

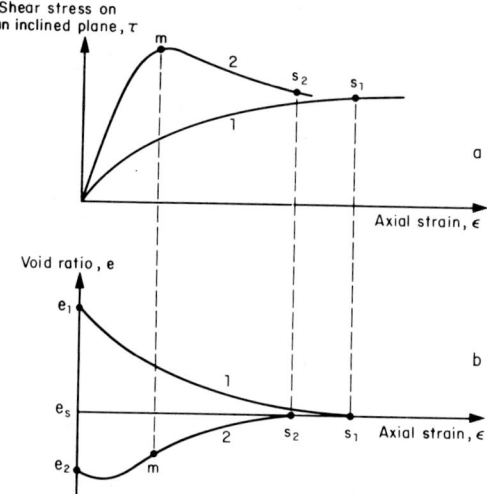

Fig. 4.1. Shear stress (a), and volume change (b) vs. axial strain in loose (*1*) and dense (*2*) free-draining specimens of a cohesionless soil.

remains constant thereafter (Fig. 4.1a). In the same specimen, the void ratio decreases continuously during the test until a final void ratio $e_s$ (steady-state void ratio; Poulos, 1971) is attained which remains unchanged with further strain (Fig. 4.1b). Specimen 2, which is much denser, exhibits a peak at point $m$ in its stress—strain curve, and shows a continuous decrease in shear stress under additional strain until a final strength is reached which approximates that of specimen 1 (Fig. 4.1a). As to its volume vs. strain behavior, specimen 2 first contracts slightly and then dilates continuously until approximately the same steady-state void ratio $e_s$ is reached as for specimen 1 (Fig. 4.1b). The peak in the stress—strain curve of specimen 2 appears when the volume is increasing at the maximum rate.

It can be experimentally shown that specimens of the same sand sheared after consolidation at void ratio $e_s$ first contract slightly and then expand by the same amount so that, when subject to large shear deformation, they undergo no net change in volume. These specimens also exhibit a peak in the stress—strain curve at the particular strain level where the ratio of increase in volume is maximum.

The foregoing behavior implies that, under the same test conditions, a void ratio exists, $e_L > e_s$, such that specimens prepared at void ratios equal to or above $e_L$ reduce in volume continuously during the test. Running series of tests as described leads to the determination of $e_L$ and $e_s$ as functions of the effective consolidation pressure, $\sigma_c$, and therefore a plot similar to that shown in Fig. 4.2 can be constructed, which will be called a state diagram (Poulos, 1971). The $e_s$-line defines Casagrande's critical void ratio (Casagrande, 1936, 1976;

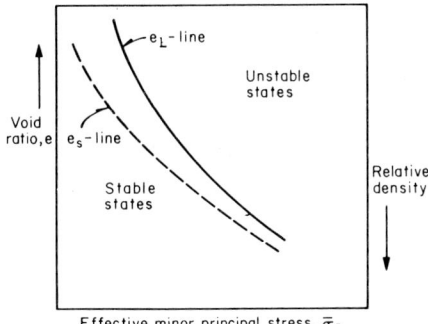

Fig. 4.2. State diagram showing the regions of stable and unstable states in cohesionless soils. The two lines represent combinations of $e$ and $\sigma_3$ associated with steady-state deformation.

Watson, 1940; Castro, 1969). Every specimen that is sheared starting from conditions represented by a point above the $e_L$-line will be continuously contractive during shear; any specimen sheared under conditions represented by a point below the $e_L$-line will show an increase in volume at some stage during the test.

The previous statements regarding the relationship between initial void ratio and tendency to volume change in cohesionless soils apply to monotonic shear and large-amplitude cyclic deformation. Under vibratory disturbances producing no shear strain or a very small one, every cohesionless soil tends to decrease in volume irrespective of density (see Section 4.4.2).

For reasons explained previously, in soil dynamics we are mainly interested in the behavior of soils under undrained conditions. A simple connection exists between drained and undrained behavior: a contractive behavior in drained conditions implies the development of positive induced pore pressure if drainage is prevented, since the tendency of the grains to accomodate in a denser state transfers stress from grain-to-grain contacts to grain-to-water contacts. Conversely, a tendency to dilate during shear results in negative induced pore pressure.

In cohesionless soils the $e_L$-line can be viewed as the boundary between two drastically different types of soil vehavior under undrained shear. Above the $e_L$-line, undrained deformation is necessary accompanied by a continuous loss of strength which can lead to liquefaction; below the $e_L$-line, liquefaction is impossible *, since the soil will stiffen, i.e. develop negative pore pressure, when sheared beyond a certain strain regardless of the stress or strain path. It will be said that a soil is in *stable conditions* if it is below

---

* Casagrande's concept of liquefaction is implied. Accordingly, liquefaction may be defined as a steady state of deformation at constant volume and under a constant resistance much lower than that exhibited by the soil before perturbation (Castro, 1969).

the $e_L$-line in the stage diagram; otherwise, it is in *unstable conditions*.

The $e_L$-line is introduced for the sake of conceptual clarity. In practice the boundary between stable and unstable states has been determined by Castro (1969, 1972) directly from the results of consolidated-undrained tests in which liquefaction has occurred; such a line is called the $e_f$-line (see Section 4.4.3). It remains to be investigated whether the $e_f$ and the $e_L$-lines coincide.

The treatment of engineering problems involving soils in stable or unstable conditions calls for quite different approaches. In soils subject to significant shear stress under unstable conditions, slight perturbations may lead to liquefaction; their detailed stress—strain relationships do not matter and it is generally enough for engineering purposes to determine that the soil is effectively in an unstable state. On the contrary, in stable soils the stress—strain relationships are important since they are necessary to predict the response of the structure or foundation to the design excitations.

In fine-grained, cohesive soils, the same line of reasoning could be followed, except that the cohesive bonds between particles add to the strength of the soil derived from friction and effective stress. Hence, liquefaction of these soils is only conceivable under void ratios high enough to abate the cohesive shear strength component to an extremely low value, i.e. a few grams per square centimeter. Therefore, it will be postulated that unstable conditions in cohesive soils are simply defined by $e > e_{LL}$, where $e_{LL}$ is the void ratio corresponding to the liquid limit. In general, $e_{LL} \geqslant e_L$ for every state of stress. This definition is compatible with the conditions encountered in the cases where liquefaction slides have been reported in clay deposits (Rosenqvist, 1953).

## 4.2 STRESS—STRAIN RELATIONSHIPS UNDER STABLE CONDITIONS

### 4.2.1 Introduction

Soil structure, defined as the relative arrangement of solid particles, was previously pointed out as one of the main factors influencing the mechanical behavior of a soil element. Furthermore, the rather weak bonds between different phases and between the solid particles themselves allow considerable changes in structure even under small load increments. Hence, in a sense, soil changes as a material during loading so that its properties are functions of the loading process. This precludes in practice the use of simple constitutive equations of a general nature to describe soil response to dynamic loads; instead, soil properties have to be determined under laboratory or field conditions duplicating those expected in the prototype. These conditions are fixed in terms of the significant variables discussed in Section 4.1.3. Since soils in unstable conditions are unable to withstand combinations of static and dynamic shear stress without large deformation, only the stress—strain relationships of stable soils will be discussed.

Four properties that fully characterize soil behavior under dynamic loads will be analyzed, namely: shear modulus $G$ (or modulus of elasticity $E$) for small-amplitude cyclic deformation, internal damping, stress—strain relationships for large-amplitude cyclic deformation, and strength under cyclic loading. Poisson's ratio is another property required for description of dynamic soil response; however, it varies within relatively close limits and affects but slightly the seismic response, so that its detailed investigation is seldom justified. *

Table 4.I is a summary of laboratory techniques to determine the dynamic properties of interest. Field techniques are reviewed in Section 4.5.2. In each practical case, the best technique is the one that closest approximates field conditions as to initial state, dynamic stress and strain amplitude (representativeness of the specimen as to its intrinsic characteristics is obviously assumed). Generally, no single test satisfies all the requirements; for instance, if the stress—strain curve for large cyclic deformations is required, it may be convenient to use a low-amplitude test to define the initial modulus and a large-amplitude test to determine the rest of the curve. We shall use Table 4.I as a background reference in the following discussion.

### 4.2.2 Shear modulus for small-amplitude vibration

Under first loading, a soil specimen subject to shear undergoes deformations that are partially irreversible, irrespective of strain amplitude; hence, stress—strain curves in loading and reloading do not coincide. If strain amplitude is small, the difference between successive reloading curves tends to disappear after a few cycles of similar amplitude and the stress—strain curve becomes a closed loop that can be described by two parameters: the average slope and the enclosed area. The first parameter defines the shear modulus, and the second the internal damping.

The main factors affecting the shear modulus of soils in general are shear strain amplitude, initial effective stress, void ratio, and shear-stress level. In addition, stress history, degree of saturation, load frequency, temperature and thixotropy have various degrees of influence in cohesive soils.

#### 4.2.2.1 Cohesionless soils

The dominant factors are strain amplitude, mean effective stress, $\bar{\sigma}_m$, and void ratio.

For strain amplitudes below $10^{-4}$, these soils exhibit a nearly constant

---

* It has been found that Poisson's ratio under dynamic loads varies between 0.25 and 0.35 for cohesionless soils and between 0.4 and 0.5 for cohesive soils. It is independent of frequency in the range of interest of earthquake engineering and, in contrast with $E$ and $G$, is insensitive to thixotropic effects (Crandall et al., 1970).

TABLE 4.I

Laboratory methods for determining dynamic soil properties

| Test conditions | Test and reference | Properties measured | Stress or strain conditions | | Strain amplitude |
|---|---|---|---|---|---|
| | | | initial | dynamic | |
| Low frequency | cyclic triaxial (Seed and Chan, 1966; Castro, 1969) | $E, D$; stress vs. strain; strength | axisymmetric consolidation | pulsating axial or confining stress; constant-amplitude stress | $10^{-4}$ to $10^{-1}$ |
| | cyclic torsion (Zeevaert, 1967; Hardin and Drnevich, 1972a and b) | $G, D$; stress vs. strain | axisymmetric consolidation | pulsating shear stress; constant-amplitude stress or free vibration | $10^{-4}$ to $10^{-2}$ |
| | cyclic simple shear (Seed and Wilson, 1967) | $G, D$; stress vs. strain; strength | $K_0$-consolidation | pulsating shear stress; constant-amplitude stress (strain) or free vibration | $10^{-4}$ to $3 \cdot 10^{-2}$ |
| High frequency | ultrasonic (Lawrence, 1965; Nacci and Taylor, 1967) | $c_p$ or $c_s$ | axisymmetric consolidation | dilation or shear; single pulse wave | $10^{-6}$ |
| | resonant column (Afifi, 1970) | $c_p$ or $c_s$; $E$ or $G$; $D$ | uniform or axisymmetric consolidation | pulsating axial or shear (torsional) stress; constant-amplitude strain | $10^{-6}$ to $10^{-2}$ |

shear modulus, $G_{max}$. Hardin and Richart (1963) and Hardin and Black (1968) propose eqs. 4.1 for $G_{max}$, in terms of void ratio, $e$, and mean effective stress, $\bar{\sigma}_m$. Equation 4.1a applies to soils with rounded grains and eq. 4.1b to those with angular grains; $G_{max}$ and $\bar{\sigma}_m$ are in psi.

$$G_{max} = 2630 \frac{(2.17 - e)^2}{1 + e} \bar{\sigma}_m^{0.5} \tag{4.1a}$$

$$G_{max} = 1230 \frac{(2.97 - e)^2}{1 + e} \bar{\sigma}_m^{0.5} \tag{4.1b}$$

Although Richart et al. (1970) claim that a number of independent investigations confirm eqs. 4.1 within ±10%, it is conceivable that the intrinsic soil characteristics, together with the effect of test conditions other than $e$ and $\bar{\sigma}_m$, give rise to larger deviations. Therefore, it is advisable to retain the portions of eqs. 4.1 expressing the influence of $e$ and $\bar{\sigma}_m$, while determining the numerical coefficient by means of one of the last two tests in Table 4.I.

The coefficients in eqs. 4.1 decrease appreciably for shear-strain amplitudes greater than $10^{-4}$ (Hardin and Richart, 1963; Hardin and Black, 1966). Also, a minor effect of stress history has been observed (Drnevich et al., 1967).

*4.2.2.2 Cohesive soils*

For shear-strain amplitudes lower than approximately $10^{-5}-10^{-4}$, effects of void ratio, mean effective stress, and stress history (the latter represented by the overconsolidation ratio $OCR$) can be expressed by eq. 4.2, where $G_{max}$ and $\bar{\sigma}_m$ are in psi and $K$ is a function of the plasticity index $I_p$, as shown in Fig. 4.3 (Hardin and Black, 1969):

$$G_{max} = 1230 \frac{(2.97 - e)^2}{1 + e} (OCR)^K \bar{\sigma}_m^{0.5} \tag{4.2}$$

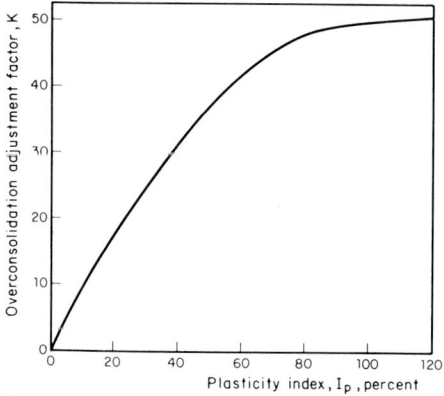

Fig. 4.3. Overconsolidation adjustment factor $K$, used in eq. 4.2. After Hardin and Black (1969).

As in the case of eqs. 4.1, the coefficient in eq. 4.2 should be determined from tests on representative samples.

The initial shear-stress level seems to have no significant effect on $G_{max}$ (Hardin and Black, 1966 and 1968). However, as in cohesionless soils, the vibration amplitude causes the shear modulus to decrease. Seed and Idriss (1970c) propose the empirical relationship shown in Fig. 4.4 to describe the variation of shear modulus with shear-strain amplitude. $G$ is normalized with respect to undrained strength, $s_u$, so that the absolute variations due to intrinsic soil properties are eliminated, and the logarithm of the ratio $G/s_u$ is plotted vs. the logarithm of the shear-strain amplitude $\gamma$. Test data obtained by other investigators are included in Fig. 4.4.

Due to thixotropy, several time-dependent effects should be taken into account when determining the shear modulus in cohesive soils.

Crandall et al. (1970) found that $G_{max}$ depends on the deformation history of the specimen in such a way that, immediately after inducing shear deformations larger than $10^{-4}$, $G_{max}$ decreases as much as 20% and then progressively increases with time up to its original value; a similar finding has been reported by Anderson (1974). Also, Afifi (1970) reported losses of 15—20% in the stiffness developed under a constant effective stress acting $10^4$ min. after a sudden increase in confining pressure of 10 psi; stiffness was regained with time under constant effective stress.

Elapsed time after primary consolidation also affects $G$. Marcuson and Wahls (1972) propose eqs. 4.3 to account for this effect in kaolinites and bentonites, respectively:

$$G_r = 1.0 + 0.046\, T_r \tag{4.3a}$$

$$G_r = 1.0 + 0.242\, T_r \tag{4.3b}$$

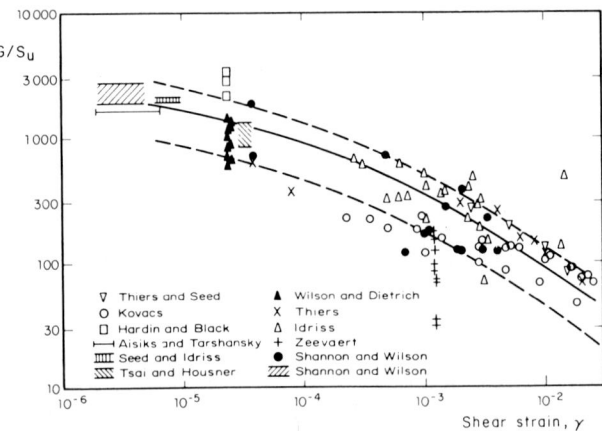

Fig. 4.4. Normalized shear modulus, $G/s_u$, vs. shear strain in cohesive soils. After Seed and Idriss (1970c).

$G_r$ is the ratio of the shear modulus at the consolidation time of interest to the shear modulus at 100% primary consolidation, and $T_r$ is the ratio of the consolidation time of interest to the time of 100% consolidation.

The last effect implies that, when laboratory tests are performed to determine $G$ in freshly consolidated samples, an extrapolation is necessary, using eqs. 4.3 or other appropriate means, in order to estimate the value of $G$ that applies to field conditions. Stokoe and Woods (1972) and Stokoe and Richart (1973) report that, when a value of $T_r$ corresponding to the age of the deposit is used, the discrepancy between resonant-column and cross-borehole test results are small. Anderson (1974) has found that a 20-year extrapolation of laboratory results gives, in most cases, good estimates of field values.

Data cited by Hardin and Drnevich (1972a) are indicative as to the effects of degree of saturation. They report a case where $G$ decreases by more than 50% when the degree of saturation was increased from 70 to 100%.

The same authors (1972a) have also found that frequency has relatively minor effects on $G$ when data are analyzed properly. *

### 4.2.3 Internal damping

Evidence of energy dissipation during dynamic loading of a soil element is provided by the hysteresis loops of the stress—strain diagram, the energy that has to be fed into the soil to keep a steady state of free vibration, the finite resonance amplitude, and the amplitude decay in free vibration. In turn, each of these observations provides a way to measure internal damping.

Some of the parameters used to measure internal damping are the specific damping $\psi$; the logarithmic decrement $\delta$; the phase angle between force and deformation $\varphi$; and the damping ratio $D$. The most common are the damping ratio (ratio of viscous to critical damping) and the logarithmic decrement (decrease of the logarithm of vibration amplitude in a cycle of free vibration). Under resonance or free-vibration conditions, all these parameters are related through the expression:

$$\psi = 2\delta = 2\pi\varphi = 4\pi D/(1 - D^2)^{1/2} \tag{4.4}$$

Damping ratio is independent of frequency in dry sands (Hardin, 1965), rocks (White, 1965), and plasticine (Crandall et al., 1970), thus indicating that the fundamental mechanism of energy dissipation in these materials is hysteretic rather than viscous. Shape and area of the hysteresis loop do not depend on loading rate; hence, $D$ does not depend on frequency. However, they are sensitive to strain amplitude; in addition, $D$ is affected by the state of effective stress, the water content, and, in clays, by the load history (Crandall et al., 1970).

---

* Taylor and Hughes (1965) report a significant effect of frequency on $G$, but do not provide conclusive data.

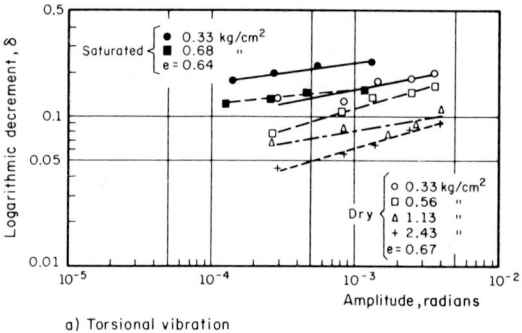
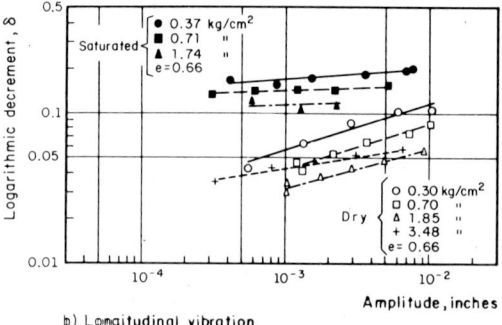

Fig. 4.5. Effects of vibration amplitude, confining pressure and saturation on logarithmic decrement of Ottawa sand. After Hall and Richart (1963).

Effects of vibration amplitude, confining pressure, and water content on the logarithmic decrement of a sand are shown in Fig. 4.5. Note that $\delta$ increases with strain amplitude and degree of saturation but ordinarily decreases with confining pressure. However, $\delta$ increases with confining pressure for very small values of the latter parameter (Biot, cited by Dobry, 1970). This change of behavior seems to result from the influence of two opposite phenomena, namely (Dobry, 1970): (1) the larger the intergranular forces, the greater the energy dissipation occurring at a given deformation; and (2) the larger the intergranular forces, the greater the rigidity and the smaller the deformation induced by a given stress increment.

*4.2.4 Stress—strain relationships in large-amplitude cyclic deformation*

When large cyclic deformations are expected, one may be interested in a realistic description of complete stress—strain relationships. This can be accomplished by first obtaining the stress—strain curve under monotonic loading and then defining the shape of the hysteresis loops by simple rules in accordance with observed experimental facts. In particular, the hysteretic loops should have a secant modulus $G$, that varies with strain amplitude, and a damping ratio $D$, independent of frequency. If an equivalent viscous model is used, the requirement of constant $D$ is fulfilled by a viscosity coefficient inversely proportional to frequency.

Dobry (1970) discusses several hysteretic models made up by their basic stress—strain curves and the corresponding geometric rules to construct hysteresis loops. He concludes that the Ramberg-Osgood model, illustrated in Fig. 4.6, shows both analytical advantages and good agreement with experiments. This model is characterized by a yield point $(\tau_y, \gamma_y)$ defining the limit of linear behavior, an initial shear modulus $G_{max}$, and two parameters $\alpha$ and $r$. The linear elastic model ($\alpha = 0$) and the perfect elasto-plastic one ($r \to \infty$) are included as limiting cases.

The shear modulus in the Ramberg-Osgood model decreases monotonically with strain amplitude beyond the yield point, whereas the damping ratio increases asymptotically toward an upper value. The degree of approximation of the model to experimental results is shown in Fig. 4.7 for dry sands and in Fig. 4.8 for the very compressible Mexico City clays.

*4.2.5 Strength under cyclic loading*

Several investigators have studied the effects of impulsive loading on soil strength. They have found that, when time to failure is reduced from a few minutes to several milliseconds, the strength of cohesionless soils increases by 10—20% (Casagrande and Shannon, 1948; Seed and Lundgren, 1954) and that of cohesive soils by 140—260%. Results of this type of tests, however, are of limited significance in earthquake engineering, since they do not simulate the type of loading occurring in reality. Seed (1960) and Seed and Chan (1966) found that stress cycling and rotation of principal stress directions occurring during earthquakes actually tend to reduce shear strength. Therefore, for earthquake engineering applications, strength has to be investigated

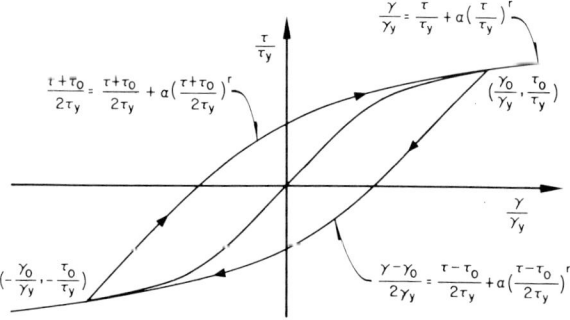

Fig. 4.6. Ramberg-Osgood constitutive model.

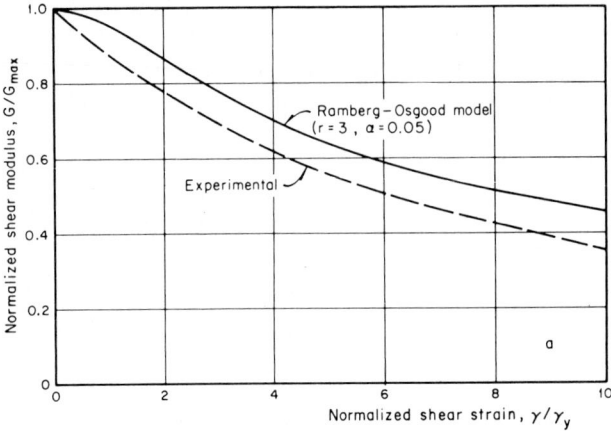
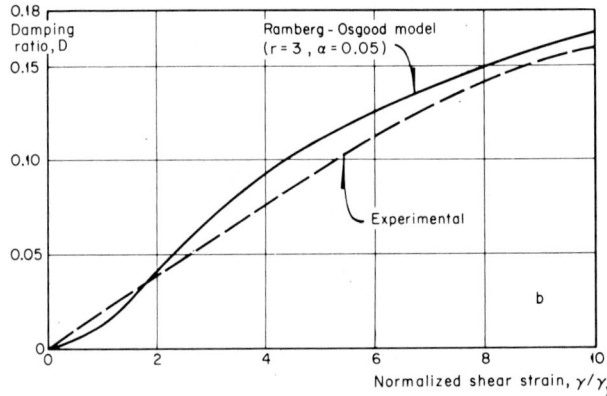

Fig. 4.7. Comparison of experimental shear modulus (a) and damping ratio (b) of dry sands with Ramberg-Osgood model. After Dobry (1970).

reproducing the salient features of the load process acting on soil elements during earthquakes. Seed and co-workers have shown that at least the following variables should be approximately reproduced: stress level, principal stress ratio, amplitude of the pulsating deviator stress and, to a lesser extent, load frequency.

The recommended procedure to determine the strength of a soil under cycling loading in a triaxial apparatus is as follows (Seed and Chan, 1966). For each set of values of effective confining pressure, principal-stress ratio and load frequency, specimens are subject to pulsating deviator stress of several amplitudes, $\Delta\sigma_d$. In each test, the axial strain is plotted as a function of the number of pulses, as shown in Fig. 4.9. From this diagram, the pulsating

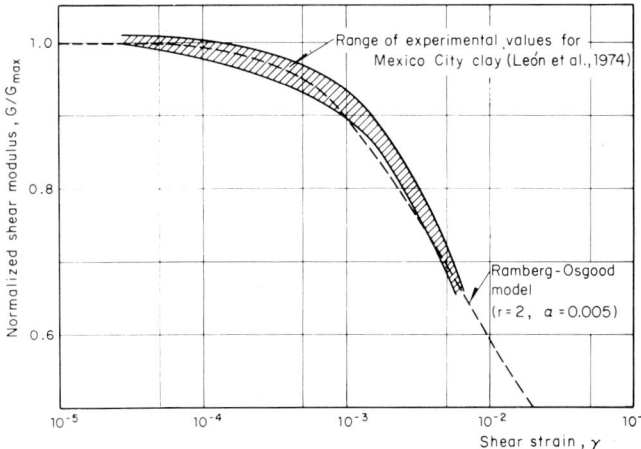

Fig. 4.8. Comparison of experimental shear modulus of a very compressible clay with Ramberg-Osgood model.

deviator stress producing failure after $N$ cycles can be determined. In this case, failure is defined as a certain level of axial strain of the specimen. From test results corresponding to several combinations of static and pulsating deviator stress level, a diagram showing combinations that produce failure can be obtained (Fig. 4.10).

It was pointed out that high strain rates cause strength to increase, whereas pulsating stresses produce strength reductions. The combined effect may be in either sense, depending on soil sensitivity. However, Seed and Chan (1966) report that, in many practical cases, the normal strength in consolidated-undrained tests can be used as a good approximation to the strength under seismic conditions (Figs. 4.11 and 4.12). In addition, it has been observed that, if static and cyclic deviator stress levels are normalized

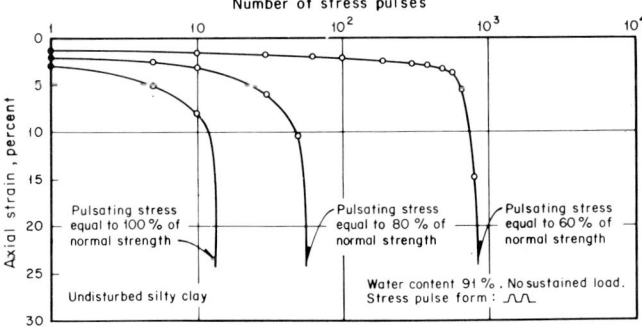

Fig. 4.9. Deformation of undisturbed silty clay under pulsating loading (Seed and Chan, 1966).

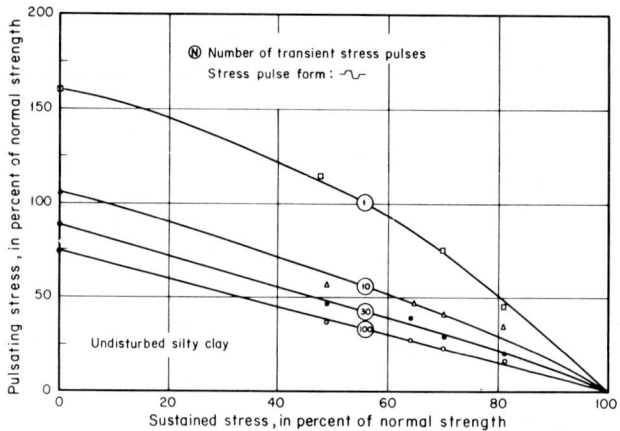

Fig. 4.10. Combinations of sustained and pulsating stress intensities causing failure of an undisturbed silty clay (Seed and Chan, 1966).

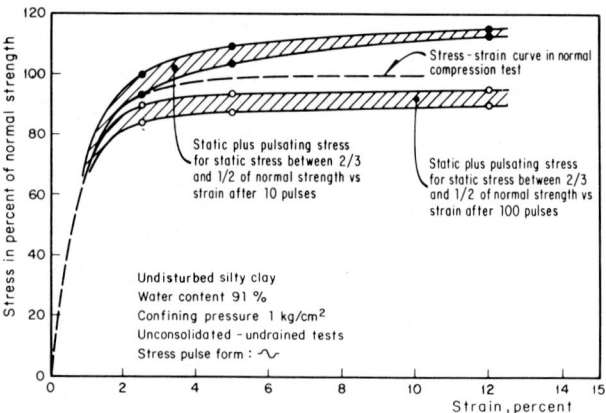

Fig. 4.11. Comparison of normal and pulsating load strength of an undisturbed silty clay. After Seed and Chan (1966).

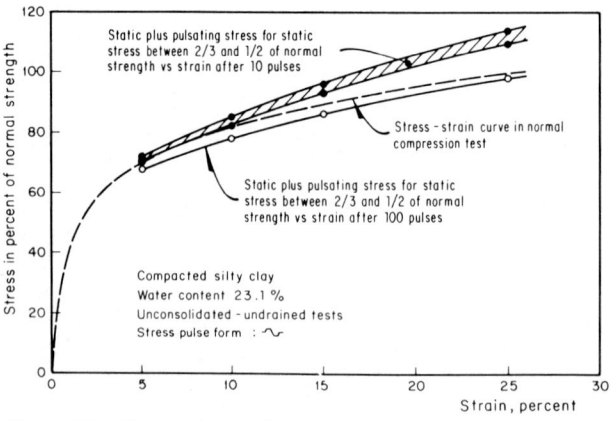

Fig. 4.12. Comparison of normal and pulsating load strength of a compacted silty clay. After Seed and Chan (1966).

dividing them by the normal consolidated-undrained compression strength, the combinations of stresses producing failure are nearly independent of consolidation pressure (Seed and Chan, 1966).

## 4.3 LOCAL AMPLIFICATION

### 4.3.1 Nature of the phenomenon. Problems of interpretation

The influence of local subsoil on earthquake ground motions is the final event of a complex propagation process in which source-mechanism and transmission-path characteristics constitute the preceding stages. Under stable soil conditions, this influence can take the form of dynamic amplification. By way of an example, acceleration signals simultaneously recorded at the surface and at different depths of stratified deposits are shown in Figs. 4.13 and 4.14. Both sets of records are of small events but clearly illustrate the modifications produced by local subsoil on incoming seismic motions, including an increase of peak amplitudes at the surface or within a specific layer, and longer duration of significant shaking. Filtering by the soil may considerably enhance wave amplitudes in certain conditions and at certain frequencies and reduce them in others.

The problem of identifying local amplification effects in instrumental records is of considerable interest and has been extensively investigated.

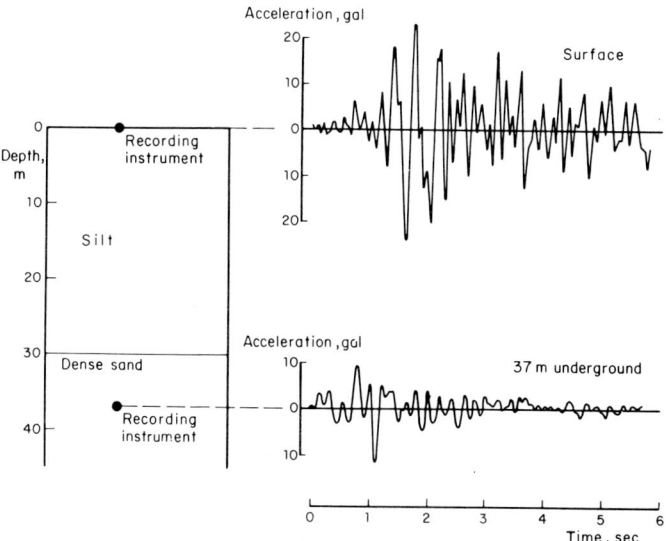

Fig. 4.13. Surface and underground acceleration records obtained at Urayasu, Japan, showing local amplification effects. After Okamoto (1973).

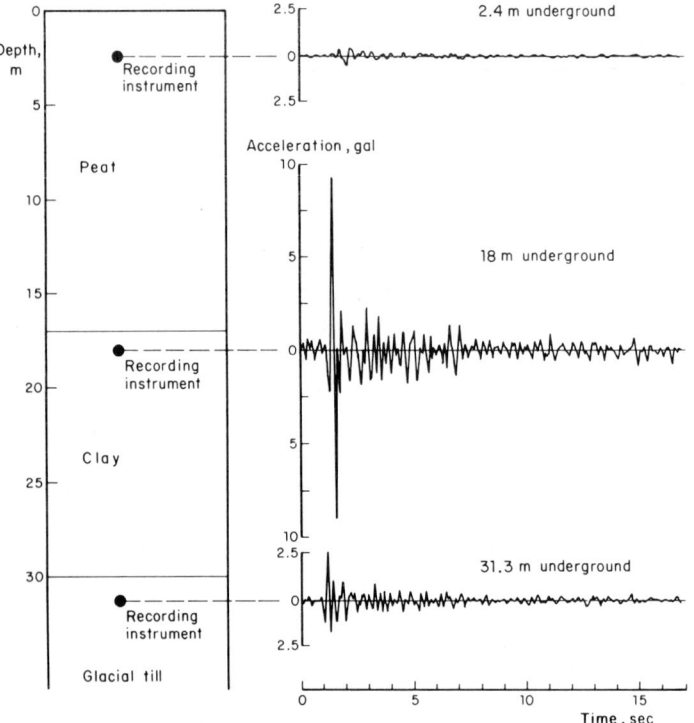

Fig. 4.14. Underground acceleration records of N—S component of small local earthquake at Union Bay, Seattle (U.S.A.). After Tsai (1969).

Under linearity assumptions, the filtering effect of a soil deposit is measured by its transfer function, i.e. by a quantity depending only on the dynamic properties and geometry of the deposit. Some of the available data actually demonstrate that amplification characteristics determined from low-intensity shaking sharply differ from those observed under strong ground motions. Although nonlinearity of soil response provides a partial explanation, there is evidence suggesting that the effects of surface geology change with different transmission paths and predominate over the influence of local subsoil. In other cases it is the source mechanism that plays a primary role. Two examples are chosen to illustrate these points. The first is a case where several strong motion records under different site conditions were obtained for the same event and local response characteristics had been previously determined from smaller-intensity earthquakes. Figure 4.15a shows amplification data calculated from seismograms of small earthquakes recorded in Pasadena, California (Gutenberg, 1957). Amplitude ratios are those at a station located on about 300 m of consolidated alluvium (CIT) with respect to a reference station on crystalline rock (SL); approximate distance be-

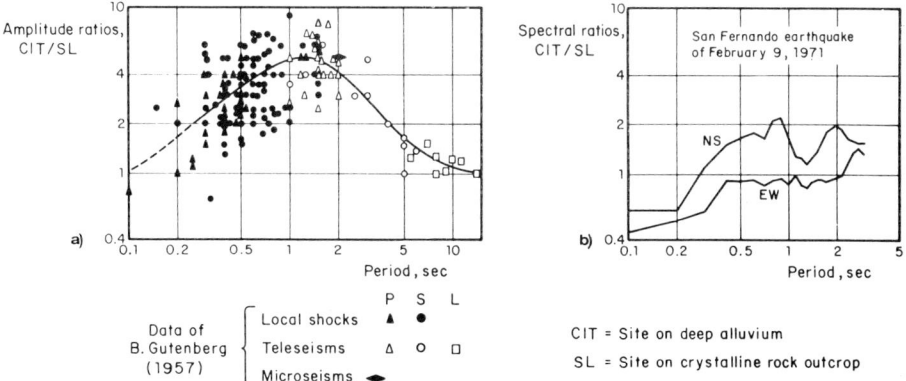

Fig. 4.15. Amplification response of deep alluvium site (California Institute of Technology campus: CIT) with respect to an outcropping bedrock site (Seismological Laboratory: SL) in the city of Pasadena, California.

tween the two sites is 4.5 km. Average amplitudes on alluvium are seen to be several times larger than on rock. Curves in Fig. 4.15b represent ratios of velocity response spectra calculated from the accelerograms recorded at CIT and SL during the San Fernando earthquake of 9 February, 1971 (Hudson, 1972). Epicentral distance from SL was of 34 km, so that attenuation and transmission-path effects should have played a minor role in the differences of ground shaking at the two stations. Motions in the N—S direction were considerably weaker than in the E—W direction. The difference in intensity seems an important cause for the discrepancy with respect to the data of Fig. 4.15a. The complex pattern of local responses for the San Fernando earthquake was in all likelihood the result of unknown three-dimensional geology effects.

The second example is the series of records obtained from 1934 to 1968 at the single accelerograph station of El Centro, California (Udwadia and Trifunac, 1973). The group of fifteen records can be arranged in different subsets, each containing earthquakes having approximately the same epicenter, at several distances and azimuthal directions. El Centro lies in a valley of well consolidated sediments, where the depth to the basement is about 6 km. Figure 4.16 shows the Fourier acceleration amplitude spectra in the E—W and vertical directions for three different earthquakes originating near the same point and therefore having the same transmission path to the recording station. No clearly defined peak attributable to local site response and involving all or some of the curves is apparent, thus indicating that the spectral content of records is largely the result of source-mechanism effects. Comparisons of other groups of spectra lead to the same conclusion. The first three peaks of the theoretical amplification curve calculated for the El Centro site under the assumption of horizontally layered soil system and

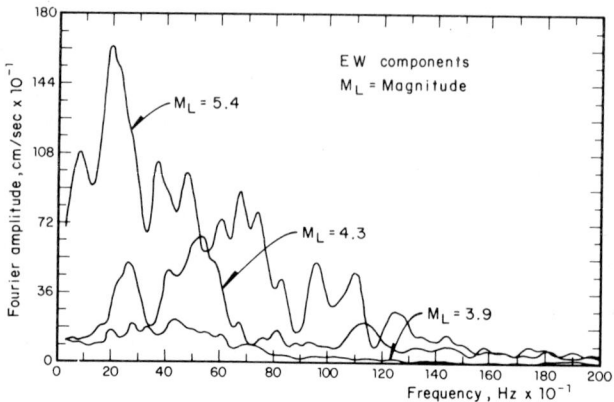

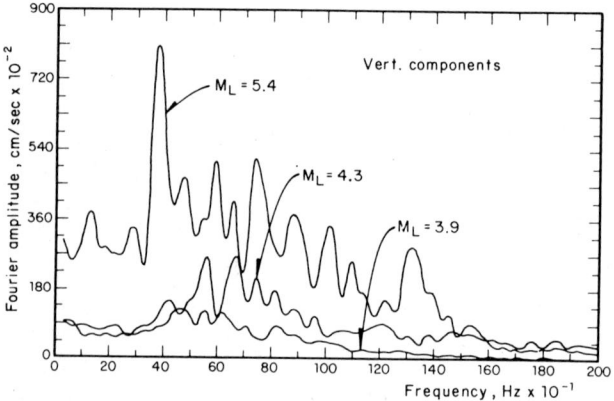

Fig. 4.16. Fourier amplitude spectra of acceleration records obtained at El Centro, California, for three different earthquakes having approximately the same location and an epicentral distance of 27 km. After Udwadia and Trifunac (1973).

vertically propagating shear waves fall at about 1.3, 3.8 and 6.3 Hz (Matthiesen et al., 1964). None of these values coincides with the spectral peaks of the curves of Fig. 4.16.

A significant indication offered by the foregoing examples is that, for strong motions at relatively close distances from the source and under stable soil behavior, the influence of local soil conditions is often not the most important factor. Use of simple theoretical models, such as the plane horizontal soil system, appears then inadequate, and other models capable of accounting for nonuniformities in the local topographic and geologic structure should be tested and applied in order to clarify several important aspects of the problem. The situation can be quite different for motions of smaller intensity, larger epicentral distances and stratigraphies characterized by sharp contrasts of seismic impedance. Experience gained from several earthquakes and results of case studies (Rosenblueth, 1960; Borcherdt, 1970; Seed et al., 1972; Okamoto, 1973; Tezcan and Ipek, 1973) demonstrate that local amplification can be considerable and become a major cause of building damage when the preceding conditions are fulfilled. In such cases one-dimensional models have proved adequate tools for the assessment of local seismic risk (Herrera et al., 1965; Idriss and Seed, 1968a; Esteva et al., 1969; Seed and Idriss, 1970a). A classical example is offered by the response of soft soil deposits in the Mexico City area; Fig. 417a shows six spectral amplification curves (ratios of velocity response spectra with 5% damping) obtained from the horizontal components of three distant earthquakes simultaneously recorded on firm ground, at the National University campus, and on the downtown lacustrine clay deposits, at Nonoalco-Tlatelolco. In Fig. 4.17b the average amplification curve derived from the data of Fig. 4.17a is compared with the theoretical amplification curve computed with a one-dimensional model. In a. profile of this nature, slight variations of soil parameters in the upper layers produce significant changes of peak amplification and fundamental frequency.

*4.3.2 Analytical models*

The model commonly used for one-dimensional amplification analysis takes into account only vertically travelling S waves: the assumptions leading to this simplified picture are extensively discussed in the literature (e.g. Newmark and Rosenblueth, 1971). In a deposit with horizontal layers, surface motions caused by incident plane waves are the result of vertical interference between uptravelling and downtravelling waves having the same phase velocity in the horizontal direction. Irregularities in the interfaces give rise to scattered waves with horizontal phase velocities different from that of the incident wave and the lateral interference becomes more important.

We shall first consider one-dimensional models with linear and nonlinear soil behavior. Some two-dimensional models will be subsequently discussed.

*4.3.2.1 One-dimensional propagation with linear soil behavior*
Two different approaches are basically available for the treatment of this problem: wave-equation methods and lumped-mass, or finite-element, methods. With the first approach the soil system is treated as a continuum and the only internal source of errors lies in the use of numerical transform techniques. For lumped-mass or finite-element methods, discretization effects must be additionally considered. The model used in wave-equation methods, assuming viscoelastic soil behavior, is shown in Fig. 4.18. Material properties are uniform within each layer of thickness $H_j$ and are represented by shear modulus $G_j$, mass density $\rho_j$, and a viscosity $\eta_j$ (or damping coefficient $D_j$). The underlying bedrock is idealized as a homogeneous elastic half-space with properties $G_r$, $\rho_r$, and $\eta_r$. Under the assumption of vertically traveling shear

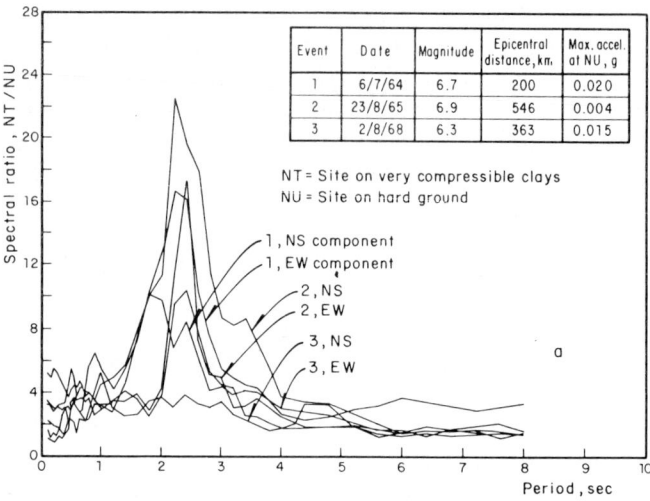

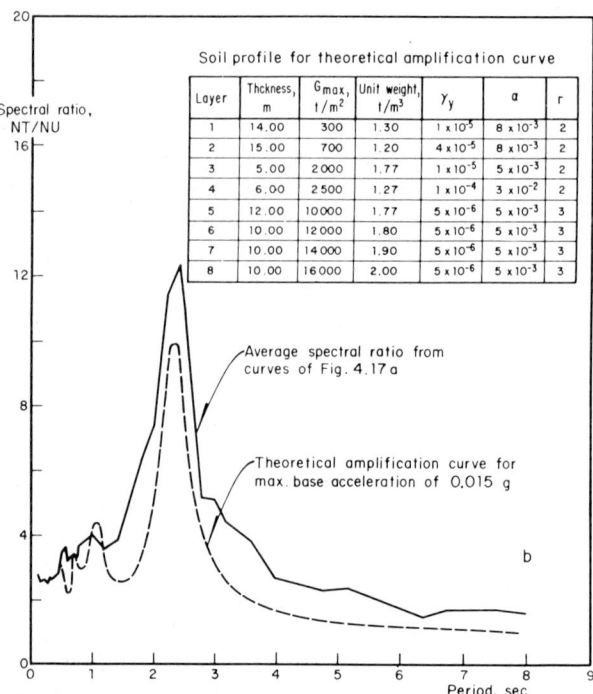

Fig. 4.17. Amplification response of central Mexico City soil. Curves in (a) are ratios of velocity response spectra from simultaneous accelerograph records at Nonoalco-Tlatelolco (NT) and National University campus (NU). Distance between the two sites is about 15 km. Parameters of the soil profile in (b) are those of Ramberg-Osgood constitutive model and were partly obtained from data by León et al. (1974).

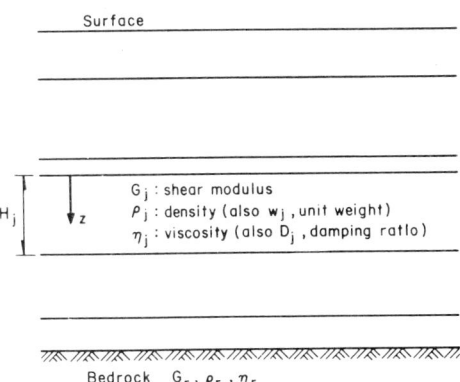

Fig. 4.18. One-dimensional model of horizontally stratified soil deposit with basic material parameters.

waves, there is only one horizontal component of displacement, $u = u(z,t)$; this must satisfy within each layer the wave equation:

$$\rho_j \ddot{u} = G_j \frac{\partial^2 u}{\partial z} + \eta_j \frac{\partial^2 \dot{u}}{\partial z^2} \tag{4.5}$$

where $t$ is time, $z$ is the depth within the $j$th layer, and the dot denotes time derivative. The free surface is assumed to be stress-free and the wave input is ordinarily specified as a seismic signal arriving from the base rock. The solution is usually sought in the form of a response at the free surface of the soil deposit. In principle, however, the input can be assigned at any other point and one may wish to compute the output at another arbitrary point; the problem of obtaining the bedrock response from a given surface signal is an especially important example. Several different solution methods have been proposed, either in the frequency domain via Fourier transform techniques (Haskell, 1960; Herrera and Rosenblueth, 1965; Roesset and Whitman, 1969; Lysmer et al., 1971; Schnabel et al., 1972a); or by multiple-reflections techniques using the simple wave equation in the time domain (Newmark and Rosenblueth, 1971; Okamoto, 1973). Due to the possibility of introducing arbitrary variations of viscosity in the soil layers and to the efficiency afforded by the Fast Fourier Transform computational algorithm (Cooley and Tukey, 1965), the first type of methods seems preferable. Under harmonic excitation of frequency $\omega$, the solution to eq. 4.5 has the form:

$$u = U_j(z) \exp(i\omega t) \qquad (i = \sqrt{-1}) \tag{4.6}$$

with:

$$U_j(z) = E_j \exp(ip_j z) + F_j \exp(-ip_j z) \tag{4.7}$$

and:

$$p_j^2 = \omega^2 \rho_j/(G_j + i\omega\eta_j) \tag{4.8}$$

The first term of eq. 4.7 represents a shear wave travelling in the negative $z$-direction with complex amplitude $E_j$, whereas the second term represents a wave travelling in the positive $z$-direction with complex amplitude $F_j$. Continuity of stresses and displacements at the interfaces yields recursive formulas relating the amplitudes $E_{j+1}$, $F_{j+1}$ and $E_j$, $F_j$ in two adjacent layers. On the free surface one obtains:

$$E_1 = F_1 = \tfrac{1}{2}A_0 \tag{4.9}$$

$A_0$ being the total displacement amplitude at the surface. Application of the recursive formulas starting from the top layer yields the relations:

$$E_r = e_r(\omega)\, E_1 \tag{4.10}$$

$$F_r = f_r(\omega)\, E_1 \tag{4.11}$$

for the wave amplitudes in the base rock, where the functions $e_r(\omega)$ and $f_r(\omega)$ are determined using the conditions $E_1 = F_1 = 1$. It is actually possible to define two different transfer functions for the soil system. One relates the surface amplitudes to those in the underlying rock and is given by:

$$H_1(\omega) = 2E_1/(E_r + F_r) = 2/[e_r(\omega) + f_r(\omega)] \tag{4.12}$$

The second relates the surface amplitudes to those which would occur in the base rock if the soil deposit were not present and is given by:

$$H_2(\omega) = 2E_1/2E_r = 1/e_r(\omega) \tag{4.13}$$

The moduli of these transfer functions are the amplification functions of the soil system. Since $H_1(\omega)$ is independent from the elastic properties of the rock, it is also called "rigid rock" transfer function. Such properties enter in the expression of $H_2(\omega)$, which is therefore termed "elastic rock" transfer function. Use of $H_2(\omega)$ is recommended since reference motions are usually assigned on outcropping bedrock and not under the deposit. Furthermore, the assumption of infinitely rigid bedrock disregards the effects of "radiation damping", i.e. of the energy carried back into the base rock because of partial refraction of the waves from the soil. This omission leads to excessively high amplification factors. A comparison of the two amplification functions for a single layer over a half-space is shown in Fig. 4.19.

The foregoing treatment applies only to harmonic motions. Transient motions are conveniently handled by Fourier decomposition (Schnabel et al., 1972b); the output time response is obtained from the inversion of the Fourier transform of surface motions.

If the viscosity coefficient $\eta_j$ is assumed constant (viscous damping), a strong reduction in the amplitude of the higher modes of the amplification

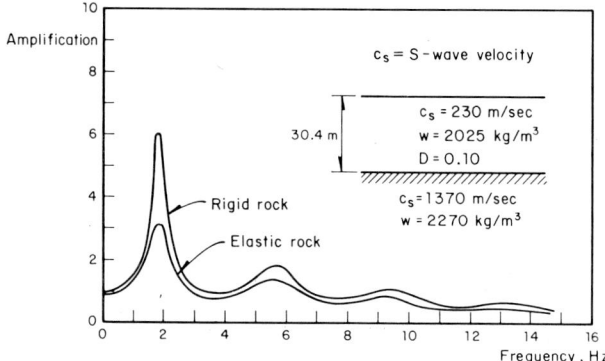

Fig. 4.19. Amplification functions of soil layer on rigid and elastic rock. After Roesset (1970).

function will result. A considerably smaller reduction is obtained if $\eta_j$ is inversely proportional to frequency (hysteretic damping), so that the damping ratio:

$$D_j = \eta_j \omega / 2 G_j \qquad (4.14)$$

is a constant. Comparison of the effects of hysteretic vs. viscous damping is made easier by assuming that the ratio $\eta_j/G_j$ is equal for all layers. Under the hypothesis of viscous damping the amplification function will have a damping ratio increasing linearly at each natural frequency $\omega_n$ and the amplification itself will be proportional to $1/\omega_n^2$. Hysteretic damping leads to a damping ratio equal for all modes and the amplification will therefore be proportional to $1/\omega_n$.

If response spectra of the surface motion are of more direct interest than, say, the acceleration history, a random vibration approach can be employed. Methods for assigning the spectral density function of bedrock excitation and calculating peak surface response statistics are the same as those used in structural response analysis (Chapter 8). Only the hypothesis of light damping is often not verified in soil systems.

As for wave-equation methods, several approaches using lumped-mass or finite-element techniques have been proposed (Idriss and Seed, 1968b; Roesset and Whitman, 1969; Roesset, 1970; Tsai, 1969; Idriss et al., 1973). When the base-rock is infinitely rigid all these methods lead to a classical system of second-order dynamic equations. For a given number of degrees of freedom, choosing the lumped masses so as to diagonalize the mass matrix is usually convenient from a computational viewpoint, although the use of a consistent mass matrix is likely to give better results (Kuhlemeyer and Lysmer, 1973). The minimum number of masses required to obtain a certain period $T$ in the response with reasonable accuracy has been discussed by Roesset (1970) and by Ohsaki and Sakaguchi (1973). When finite elements are used meaningful results are obtained when the element length is smaller than about one-sixth of the shortest seismic-wave length one wishes to propagate in the system (Kuhlemeyer and Lysmer, 1973).

When the viscosities of the layers are arbitrary, a normal mode solution of the equations of motion is not possible and step-by-step numerical integration has to be performed. A necessary and sufficient condition for modal decomposition is that the damping matrix can be expressed as a Caughey series (O'Kelly, 1964). When differences among damping ratios in the different layers are not large, average or weighted modal damping coefficients can be introduced through energy considerations. The resulting normal mode solutions furnish reasonable approximations (Whitman et al., 1972). Radiation damping effects can be accounted for by placing a massless dashpot with constant $(G_r \rho_r)^{1/2}$ between the bottom mass of the system and the halfspace, in the hypothesis that the latter is elastic, homogeneous and has zero viscosity (Tsai, 1969; Rosenblueth and Elorduy, 1969). In this case, an additional first-order differential equation describing the motion of the dashpot must be introduced. This renders a classical normal mode solution impossible in all cases. Radiation effects can otherwise be approximately described by adding to the modal damping ratio a fictitious component proportional to the ratio between the average seismic impedance of the deposit and that of bedrock (Roesset and Whitman, 1969).

The stability and efficiency of numerical methods for step-by-step integration in the time domain has been extensively investigated (Nickell, 1971; Bathe and Wilson, 1973; Goudreau and Taylor, 1973; Ayala and Brebbia, 1973). In spite of heavier requirements in computer time, use of unconditionally stable, implicit schemes is normally preferable, and Newmark's beta method with constant acceleration is especially appealing because of the absence of spurious damping.

*4.3.2.2 One-dimensional propagation with nonlinear behavior*

The strain-dependent, hysteretic characteristics of soil behavior under dynamic loading are put in clear evidence by laboratory test results of several types. Field evidence of nonlinearity effects is usually indirect and often suggests that significant yielding has occurred: such is the case with short open cracks and compression ridges encountered on flat ground in the epicentral area of strong earthquakes (Ambraseys, 1970). However, for dense materials showing no evidence of cracking or failure, recent instrumental data (notably those from the 1971 San Fernando earthquake) indicate considerably less amplification than predicted by linear theories. In addition to transmission-path effects, this can be the result of nonlinearity in soil response.

In the absence of pertinent strong-motion data and experimental field information at large strains, characteristics of nonlinear response have to be discussed on the basis of numerical simulation and laboratory test results. A systematic feature consists of a sizable reduction of the peak amplification factor and of the associated frequency with increasing excitation intensity. Strong damping produced by hysteresis is the major factor contributing to the observed reductions. This is important when one tries to model the primary features of nonlinear response by means of simple equivalent viscoelastic systems.

When a fully nonlinear and hysteretic constitutive law is assumed for the soil, direct numerical integration of the equations of motion in the time domain with step-by-step linearization of the stress—strain law is the most common approach. Numerical solutions based on the Ramberg-Osgood model have been produced for several case problems using finite elements (Cervantes et al., 1973), lumped masses (Constantopoulos, 1973) or the method of characteristics (Streeter et al., 1974; Papadakis, 1973). Stability of integration schemes becomes a much heavier computational requirement for discrete nonlinear models. Within the method of characteristics, the stability condition is expressed by the Courant-Levy-Friedrichs inequality:

$$c_s \Delta t \leq H \qquad (4.15)$$

to be satisfied in each layer of thickness $H$. After assigning $H$ on the basis of the highest frequency to be retained in the response, $\Delta t$ is determined by the equality sign in eq. 4.15 and by choosing the value of the shear-wave velocity $c_s$ that corresponds to the highest value of the shear modulus. The most important advantage of the method of characteristics consists in reducing the problem to the numerical solution of two first-order differential equations, one along the positive and one along the negative characteristic, containing the horizontal particle velocity and the shear stress as unknowns.

Direct determination of equivalent amplification functions can be achieved via random vibration analysis, assuming a stationary excitation process with smooth spectral-density function and representing soil nonlinearity by an arbitrary hysteretic model (Faccioli et al., 1973; Faccioli and Ramírez, 1976). Nonlinear response characteristics are modelled by an equivalent viscoelastic system through an iterative procedure. Equivalent linearization is achieved by the describing-function technique (Šiljak, 1969), and the equivalent stiffnesses and damping ratios depend only on the excitation amplitude. Since the describing-function technique strictly applies to sinusoidal motions, the method determines a harmonic response which is, in some average sense, equivalent to the actual stochastic response. A converging solution for the amplification function of the system is obtained when the equivalent properties do not change in two successive iterations. The validity of the method critically depends on the hypothesis that the output be narrow-band; most of the available numerical results show that this assumption is reasonable. Figure 4.20 illustrates ratios of acceleration response spectra for 2% damping calculated by Constantopoulos (1973) at two different excita-

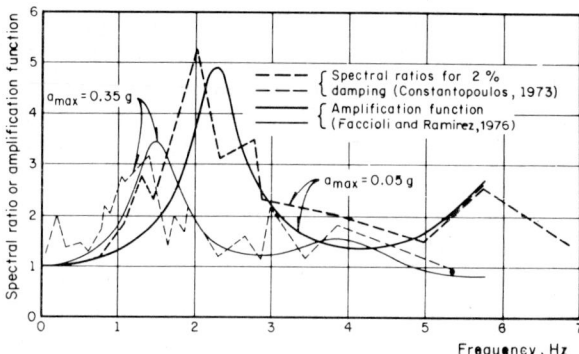

Fig. 4.20. Influence of input acceleration intensity ($a_{max}$ = peak value) on maximum amplification factor and associated frequency of a multilayered soil profile governed by Ramberg-Osgood constitutive laws.

tion intensities for an assigned nonlinear system having nine degrees of freedom; the base motion is the S69E component of the July 21, 1952 Taft record. Continuous curves are the amplification functions given by the random vibration method.

Nonlinear dependence of shear modulus and damping factor on shear-strain amplitude can be taken into account without resorting to a fully nonlinear material description. This is achieved by the "strain-compatible" method, an iterative procedure with empirically adjusted viscoelastic properties. Each iteration consists of a full linear numerical solution for the given soil profile and can be carried out either by wave-equation or lumped-mass methods (Idriss and Seed, 1968b). Shear moduli and damping coefficients are selected according to the largest value of the shear strain computed in the preceding iteration, on the basis of available experimental curves. A "representative" fraction of the largest strain, usually between 0.6 and 0.7, is adopted for this purpose. Convergence is attained when variations of the shear strains in two successive iterations become negligible. This method cannot be used for calculating residual strains or displacements and does not appear to provide substantial reductions in computational work with respect to fully nonlinear solutions. If the equivalent properties are calculated from a Ramberg-Osgood model, the iterative procedure underestimates maximum response displacements and overestimates maximum accelerations in comparison to direct numerical simulation (Constantopoulos, 1973).

### 4.3.2.3 Other propagation models

Steady-state analytical solutions have been obtained for a few cases where the configuration of the local deposit is no longer horizontal and the soil is treated as a perfectly elastic material. Useful models are represented by the semi-cylindrical and semi-elliptical alluvial valleys with incident plane SH-waves (Trifunac, 1971; Wong and Trifunac, 1974). The solution gives the total displacement field as a function of angle of incidence and position. Surface displacement amplitudes for different angles of incidence are shown in Fig. 4.21 for the semi-cylindrical configuration; the seismic impedance ratio is in this case 0.33 and the amplitude of excitation is unity. The complexity of resulting amplification patterns and their sensitivity to the direction of approach warns against simplified interpretations of actual strong-motion accelerograms when the angle of incidence and details of local geology are unknown. Only when wavelength is very long in comparison with the cross-section of the valley do effects of irregularity become negligible. As wavelength decreases, interference and local amplification caused by the valley are more significant.

A method based on integral representations of the displacement fields has been successfully applied to the more general two-dimensional problem of determining the motions of a layer having an irregular interface with an underlying half-space (Aki and Larner, 1970). Also in this case the excitation is assumed in the form of SH waves with an arbitrary angle of incidence. The method, although primarily intended for the study of large-scale inhomogeneities, is well suited for the analysis of ground motion in a local soil basin. Effects on surface motions produced by two different basin configurations with vertical wave incidence are illustrated in Fig. 4.22. Seismic impedance ratio is 1/7 in both cases, whereas the length of the wave in the layer is a different multiple of the maximum interface depth. The dotted curves represent the one-dimensional solution assuming that the basin con-

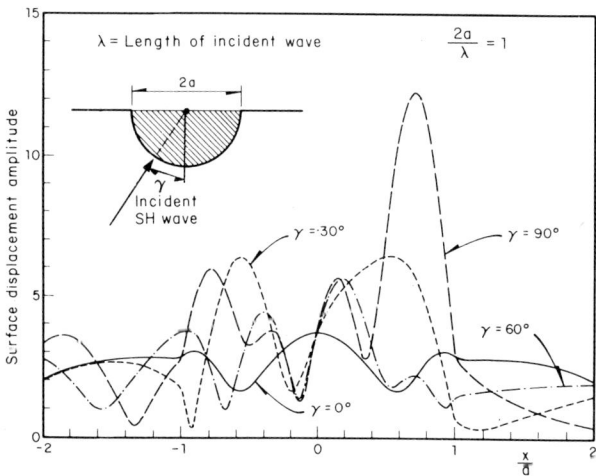

Fig. 4.21. Surface displacement amplitudes for elastic model of semi-cylindrical valley excited by SH waves at different angles of incidence. Center of the valley corresponds to $x/a = 0$ and edges to $x/a = \pm 1$. After Trifunac (1971).

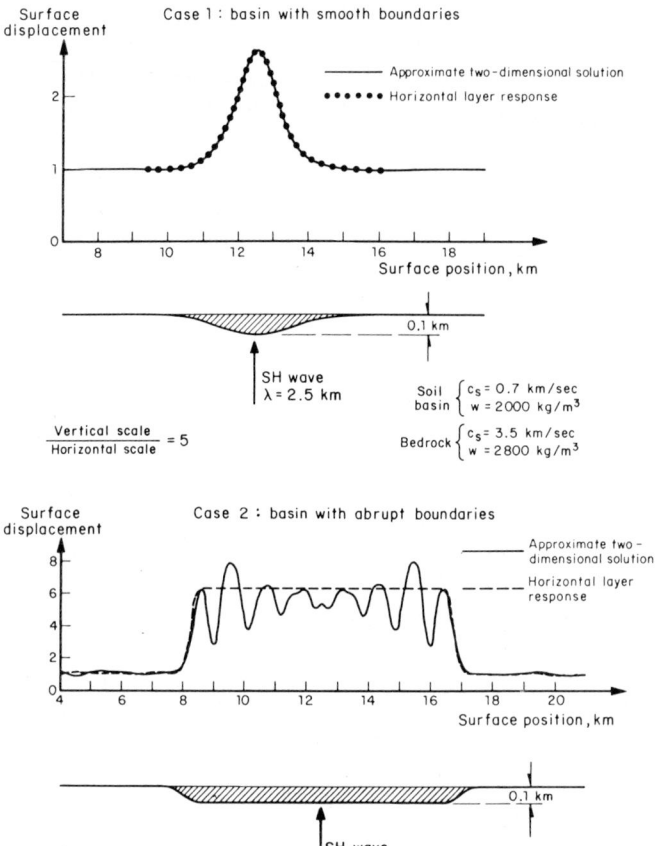

Fig. 4.22. Effects of boundary configuration on amplitude response of a soil basin excited by vertically incident SH waves. After Aki and Larner (1970).

sists of a horizontal layer having a constant thickness equal to that directly below the point. The first example shows that flat-layer analysis is adequate when the interface is smooth throughout. In the second example, however, presence of relatively steep boundaries produces significant lateral interference and use of one-dimensional models is therefore bound to introduce large errors in the estimation of local amplitudes. In the second case, the wavelength in the basin satisfies the resonance condition predicted by horizontal-layer theory.

Local surface inhomogeneities embedded in horizontally layered infinite media resting on a rigid half-space have been treated by a mixed finite-element standard matrix approach (Lysmer and Drake, 1972). The irregular zones involving complicated soil structures are analyzed with finite elements and the coupling with adjacent flat-layered regions is achieved through continuity requirements imposed on stress and displacement components. Results for case problems include propagation of surface waves across large alluvial valleys; an example for incident Love wave motion with a period of 2 sec is illustrated in Fig. 4.23. The local surface amplifications are in this case due to the large amplitude of the fundamental Love wave mode in the valley and the refraction caused by the sloping boundaries is found to be negligible, in contrast with the case of vertical SH wave incidence shown in Fig. 4.22. The small magnitude of boundary effects on peak amplification values is confirmed by comparing with the solution computed when the alluvial section extends to infinity.

Finite-element techniques have been used to investigate the seismic response of soil deposits with sloping rock boundaries subjected to horizontal acceleration histories, adopting a strain-compatible material description (Dezfulian and Seed, 1970). An improved formulation, with damping allowed to vary from element to element, has been proposed which yields results in good agreement with data from shaking table tests on small-scale clay banks (Idriss et al., 1973). Special analyses have also been conducted for rigid base motions with both vertical and horizontal acceleration components and for elastic base with travelling wave excitations (Dibaj and Penzien, 1969; Dezfulian and Seed, 1971).

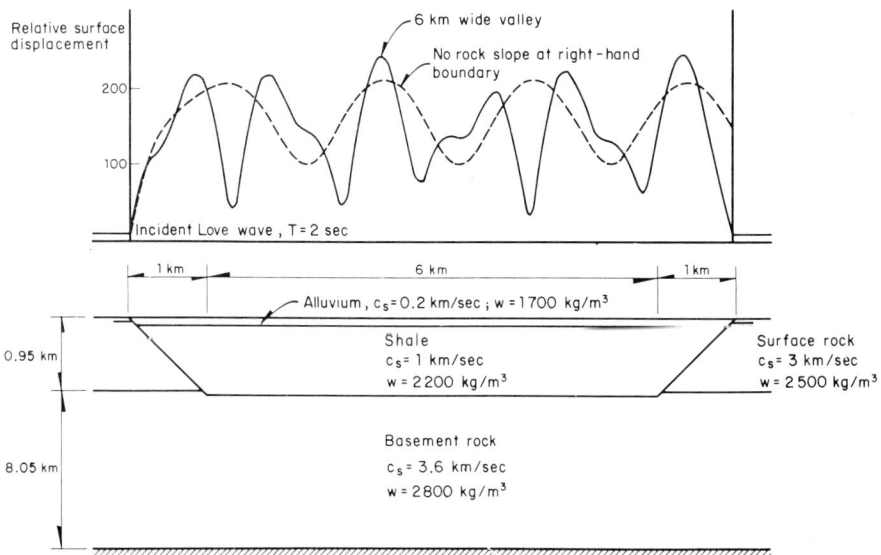

Fig. 4.23. Amplitude response of two-dimensional finite-element model of alluvial valley excited by a Love wave. After Lysmer and Drake (1972).

## 4.3.3 Effects of weak interbedded layers

Lenses and thin layers of weaker materials, typically of loose sand and silt, are frequently encountered in local surface deposits. Even if the strength of the predominant soils in the deposit is high, response to a strong earthquake can be affected to a considerable extent by the behavior of the weaker layers. In sloping surfaces, liquefaction of saturated silt and sand seams embedded in large clay masses can become the principal cause of major slide movements (Seed, 1970). On flat ground, early failure of a layer at depth may have positive or negative consequences on the stability of overlying structures but it is likely that significant changes in amplitude and frequency content of the surface response will ensue. This has been demonstrated by numerical solutions of case problems for horizontal deposits, assuming elasto-plastic soil behavior and introducing thin layers having a yield limit smaller than that of the surrounding material (Ambraseys, 1970 and 1973). Yielding of a deep stratum creates a discontinuity and in turn causes the overlying soil to oscillate freely, and introduces a higher frequency modulation in the response. This could explain some features of the strong-motion accelerogram recorded during the 1964 Niigata earthquake at Kawagishi-Cho, where underground liquefaction started presumably six or seven seconds after the onset of motion (Fig. 4.24). It can be assumed that the yield limit of the weakest material imposes an upper bound on the magni-

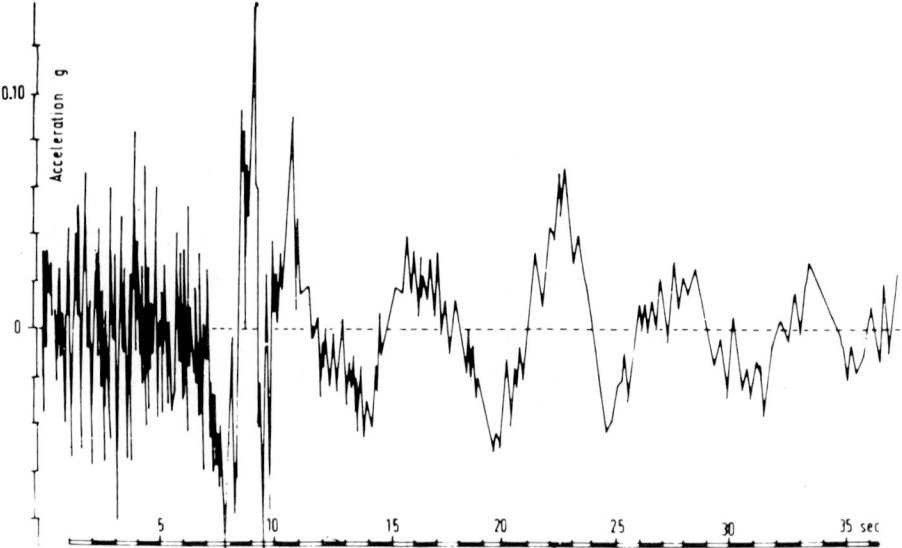

Fig. 4.24. Component of strong-motion accelerogram recorded at Kawagishi-Cho during the 16 June 1964 Niigata earthquake. Courtesy N.N. Ambraseys.

tude of the stresses, and hence also of the accelerations and velocities that can be transmitted to the surface of the deposit. The bounds would depend, for clays, on the value of consolidated-undrained shear strength under field stress conditions, and, for sands, on the effective stress and the effective angle of friction. It follows that, when deposits containing weak materials predominate, the pattern of surface ground motion in the epicentral area of strong earthquakes could be governed by the dynamic behavior of local soils to a larger extent than by source-mechanism and transmission-path characteristics. Evidence of yielding in near-surface soils has been observed in the near-field region of several earthquakes, and the slow and erratic attenuation pattern of peak accelerations at close distances from the source could then significantly depend on the strength distribution of the soils.

*4.3.4 Modification of design spectra according to soil profile*

At intermediate and large focal distances, several strong-motion records document the effects of the local soil profile on the shape of the response spectrum. Seed and Idriss (1969) discussed the five spectra illustrated in Fig. 4.25 together with soil profiles at the recording sites. Standard penetration values are also displayed. Four records are of the same earthquake and in the same city (Hisada et al., 1965) and all records are representative of motions at large epicentral distances. The five spectra are arranged according to increasing degrees of softness of the subsoil at the recording station and for each the period associated with the largest normalized spectral acceleration is given. Note that the dominant period increases with the softness of the profile.

Ohsaki (1969) selected from the catalogue of Japanese accelerograms groups of records obtained on hard and soft soil. Acceleration spectra of records in the basement of buildings having spread foundations on rock and hard soils and on multilayered soft soils are illustrated in Figs. 4.26a and b, respectively. As in Fig. 4.25, spectral ordinates are normalized with respect to maximum ground acceleration. The spectra of Fig. 4.26a have their peaks at periods of 0.2—0.3 sec and the decrease of spectral ordinates with increasing period is very sharp. On the other hand, the spectra of Fig. 4.26b possess several peaks and the range of periods in which large seismic forces act on the structure is very wide.

Turning now to the problem of establishing a design spectrum at the surface of a local soil deposit, we note that reference seismic motions in a region are generally assigned for hard ground conditions. Such motions can be directly specified in the form of a design spectrum or in the form of one or more design accelerograms. When no local records are available for the region under study, bedrock accelerograms can be selected from the existing catalogue of strong-motion data or generated by means of stochastic simulation models (Rascón and Chávez, 1973; also Chapter 8). If local records are

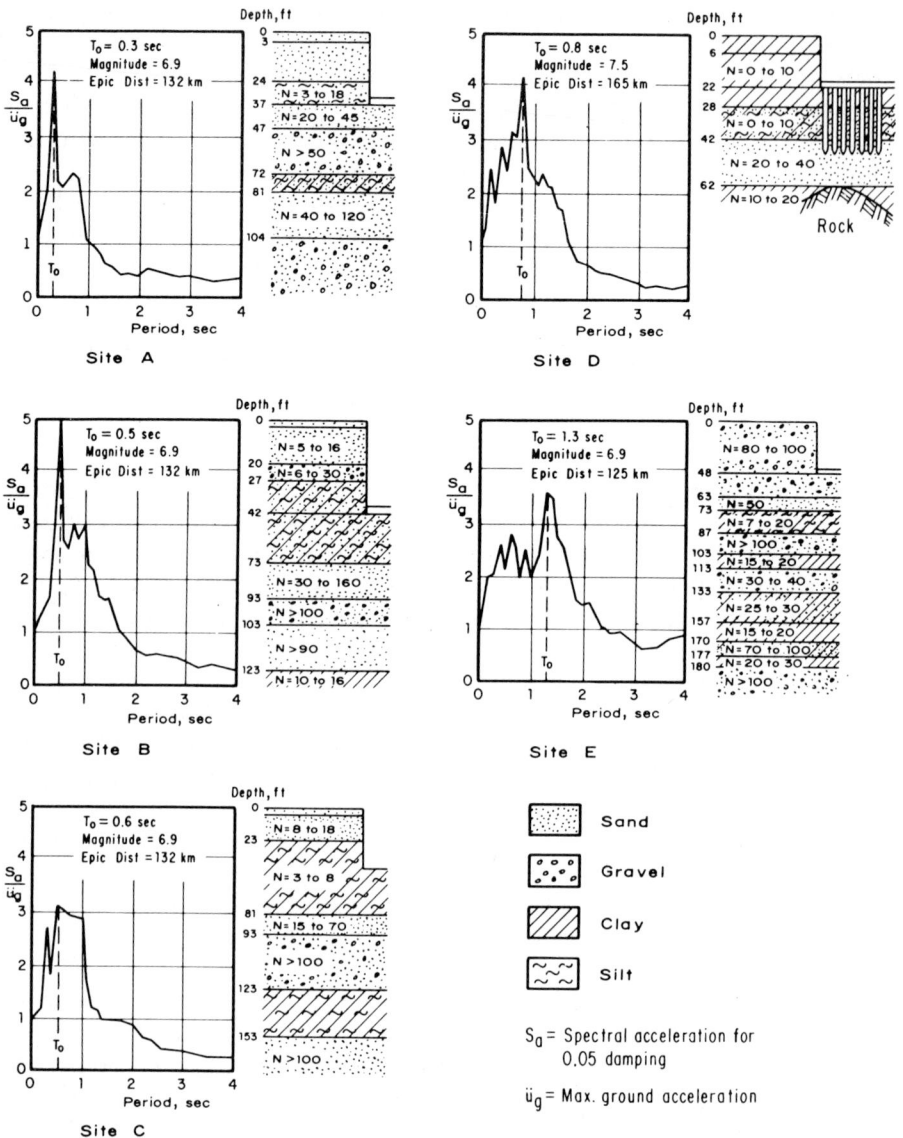

Fig. 4.25. Effect of local soil conditions on shape of response spectra. After Seed and Idriss (1969).

absent and reference spectral parameters are difficult to establish we can proceed by analogy and choose as design spectrum on the soil a normalized average or envelope spectrum from records obtained at other sites with similar soil profiles. The envelope of the data of Fig. 4.26b may be used for this purpose. Although this criterion is likely to lead to a conservative design, it

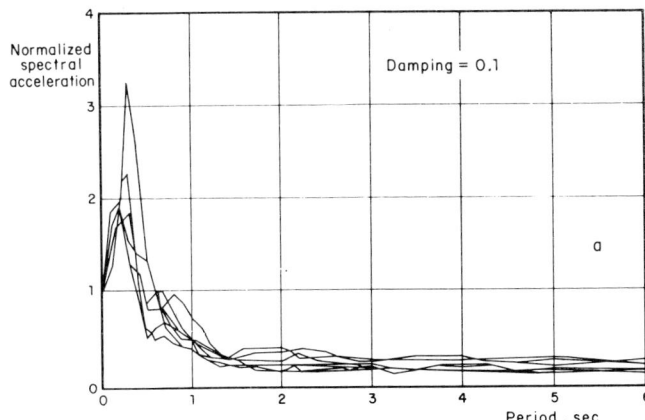

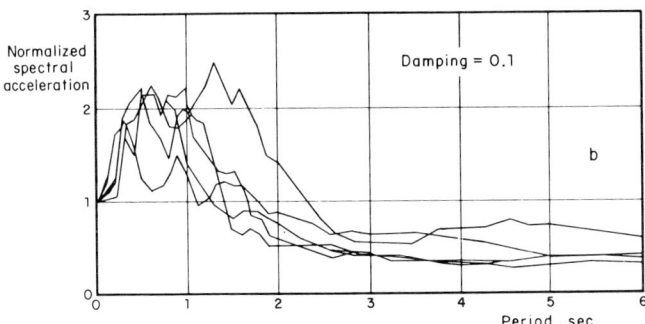

Fig. 4.26. Response acceleration spectra for rock and hard ground (a), and for soft multilayered soils (b). After Ohsaki (1969).

may represent a reasonable option in a number of practical cases. Otherwise, we can proceed via numerical simulation and directly determine the filtering effects of the soil on the reference motions. For complicated boundary configurations and irregular stratification, this is best achieved by numerical calculation of the response to a specific earthquake input using one of the previously described two-dimensional models. If this is to provide statistically significant results, response calculations have to be carried out for several different excitation samples. A representative soil spectrum can then be obtained by taking the envelope, or some average, of the resulting family of response spectra. This is generally a long and costly procedure, especially when a fine discretization of the soil system is required and nonlinear properties must be taken into account.

Considerable simplifications are introduced when the deposit can be treated by infinite horizontal layer theory. If we take out the very short period range, the theoretical amplification function of the system provides an acceptable approximation to the ratio of undamped response spectra. In the case of viscoelastic behavior, the amplification function can be exactly calculated; a more refined method taking into account statistically assessed differences between the spectral ratio and the amplification function is also available (Roesset, 1970). If a nonlinear hysteretic material description is introduced, the equivalent amplification function method provides a convenient approach, considerably more economic than numerical simulation. Stochastic properties of reference motions are in this case directly included in the spectral-density representation of the input (Chapter 8); methods for computing spectral densities from velocity response spectra are available in the literature (Jennings, 1963). A probabilistic formulation, based on equivalent transfer functions for nonlinear response and leading to a conditional estimation of the mean annual number of earthquakes whose intensity on soil exceeds a given value, has also been suggested (Esteva and Villaverde, 1973).

Details of local geology and dynamic properties of subsoil materials are often poorly known. The convenience of determining the modification to design spectra by numerical simulation, using more or less sophisticated multidimensional models, should then be assessed with extreme care. Additional large uncertainties affect also the earthquake input characteristics. In many cases, it seems more useful to perform parametric studies with one-dimensional models, since these provide some insight on the extent to which soil responses are influenced by the most critical uncertainties of the problem.

## 4.4 COMPACTION AND LOSS OF STRENGTH

### 4.4.1 Introductory remarks

Dynamic behavior of unstable foundation soils constitutes one of the least understood, and hence potentially more fertile areas of earthquake engineering research. Despite recent advances, an uncontroversial explanation of the mechanism governing liquefaction of saturated sands and cyclic loss of strength of soft cohesive soils is still lacking and the applicability of laboratory findings to field conditions is the object of a continuing debate. Displacements caused by local soil failures or compaction are especially important for engineering structures sensitive to differential settlements such as bridges, nuclear plants, utility networks, and earth and rockfill dams. Although such structures represent only a small fraction of the total, their failure implies very high costs to a community and their performance under severe ground shaking should be assessed with care.

*4.4.2 Volume change under vibration*

Vibration has been long recognized as an effective way of densifying cohesionless soils, and several laboratory and field techniques are available for this purpose (Broms and Forssblad, 1969). Studies of densification caused by vertical accelerations and alternating vertical stresses in shaking-table and one-dimensional cyclic load tests have shown that the density changes are very small for accelerations below $1g$ (Figs. 4.27 and 4.28), shaking-table tests produce significant compaction only at acceleration levels that cause vertical stresses to vanish at some stage of a load cycle (D'Appolonia and D'Appolonia, 1967; Whitman and Ortigosa, 1968).

Changes in volume and in frictional characteristics of dry sand induced by high horizontal accelerations and a large number of cycles at small amplitude were investigated by means of shear boxes mounted on shaking tables (Barkan, 1962; Youd, 1970). For initially loose samples it was found that the final density increases with increasing acceleration and that, for a given acceleration level, an equilibrium state is attained which is a function of the applied vertical pressure (Fig. 4.29). If, after reaching the equilibrium state, the vibrational excitation is maintained and, at the same time, direct shear is applied to the sample, no further volume change occurs. Additional densification of a specimen in an equilibrium state can be caused by an acceleration or a vertical pressure exceeding the corresponding equilibrium value.

Other studies based on cyclic simple-shear and shaking-table tests with acceleration and strain amplitudes in the same range as those expected in strong-motion earthquakes show that shear-strain magnitude, relative density and number of loading cycles are the primary factors governing compaction

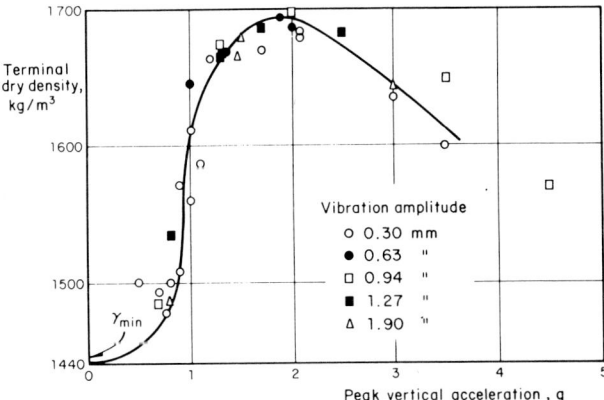

Fig. 4.27. Effect of acceleration intensity on densification of sand in vertical shaking-table tests. After D'Appolonia and D'Appolonia (1967).

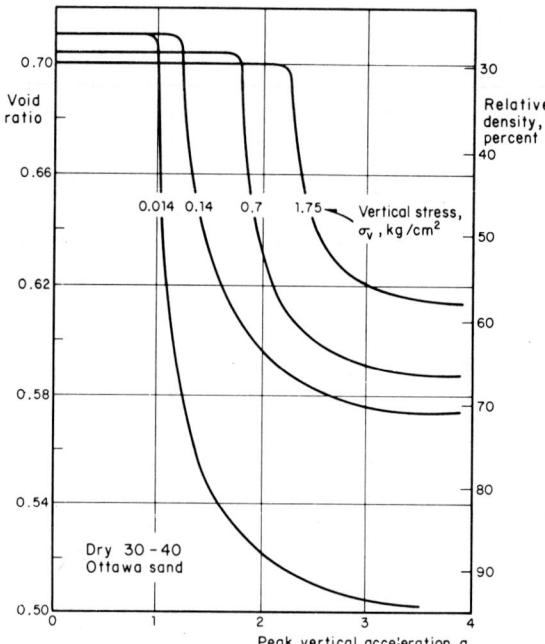

Fig. 4.28. Effects of acceleration intensity and confining stress on densification of dry Ottawa sand in vertical shaking-table tests. After Whitman and Ortigosa (1968).

of dry and saturated, drained cohesionless soils (Silver and Seed, 1969; Youd, 1972). Typical behavior of loose Ottawa sand is displayed in Fig. 4.30. Each cycle exhibits a contraction—dilation sequence similar to that observed in monotonic shear, but a finite amount of volume reduction is pro-

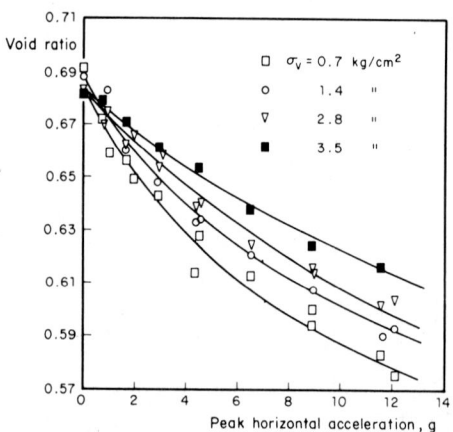

Fig. 4.29. Effects of acceleration intensity and confining stress on final void ratio of dry Ottawa sand in shear vibration tests. After Youd (1970).

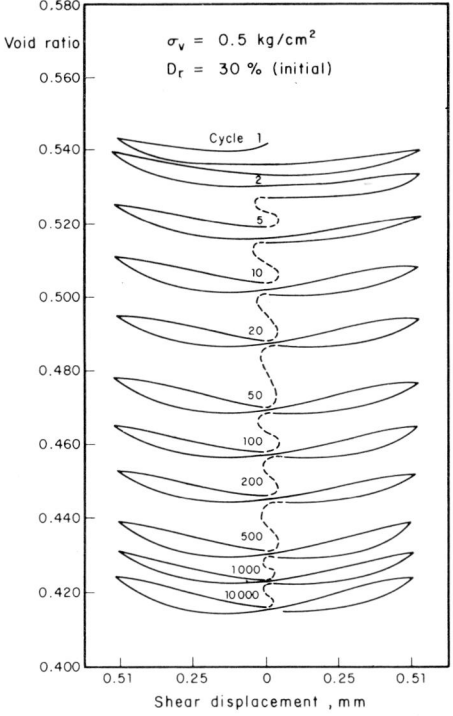

Fig. 4.30. Compaction vs. shear-strain history in a cyclic shear test on saturated, drained Ottawa sand. After Youd (1972).

gressively accumulated until a limiting density is reached, considerably greater than 100% relative density as defined by ASTM D-2049-69. The combined effects of shear-strain amplitude and number of loading cycles on the compaction rate are shown in Fig. 4.31, while the influence of initial relative density on the permanent vertical strain accruing in ten cycles is illustrated in Fig. 4.32. Note that the compaction rate decreases as the number of cycles increases. This suggests that settlements in the field can be overestimated by laboratory tests on undisturbed samples if there has been some prior shaking causing significant shear strains.

Shaking-table tests on large samples appear especially suitable for reducing the influence of stress concentrations and non-uniformities in density which may develop in small-scale tests. Even more important, the effect of multi-directional shaking can be evaluated if table motions can be programmed for more than one degree of freedom. One-directional shaking-table tests with horizontal base motion on thin layers of uniform dry sand qualitatively confirm the findings of cyclic shear tests (Silver and Seed, 1969). Vertical settlements for maximum base accelerations of $0.3g$ are of the order of 1—2% in

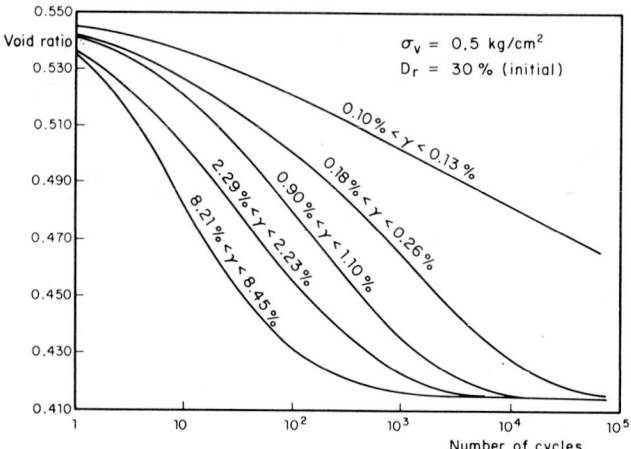

Fig. 4.31. Effects of shear-strain amplitude and number of cycles on compaction of saturated, drained Ottawa sand in cyclic shear tests. After Youd (1972).

the worst cases and considerably less for more favorable situations.* Influence of two-dimensional horizontal excitation and also effects of superimposing a vertical component of motion were studied by Pike et al. (1974), and compared with the results of one-dimensional horizontal shaking. Circular layers of uniform dry sand, having a thickness of 7.5 cm and a diameter

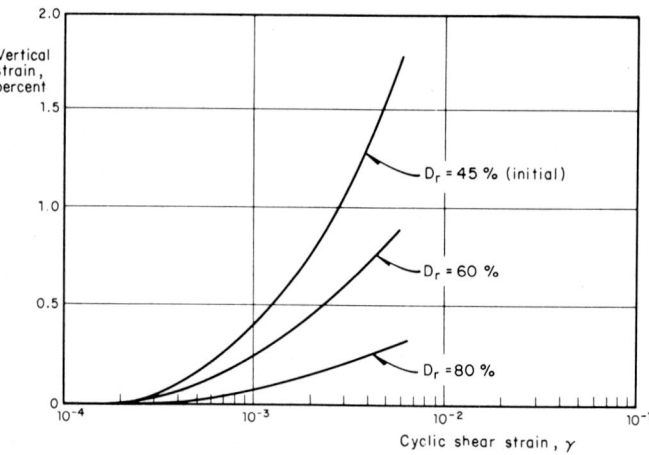

Fig. 4.32. Effects of relative density and shear-strain amplitude on total vertical settlement of a dry sand in cyclic shear tests. After Silver and Seed (1969).

---

* Settlements of 0.5—1.0% were observed in unsaturated sand fills after the San Fernando earthquake (Lee and Albaisa, 1974).

of 122 cm, were tested at different relative densities under largest horizontal accelerations of the order of $1g$ and vertical accelerations up to about $0.3g$ at a typical frequency of 6 Hz. Stresses and strains generated in the sample were assumed to be comparable to those developed in the field at a depth of about 1.5—3.0 m. Two-dimensional horizontal base motions with independent $X$ and $Y$ random acceleration histories produced the effects illustrated in Fig. 4.33, where the vertical settlement for the combined motion is compared with that caused by each component acting alone. It is seen that the settlement under combined motion approximately equals the sum of the independent settlements. This relation was found to be valid over a wide range of densities. The results of Fig. 4.34 were obtained superimposing sinusoidal vertical accelerations of increasing intensity on the combined random horizontal motion.

In conclusion, the most important observations concerning volume changes of dry cohesionless soils at vibration levels of earthquake engineering interest can be summarized as follows:

(1) Cyclic shearing, such as may be associated with the passage of seismic S-waves through surface soil strata, constitutes the most effective densification process.

(2) For a given initial density, cyclic shearing strain amplitude is the single most important parameter affecting the rate and the total amount of densification.

(3) If a large number of small-amplitude shear-strain cycles are applied and the initial conditions of the material lie above a certain critical line in the state diagram, an equilibrium void ratio is attained. Under the same initial conditions, this must be lower than the $e_s$-line void ratio (see Section 4.1.4) since it corresponds to a very dense condition that can only be attained by cyclic loading.

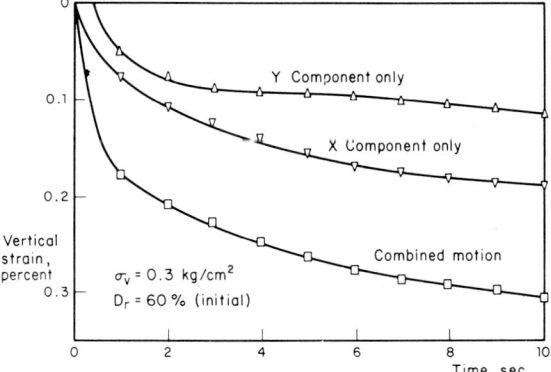

Fig. 4.33. Vertical settlements of a dry sand observed in shaking-table tests with independent and combined horizontal random motions. After Pyke et al. (1974).

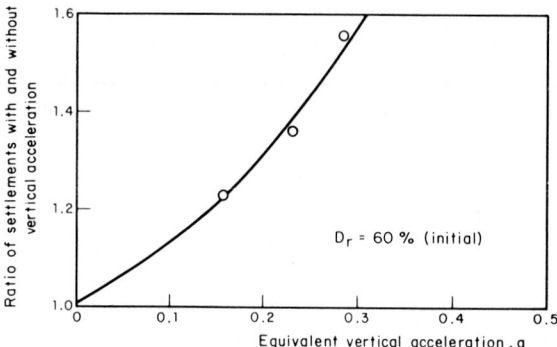

Fig. 4.34. Effects of vertical motion superimposed on horizontal motion on settlements of a dry sand in shaking-table tests. After Pyke et al. (1974).

(4) Higher surcharge pressures will generally make the rate of densification slower since the shear strains will be smaller if the excitation remains the same.

(5) Densification produced in a given number of cycles of specified intensity is largely frequency-independent.

(6) The basic characteristics of compaction of sand in cyclic simple shear, unidirectional shaking-table tests, and multidirectional shaking-table tests are similar, but quantitative effects due to multidirectional shaking are significant (Figs. 4.33 and 4.34).

### 4.4.3 Loss of strength of loose saturated sands and soft cohesive soils

Presence of water in the pores of cohesionless soil does not significantly affect its behavior under cyclic loading if drainage can freely occur. When drainage is prevented or is not instantaneous, the natural tendency of the soil to volume decrease under vibrations causes an increase in pore pressure (cf. Section 4.1.4); the magnitude of the increase depends on the initial state of the material, the state of stress, and the characteristics of the vibratory process.

If, as a consequence of the development of high pore pressures, an important portion of a soil mass suffers a substantial reduction of shear strength and sudden large deformations set in under the acting state of stress, liquefaction failure proper is said to occur. This behavior is typical of saturated sand in a flow slide: the soil flows until the shear stresses within the mass become compatible with its reduced strength. Consolidated-undrained triaxial tests show that this phenomenon is governed by the $e_f$ critical state line (see Section 4.1.4) (Castro, 1969). Figure 4.35 illustrates the $e_f$-lines of several different sands, briefly described in Table 4.II. Field experience and laboratory tests indicate that liquefaction in the foregoing sense can only de-

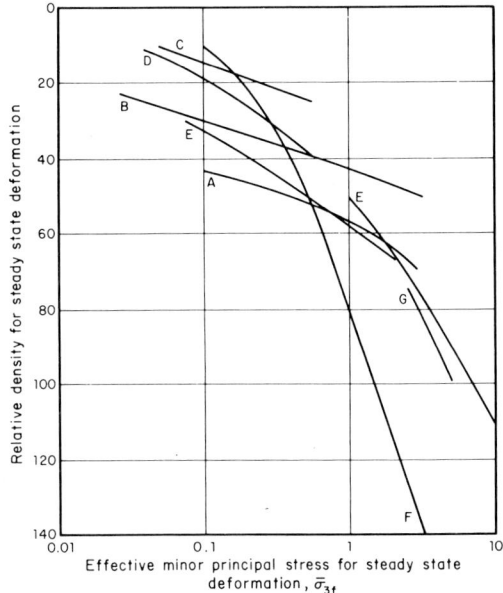

Fig. 4.35. Critical-state lines ($e_f$-lines) of sands described in Table 4.II (Castro, 1972).

velop in loose or very loose sands when there exists a driving force producing high shear stresses, such as at the base of tall slopes and under the foundation of heavy structures (Castro, 1975). In these circumstances, small static or dynamic perturbations can suffice to trigger liquefaction. Note that, for a given initial density, susceptibility to liquefaction increases with confining pressure.

The $e_f$-line can be determined in either cyclic or monotonic tests (Castro, 1969 and 1975). A typical feature of both types of test on unstable sands is that, after reaching a well-defined peak strength, the specimens suddenly undergo a large decrease in resistance and strains of the order of 20% are reached in a fraction of a second. Progressive increase of pore pressure in the cyclic tests follows the same general pattern of volume reduction observed in drained tests (Fig. 4.30).

Pore-pressure effects in saturated cohesive soils are more difficult to analyze experimentally and have been less extensively investigated. On the one hand, cohesive soils are more sensitive than sands to strain-rate effects and show a considerable increase in stiffness and strength above static values when subjected to rapid transient loads. On the other, there are few data showing that, in undrained conditions, a gradual load transfer from soil to water occurs in a fashion qualitatively similar to that observed in sands. Quasi-static cyclic triaxial tests on normally consolidated clay suggest the existence of a critical level of repeated deviator stress below which failure

TABLE 4.II

Brief description of the sands corresponding to the critical-state lines of Fig. 4.35 (Castro, 1972)

| Sand type | Description | Grain size (mm) | Uniformity coefficient, $C_u$ |
|---|---|---|---|
| A | slightly silty fine to medium sand with subangular to angular grains containing some shell fragments | $D_{10} = 0.13$<br>$D_{60} = 0.40$ | 3.08 |
| B | uniform clean fine quartz sand with sub-rounded grains | $D_{10} = 0.097$<br>$D_{60} = 0.17$ | 1.8 |
| C | uniform fine to medium sand with angular grains | $D_{10} = 0.14$<br>$D_{60} = 0.33$ | 2.3 |
| D | clean medium to fine sand | $D_{10} = 0.15$<br>$D_{60} = 0.90$ | 6.0 |
| E | uniform fine sand, angular grains, slightly micaceous | $D_{10} = 0.080$<br>$D_{60} = 0.17$ | 2.1 |
| F | uniform fine to medium sand with angular grains, containing plate-shaped grains corresponding to glass shards of volcanic origin. Some grain breakage occurred during the tests. | $D_{10} = 0.117$<br>$D_{60} = 0.23$ | 2.0 |
| G | silty fine sand with angular grains | $D_{60} = 0.15$<br>26% passing the no. 200 sieve (0.074 mm) | — |

under cyclic loading cannot be obtained (Sangrey et al., 1969). However, any repeated stress level above the critical limit may lead to large deformations and to a progressive build-up of pore pressure. The ratio of the critical level of repeated stress to the standard undrained strength seems to depend mainly on the sensitivity of the clay. In very sensitive clays the loss of resistance caused by repeated loading may be substantial and the pore pressure may increase to such an extent that it appears appropriate to speak of liquefaction failure in a broad sense. Pore-pressure histories observed by Seed and Chan (1966) in medium-sensitive clays subjected to several separated trains of stress pulses after application of a sustained deviator stress lend support to this hypothesis. Their data further show that earthquake-type vibrations may initiate increases in pore pressure and creep deformations leading to failure after some time.

## 4.4.4 Cyclic mobility of cohesionless soils in laboratory tests

Large-scale damage caused by failures of saturated foundation soils, and backfills of quay walls and waterfront bulkheads has occurred in strong earthquakes at Puerto Montt (1960), Niigata (1964), Anchorage and several other locations in Alaska (1964), Hachinohe (Tokachioki earthquake, 1967), and San Fernando (1971), to quote only the most important cases. The variety of combinations of seismic excitation characteristics, soil conditions, surface topography and type of structure affected discourages any attempt at quantitative predictions of the phenomenon taking all factors into account. A large amount of laboratory data on small samples has been obtained in recent years; yet, it is not clear whether complex scale effects associated with actual boundary and drainage conditions in the field can be approximated by these tests.

Pioneering work on the behavior of uniform, fine saturated sands in undrained stress-controlled, cyclic triaxial and simple-shear tests has been carried out by Seed and co-workers at the University of California in Berkeley (Seed and Lee, 1966; Lee and Seed, 1967, Peacock and Seed, 1968). These investigators defined as initial liquefaction the instant when the pore pressure reaches the value of the confining pressure for the first time during the test and as complete liquefaction the state corresponding to a strain amplitude of 20%. In loose sands initial and complete liquefaction occur almost simultaneously, whereas in dense sands several tens or hundreds of loading cycles may be required to achieve the transition. The number of cycles necessary to cause initial liquefaction increases with relative density and decreases with the ratio of pulsating deviator stress to effective confining pressure. Assuming that the most important densification effects in surface soils are attributable to vertically propagating shear waves, the cyclic simple-shear test appears more suitable than the triaxial test for reproducing the state of stress on a soil element in the field. If the alternating shear stress in the simple-shear test is taken to correspond directly to that in the field, it is found that the triaxial test overestimates the liquefaction resistance in terms of deviator stress intensity by a factor of about 1.3—1.4. The comparability of the two types of tests was investigated by Finn et al. (1969), who showed that they give identical results if the difference between the initial confining states of stress is taken into account. More sophisticated torsional shear devices, allowing soil samples to consolidate at different values of the intermediate principal stress, have also been introduced in order to study the effect of initial anisotropy on pore-pressure development and liquefaction resistance (Ishihara and Li, 1972; Ishibashi and Sherif, 1974). Gyratory and biaxial loading effects are being investigated by means of special cyclic shear apparatuses developed at Harvard (Rendón, 1973) and at the National University of Mexico (Jaime, 1975).

Cyclic tests by Castro (1969) confirm the findings of Seed and co-workers and partially clarify the meaning of their results. Castro tested samples of the same sand at relative densities falling both above and below the $e_f$-line determined from monotonic tests. Typical stress—strain curves for a sample prepared at an unstable state (above the $e_f$-line) are illustrated in Fig. 4.36. The stress—strain behavior of Fig. 4.36 is essentially the same as in a monotonic test starting from identical initial conditions. Results of a cyclic test on a sample prepared at a stable state (below the $e_f$-line) are shown in Fig. 4.37. During the first few cycles the sample behaved quasi-elastically but after the thirteenth cycle, i.e. after the effective stress vanished for the first time, the sample deformed over a wide strain range without mobilizing any resistance. Only at strain amplitudes greater than about 3% the specimen started developing new resistance as the pore pressure dropped. We have again a case of failure due to the build-up of high pore pressure and large cyclic strains, but of a quite different nature from that illustrated in Fig. 4.36. The latter would be associated with the occurrence of flow-slides (liquefaction in a strict sense). The behavior of Fig. 4.37 would correspond to slumping of a slope under earthquake but not to flow of the soil. Results of a test such as that of Fig. 4.37 strongly underestimate the ability of a dense sand to resist cyclic loading because a substantial redistribution of water occurs, beginning from relatively small strain amplitudes and leading to the formation of a layer of very loose sand and even clean water on top of the specimen. The presence of this layer is essentially responsible for the high recorded pore pressure and axial strains and, ultimately, for collapse. Following a suggestion by Casagrande, the term "cyclic mobility" can be used for the large strains occurring in this type of test; development of cyclic mobility is intimately related to the specific experimental conditions involved.

Since cyclic mobility observed in the laboratory occurs even in very dense

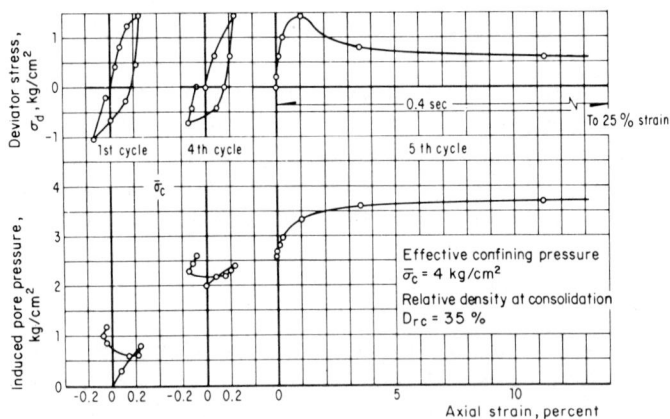

Fig. 4.36. Stress—strain curves from cyclic triaxial test on a saturated sand in unstable initial conditions. After Castro (1969).

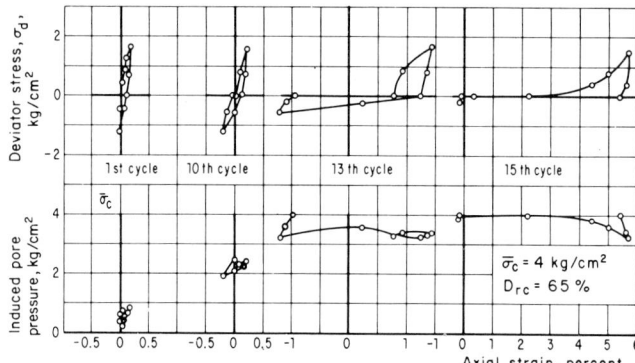

Fig. 4.37. Stress—strain curves from cyclic triaxial test on a saturated sand in stable initial conditions. After Castro (1969).

sands and, in contrast to liquefaction proper, resistance to cyclic mobility increases with increasing confining pressure, it would be useful to determine the conditions and extent to which this phenomenon is affected by water redistribution. Whether the inhomogeneities observed in laboratory tests may develop under field conditions remains in most cases an open question. It appears, however, unlikely that any saturated cohesionless soil, under a sufficiently intense and long earthquake, would suffer large strains as implied by the data of Seed and co-workers. Given the present state of knowledge, one must consider the critical-state theory as an acceptable postulate for analyzing liquefaction and admit that cyclic mobility phenomena may be governed by a different mechanism, still partially unclear. It should also be recognized that laboratory test results on medium and dense sands furnish conservative estimates of the resistance to cyclic mobility under field conditions. A comparative analysis of cyclic triaxial test data and field evidence of the behavior of saturated sands during earthquakes lends substantial support to the latter statement (Castro, 1975). The same analysis shows that an important reason for the conservatism of cyclic tests is the loosening of medium dense and dense sands produced by current procedures for extracting samples below the water table; the reverse is true for loose sands. Systematic investigations on the effects of different sampling techniques will probably bring significant improvements in liquefaction prediction.

As it is improbable that water redistribution can be eliminated from small-scale tests, better understanding of cyclic mobility is likely to come from results of large-scale tests, such as those performable on shaking tables. A step in this direction has been undertaken by Finn and co-workers at the University of British Columbia by studying the behavior of loose saturated sand in a rigid container of $180 \times 45 \times 17.5$ cm under horizontal shaking-table excitation at acceleration levels and frequencies of earthquake engineering interest (Finn et al., 1971). Typical results (Fig. 4.38) display the same qualitative trends as small-scale cyclic tests but a detailed analysis on the comparability of the two types of test and data on the relative density distribution within liquified large samples are not yet available. The role played by cyclic shear stress magnitude is difficult to assess in these tests because most of the shear force is carried by the rigid container walls rather than by the soil.

### 4.4.5 Effects of permeability, drainage path and boundary conditions

Fine sands and silty sands are the soils with highest probability of failure due to liquefaction or cyclic mobility (Fig. 4.39). Clay fines, being plastic, are less readily liquifiable although, as we have seen, substantial increases of pore pressure associated with large deformations may be induced even in clays.

Although data of Fig. 4.39 give an indirect idea of the influence of the permeability coefficient, it is unrealistic to try to evaluate the role of the latter factor separately from those of drainage path, boundary conditions

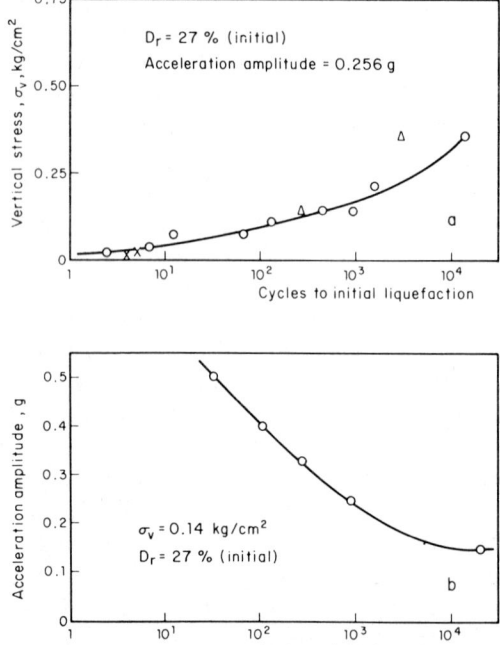

Fig. 4.38. Effects of confining pressure (a) and acceleration amplitude (b) on liquefaction (or cyclic mobility) resistance of saturated Ottawa sand in horizontal shaking table tests. After Finn et al. (1971).

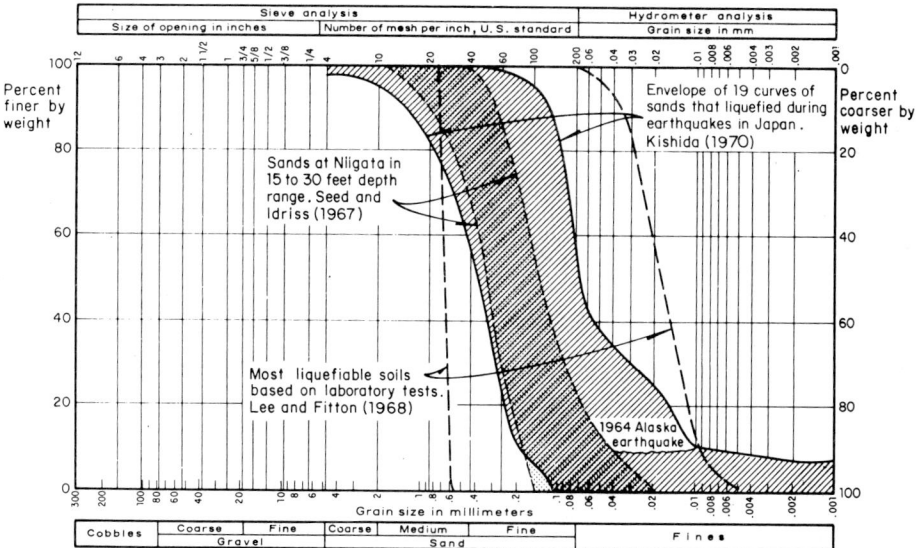

Fig. 4.39. Effect of grain size on susceptibility to liquefaction or cyclic mobility. After Shannon and Wilson et al. (1971).

and in-situ density. When the drainage path imposed by the boundary conditions is excessively long, as in deep deposits or under large impervious foundations, and permeability is not high, the ability to develop high pore pressures depends almost exclusively on soil density. Potentially dangerous configurations are also associated with lenses and seams of saturated sand or silt imprisoned in masses of soil of substantially lower permeability. Drainage can then be prevented and liquefaction or cyclic mobility of the sand lenses may trigger catastrophic collapses of the surrounding soil mass if the ground surface is sloping. Quantitative analysis of pore-pressure dissipation can be performed in simple cases of horizontal surface deposits. In a typical two-layer soil model it is assumed that pore pressure produced by earthquake vibrations or some other dynamic load have led to vanishing of effective stresses throughout the lower layer and, starting from this initial condition, the static redistribution of excess pressures in the system is studied (Housner, 1958; Ambraseys and Sarma, 1969; Yoshimi and Kuwabara, 1973). This model predicts that loss of strength of a loose layer at some depth can cause foundation failures in an initially stable surface layer with a delay of several minutes. Such negative effect can be reduced by substantially increasing the permeability and initial strength of the surface soil.

*4.4.6 Evaluation of liquefaction and cyclic mobility potentials*

It follows from the critical-state theory that proper evaluation of the liquefaction potential requires the determination of the $e_f$-line of the cohesionless soils at the site under consideration. This is hardly a trivial matter but the prerequisite that the material be in a loose condition can often be ascertained from penetration tests in situ. Estimating the level of the confining stress and of the shear stresses induced by existing static forces is also feasible in several cases without excessive difficulties. Data of Fig. 4.35 may in some cases give useful indications as to the convenience of performing independent tests for defining a specific critical state curve. As to the character and intensity of the perturbation, a relatively small earthquake could suffice for inducing liquefaction in very loose sands under high shear stresses.

The experimental findings mentioned in Section 4.4.4 indicate that far greater uncertainties surround the assessment of the cyclic mobility potential in the field. The methods summarized below should be considered with caution, recognizing that they are only rough tools for obtaining conservative estimates, especially when dealing with medium-dense and dense sands. Empirical methods based on field observations will be discussed in connection with the measurement of soil properties in situ.

Among the semi-analytical methods, one-dimensional stochastic approaches, in which the phenomenon of pore-pressure accumulation is treated as a low-cycle fatigue process, have been independently proposed by Donovan (1971) and Faccioli (1973). Statistical estimates of stress response at different depths are obtained either from surface acceleration response or from an assumed spectral-density function of bedrock excitation. Assuming deterministic properties for the soil system, a linear fatigue criterion, and a suitable expression fitting cyclic resistance laboratory data, the time and depth of first vanishing of the effective stress within the deposit can be estimated by techniques of random vibration theory (Lin, 1967).

A more widely used deterministic approach, originally applied by Seed and Idriss (1967) to the analysis of soil collapse at Niigata and later proposed by the same authors in a simplified version (1970b), consists of the following steps:

(1) Calculate the shear-stress response induced by the design earthquake at different depths in the deposit. In the simplified version the maximum shear stress near the surface is estimated from the maximum ground surface acceleration.

(2) Transform each stress history into an equivalent number of uniform stress cycles. The equivalent uniform stress amplitude may be taken as two thirds of the maximum stress. The number of equivalent stress cycles depends on the earthquake magnitude: a representative number of 10, 20, and 30 significant cycles is suggested for earthquake magnitude of 7, 7½, and 8 respectively.

(3) Determine at various depths the cyclic shear-stress amplitudes required to cause large strains in the same number of uniform cycles as calculated in step (2). This information is derived from cyclic triaxial or simple-shear tests on representative soil samples under different confining pressures. The shear

stress causing large strains in a given number of cycles can be alternatively estimated from empirical correlations based on relative density values.

(4) Compare at each depth the shear stress generated by the earthquake with that required to cause large strains. In zones where the latter is exceded by the former, danger of cyclic mobility exists.

The sources of the largest uncertainties are the selection of a deterministic earthquake input and the variability of soil properties in the field. Furthermore, the assumption of constant water content deformation may lead to unacceptably conservative conclusions in certain cases, as discussed in Section 4.1.2. Material descriptions including two phases, i.e. a solid skeleton and a fluid filling the pores, appear in such circumstances more appropriate and their applicability should be more thoroughly investigated. Since the pore-pressure history can be followed independently in a two-phase description, the points where this variable first reaches the level of the confining pressure would indicate the potentially liquifiable zones under a given excitation. Ghaboussi and Wilson (1973) have applied Biot's (1961) linear dynamical theory of saturated porous elastic solids to the seismic finite-element analysis of an earth dam-reservoir. Although, due to the linearity assumption, progressive pore-pressure accumulation phenomena are neglected by this model, interesting qualitative effects introduced by the two-phase description can be appreciated in the pore pressure and effective stress histories (Fig. 4.40). The dam core has a coefficient of permeability $10^4$ times smaller than that of the upstream and downstream shells. Note that the major part of the oscillations is carried by the fluid and that the stresses in the solid change in a much smoother fashion.

In addition to the previous suggestions for future experimental research, improvements on the current criteria require that more effort be directed to detailed studies of subsoil conditions at sites where liquefaction or cyclic mobility has actually occurred in recent earthquakes. Representative sites located in regions of intense seismicity and having a subsoil with predominance of fine, uniform saturated sands should be thoroughly instrumented for pore pressure, acceleration and possibly strain measurements. Pore-pressure measurements at different depths appear especially important for clarifying accumulation and dissipation processes, as well as drainage-path effects.

A question of importance is the possibility of reliquefaction. If a saturated sand deposit has experienced liquefaction or cyclic mobility during a past earthquake, how likely is this phenomenon to occur again during a future earthquake? Given the present state of knowledge, only some educated guesses can be offered. First, the age of the deposits should be estimated as precisely as possible and related to the seismic history of the site. Deposits of fine, uniform cohesionless soils are often the product of erratic transport and sedimentation operated by rivers which have been continuously changing their course with time. They are further found in coastal

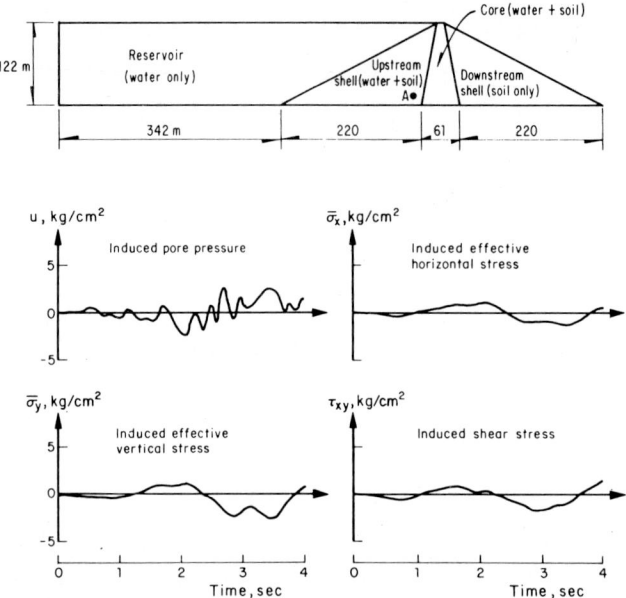

Fig. 4.40. Histories of induced pore pressure and effective stresses at point A of two-dimensional dam-reservoir system shown on top of figure, analyzed by a two-phase finite-element model. Base of system is subjected to 4 sec of the N—S component of the 1940 El Centro earthquake. After Ghaboussi and Wilson (1973).

areas where the elevation above sea level may have varied in time. Quite common is also the case of reclaimed land or artificial fills where portions of cities have been erected. In all these circumstances it is conceivable that the deposit in its present configuration is recent and that it has never suffered before severe or moderate shaking. Such seems to have been the case with some sand fills in Niigata: the earthquake of 1964 was the first strong motion in 130 years (Koizumi, 1966). Evidence of past liquefaction or cyclic mobility seems difficult to ascertain, unless caused by a very recent earthquake. In free-field liquefaction the final equilibrium configuration of the deposit should be characterized by low shear stresses, so as to ensure compatibility with the residual shear strength of the material; in loose sands the latter is usually very small. This fact suggests that reliquefaction of the same deposit is a rather unlikely event, unless a new driving force has been applied after the first liquefaction and/or its current equilibrium configuration favors further flow of the soil.

On the other hand, some laboratory data show that after cyclic mobility has been induced once in a sand specimen, it can be induced for the second time in a substantially smaller number of cycles if the specimen is reconsolidated at the same initial stress (Finn et al., 1970). It is presently impossible to say how much of this is strictly a laboratory phenomenon and how much is representative of a different field behavior, possibly related with reductions in density. Again, this seems to justify the high degree of conservatism currently adopted in problems related with cyclic mobility.

*4.4.7 Loss of strength and fatigue effects in cohesive soils*

Loss of strength in clays of different sensitivities has been investigated by Seed and Chan (1966) and Thiers and Seed (1968a,b) with cyclic triaxial and simple-shear tests. The strength developed by applying a sustained deviator stress plus a pulsating load was found to be considerably lower than the transient strenght, but often near to the value of normal undrained strenght (see Section 4.2.5).

Typical failure curves (Fig. 4.41) for an undisturbed silty clay (San Francisco bay mud) suggest the onset of low-cycle fatigue phenomena. As expected, compacted clays are much less affected by an increasing number of stress cycles. No direct evidence of the existence of a critical level of repeated stress is provided by these experiments but it might be speculated that this level coincides with the asymptotic value of cyclic failure curves. The overall strength reduction with respect to the normal undrained value seems to depend primarily on the cyclic strain amplitude. Thus, even in a soft medium-sensitive clay the normal strength remains unaffected by two hundred load cycles of strain magnitude not exceeding 1.5%, values that may represent upper bounds for strong earthquakes.

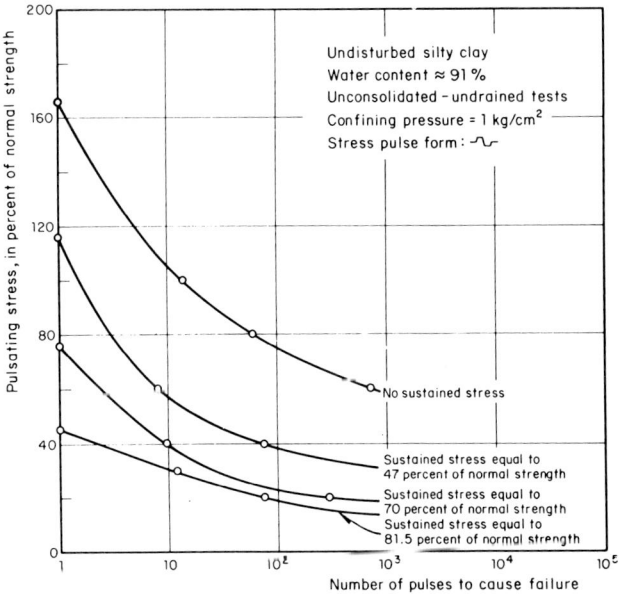

Fig. 4.41. Relationship between stress amplitude and number of pulses causing failure in an undisturbed silty clay (Seed and Chan 1966).

On the basis of available data it appears that the failure of soft cohesionless soils under earthquake loading conditions could be investigated by the methods described in Section 4.4.6. The stochastic fatigue model (Faccioli, 1973) seems suitable if one does not wish to analyze directly the pore-pressure effects. Two-dimensional failure criteria, accounting for the presence of a sustained stress, can also be introduced (see Section 4.2.5).

4.5 SOIL EXPLORATION

*4.5.1 Considerations on field exploration programs*

Since seismic risk partly depends on local geology and subsoil, the question arises of how accurately the influence of site conditions can be assessed on the basis of field data. Depending on the answer and on the project under consideration, decisions can be taken as to the type and extent of field exploration work and of laboratory tests that should be performed. This leads to a sequential decision process in which each step will depend on the information gained from the previous step and where, in general, professional experience and sound engineering judgement are hardly replaceable by mathematical rules. In some cases, however, parts of this process can be handled by probabilistic decision models (Baecher, 1972; Benjamin and Cornell, 1970).

Subsurface data required for foundation and settlement analysis under static loads usually provide a preliminary answer to the question posed at the beginning. If stable soil conditions predominate, attention can be confined to the occurrence of dynamic amplification and the shear moduli and damping factors of the different soil materials should be primarily evaluated. The number of field measurements for the determination of shear moduli will depend on the extension of the area under study, the degree of uniformity in the stratigraphic profile and the structures to be built. For large areas with relatively uniform soil conditions a few measurement points on a regular space grid will generally suffice. For special projects involving different types of structures, such as a nuclear plant, at least one or two measurements under each major structure should be performed and more if the structure is large and the soil profile nonuniform.

When penetration resistance values disclose the presence of weak soils, such as soft clays or fine loose sands, the occurrence of direct foundation damage as a result of local compaction or loss of shear strength will be of primary concern.

In this case the extension and thickness of the potentially unstable layers should be determined with good approximation. If the weak soils occur in lenses of limited extension and some indications as to their size and frequency are available, their search by means of borings arranged in regular space grids can be rationalized by means of probabilistic techniques of the "target hitting" type used in statistical geology (Koch and Link, 1970). Once the target has been hit, the geotechnical engineer is confronted with the problem of deciding the location and number of additional borings necessary to better define the shape and extension of the lens: the optimal decision can in some cases be attained using Bayes theorem (Wu, 1974).

In large-scale projects, such as the microregionalization of an urban area, one of the first decisions in soil exploration concerns the design of a boring program for an initial determination of the subsurface stratigraphy. Standard penetration borings are frequently used for this purpose. An important design variable is often represented by depth to bedrock or sufficiently hard ground. In addition to influencing the static foundation design, the hard ground profile may be needed in establishing the boundary conditions of discrete mathematical models for seismic soil-response analysis. If the stratigraphy is not very erratic, optimal design of boring programs for determining the depth of firm ground through regular or quasi-regular space grids can be achieved by a Bayesian probabilistic approach (Veneziano and Faccioli, 1975).

In conclusion, it is felt that selection of field exploration strategies could benefit from an increased application of probabilistic decision methods and from research efforts directed to a suitable quantification of concepts such as "field experience" or "engineering judgement" whenever possible.

*4.5.2 Exploration methods*

*4.5.2.1 Geophysical methods*
For shear-strain amplitudes below about $10^{-5}$ most soils can be approximately treated as linear elastic media; their material constants can therefore be calculated from the values of the longitudinal ($c_p$) and shear ($c_s$) wave velocities through the relations:

$$G = \rho c_s^2 \tag{4.16}$$

$$E = \rho(3c_p^2 - 4c_s^2)/(c_p^2/c_s^2 - 1) \tag{4.17}$$

$$\mu = (c_p^2/2c_s^2 - 1)/(c_p^2/c_s^2 - 1) \tag{4.18}$$

$$K = \rho(c_p^2 - 4c_s^2/3) \tag{4.19}$$

where $\mu$ is Poisson's ratio, and $K$ the bulk modulus. Field techniques for the determination of $c_p$ will not be treated here; their characteristics and limitations are discussed in several references (e.g. White, 1965; Richart et al., 1970). Considerable effort is currently being devoted to improving field methods for the determination of $c_s$. As in most geoseismic methods, the problem is one of generating and propagating through the soil as clean as possible shear-wave signals, recording their arrival at a suitable point, and measuring their propagation time between the energy source and the recording point along an appropriate path. Whereas field methods have the intrinsic advantage of directly operating on the actual configuration and boundary conditions, a serious shortcoming stems from the limitations of the energy input at the source and the ensuing low level of induced strains. Shear-strain amplitudes produced by strong-motion earthquakes in stable soils may exceed $10^{-3}$ whereas the strains generated during typical S wave field tests hardly exceed levels of $10^{-6}$. Actual measurements during such tests are very few; strain values recorded in down-hole tests using SH signals generated on the ground surface are illustrated in Fig. 4.42. Because of the limitations in strain amplitude, field measurements can only be used for establishing values of $G_{max}$ (see Section 4.2.2). The variation of $G$ as a function of shear-strain amplitude up to levels of strong-motion interest must be determined from laboratory tests.

As regards the effectiveness of different methods, it may be observed that a source will generate S waves to the extent that it is directional, unbalanced and asymmetric. Experiments have shown that most seismic sources possess such properties to some extent. The problem therefore consists in eliminating unwanted wave types as much as possible. Horizontally polarized shear wave (SH) type sources are particularly attractive since the radiation pattern is simple and, in horizontally layered structures, reflection and refraction of SH waves at an interface will only give rise to waves of the same type. If the source consists of a horizontal force at the surface, the SH wave radiation pattern is as illustrated in Fig. 4.43. A convenient source is represented by a wooden plank pressed firmly on the ground and impacted horizontally at

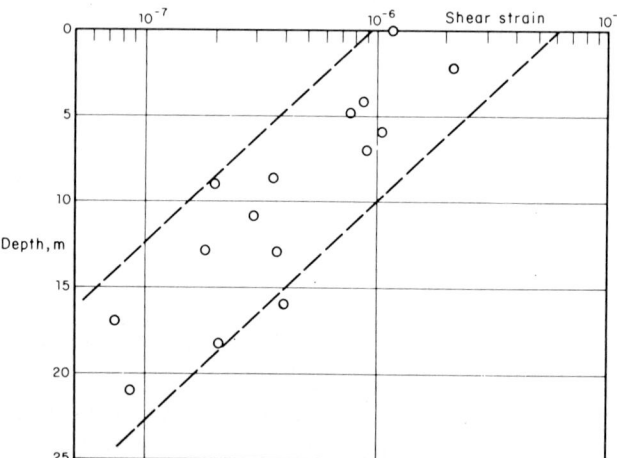

Fig. 4.42. Induced shear-strain levels in seismic down-hole tests for measuring S wave propagation velocity. After Hara (1972), cited by Ohsaki and Iwasaki (1973).

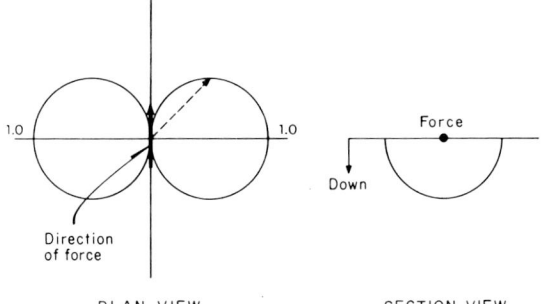

Fig. 4.43. Radiation pattern for SH wave amplitude from horizontal surface force.

one end with a hammer or a swinging mass. Measurements are taken in a borehole with the axis near the center of the plank: the direct SH wave arrival is detected by the horizontal transducers of a suitable seismometer which can be moved along the hole and clamped to its wall at the desired depth. For maximum measurement effectiveness one of the two horizontal sensing components should be oriented normally to the force. With this method, the SH wave arrival at a given depth can be easily identified by hitting the plank first at one end and then at the opposite one: two wave forms in which SH motion is exactly reversed must be obtained. Due to the ambient noise and the small energy input at the source, arrival times tend to be poorly defined at depths greater than 20—30 m. This technique, introduced by Kobayashi (1959), was improved and systematically employed by Shima and co-workers for determining the dynamic properties of the subsoil in the Tokyo area (Kawasumi et al., 1966; Shima et al., 1968a,b, 1969). A more powerful SH wave generator, using explosives, has been tested by Shima and Ohta (1967).

Although shear waves of the vertically polarized type exhibit a more complicated radiation pattern, they have also been employed for shear modulus measurements (Ohta and Shima, 1967); effectiveness of SV sources using explosives has been studied by Hattori (1972). Reviews of field methods of engineering interest, covering the often far from trivial aspects of identification and interpretation, are due to Duke (1969) and Mooney (1974).

Field measurements of soil damping are still at an experimental stage (Kudo and Shima, 1970) and considerable further research seems necessary for developing simple and reliable techniques for earthquake engineering applications.

*4.5.2.2 Penetration tests*

Penetration resistance represents a convenient property for field classification purposes and for estimating bearing capacities to be used in foundation design. Moreover, a distinct advantage of penetration tests lies in their simplicity and low cost. Penetration resistance can be measured by a wide variety of dynamic or static cone penetrometers (Sanglerat, 1972), or by the standard penetration test (SPT), as defined by the norm ASTM D-1586. An exhaustive state-of-the-art review if the SPT is due to de Mello (1971). The popularity of SPT in earthquake engineering applications derives from its diffusion as a standard exploration tool in seismically active countries such as the U.S. and Japan. As a result, its data have been correlated with dynamic soil properties. In sands, SPT has acquired an importance perhaps exaggerated because of the difficulty, cost and frequent unreliability of special techniques for obtaining and testing undisturbed samples, particularly at large depths and below the groundwater level.

A commonly used index property for defining the state of granular soils in situ is the relative density $D_r$; several laboratory and field investigations have been performed in order to establish quantitative correlations between $D_r$ and the blowcount $N$ of the SPT (Gibbs and Holtz, 1957; Schultze and Menzenbach, 1961; Schultze and Melzer, 1965; Bazaraa, 1967). The correlation by Gibbs and Holtz, obtained from laboratory experiments on two different sands, is the most popular in seismic site evaluation problems and is illustrated in Fig. 4.44 together with other empirical curves as a function of the overburden pressure. It is common practice, especially in the U.S., to compute $D_r$ from the in-situ value of $N$ using the Gibbs-Holtz correlation and to prepare specimens of the sand in question at the same value of $D_r$ in order to perform dynamic laboratory tests. This procedure is open to criti-

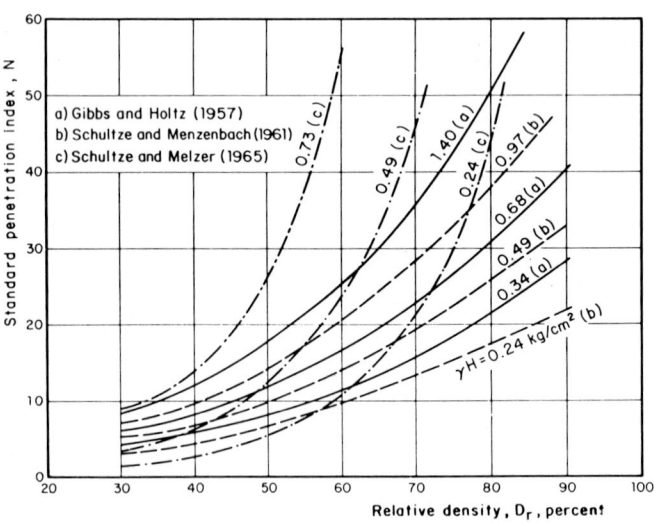

Fig. 4.44. Empirical correlations between standard penetration index $N$, overburden pressure ($\gamma H$) and relative density $D_r$ in sands. After Tavenas (1971).

cism. First of all, the relative density concept may not be applicable for defining the state of natural sand deposits. In the words of R.B. Peck (1971): "All alluvial and all windblown deposits are sorted by the transporting agent and are deposited in sedimentation units within each of which the grain size is likely to be uniform. Because of rapid fluctuations in velocity of either water or wind, the sedimentation units deposited under a given set of conditions may be small. (...) To be meaningful, the results of a determination of density index must be obtained on samples from a single sedimentation unit. If the materials from even two sedimentation units are mixed, the grain-size distribution of the mixture becomes radically different from that of each contributing constituent. (...) Neither the densest state nor the loosest state of a mixture has any relationship to the densest state or loosest state of the materials in the original sedimentation units. Since it is impractical, or even impossible, to subdivide the deposit samples into representatives of individual sedimentation units, and since it is unavoidable that several units should become mixed, it is impossible to determine a meaningful value of density index for any natural material deposited in such fashion. For these reasons, density index (...) cannot be expected to correlate with the strength or compressibility of the actual deposit, or with the tendency towards liquefaction. (...) The use of density index as an index to the (...) tendency to liquefaction of granular materials should be limited to artificially placed deposits in which materials are well mixed, so that the laboratory samples have the same grain-size characteristics as the significant field materials." Additional sources of error likely to affect any $N-D_r$ correlation include dispersion in the $N$-value, in the minimum and maximum dry weight needed for the determination of $D_r$ (Tavenas, 1971), as well as the effects of introducing large amounts of data from different sands into the analysis (de Mello, cit.). If e.g. the Gibbs and Holtz correlation is considered, the limits of the zone of possible scattering for both $N$ and $D_r$ are such that the correct statistical use of the correlation may yield information of little practical use (Fig. 4.45).

It thus appears that the introduction of $D_r$ as an intermediate variable to describe the natural state of soil is bound to introduce large, often unacceptable, errors. As a consequence, correlation with properties such as the resistance to cyclic mobility or liquefaction of saturated sands should be sought directly in terms of the blowcount $N$; related laboratory tests should be performed as much as possible on undisturbed samples.

Analyses of standard penetration resistance in saturated sand deposits which actually liquefied have been performed by Japanese investigators, especially in connection with the Niigata and Tokachioki earthquakes. In Niigata, in order to study the changes in subsoil conditions produced by seismic motions, several penetration tests were carried out after the earthquake at locations where SPT profiles had been previously determined (Ohsaki, 1966; Koizumi, 1966). Results for one location are shown in Fig.

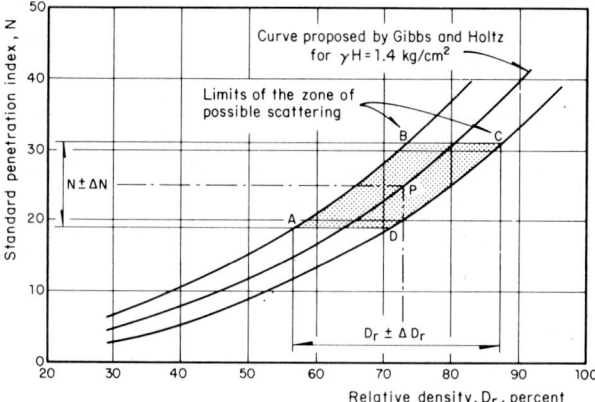

Fig. 4.45. Influence of experimental errors on the applicability of the Gibbs and Holtz correlation between standard penetration index and relative density in sands, according to Tavenas (1971).

4.46: note that the looser sands were compacted by the earthquake whereas the denser sands were loosened. It follows that at each depth a critical value of the penetration resistance, $N_{cr}$, exists that did not vary during the earth-

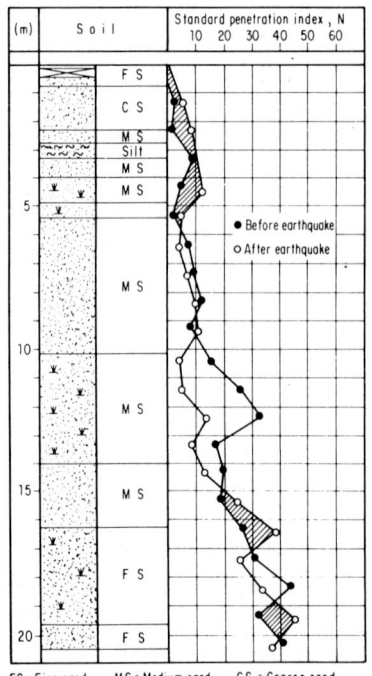

Fig. 4.46. Variations in standard penetration index $N$ caused at one location by the 1964 Niigata earthquake. After Ohsaki (1966).

quake. Ohsaki makes the reasonable assumption that the value of $N_{cr}$ corresponds to a critical void ratio of the sand at the given depth. A sand having $N$-values below $N_{cr}$ is potentially unstable and, if subjected to shear strains by earthquake vibration, will undergo volume contraction and pore-pressure increase. The curve of $N_{cr}$ vs. depth obtained by Koizumi for the Niigata subsoil is illustrated in Fig. 4.47. For practical use, Ohsaki (1970) suggested the approximate expression:

$$N_{cr} = 2z \tag{4.20}$$

where $z$ denotes the depth in m. The confidence that can be placed in the foregoing criterion depends mainly on the dispersion one is willing to assume for the values of $N$ in the SPT; doubts may arise as to the possibility of a soil being loosened by an earthquake against several meters of overburden plus negative pore pressure. However, although crude and based only on one set of observations, the criterion has the great merit of using quantities directly measurable in the field and of being consistent with critical-state theory predictions.

Since standard penetration tests are performed much more often than field measurements of the shear modulus, a study of the possible correlations between $G$ and $N$ based on in-situ data for typical subsoil conditions is of considerable utility. From a wide spectrum of data obtained in Japan at sites of construction of high-rise buildings, Ohsaki and Iwasaki (1973) performed a statistical analysis for typical foundation soils. The assumed correlation is of the form:

$$G = aN^b \tag{4.21}$$

and the numerical values in Table 4.III have been calculated by least-square fitting and expressing $G$ in $t/m^2$.

The correlation appears especially good for cohesive soils. Other proposed correlations are discussed in the same paper.

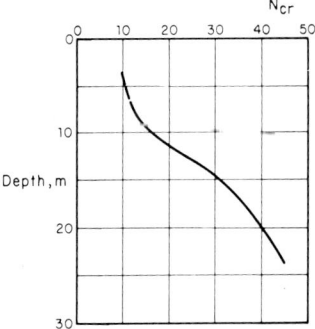

Fig. 4.47. Relation between critical value of standard penetration index, $N_{cr}$, and depth for Niigata subsoil (Koizumi, 1966).

TABLE 4.III

Numerical values of parameters of eq. 4.21 (Ohsaki and Iwasaki, 1973)

| Soil type | $a$ | $b$ | Correlation coefficient |
| --- | --- | --- | --- |
| Sandy soils | 650 | 0.94 | 0.852 |
| Intermediate soils | 1182 | 0.76 | 0.742 |
| Cohesive soils | 1400 | 0.71 | 0.921 |
| All soils | 1218 | 0.78 | 0.888 |

*4.5.2.3 Sampling*

Undisturbed sampling is a crucial step in any exploration program involving the performance of representative laboratory tests. In soil dynamics applications, undisturbed samples are usually required for the determination of shear modulus and damping, or else to investigate the susceptibility of cohesionless soils to liquefaction and cyclic mobility. A standard reference on sampling techniques is represented by the work of Hvorslev (1949). Whereas the extraction of undisturbed samples from cohesive soil deposits is more or less routine matter, the same amount of confidence cannot be placed on the techniques employed for saturated sands. A supposedly reliable sampler for saturated sands is the one designed by Bishop (1948); currently used procedures also include pitcher or piston-type samplers (Castro, 1975). No single sampler appears to be best for all cases; tools best suited to the specific conditions must be determined in each case by the soils engineer.

If dynamic tests for investigating cyclic mobility must be performed, differences in sampling techniques and details of handling can lead to significant differences in test results. Although badly needed, no systematic investigations on this important aspect of the problem have yet been undertaken. Preliminary results by Castro (1975) indicate that most current procedures for sampling sands below the water table produce significant loosening of medium dense and dense materials, and thus lead to conservative estimates of their cyclic resistance.

REFERENCES

Afifi, S.E.A., 1970. *Effects of Stress History on the Shear Modulus of Soils.* Thesis, Dep. Civ. Eng., Univ. Michigan, Ann Arbor, Michigan.

Aki, K. and Larner, K.L., 1970. Surface motion of a layered medium having an irregular interface due to incident plane SH waves. *J. Geophys. Res.*, 75: 933—954.

Ambraseys, N.N., 1970. Factors controlling the earthquake response of foundation materials. *Proc. 3rd European Symp. Earthquake Eng., Sofia*, pp. 309—323.

Ambraseys, N.N., 1973. Dynamics and response of foundation materials in epicentral regions of strong earthquakes. *Proc. 5th World Conf. Earthquake Eng., Rome*, pp. CXXVI—CXLVIII.

Ambraseys, N.N. and Sarma, S., 1969. Liquefaction of soils induced by earthquakes. *Bull. Seismol. Soc. Am.*, 59: 651—664.
Anderson, D.G., 1974. *Dynamic Modulus of Cohesive Soils.* Thesis, Dep. Civ. Eng., Univ. Michigan, Ann Arbor, Michigan.
Ayala, G. and Brebbia, C.A., 1973. A survey of numerical integration in time domain. *Report CE/4/73, Dep. Civ. Eng., Univ. Southampton,* Southampton.
Baecher, G.B., 1972. *Site Exploration: A Probabilistic Approach.* Thesis, Massachusetts Institute of Technology, Cambridge, Massachusetts.
Barkan, D.D., 1962. *Dynamics of Bases and Foundations.* McGraw-Hill, New York, 434 pp.
Bathe, K.J. and Wilson, E.L., 1973. Stability and accuracy analysis of direct integration methods. *Int. J. Earthquake Eng. Struct. Dyn.*, 1: 283—291.
Bazaraa, A., 1967. *Use of the Standard Penetration Test for Estimating Settlements of Shallow Foundations on Sand.* Thesis, Dep. Civ. Eng., Univ. Illinois, Urbana, Illinois.
Benjamin, J.R. and Cornell, C.A., 1970. *Probability, Statistics and Decision for Civil Engineers.* McGraw-Hill, New York, 684 pp.
Biot, M.A., 1961. Mechanics of deformation and acoustic propagation in porous media. *J. Appl. Phys.*, 33: 1482—1498.
Bishop, A.W., 1948. A new sampling tool for use in cohesionless sands below ground water level. *Geotechnique*, 1: 125—131.
Bishop, A.W. and Eldin, G., 1950. Undrained triaxial tests on saturated sands and their significance in the general theory of shear strength. *Geotechnique*, 2: 13—32.
Borcherdt, R.D., 1970. Effects of local geology on ground motion near San Francisco bay. *Bull. Seismol. Soc. Am.*, 60: 29—61.
Broms, B.B. and Forssblad, L., 1969. Vibratory compaction of cohesionless soils. *Proc. Specialty Session No. 2, 7th Int. Conf. Soil Mech. Found. Eng., Mexico City*, pp. 101—118.
Casagrande, A., 1936. Characteristics of cohesionless soils affecting the stability of slopes and earth fills. *J. Boston Soc. Civ. Eng.*, January, pp. 257—276.
Casagrande, A., 1976. Liquefaction and cyclic deformation of sands. A critical review. *Rep. 88, Harvard Soil Mechanics Series, Harvard Univ.*, Cambridge, Massachusetts.
Casagrande, A. and Shannon, W.L., 1948. Stress-deformation and strength characteristics of soils under dynamic loads. *Proc. 2nd Int. Conf. Soil Mech., Rotterdam*, V, pp. 29—34.
Castro, G., 1969. Liquefaction of sands. *Rep. 81, Harvard Soil Mechanics Series, Harvard Univ.*, Cambridge, Massachusetts.
Castro, G., 1972. Liquefaction and cyclic mobility. Given at a seminar on *Earthquake Response of Subsoils* presented by Geotechnical Engineers Inc., July 1972, Winchester, Massachusetts.
Gastro, G., 1975. Liquefaction and cyclic mobility of saturated sands. *Proc. ASCE, J. Geotech. Eng. Div.*, 101 (GT6): 551—570.
Cervantes, R., Esteva, L. and Alduncin, G., 1973. Riesgo sísmico en formaciones estratificadas. *Intern. Rep., Instituto de Ingeniería, Universidad Nacional Autónoma de México*, Mexico City.
Constantopoulos, I.V., 1973. Amplification studies for a nonlinear hysteretic soil model. *Rep. R73-46, Dep. Civ. Eng., Massachusetts Institute of Technology*, Cambridge, Massachusetts.
Cooley, J.W. and Tukey, J.W., 1965. An algorithm for the machine calculation of complex Fourier series. *Math. Comput.* 19: 297—301.
Crandall, S.H., Kurzweil, L.G., Nigam, A.K. and Remington, P.J., 1970. Dynamic properties of modelling clay. *Rep. 76205-3, Acoustics and Vibration Laboratory, Massachusetts Institute of Technology*, Cambridge, Massachusetts.

D'Appolonia, D.J. and D'Appolonia, E., 1967. Determination of the maximum density of cohesionless soils. *Proc. 3rd Asian Regional Conf. Soil Mech. Found. Eng., Haifa*, pp. 266—268.

de Mello, V.B., 1971. The standard penetration test. *Proc. 4th Pan-Am. Conf. Soil Mech. Found. Eng., San Juan*, 1, pp. 1—86.

Dezfulian, H. and Seed, H.B., 1970. Seismic response of soil deposits underlain by sloping rock boundaries. *Proc. ASCE*, 96 (SM6): 1893—1916.

Dezfulian, H. and Seed, H.B., 1971. Response of nonuniform soil deposits to travelling seismic waves. *Proc. ASCE*, 97 (SM1): 27—46.

Dibaj, M. and Penzien, J., 1969. Response of earth dams to travelling seismic waves. *Proc. ASCE*, 95 (SM2): 541—560.

Dobry, R., 1970. Damping in soils: its hysteretic nature and the linear approximation. *Rep. R70-14, Dep. Civ. Eng.*, Massachusetts Institute of Technology, Cambridge, Massachusetts.

Donovan, N.C., 1971. A stochastic approach to the seismic liquefaction problem. Presented at *1st Int. Conf. Applications of Statistics and Probability to Soil and Structural Engineering, Hong-Kong*.

Drnevich, V.P., Hall, J.R. and Richart, F.E., 1967. Effects of amplitude of vibration on the shear modulus of sand. *Proc. Int. Symp. Wave Propagation and Dynamic Properties of Earth Materials*, University of New Mexico, Albuquerque, New Mexico, pp. 189—199.

Duke, C.M., 1969. Techniques for field measurement of shear waves in soils. *Proc. 4th World Conf. Earthquake Eng., Santiago*, III (A5), pp. 39—54.

Esteva, L. and Villaverde, R., 1973. Seismic risk, design spectra and structural reliability. *Proc. 5th World Conf. Earthquake Eng., Rome*, pp. 2586—2596.

Esteva, L., Rascón, O. and Gutierrez, A., 1969. Lessons from some recent earthquakes in Latin America. *Proc. 4th World Conf. Earthquake Eng., Santiago*, III (J2), pp. 58—73.

Faccioli, E., 1973. A stochastic model for predicting seismic failure in a soil deposit. *Int. J. Earthquake Eng. Struct. Dyn.*, 1: 293—307.

Faccioli, E. and Ramírez, J., 1976. Earthquake response of nonlinear hysteretic soil systems. *Int. J. Earthquake Eng. Struct. Dyn.*, 4: 261—276.

Faccioli, E., Esteva, L. and Cervantes, R., 1973. Probabilistic analysis of nonlinear seismic response of stratified soil deposits. *Proc. 5th World Conf. Earthquake Eng., Rome*, pp. 414—418.

Finn, W.D.L., Pickering, D.J. and Bransby, P.L., 1969. Sand liquefaction in triaxial and simple shear tests. *Rep. 11, Soil Mechanics Series, Dep. Civ. Eng., Univ. British Columbia*, Vancouver.

Finn, W.D.L., Bransby, P.L. and Pickering, D.J., 1970. Effect of strain history on liquefaction of sand. *Proc. ASCE*, 96 (SM6): 1917—1934.

Finn, W.D.L., Emery, J.J. and Gupta, Y.P., 1971. Liquefaction of large samples of saturated sand on a shaking table. *Proc. 1st Can. Conf. Earthquake Eng., Vancouver*, pp. 97—110.

Ghaboussi, J. and Wilson, E.L., 1973. Seismic analysis of earth dam-reservoir system. *Proc. ASCE*, 99 (SM10): 849—862.

Gibbs, H.J. and Holtz, W.G., 1957. Research on determining the density of sands by spoon penetration testing. *Proc. 4th Int. Conf. Soil Mech. Found. Eng., London*, 1, pp. 35—39.

Goudreau, G.L. and Taylor, R.L., 1973. Evaluation of numerical integration methods in elastodynamics. *J. Comput. Methods Appl. Mech. Eng.*, 2: 69—98.

Gutenberg, B., 1957. Effect of ground on earthquake motion. *Bull. Seismol. Soc. Am.*, 47: 221—250.

Hall, J.R. and Richart, F.R., 1963. Dissipation of elastic wave energy in granular soils. *Proc. ASCE*, 89 (SM6): 27—56.

Hardin, B.O., 1965. The nature of damping in sands. *Proc. ASCE*, 91 (SM1): 63—97.
Hardin, B.O. and Black, W.L., 1966. Sand stiffness under various triaxial stresses. *Proc. ASCE*, 92 (SM2): 27—42.
Hardin, B.O. and Black, W.L., 1968. Vibration modulus of normally consolidated clay. *Proc. ASCE*, 94 (SM2): 353—368.
Hardin, B.O. and Black, W.L., 1969. Closure to: Vibration modulus of normally consolidated clay. *Proc. ASCE*, 95 (SM6): 1531—1539.
Hardin, B.O. and Drnevich, V.P., 1972a. Shear modulus and damping soils: measurement and parameter effects. *Proc. ASCE*, 98 (SM6): 603—624.
Hardin, B.O. and Drnevich, V.P., 1972b. Shear modulus and damping soils: design equation and curves. *Proc. ASCE*, 98 (SM7): 667—692.
Hardin, B.O. and Richart, F.E., 1963. Elastic wave propagation in granular soils. *Proc. ASCE*, 89 (SM1): 33—65.
Haskell, N.A., 1960. Crustal reflections of plane SH waves. *J. Geophys. Res.*, 65: 4147—4150.
Hattori, S., 1972. Investigation of seismic waves generated by small explosions. *Bull. Int. Inst. Seismol. Earthquake Eng.*, 9: 27—106.
Herrera, I., and Rosenblueth, E., 1965. Response spectra on stratified soil. *Proc. 3rd World Conf. Earthquake Eng., Auckland and Wellington*, pp. 1.44—60.
Herrera, I., Rosenblueth, E. and Rascón, O., 1965. Earthquake spectrum prediction for the valley of Mexico. *Proc. 3rd World Conf. Earthquake Eng., Auckland and Wellington*, pp. 1.61—74.
Hisada, T., Nakagawa, K. and Izumi, M., 1965. Normalized acceleration spectra for earthquakes recorded by strong-motion accelerographs and their characteristics related with subsoil conditions. *BRI Occas. Rep. No. 23, Build. Res. Inst.*, Tokyo.
Housner, G.W., 1958. The mechanism of sandblows. *Bull Seismol. Soc. Am.*, 48: 155—161.
Hudson, D.E., 1972. Local distribution of strong earthquake ground motions. *Bull. Seismol. Soc. Am.*, 62: 1765—1786.
Hvorslev, M.J., 1949. Subsurface exploration and sampling of soils for civil engineering purposes. *Report* edited by Waterways Experiment Station, Vicksburg, Mississippi.
Idriss, I.M. and Seed, H.B., 1968a. An analysis of ground motions during the 1957 San Francisco earthquake. *Bull Seismol. Soc. Am.*, 58: 2013—2032.
Idriss, I.M. and Seed, H.B., 1968b. Seismic response of horizontal soil layers. *Proc. ASCE*, 94 (SM4): 1003—1031.
Idriss, I.M., Lysmer, J., Hwang, R. and Seed, H.B., 1973. Quad-4. A computer program for evaluating the seismic response of soil structures by variable damping finite element procedures. *Rep. 73-16, Earthquake Eng. Res. Center, Univ. California*, Berkeley, California.
Ishibashi, I. and Sherif, M.A., 1974. Soil liquefaction by torsional simple shear device. *Proc. ASCE*, 100 (GT8): 871—888.
Ishihara, K. and Li, S., 1972. Liquefaction of saturated sand in triaxial torsion shear test. *Soils Found.*, 12(2): 19—40.
Jaime, A., 1975. Aparato de corte simple cíclico bidireccional. *Proc. 5th Pan-Am. Conf. Soil Mech. Found. Eng., Buenos Aires*, II, pp. 395—402.
Jennings, P.C., 1963. *Response of Simple Yielding Structures to Earthquake Excitation*. Thesis, Earthquake Eng. Res. Lab., California Institute of Technology, Pasadena, California.
Kawasumi, H., Shima, E., Ohta, Y., Yanagisawa, M., Allam, A. and Miyakawa, K., 1966. S wave velocities of subsoil layers in Tokyo, 1. *Bull. Earthquake Res. Inst.*, 44: 731—747 (in Japanese).
Kishida, H., 1970. Characteristics of liquefaction of level sandy ground during the Tokachioki earthquake. *Soils Found.* 10(2): 103—111.

Kobayashi, N., 1959. A method of determining the underground structure by means of SH wave. *Zisin*, 12: 19—24 (in Japanese).
Koch, G.S. and Link, R.F., 1970. *Statistical Analysis of Geological Data*. Wiley, New York, Vol. 2: 210—216.
Koizumi, Y., 1966. Changes in density of sand subsoil caused by the Niigata earthquake. *Soils Found.* 6 (2): 38—44.
Kudo, K. and Shima, E., 1970. Attenuation of shear waves in soil. *Bull Earthquake Res. Inst.*, 48: 145—158.
Kuhlemeyer, R.L. and Lysmer, J., 1973. Finite-element method accuracy for wave propagation problems. *Proc. ASCE*, 99 (SM5): 421—426.
Lawrence, F.V., 1965. Ultrasonic wave velocities in sand and clay. *Rep. R65-05, Dep. Civ. Eng., Massachusetts Institute of Technology*, Cambridge, Massachusetts.
Lee, K.L. and Albaisa, A., 1974. Earthquake-induced settlements in saturated sands. *Proc. ASCE*, 100 (GT4): 387—406.
Lee, K.L. and Fitton, J.A., 1968. Factors affecting the cyclic loading strength of soil. In: *Symposium on Vibration Effects of Earthquakes on Soils and Foundations, ASTM Spec. Tech. Publ.*, 450: 71—95.
Lee, K.L. and Seed, H.B., 1967. Cyclic stress conditions causing liquefaction of sand. *Proc. ASCE*, 93 (SM1): 47—70.
León, J.L., Jaime, A. and Rábago, A., 1974. Propiedades dinámicas de los suelos. Estudio preliminar. *Intern. Rep., Instituto de Ingeniería, Universidad Nacional Autónoma de México*, Mexico City.
Lin, Y.K., 1967. *Probabilistic Theory of Structural Dynamics*. McGraw-Hill, New York, 366 pp.
Lysmer, J. and Drake, L.A., 1972. A finite-element method for seismology. In: B.A. Bolt (Editor), *Seismology: Surface Waves and Earth Oscillations. Methods in Computational Physics, Volume 11*. Academic Press, New York, pp. 181—216.
Lysmer, J., Seed, H.B. and Schnabel, P.B., 1971. Influence of base-rock characteristics on ground response. *Bull. Seismol. Soc. Am.*, 61: 1213—1232.
Marcuson, W.F. and Wahls, H.E., 1972. Time effects on dynamic shear modulus of clays. *Proc. ASCE*, 98 (SM12): 1359—1373.
Matthiesen, R.B., Duke, C.M., Leeds, D.J. and Fraser, J.C., 1964. Site characteristics of Southern California strong-motion earthquake stations, 2. *Rep. 64-15, Dep. Eng., Univ. California*, Los Angeles, California.
Mooney, H.M., 1974. Seismic shear waves in engineering. *Proc ASCE*, 100 (GT8): 905—924.
Nacci, V.A. and Taylor, R.J., 1967. Influence of clay structure on elastic wave propagation. *Proc. Int. Symp. Wave Propagation and Dynamic Properties of Earth Materials*, University of New Mexico, Albuquerque, New Mexico, pp. 491—501.
Newmark, N.M. and Rosenblueth, E., 1971. *Fundamentals of Earthquake Engineering*. Prentice-Hall, Englewood Cliffs, New Jersey, 640 pp.
Nickell, R.E., 1971. On the stability of approximation operators in problems of structural dynamics. *Int. J. Solids Struct.*, 7: 301—319.
Ohsaki, Y., 1966. Niigata earthquakes, 1964, building damage and soil condition. *Soils Found.* 6 (2): 14—37.
Ohsaki, Y., 1969. The effects of local soil conditions upon earthquake damage. *Proc. Specialty Session No. 2, 4th Int. Conf. Soil Mech. Found. Eng.*, Mexico City, pp. 3—32.
Ohsaki, Y., 1970. Effects of sand compaction on liquefaction during the Tokachioki earthquake. *Soils Found.*, 10 (2): 112—128.
Ohsaki, Y. and Iwasaki, R., 1973. On dynamic shear moduli and Poisson's ratios of soil deposits. *Soils Found.*, 13 (4): 61—73.

Ohsaki, Y. and Sakaguchi, O., 1973. Major types of soil deposits of urban areas in Japan. *Soils Found.*, 13 (2): 49—65.

Ohta, Y. and Shima, E., 1967. Experimental study on generation and propagation of S-waves: II. Preliminary experiments on generation of SV-waves. *Bull. Earthquake Res. Inst.*, 45: 33—42.

Okamoto, S., 1973. *Introduction to Earthquake Engineering*. University of Tokyo Press, Tokyo, 571 pp.

O'Kelly, M., 1964. *Vibration of Viscously Damped Linear Dynamic Systems*. Thesis, California Institute of Technology, Pasadena, California.

Papadakis, C.N., 1973. *Soil Transients by Characteristics Method*. Thesis, Dep. Civ. Eng. Univ. Michigan, Ann Arbor, Michigan.

Peacock, W.H. and Seed, H.B., 1968. Sand liquefaction under cyclic loading simple-shear conditions. *Proc ASCE*, 94 (SM3): 689—708.

Peck, R.B., 1971. The standard penetration test. Discussion, *Proc. 4th Panam. Conf. Soil Mech. Found. Eng.*, San Juan, III, pp. 59—61.

Poulos, S.J., 1971. The stress—strain curves of soils. Mimeographed pamphlet.

Pike, R., Chan, C.K. and Seed, H.B., 1974. Settlement and liquefaction of sands under multi-directional shaking. *Rep. EERC 74-2, Earthquake Eng. Res. Center, Univ. California*, Berkeley, California.

Rascón, O. and Chávez, M., 1973. On an earthquake simulation model. *Proc. 5th World Conf. Earthquake Eng.*, Rome, pp. 2899—2907.

Rendón, F., 1973. Summary on liquefaction and cyclic mobility research performed on gyratory shear apparatus. *Intern. Rep., Soil Mechanics Laboratory, Harvard University*, Cambridge, Massachusetts.

Richart, F.E., Hall, J.R. and Woods, R.D., 1970. *Vibrations of Soils and Foundations*. Prentice-Hall, Englewood Cliffs, New Jersey, 414 pp.

Roesset, J.M., 1970. Fundamentals of soil amplification. In: R.J. Hansen (Editor), *Seismic Design for Nuclear Power Plants*. M.I.T. Press, Cambridge, Massachusetts, pp. 183—244.

Roesset, J.M. and Whitman, R.V., 1969. Theoretical background for soil amplification studies. *Rep. R69-15, Dep. Civ. Eng., Massachusetts Institute of Technology*, Cambridge, Massachusetts.

Rosenblueth, E., 1960. The earthquake of 28 July 1957 in Mexico City. *Proc. 2nd World Conf. Earthquake Eng., Tokyo and Kyoto*, pp. 359—379.

Rosenblueth, E. and Elorduy, J., 1969. Características de los temblores en la arcilla de la Ciudad de México. In: *Nabor Carrillo. El Hundimiento de la Ciudad de México y Proyecto Texcoco*, contribución de Proyecto Texcoco al Séptimo Congreso Internacional de Mecánica de Suelos e Ingeniería de Cimentaciones. Secretaría de Hacienda y Crédito Público, Fiduciaria: Nacional Financiera, S.A., México, pp. 287—328.

Rosenqvist, I.T., 1953. Considerations on the sensitivity of Norwegian quick-clays. *Geotechnique*, 3: 195—200.

Sanglerat, G., 1972. *The Penetrometer and Soil Exploration*. Elsevier, Amsterdam, 464 pp.

Sangrey, D., Henkel, D. and Esrig, M., 1969. The effective stress response of a saturated clay soil to repeated loading. *Can. Geotech. J.*, 6: 241—252.

Schnabel, P.B., Seed, H.B. and Lysmer, J., 1972a. Modification of seismograph records for effects of local soil conditions. *Bull Seismol. Soc. Am.*, 62: 1649—1664.

Schnabel, P.B., Lysmer, J. and Seed, H.B., 1972b. Shake. A computer program for earthquake response analysis of horizontally layered sites. *Rep. EERC 72-12, Earthquake Eng. Res. Center, Univ. California*, Berkeley, California.

Schultze, E. and Melzer, K., 1965. The determination of the density and the modulus of compressibility of non cohesive soils by soundings. *Proc 6th Int. Conf. Soil Mech. Found. Eng., Montreal*, 1, pp. 354—358.

Schultze, E. and Menzenbach, E., 1961. Standard penetration test and compressibility of soils. *Proc. 5th Int. Conf. Soil Mech. Found. Eng., Paris*, 1, pp. 527—532.

Seed, H.B., 1960. Soil strength during earthquakes. *Proc. 2nd World Conf. Earthquake Eng., Tokyo and Kyoto*, 1, pp. 183—194.

Seed, H.B., 1970. Soil problems and soil behavior. In: R.L. Wiegel (Editor), *Earthquake Engineering*. Prentice-Hall, Englewood Cliffs, New Jersey, pp. 227—252.

Seed, H.B. and Chan, C.K., 1966. Clay strength under earthquake loading conditions. *Proc. ASCE*, 92 (SM2): 53—78.

Seed, H.B. and Idriss, I.M., 1967. An analysis of the soil liquefaction in the Niigata earthquake. *Proc. ASCE*, 93 (SM3): 83—108.

Seed, H.B. and Idriss, I.M., 1969. Influence of soil conditions on ground motions during earthquakes. *Proc. ASCE*, 95 (SM1): 99—137.

Seed, H.B. and Idriss, I.M., 1970a. Analyses of ground motions at Union Bay, Seattle, during earthquakes and distant nuclear blasts. *Bull. Seismol. Soc. Am.*, 60: 125—136.

Seed, H.B. and Idriss, I.M., 1970b. A simplified procedure for evaluating soil liquefaction potential. *Rep. EERC 70-9, Earthquake Eng. Res. Center, Univ. California*, Berkeley, California.

Seed, H.B. and Idriss, I.M., 1970c. Soil moduli and damping factors for dynamic response analyses. *Rep. EERC-70-10, Earthquake Eng. Res. Center, Univ. California*, Berkeley, California.

Seed, H.B. and Lee, K.L., 1966. Liquefaction of saturated sands during cyclic loading. *Proc. ASCE*, 92 (SM6): 105—134.

Seed, H.B. and Lundgren, R., 1954. Investigation of the effect of transient loading on the strength and deformation characteristics of saturated sands. *Proc. ASTM*, 54: 1288—1306.

Seed, H.B. and Wilson, S.D., 1967. The Turnagain Heights landslide, Anchorage, Alaska. *Proc. ASCE*, 93 (SM4): 325—353.

Seed, H.B., Whitman, R.V., Dezfulian, H., Dobry, R. and Idriss, I.M., 1972. Soil conditions and building damage in 1967 Caracas earthquake. *Proc. ASCE*, 98 (SM8): 787—806.

Shannon & Wilson, Inc., and Agbabian-Jacobsen Associates, 1971. Soil behavior under earthquake loading conditions. *Report prepared for U.S.A.E.C.*, Contract W-7405-eng-26.

Shima, E. and Ohta, Y., 1967. Experimental study on penetration and propagation of S-waves: 1. Designing of SH-wave generator and its field tests. *Bull. Earthquake Res. Inst.*, 45: 19—31.

Shima, E., Ohta, Y., Yanagisawa, M., Allam, A. and Kawasumi, H., 1968a. S-wave velocities of subsoil layers in Tokyo, 2. *Bull. Earthquake Res. Inst.*, 46: 759—772 (in Japanese).

Shima, E., Ohta, Y., Yanagisawa, M., Kudo, K. and Kawasumi, H., 1968b. S-wave velocities of subsoil layers in Tokyo, 3. *Bull. Earthquake Res. Inst.*, 46: 1301—1312 (in Japanese).

Shima, E., Ohta, Y., Yanagisawa, M., Kudo, K. and Kawasumi, H., 1969. S-wave velocities of subsoil layers in Tokyo, 4. *Bull. Earthquake Res. Inst.*, 47: 819—829.

Šiljak, D., 1969. *Nonlinear Systems*. Wiley, New York, 618 pp.

Silver, M.L. and Seed, H.B., 1969. The behavior of sands under seismic loading conditions. *Rep. EERC 69-16, Earthquake Eng. Res. Center, Univ. California*, Berkeley, California.

Stokoe, K.H. and Richart, F.E., 1973. In-situ and laboratory shear-wave velocities. *Proc. 8th Int. Conf. Soil Mech. Found. Eng., Moscow*, 1 (2): 403—409.

Stokoe, K.H. and Woods, R.D., 1972. In-situ shear-wave velocity by cross-hole method. *Proc. ASCE.*, 98 (SM5): 443—460.

Streeter, V.L., Wylie, E.B. and Richart, F.E., 1974. Soil motion computations by characteristics method. *Proc. ASCE*, 100 (GT3): 247—253.

Tavenas, F.A., 1971. The standard penetration test. Discussion, *Proc. 4th Panam. Conf. Soil Mech. Found. Eng.*, San Juan, III, pp. 64—70.

Taylor, P. and Hughes, J., 1965. Dynamic properties of foundation subsoils as determined from laboratory tests. *Proc. 3rd World Conf. Earthquake Eng., Auckland and Wellington*, pp. 1.196—212.

Tezcan, S.S. and Ipek, M., 1973. Long distance effects of the 28 March 1970 Gediz Turkey earthquake. *Int. Earthquake Eng. Struct. Dyn.*, 1: 203—215.

Thiers, G.R. and Seed, H.B., 1968a. Cyclic stress—strain characteristics of clay. *Proc. ASCE*, 94 (SM2): 555—569.

Thiers, G.R. and Seed, H.B., 1968b. Strength and stress—strain characteristics of clays subjected to seismic loading conditions. In: *Symp. on Vibration Effects of Earthquakes on Soils and Foundations. ASTM Spec. Tech. Publ. 450*, pp. 3—56.

Trifunac, M.D., 1971. Surface motion of a semi-cylindrical alluvial valley for incident plane SH-waves. *Bull. Seismol. Soc. Am.*, 61: 1755—1770.

Tsai, N.C., 1969. *Influence of Local Geology on Earthquake Ground Motion*. Thesis, Earthquake Eng. Res. Lab., California Institute of Technology, Pasadena, California.

Udwadia, F.E. and Trifunac, M.D., 1973. Comparison of earthquake and microtremor ground motions in El Centro, California. *Bull. Seismol. Soc. Am.*, 63: 1227—1253.

Veneziano, D. and Faccioli, E., 1975. Bayesian design of optimal experiments for the estimation of soil properties. *Proc. 2nd Int. Conf. on Applications of Statistics and Probability in Soil and Structural Engineering, Aachen*, II, pp. 191—213.

Watson, J.D., 1940. *Stress-Deformation Characteristics of Cohesionless Soils from Triaxial Compression Tests*. Thesis, Harvard University, Cambridge, Massachusetts.

White, J.E., 1965. *Seismic Waves*. McGraw-Hill, New York, 302 pp.

Whitman, R.V. and Ortigosa, P., 1968. Densification of sand by vertical vibrations. *Tech. Pap. T68-5, Dep. Civ. Eng., Massachusetts Institute of Technology*, Cambridge, Massachusetts.

Whitman, R.V., Roesset, J.M., Dobry, R. and Ayestaran, L., 1972. Accuracy of modal superposition for one-dimensional soil amplification analysis. *Proc. Int. Conf. on Microzonation, Seattle*, pp. 483—498.

Wong, H.L. and Trifunac, M.D., 1974. Surface motion of a semi-elliptical alluvial valley for incident plane SH waves. *Bull. Seismol. Soc. Am.*, 64: 1389—1408.

Wu, T.H., 1974. Uncertainty, safety and decision in soil engineering. *Proc. ASCE*, 100 (GT3): 329—348.

Yoshimi, Y. and Kuwabara, F., 1973. Effects of subsurface liquefaction on the strength of surface soil. *Soils Found.*, 13 (2): 67—82.

Youd, T.L., 1970. Densification and shear of sand during vibration. *Proc ASCE*, 96 (SM3): 863—880.

Youd, T.L., 1972. Compaction of sands by repeated shear straining. *Proc. ASCE*, 98 (SM7): 709—726.

Zeevaert, L., 1967. Free vibration torsion tests to determine shear modulus of elasticity of soils. *Proc. 3rd Panam. Conf. Soil Mech. Found. Eng.*, Caracas, 1, pp. 111—129.

*Chapter 5*

THE PHYSICS OF EARTHQUAKE STRONG MOTION

JAMES N. BRUNE

*University of California, San Diego Scripps Institution of Oceanography, Institute of Geophysics and Planetary Physics, La Jolla, Calif. U.S.A.*

5.1 INTRODUCTION

This review of the physics of earthquake source is aimed at an intuitive physical understanding of the factors which determine the intensity of strong ground motion. It does not cover the development of mathematical techniques needed in quantitative calculations of ground motion for specific cases.

Our understanding of strong motion is primarily data-limited, and it is not surprising that there are a number of theories, models and mathematical techniques available, with little assurance that they will yield reliable results. Only a few measurements of strong ground motion are available from near-source accelerographs. Each new such record has given surprising results. Thus, we have little statistical basis for defining the maximum expected near-source ground motion, or what is typical.

The data limitation cannot be corrected in a short time because of the relative infrequency of strong earthquakes, although the deployment of massive numbers of strong-motion instruments promises to increase greatly the rate of acquisition of critical data. Because the problem is so important, we must make maximum use of theoretical modeling, even with its uncertainties, until an adequate strong-motion data base is obtained.

Our present understanding of the physics of earthquakes indicates that accelerations greater than $1g$ and velocities greater than 100 cm/sec are possible on solid rock in the near field of earthquakes. Unfortunately, we cannot adequately describe the probabilities of such strong ground motion because we lack knowledge about the details of the faulting mechanism in nature. The purpose of this chapter is to present the main physical concepts needed to estimate such probabilities.

5.2. PHYSICAL PARAMETERS OF THE EARTHQUAKE SOURCE

*5.2.1 The elastic rebound mechanism*

The physical mechanism of earthquakes as now understood was first clearly outlined by Reid (1910/1969) in his study of the great San Francisco earthquake of 1906. Earthquakes occur when parts of the earth elastically rebound from a state of high stress and convert some fraction of the potential elastic energy into kinetic energy of wave propagation (cf. Chapter 1). The energy conversion results from failure of the rocks along a fault surface. The gradual increase in stress preceding failure is brought about by long-term tectonic motions. The long-term motions can be described as motions of relatively rigid plates sliding on a weak asthenosphere. The fault surfaces on which large earthquakes occur represent the boundaries between plates or smaller blocks which slide past one another. Zones of transcurrent or strike-slip motion occur where plates slide horizontally past one another, zones of thrust faulting occur where plates collide and one is thrust under the other, and zones of normal faulting occur where plates are pulling apart and material from the mantle flows into the gap.

Friction along fault boundaries causes the motion to be irregular, to "chatter", i.e., alternately lock and slip as elastic energy is repeatedly stored and released. Thus, earthquakes result from "stress drop" along a fault surface following a period of gradual stress increase before the earthquake. The stress drop occurs over a limited part of the fault plane and its time history is a function of the physical properties of the fault surface, i.e., its frictional properties, roughness, temperature, etc.

*5.2.2 Point source theory*

For long wavelengths and at large distances relative to the source dimensions, earthquakes may be approximated by point sources. The theory of point sources stems from the early work of Love (1927). Forces applied to the medium are zero everywhere except at certain points where they become infinite in such a way that their volume integral preserves net moment and force balance.

An earthquake may be considered a confined source with characteristic spatial dimensions $l$ and time duration approximately $l/\beta$, where $\beta$ denotes the velocity of shear waves. For relatively long wavelengths and long periods, the source time function may be approximated by a constant times the unit Heaviside function, $H(t)$ [$H(t) = 0, t < 0, H(t) = 1, t > 0$].

Since the earthquake-producing forces are at equilibrium before and after the earthquake (after the released energy has radiated from the vicinity of the source), the point source (long period) representation of the earthquake must preserve equilibrium and thus have neither net force nor net moment. Possible point source representations are illustrated schematically in Fig. 5.1.

The single unbalanced force and single unbalanced couple are not possible equilibrium point sources, although any source can be represented by superimposing them.

The initial formulation by Love (1914, 1927) has been converted by Keylis-Borok (1960) to elucidate the separate character of the shear and com-

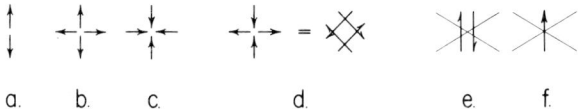

Fig. 5.1. Examples of various types of point sources. In the low-frequency limit, most earthquakes can be approximated by the double couple d. Sources e and f are not possible equilibrium sources.

pressional fields, and the results immediately following are taken from his work. The displacements caused by a simple force $K(t)$ located at the center of a rectangular system of coordinates $(x, y, z)$ and directed along the $x$-axis are presented first. Let us designate $x, y, z$ as the coordinates of the point of observation; $u_\alpha^{(q)}$, $u_\beta^{(q)}$ as the components of displacement for longitudinal and transverse waves, respectively, along the $q$-axis ($q$ being $x, y$, or $z$); $\alpha, \beta$ as the velocities of the longitudinal and transverse waves; $\rho$ as the density of the elastic medium, and $R$ as the distance from the point of origin to the point of observation, $R = \sqrt{(x^2 + y^2 + z^2)}$. $\mathcal{R}_{\theta,\phi}^{N;\alpha}$ is the geometrical radiation pattern (a product of sines and cosines of multiples of $\theta$ and $\phi$) for the near-field (N) compressional field ($\alpha$) and similarly for other geometrical factors (i.e., $\mathcal{R}_{\theta,\phi}^{F;\beta}$ = radiation pattern for far-field shear waves, etc.). Thus:

$$u_\alpha^{(q)} = (4\pi\rho R^3)^{-1} \mathcal{R}_{\theta,\phi}^{N;\alpha} [F(t - R/\alpha) + (R/\alpha)F'(t - R/\alpha)] + (4\pi\rho R\alpha^2)^{-1}$$
$$= \mathcal{R}_{\theta,\phi}^{F;\alpha} K(t - R/\alpha) \tag{5.1}$$

$$u_\beta^{(q)} = (4\pi\rho R^3)^{-1} \mathcal{R}_{\theta,\phi}^{N;\beta} [-F(t - R/\beta) - (R/\beta)F'(t - R/\beta)] + (4\pi\rho R\beta^2)^{-1}$$
$$= \mathcal{R}_{\theta,\phi}^{F;\beta} K(t - R/\beta) \tag{5.2}$$

where:

$$F' = \frac{\partial F}{\partial t}; \quad F(t) = \int_0^t \int_0^{t'} K(t')dt'dt \tag{5.3}$$

Various other types of point sources may be obtained by superposition of the simple point force. For the case of a couple we let forces $K(t)$ and $-K(t)$ be active at points $(0, 0, 0)$ and $(0, -h, 0)$ respectively. If we resolve the formulas for displacements caused by the second of these forces into the Taylor series about the coordinate of origin $y = 0$, and eliminate terms of the order of $h^2$ and higher, the total displacement will read:

$$u_v^{(q)} = h\frac{\partial}{\partial y_0} \cdot u_{v,1}^{(q)}\Big|_{y=h} \quad (v = \alpha, \beta) \tag{5.4}$$

A double couple is obtained by superposition of two couples of opposite sign, one rotated by 90° relative to the other (Fig. 5.1d). We can apply an analogous method to higher-order multipoles. The double couple produces a static displacement field which falls off with distance proportional to $R^{-2}$. The dynamic field at large distance falls off as $R^{-1}$ (so that energy per unit area of the wave front is preserved).

For a simple point source the far-field displacement is in phase with the forcing function, whereas the near-field displacement is in phase with the double time integral of the forcing function (because of the inertia of the mass near the source). For the double couple, the displacement field is differentiated according to eq. 5.4 and hence, the far-field displacement is in phase with the derivative of the forcing function and the near-field displacement in phase with the single time integral of the forcing function.

### 5.2.3 Far-field approximation

In strong-motion studies, it is important to know under what conditions we may assume we are in the far field. The term $F'(t - R/\beta)$ in eq. 5.2 falls off as $R^{-2}$ whereas the last term falls off as $R^{-1}$. We define the far field as distances beyond which the $R^{-2}$ term can be ignored with respect to the $R^{-1}$ term. The $R^{-2}$ term is proportional to $\omega^{-1}$ in its spectrum. The $R^{-1}$ term is constant in its spectrum with an additional $\beta^{-1}$ factor. Thus, the ratio of the two terms is proportional to $\beta/\omega R$, and the far-field approximation for a point source is valid if:

$$\frac{\beta}{\omega R} \ll 1 \tag{5.5}$$

For a finite source, the far-field approximation also requires that the distance be large compared to the source dimension.

### 5.2.4 Far-field radiation from a double-couple point source

The spectrum of the far-field radiation from a double couple acting as a step (Heaviside) function in time is independent of frequency. This follows because for a simple point force (first order) the far-field spectrum is proportional to the spectrum of the forcing function, $K(t) \propto H(t)$ with far-field spectrum proportional to $\omega^{-1}$, and the double couple (second order) may be obtained by differentiation. Thus, the far-field radiation for a double couple with $K(t) = H(t)$ has the shape of $H'(t) = \delta(t)$, the delta function, and hence, its spectrum is constant and it is phase-shifted by $\pi/2$ from the phase of the forcing function. For compressional waves, the displacement is given by:

$$u_\alpha(t,\overline{R}) = \frac{M_0 \delta(t - R/\alpha)}{4\pi\rho\alpha^3 R} \sin 2\theta \sin^2\phi \tag{5.6}$$

where $u_\alpha$ is the radial displacement, $M_0$ is the moment of one couple of the double couple, $\theta$ is the azimuthal angle of the station relative to the fault plane in the plane containing the couples, and $\phi$ is the co-latitudinal angle of the station relative to the plane of the couples (White, 1965).

For the S-waves:

$$u_\beta^{S\theta}(t,\overline{R}) = \frac{M_0 \delta(t - R/\beta)}{4\pi\rho\beta^3 R} \cos 2\theta \sin \phi \tag{5.7}$$

$$u_\beta^{S\phi}(t,\overline{R}) = \frac{M_0 \delta(t - R/\beta)}{8\pi\rho\beta^3 R} \cos 2\theta \cos 2\phi \tag{5.8}$$

The equations for the spectra are:

$$\Omega_\alpha(\omega) = \frac{M_0}{4\pi\rho\alpha^3 R} \sin 2\theta \sin^2\phi \tag{5.9}$$

$$\Omega_\beta^{S\theta}(\omega) = \frac{M_0}{4\pi\rho\beta^3 R} \cos 2\theta \sin \phi \tag{5.10}$$

$$\Omega_\beta^{S\phi}(\omega) = \frac{M_0}{8\pi\rho\beta^3 R} \cos 2\theta \cos 2\phi \tag{5.11}$$

the phase being taken relative to the wave arrival time in each case (White, 1965).

## 5.2.5 Seismic moment and fault slip

If earthquakes are essentially a pure-shear rebound phenomenon, then there is no net dilatation, and we may infer that in the low-frequency limit, the equivalent point source is a double couple. The preponderance of evidence indicates that this is a good approximation, although some studies have indicated that deep earthquakes may have a net dilatation or compression superposed (Dziewonski and Gilbert, 1974; Randall, 1968; Randall and Knopoff, 1970).

The scale parameter of the double-couple point source is its seismic moment $M_0$. This can be related to the dimensions and average offset of the fault surface (Aki, 1966). The following is an enlarged view of the derivation. Consider a very thin slab-shaped zone representing the fault plane (Fig. 5.2). As a result of a change in shear stress along the fault surface, the two sides will be offset relative to one another. This may be thought of as the result of a "machine" inside the fault zone which exerts tangential stresses on the walls of the slab. In order for the system to be in equilibrium, counterbalancing torques are required at the ends of the slab. Suppose the fault zone is filled with material (of the same rigidity) and that this inner material is sheared so as to provide continuity of displacement at its boundary. The inner material is welded to the outer material. For each local section of sur-

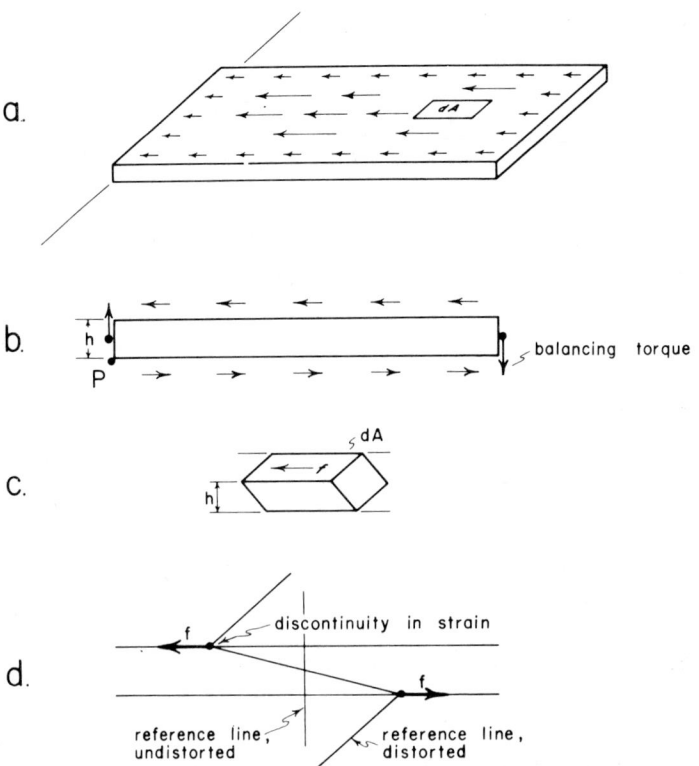

Fig. 5.2. Diagram illustrating the derivation of the relation $M_0 = \mu A \bar{u}$.

face area d$A$, a shearing force is applied, and for the interior tablet the local displacement is related to the force $f$ and rigidity $\mu$ by $u = hf/\mu dA$ (see Fig. 5.2c). The moment about point $P$ for the small block with fault area d$A$ is then $hf$. Since the material is welded together the situation is equivalent to a distribution of body forces in a continuous medium such as indicated in Fig. 5.2d. The total moment exerted by these forces is given by integration over the whole fault surface:

$$M_0 = \int \frac{hf}{dA} \, dA = \int \mu u \, dA = \mu A \bar{u} \qquad (5.12)$$

Since the material is in equilibrium, there is an opposing equal couple from forces acting on the orthogonal surface. One can think of the earthquake as occurring when the body forces which hold this dislocation are suddenly released. Since we are looking for a long-period, point source approximation relating seismic moment to the area and average offset, it is not important that the actual fault zone does not have this particular shape. For high frequencies, a general source would have to be represented by a complex distribution of multipoles.

In the limit of an infinitely thin slab ($h \to 0$) the energy in the interior slab is infinite since the strain energy goes as the square of the strain (as $h$ decreases, strain increases as $h^{-1}$ and volume decreases as $h$). Thus, the dislocation model has an infinite supply of potential energy. How much of this energy is radiated, of course, depends on how the displacement is relaxed. For certain types of displacement time functions with rupture velocities greater than or equal to the shear wave velocity, the radiated energy from a forced dislocation is infinite. This can be seen from the equations of Haskell (1964, 1966), which contain factors like $(\beta/v - \cos\theta)^{-1}$. This factor is singular for certain values of $\theta$ if the rupture velocity $v \geq \beta$. On the other hand, if the displacement is released quasi-statically, the radiated energy is zero. This peculiar property of the dislocation illustrates why it can only be used as a kinematic model, and thus, may not represent how an actual dislocation occurs in nature.

## 5.2.6 Energy and stress

Since the earthquake is an elastic rebound phenomenon the energy for the faulting process comes from stored elastic energy. Total energy released in the faulting process can be equated to the difference between the total elastic potential energy before and after the earthquake. The total energy release can be approximately divided into seismic wave energy and frictional heating on the fault. If we assume that the initial state of strain was pure shear, after the earthquake the rocks will be in a more complex state of strain involving both shear and compressional strain energy. In its general form, the total energy stored in the rocks is:

$$E = \tfrac{1}{2} \int_v p_{ij} e_{ij} \, dV \tag{5.13}$$

where $p_{ij}$ and $e_{ij}$ are the stress and strain tensors respectively and $v$ is the volume. If the medium is isotropic this may be written:

$$E = \tfrac{1}{2} \int_v (\lambda_I \theta^2 + 2\mu e_{ij}^2) \, dV \tag{5.14}$$

where $\lambda_I$ is the isothermal Lamé parameter. The Lamé parameter $\mu$ is the same for the isothermal and adiabatic cases. The adiabatic heating due to an earthquake includes both increases and decreases in temperature (in zones of compression and dilatation respectively). This causes small increases and decreases in surface heat flow, but averaged over geologic time and many earthquakes, the difference between the isothermal and adiabatic energy release will be small.

Most of the energy release comes from the strained volume near the fault. Near the center of the fault plane (away from the edges) the stress field is approximately plane shear strain before and after the earthquake, and we can write for a small volume there:

$$\Delta E \simeq \frac{1}{2}\mu(\epsilon_1^2 - \epsilon_2^2)dV \tag{5.15}$$

where $\epsilon_1$ and $\epsilon_2$ refer to the shear strain (derivative of parallel displacement with respect to the transverse coordinate) before and after the earthquake, respectively. This expression may be factored to illustrate the role of the stress drop $\Delta\sigma$ and the prestress (or average stress):

$$\Delta E = \mu dV(\epsilon_1 - \epsilon_2)\frac{(\epsilon_1 + \epsilon_2)}{2} = \frac{1}{\mu}dV\Delta\sigma\bar{\sigma} \tag{5.16}$$

where $\bar{\sigma}$ is the time average shear stress during the faulting process. This illustrates the crucial role of prestress in determining the total energy release. Later we describe how stress drop can be determined from measurements of fault offset in the field. Equation 5.16 shows that even so we cannot determine the energy release because it is determined not only by the stress drop, but also by the prestress (or average stress). As an example, if the stress drops from 1000 bars to 900 bars the energy release is 950/50 = 19 times greater than if the stress drops from 100 bars to zero, even though the permanent displacements observed in the field (and the seismic moment) are the same.

To illustrate the connection between the elastic energy release in the volume around the earthquake and the stresses acting on the fault we may consider the case of quasi-static stress release on the fault surface. Since the fault surface is by hypothesis the only place where the boundary conditions change, this approach is equivalent to considering the volume integrals above. In the quasi-static process we imagine a "machine" inside the fault zone which allows the stress to drop at such a slow rate that all of the elastic energy released goes into heat in the "machine". The energy change is then the work done on the machine. For a unit area of the fault $\Delta E = \sigma(u)du\Delta A$, or since $\sigma$ is linearly related to $u$ by the rigidity $\mu$ and some length parameter $l$:

$$\frac{\Delta E}{\Delta A} = \int_0^u \sigma(u)du = \int_0^u \frac{\mu}{l}udu = \frac{1}{2}\frac{\mu}{l}u^2\Big]_{u_2}^{u_1} = \frac{1}{2}\frac{\mu}{l}(u_1^2 - u_2^2)$$

$$= \frac{\mu}{l}\left(\frac{u_1 + u_2}{2}\right)(u_1 - u_2) = \left(\frac{\sigma_1 + \sigma_2}{2}\right)(u_1 - u_2) = \bar{\sigma}u \tag{5.17}$$

where $\bar{\sigma} = (\sigma_1 + \sigma_2)/2$ and $u = u_1 - u_2$.

The total work done on the machine is then:

$$\Delta E = \int_A \bar{\sigma}u dA \tag{5.18}$$

If we assume the stress drop is constant over the fault this may be written:

$$\Delta E = \bar{\sigma}\bar{u}A \tag{5.19}$$

where $\bar{u}$ is the areal average of $u$.

### 5.2.7 Frictional heat generation and seismic efficiency

In earthquakes much of the elastic potential energy goes into seismic radiation, but in general, unless the friction during sliding is nearly zero, a large amount of energy will also go into heat (or crushing of rock into gouge). Several simple possibilities are illustrated diagrammatically in Fig. 5.3. The basic assumption in simple models is that the friction force is constant as sliding along the fault progresses (independent of $u$ or $\dot{u}$, or prior history of the fault slip). Thus, the heat liberated is given by the product of the slip and the frictional stress no matter what the time history of the slip is. The remainder of the energy must go into seismic energy. The possibilities cover all ratios of seismic energy to heat energy. In the case of zero frictional stress, all of the energy goes into seismic radiation (e.g., fault melting, McKenzie and Brune, 1972). In the case of friction overshoot, stress drops below the kinetic friction (Savage and Wood, 1971). For the abrupt locking case (Housner, 1955; Brune, 1970) ($\epsilon < 1$) more high-frequency energy is radiated than would be expected on the basis of the Orowan (1960) model ($\sigma_2 = \sigma_f$). At present, we do not know the range of ratios for actual earthquakes, but evidence suggests that many earthquakes are approximated by Orowan's model. There may be some instances of overshoot and some instances of abrupt locking. The possible occurrence of abrupt locking might result from a non-constant sliding friction, e.g., resistance to sliding may

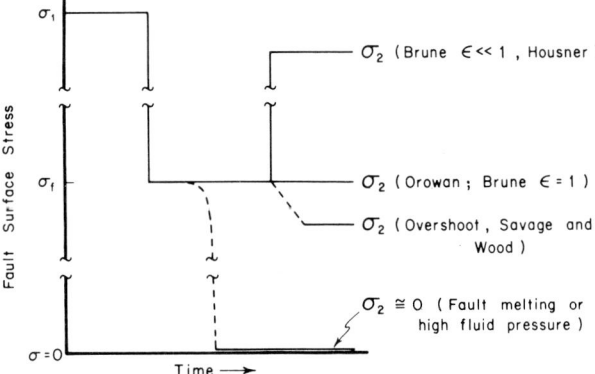

Fig. 5.3. Schematic illustrations of some possible records of average fault stress. In nature the fault stress is probably more complicated than for any of the curves shown here.

suddenly increase as the particle velocity drops below a critical limit, or as asperities are encountered; or a propagating dislocation of relatively small dimensions may lock the fault behind itself as it propagates. Savage and Wood (1971) have argued that for most earthquakes the overshoot model is appropriate. Dieterich (1973, 1974), using two- and three-dimensional finite-element models with variable friction along the fault, found the average stress drop less than the average difference between the initial stress and the frictional stress.

The record of stress on the fault is important to strong-motion seismology because it determines the relative amount of high-frequency energy and hence, the peak accelerations for a given stress drop. Stress drop (shear) is often the only stress parameter estimated for an earthquake. For the abrupt-locking model ($\epsilon \ll 1$), the radiated high-frequency energy is relatively great (relative to the Orowan model).

The dynamic frictional stress on faults is not known with certainty. Brune et al. (1969) have shown that the lack of an observed heat-flow anomaly over the San Andreas fault suggests friction stresses of less than a few hundred bars in the zone of earthquakes (averaged over geologic slip). Jeffreys (1962), Ambraseys (1969), and McKenzie and Brune (1972) have shown that if frictional stress exceeds a certain value for large earthquakes, melting should occur on fault planes. It is possible that fault-plane melting does occur, but the lack of abundant references to this phenomenon in the literature may suggest that frictional stresses on faults are relatively low (less than a few hundred bars). More study is needed on this important problem.

The parameter commonly used to represent the ratio of seismic energy to potential elastic energy release is the seismic efficiency $\eta_1$. For the case of assumed constant friction during sliding:

$$\eta_1 = \frac{\frac{1}{2}(\sigma_1 + \sigma_2) - \sigma_f}{\frac{1}{2}(\sigma_1 + \sigma_2)} = 1 - \frac{\sigma_f}{\bar{\sigma}} \tag{5.20}$$

$\eta_1$ may also be expressed in terms of seismic energy, seismic moment, and average stress, $\bar{\sigma}$:

$$\bar{\sigma} = \frac{\mu}{\eta_1} \frac{E}{M_0} \tag{5.21}$$

This has led Wyss (1970) to define the *apparent* stress as:

$$\sigma_{app} = \mu \frac{E}{M_0} \tag{5.22}$$

i.e., the estimated average stress assuming 100% seismic efficiency. For the Orowan model ($\sigma_f = \sigma_2$) the seismic efficiency is determined by the ratio of the stress drop to the sum of the frictional stress and initial stress (Wyss and Molnar, 1972):

$$\eta_1 \text{(Orowan)} = \frac{\sigma_1 - \sigma_f}{\sigma_1 + \sigma_f} = \frac{\Delta\sigma}{2\bar{\sigma}} \tag{5.23}$$

*5.2.8 Stress drop, fault displacement, and source dimension*

In a medium which is linearly elastic (except for the fault surface which can be treated as an interior boundary) the displacements are linearly related to the stress drop by the elastic moduli. Thus, stress drops can be measured by the difference in displacements before and after the earthquake. Analytical solutions for the shear-stress drop in terms of the ratio of maximum (or average) fault offset to fault dimensions have been obtained for three fault geometries shown in Fig. 5.4, the Starr (1928) solution, the Knopoff (1958) solution, and the Neuber (1958), Keylis-Borok (1959) solution. These results may be summarized by the general formula:

$$\Delta\sigma = \eta_2 \mu \frac{U_m}{W} \tag{5.24}$$

where $\Delta\sigma$ is the stress drop (constant) over the fault surface, $\mu$ is the modulus of rigidity, $U_m$ is the maximum displacement, $W$ is the fault width (or depth for the Knopoff model), and $\eta_2$ is a parameter which is a function of the geometrical shape of the fault surface and takes on values 1/2 (Knopoff), 4/3 (Starr), and $7\pi/12$ * (Neuber, Keylis-Borok). For irregularly shaped slip surfaces, the parameter could be larger depending on how the source dimension is defined. Since displacements observed in actual earthquakes are often quite complicated, some type of average is commonly used.

The importance of eq. 5.24 is that we may estimate the stress drop from

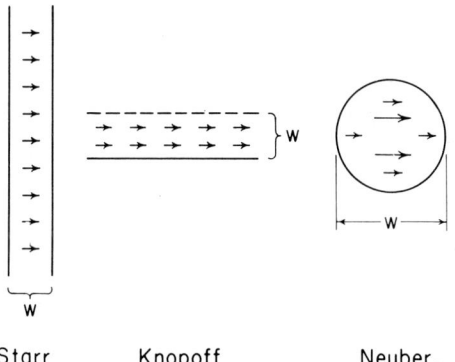

Starr      Knopoff      Neuber, Keylis – Borok

Fig. 5.4. Schematic illustration of three geometries for which solutions for constant stress drop on a surface have been obtained.

---

* This number is erroneously given as $2\pi/3$ in Brune and Allen (1967).

measurements made in the field after an earthquake. This has been done for several large earthquakes and the results suggest that stress drops averaged over the fault plane are commonly about 20—100 bars, but may range from very small values (<10 bars) to a few hundred bars (Brune and Allen, 1967; Chinnery, 1964). It should be emphasized that earthquakes in general cannot be represented by a uniform stress drop over the fault plane; the displacements at the surface often vary erratically along the fault trace and multiple breaks are common. Usually only surface observations of fault offset are available. Thus, the statement that stress drops are commonly about 20—100 bars is gross simplification.

Stress drops may also be estimated by direct measurement of the strain change with geodetic surveys or strain meters. Reid (1910/1969) found a geodetically measured stress drop of about 133 bars for the 1906 San Francisco earthquake.

### 5.2.9 Effective accelerating stress

The stress available to accelerate the sides of the fault, the effective accelerating stress, is the difference between the initial shear stress $\sigma$, and the frictional resisting stress $\sigma_f$ (see Fig. 5.3). In the case of instantaneous failure over a finite area, this will initially be $\sigma_1 - \sigma_f$. This will also be the approximate accelerating stress in a case of a growing rupture surface, except near the crack tip where (in the theoretical models) there can be a singularity. The very high stresses will be confined to a region close to the crack tip and thus will affect primarily high frequencies. The effective accelerating stress is approximately equal to the stress drop in Orowan's model, but is less for the stress overshoot model (Savage and Wood, 1971) and greater for the abrupt stopping model ($\epsilon < 1$, Brune, 1970). Ambraseys (1969) and Brune (1970) have pointed out that the peak particle velocities in the near field of earthquakes should be limited by the strength or maximum stresses of the rocks. They concluded that particle velocities near 100 cm/sec and particle accelerations greater than 1$g$ could be expected on the bases of current estimates of near-source stresses.

## 5.3 EARTHQUAKE MODELING

### 5.3.1 Instantaneous stress pulse model

No analytic solution has been obtained for the case of a growing and stopping rupture surface in a three-dimensional medium with a specified failure criterion. Several authors have obtained approximations and partial solutions. Before reviewing the results of these authors, we will discuss a simple model in which a stress pulse is applied instantaneously over the whole fault surface (Brune, 1970). The reasons for discussing this model even though its basic assumption, instantaneous stress drop over the fault plane, is not likely to be closely approximated in nature, is that it is physically realistic in the sense that it does not pose any forced motions with consequent non-physical stresses (as may be the case for a dislocation model or any model in which a rupture velocity is prescribed). The problem is one of relaxation from one state of equilibrium to another and a known amount of energy will be radiated (the radiation pattern, of course, will be severely affected by this simplification). If we only consider the stress above friction, the difference in energy between the initial and final stress states must go into seismic energy. Thus, we may consider RMS (root mean square) averages over the radiation pattern with some confidence. Another reason why the instantaneous stress pulse model may serve as a rough approximation to nature is that the particle velocities are much lower than the rupture velocities; thus the interaction of the particle velocity with rupture velocity will be small. The rupture will pass so quickly that the particles will be allowed to slip approximately (within a factor of about two) as if the rupture were propagating with infinite velocity (except near the fault at the time of rupture passage, where high-frequency energy may be focussed by rupture propagation, or anomalous stress concentration may be associated with the crack tip).

*5.3.1.1 Near-source displacements*

We assume that stress drop occurs instantaneously over the fault surface. In a linear system, this is equivalent to applying instantaneously a uniform stress pulse on the interior surface of the fault plane (see Fig. 5.5). During slip, the fault is equivalent to a surface totally reflecting to shear waves. Initially, the motion at a point near the center of the fault occurs as if the fault plane were infinite (before the end effects can propagate to the center of the fault). The stress pulse sends a pure shear wave propagating perpendicular to the dislocation surface. The initial time function for this pulse follows directly from the boundary conditions:

$$\sigma(x, t) = \sigma H(t - x/\beta) \tag{5.25}$$

where $H(t)$ is the Heaviside function, $\sigma$ is the effective shear stress, and $\beta$ is the shear-wave velocity.

The tangential displacement $u$ corresponding to eq. 5.25 may be obtained by integration since $\sigma = \mu \partial u / \partial x$. At $x = 0$:

$$\begin{aligned} u &= 0 & t &< 0 \\ u &= (\sigma/\mu)\beta t & 0 &< t < T \end{aligned} \tag{5.26}$$

where $T$ is the time required for elastic waves to propagate from the ends of the rupture surface.

For a point near the fault, displacement increases linearly with time until

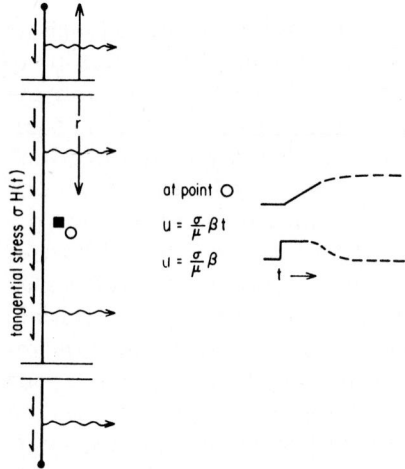

Fig. 5.5. Illustration of the stress-pulse model for an observation point O, near the dislocation surface (fault plane). In this schematic diagram the displacement u is tangential to the dislocation surface. A shear pulse on the inner right surface of the dislocation sends a shear wave propagating to the right (plus x-direction). A wave of opposite sense propagates to the left. r is the approximate radius of an equivalent circular dislocation.

the effects of the boundaries reach the observation point and stop the linear increase in displacement. Thus, for an observation point near the center of the dislocation surface, the particle displacement is initially given by eq. 5.26. Near the edge of the fault, boundary conditions require compression and rarefaction of the medium, with consequent motions normal to the fault plane. These normal motions are smaller (see discussion on p. 164) than the tangential motion, but produce an angular momentum of opposite sign, so that the net angular momentum is zero (the other couple of the equivalent double-couple point source representation). The initial stress has no angular moment since the stress is applied in a plane (i.e., there is no moment arm). The initial particle velocity is:

$$\dot{u} = \frac{\sigma}{\mu}\beta \qquad (5.27)$$

For $\sigma = 100$ bars ($=10^8$ dyn/cm$^2$), $\mu = 3 \cdot 10^{11}$ dyn/cm$^2$, and $\beta = 3$ km/sec ($=3 \cdot 10^5$ cm/sec), eq. 3 gives $\dot{u} = 100$ cm/sec. A stress of 1000 bars gives $\dot{u} = 1000$ cm/sec. 150 cm/sec appears to be a good value for the upper limit of initial velocities observed in most earthquakes. Equation 5.27 has also been derived by Jeffreys (1962, p. 359) and Ambraseys (1969).

*5.3.1.2 Maximum near-source acceleration*

Although the accelerations given by eq. 5.26 are infinite at the arrival times of the pulse, the forces remain finite because the accelerating masses are zero in the limit. We may predict the maximum accelerations expected at any finite frequency, or frequency band, by limiting the spectra of eq. 5.26. This will correspond to the maximum accelerations expected for masses with volume of the order of the cube of the wavelength. During cracking of rocks, rockbursts, etc., very high accelerations are obtained, but only at very high frequencies. For engineering seismology, these high accelerations at high frequencies are not important since they attenuate very rapidly with distance and since buildings are not very susceptible to damage at high frequencies. The frequencies of greatest interest in engineering seismology are less than 10 Hz.

First, consider the case of transonic or supersonic rupture propagation ($v > \beta$). For a small time interval $\Delta t$ as the rupture propagates along the fault plane with velocity $v$ (Fig. 5.6), a shear pulse travels away from the fault with velocity $\beta$. A small mass, $\rho V$, where $V$ is the triangular volume of dimensions one unit deep, $v\Delta t$ along the fault and approximately $\beta \Delta t$ perpendicular to the wave front, is accelerated by a force $\sigma \cdot 1 \cdot v\Delta t$. The acceleration is then given by:

$$\ddot{u} = \text{force/mass} = 2\sigma/(\rho\beta\Delta t\sqrt{1 - \beta^2/v^2}) \tag{5.28}$$

For $v >> \beta$ this approaches $2\sigma/(\rho\beta\Delta t)$. In the limit of small $\Delta t$, $\ddot{u}$ approaches infinity; however, for any finite mass and finite $\Delta t$ (finite frequency), $\ddot{u}$ is finite. For example, if we consider a volume 0.3 km in dimension along the

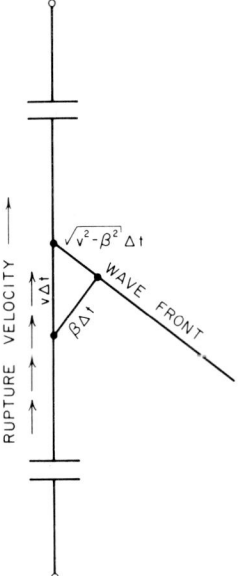

Fig. 5.6. Illustration of the finite rate of stress application for a supersonically traveling rupture.

fault ($\Delta t = 0.1$, frequency $\sim 10$ Hz) and $\sigma = 100$ bars, we have:

$$\ddot{u}(10 \text{ Hz}) \cong \sigma \cdot 2 \cdot 10^{-5} \cong 2g$$

If the rupture velocity is near or less than $\beta$, the mass will be accelerated not only by the initial stress, but also by the stress concentration near the rupture front. Accelerations near the crack tip can be much greater than those given by eq. 5.28 depending on the rupture velocity. This will be considered in the section on focussing by rupture propagation.

Alternatively, we may estimate the maximum accelerations by considering the contribution of a finite band of frequencies from 0 to some cutoff angular frequency $\omega_s$:

$$u(t) = \frac{1}{2\pi} \frac{\sigma}{\mu} \beta \text{Re} \int_{-\omega_s}^{\omega_s} -\frac{1}{\omega^2} e^{i\omega t} d\omega$$

$$\ddot{u}(t) = \frac{1}{2\pi} \frac{\sigma}{\mu} \beta \text{Re} \int_{-\omega_s}^{\omega_s} e^{i\omega t} d\omega = \frac{1}{\pi} \frac{\sigma}{\mu} \beta \omega_s \left(\frac{\sin \omega_s t}{\omega_s t}\right) \qquad (5.29)$$

For a cutoff frequency of 10 Hz and $\sigma = 100$ bars:

$$\ddot{u}(10 \text{ Hz}) \simeq 2g$$

in agreement with eq. 5.28. This result is equivalent to considering a velocity pulse with a finite rise time, i.e., $\dot{u}$ accelerating to 100 cm/sec in 0.05 sec gives about $2g$ acceleration. The largest acceleration observed to date on a strong-motion seismogram is about $1.25g$, for the San Fernando earthquake.

### 5.3.1.3 Near-field effect of finite fault dimension

As the effects of the edges of the dislocation become felt at the observation point, the particle velocity will be decreased and approach zero for times large compared to the distance to the edge divided by the shear velocity $\beta$. This effect may be approximated by replacing eq. 5.26 by:

$$u(x = 0, t) = (\sigma/\mu)\beta\tau(1 - e^{-t/\tau}) \qquad t > 0 \qquad (5.30)$$

$$\dot{u}(x = 0, t) = (\sigma/\mu)\beta e^{-t/\tau} \qquad t > 0 \qquad (5.31)$$

At $t = 0$, eq. 5.30 gives $\dot{u} = (\sigma/\mu)\beta$ as in eq. 5.27.

The Fourier transform of eq. 5.30 gives:

$$\Omega(\omega) = \frac{\sigma}{\mu} \beta \omega^{-1}(\omega^2 + \tau^{-2})^{-1/2} \qquad (5.32)$$

In this approximation $\tau = O(r/\beta)$. The result is diagrammatically illustrated in Fig. 5.7. The initial high-frequency spectrum and rise velocity are not altered, but the velocity decays to zero. The corresponding spectra are

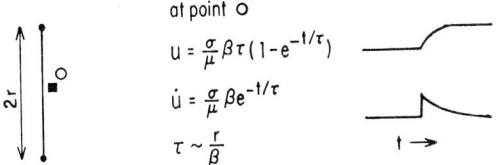

Fig. 5.7. Illustration of the approximate effect of finite source dimensions on the near-field displacement.

shown in Fig. 5.8. As slippage proceeds, perpendicular (transverse) displacements develop but these are of smaller magnitude than the tangential displacements. Trifunac (1972a) found that the Pacoima Dam strong-motion records of the San Fernando earthquake could be fit approximately by eq. 5.32 with $\sigma \simeq 85$ bars.

*5.3.1.4 Abrupt locking and partial stress drop*

If the stress does not remain at the frictional stress, but is abruptly increased during faulting so that the final stress drop is some fraction $\epsilon$ of the effective accelerating stress (abrupt locking model, Brune, 1970) the rise-time and high-frequency spectra are not drastically changed, but the long-period behavior and spectra are reduced by $\epsilon$. As a result, in the near field,

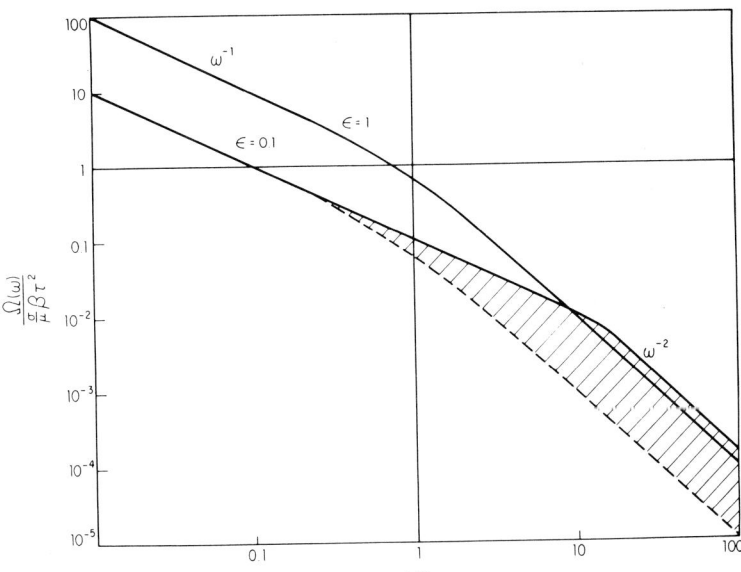

Fig. 5.8. Approximate near-field spectra for the cases $\epsilon = 1$ and $\epsilon = 0.1$. The cross-hatched area represents the part of the spectra which is important to distinguish between the partial stress drop model (e.g., $\epsilon = 0.1$) and Orowan's model ($\epsilon = 1$).

velocities and accelerations are about $1.6/\epsilon$ higher at high frequencies than would be expected for 100% fractional stress drop corresponding to the same seismic moment. The situation is equivalent to stress—time curve for $\epsilon \ll 1$ in Fig. 5.3, and the solid line in Fig. 5.9. For $\epsilon = 0.1$, the near-field spectra will be approximately as indicated in Fig. 5.8 (lower solid curve). At present, there is no conclusive evidence that this mechanism is or is not important in natural earthquakes.

### 5.3.1.5 Far-field RMS average spectra

Using the relationship between low-frequency spectra and seismic moment, and considering energy balance at high frequencies, an expression can be obtained for the far-field displacement pulse and spectra corresponding to a circular fault of radius $r$ (Brune, 1970, 1971-correction). The primary effect of moving from the near-field to the far-field is the relative attenuation of the low frequencies due to diffraction around the fault. The RMS average displacement pulse and spectra are:

$$U(R, t, \theta, \phi) = \overline{\mathcal{R}}_{\theta\phi} \frac{\sigma}{\mu} \beta r R^{-1} t'' \exp(-2.34 \, \beta r^{-1} \, t'') \tag{5.33}$$

$$\Omega(R, \omega, \theta, \phi) = \overline{\mathcal{R}}_{\theta\phi} \frac{\sigma}{\mu} \beta r R^{-1} \left[\omega^2 + (2.34 \, \beta r^{-1})^2\right]^{-1} \tag{5.34}$$

where $\overline{\mathcal{R}}_{\theta\phi}$ is the RMS radiation pattern, $R$ is the distance, and $t'' = t - R/\beta$. For the case of abrupt locking, i.e., stress drop only a fraction $\epsilon$ of the effective accelerating stress, the spectrum is multiplied by a function which reduces the low-frequency part of the spectrum relative to the average high-frequency spectrum by a factor of about $1.6\epsilon$. The spectra are plotted in Fig. 5.10. Tucker and Brune (1973) found that the majority of aftershocks of the San Fernando earthquake of February 9, 1971 had S-wave spectra which could be fit by eq. 5.34 and gave values of $\sigma$ ranging from about 0.1 bar to about 200 bars.

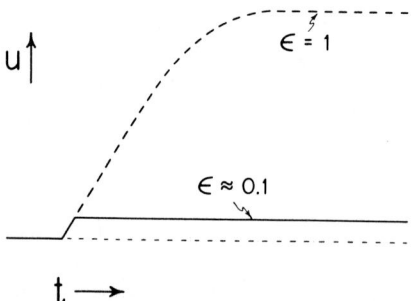

Fig. 5.9. Approximate near-source displacement time function for the cases $\epsilon = 1$ and $\epsilon = 0.1$.

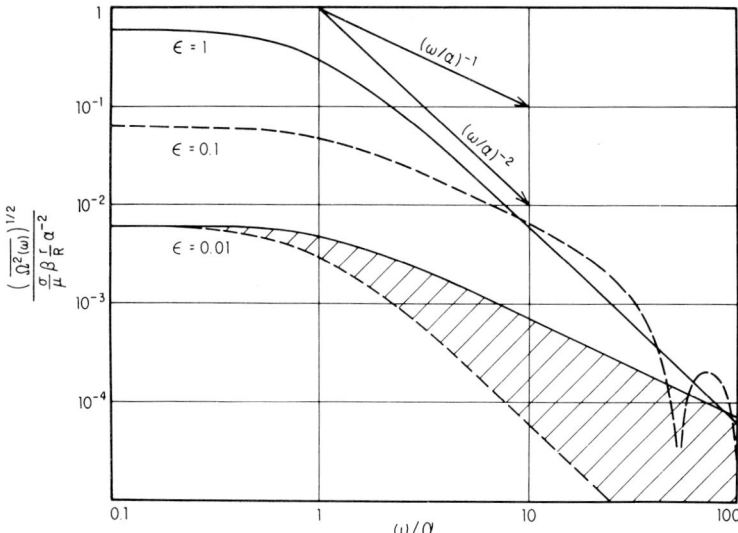

Fig. 5.10. Far-field spectra for various values of $\epsilon$. The cross-hatched area represents the part of the spectra which is important in distinguishing between the partial stress drop model (e.g., $\epsilon = 0.01$) and Orowan's model ($\epsilon = 1$).

*5.3.2 Crack tip stress singularity*

Burridge (1973) pointed out that for the case of instantaneous stress drop, stress concentration should always cause the rupture to accelerate the shear-wave velocity for screw dislocations (rupture normal to the direction of slip), or to between the Rayleigh and the P-wave velocities for edge dislocations (rupture in the direction of slip). In nature, some stress concentration will occur at the crack tip, but since the rocks cannot sustain unlimited short-term stresses, stress at the crack tip will be limited by rock failure.

Ida (1973) has made estimates for maximum particle velocity by removing the stress singularity at the crack tip with assumed inelastic behavior of the material near the fault tip. Following Barenblatt (1959), the behavior was formulated in terms of a "cohesive force" that works across the crack tip against fracture. Ida (1973) derives estimations of maximum particle velocity and acceleration near the crack tip given by:

$$\dot{u}_M \sim (\sigma_0/\mu)v \qquad (5.35)$$

$$\ddot{u}_M \sim (\sigma_0/\mu)^2 (v^2/D_0) \qquad (5.36)$$

An estimate of the distance within which large accelerations can be expected is:

$$x_M \sim (\mu/\sigma_0)D_0 \qquad (5.37)$$

An estimate of the duration of the strong shaking is:

$$t_M \sim (\mu/\sigma_0)(D_0/v) \tag{5.38}$$

In this formulation, $v$ is the rupture velocity, $\sigma_0$ is the stress of the elastic limit (approximately equal to the yield strength), and $D_0$ is the displacement required to break down cohesion. Associated with $\sigma_0$ and $D_0$ is a value $\gamma^I$ of specific cohesive surface energy. Unfortunately, it is not clear at the present time what values of $\gamma^I$ and $D_0$ should be specified for earthquakes or how they might be estimated from laboratory experiments. Using values of $\sigma_0 \simeq 1$ kbar and $D_0 \simeq 10$ cm yields velocity and acceleration estimates of 1 m/sec and 1g respectively, in good agreement with eqs. 5.27 and 5.29. The associated value of $t_M$ is about 0.1 sec and $x_M$ about 0.1 km. The implied value of specific surface energy is $10^{10}$ erg/cm$^2$. A comparable estimate of $\gamma^I$ was made independently by Kikuchi and Takeuchi (1970).

*5.3.3 Focussing of energy by rupture propagation*

The instantaneous stress pulse model discussed above gave the average energy radiation and RMS spectrum (averaged over a sphere around the source) from a simple stress-drop mechanism. This model does not predict the variations in radiation pattern due to focussing by rupture propagation. Many studies have indicated that stress release during earthquakes may be quite complex, but for wavelengths near the source dimension, earthquakes often look like simple unilateral or bilateral propagating sources. Benioff (1955) showed how source propagation would cause high amplitudes and high frequencies to be focussed in the direction of propagation (Doppler focussing). Ben-Menahem (1961) and Haskell (1964) derived equations expressing the effects of double-couple source propagation in the far field.

A coherent propagating rupture can theoretically focus high frequencies in the direction of propagation and give arbitrarily high accelerations and velocities. In nature, the focussing cannot be perfectly effective, but it is not known what the practical limits are. The coherency is degraded both by inhomogeneities in the rock and complexity in the source. Coherency at 5 Hz would require phase coherency of about 0.1 sec. Since variations in elastic wave velocities in rocks are of the order of 5% or greater, in the neighborhood of most faults this would imply that rupture wave fronts could not remain coherent at 5 Hz after travelling distances greater than about 10 km. In a similar way, lack of coherency in the faulting itself limits the likelihood of a shock wave build-up. Most faults do not remain planar for a distance of more than about 10 km (without turning, branching, etc.) and this limits coherency in approximately the same way as the effect of inhomogeneity in wave velocities.

## 5.3.4 Approximate solutions to the dynamic problem of propagating ruptures

Several authors have published partial solutions for the problem of propagating ruptures. No one has obtained an analytic solution in three dimensions for a spontaneous rupture which includes the stopping. Kostrov (1964) obtained the solution for a uniformly growing circular fault and Burridge and Willis (1969), Burridge and Levy (1974), and Richards (1973) gave solutions for a uniformly growing elliptical fault. Burridge and Halliday (1971) solved the problem of a spontaneous two-dimensional crack with prescribed static and dynamic friction stopped by variation in initial stress. A similar problem was solved by Andrews (1975).

Aki (1968) and Haskell (1969), using dislocation results of Maruyama (1963) developed formulations for computing the near-field displacements for propagating forced dislocations (kinematic models). The Haskell formulation has been widely used to study near-source motion, especially for the Parkfield and San Fernando earthquakes (Trifunac and Udwadia, 1974; Anderson, 1973; Tsai and Patton, 1973; Canitez and Toksöz, 1972; and Boore and Zoback, 1974a,b). Boore and Zoback (1974a,b) used 2-D dislocation modelling to study effects of fault orientation and rupture velocity. In such kinematic models, the dislocation along the fault is arbitrarily prescribed, usually in terms of a rise time and rupture velocity, and the resulting displacements calculated. The prescribed displacement may or may not be physically reasonable, depending on how carefully it is chosen. There is large flexibility in fitting data, but little basis for deciding uniqueness or for predicting the range of expected values of acceleration and velocity. For rupture velocities near or greater than the shear-wave velocity, arbitrarily high focussing effects can occur with consequent infinite energy radiation. In this sense, the kinematic model is not, by itself, capable of predicting maximum accelerations; we need to have some constraints from the dynamic equations to aid us in prescribing kinematic dislocations.

Numerical techniques promise to give important flexibility in dealing with dynamic propagating ruptures. Hanson et al. (1974) solved a problem of a two-dimensional propagating and stopping rupture. Dieterich (1973, 1974) studied dynamic three-dimensional numerical problems where stopping was caused by variations in frictional properties along the fault. Cherry et al. (1975) studied a finite-difference numerical model involving frictional dissipation in the elements.

Laboratory models also promise to give valuable insights into the faulting process. Wu et al. (1972) studied rupture propagation in plexiglass models and Johnson et al. (1973) studied time functions and rupture propagation in slabs of westerly granite. Brune (1973) and Archuleta and Brune (1975) studied spontaneous growing and stopping ruptures in stressed foam rubber.

*5.3.4.1 The growing elliptical crack*

Richards (1973) has extended the previous results of Kostrov (1964) and Burridge and Willis (1969) to obtain analytic expressions for particle accelerations at given directions from a uniformly growing elliptical crack. These results, though they do not include the effects of crack stoppage, are very useful in understanding the effects of focussing. Since the problem is self-similar, the results may be used to study a growing elliptical crack of any size. The model consists of a rupture surface which steadily grows from a point with two, in general different, rupture velocities. As the rupture passes any point on the plane of the fault, the stress drops a given amount. There are three general properties of the solution:

(1) There is a singularity in particle velocity and acceleration on the fault surface at the time of the crack passage (even if the rupture velocity is near zero). This results from stress concentration at the crack tip and the fact that the stress is forced to drop instantaneously.

(2) Doppler focussing of high-frequency energy occurs in the direction of fault propagation. The distribution of high particle velocities with angle away from the direction of rupture propagation depends on the rupture velocity, but is typically limited to a range of angles of about $\pm 5°$.

(3) The particle velocity asymptotically reaches the same value at all points in the medium after the rupture passes nearby because the crack grows continuously, attains infinite dimensions, and any point in the medium, no matter what its position relative to the point of rupture initiation, becomes effectively near the fault plane (relative to the scale of the fault dimensions). The model is unrealistic in this respect, because it does not include the effect of fault stoppage (finite fault dimension).

*5.3.5 Foam-rubber model*

Brune (1973) constructed a three-dimensional foam-rubber model of spontaneous stick-slip along precut surfaces in stressed foam rubber. A semicircular and an elongated cut were considered. The overall behavior of the fault slip in this model, including fault creep, peak particle velocities, and the presence of multiple events, was similar to that inferred for earthquakes. Only point measurements of particle displacement near the center of the fault trace were made in this first study, but later Archuleta and Brune (1975) measured the motion at a large number of points. Figure 5.11 shows displacement versus time plots for a number of different stick-slip events observed for the elongated fault. The time functions range from rather simple truncated ramps with rise times of about 2 msec, to complex multiple events with durations of 10 msec. The simple truncated ramps were the most common. In some cases, there is obvious pre-slip acceleration and/or post-slip deceleration. In other cases, the slip begins and/or ends abruptly. Figure 5.12

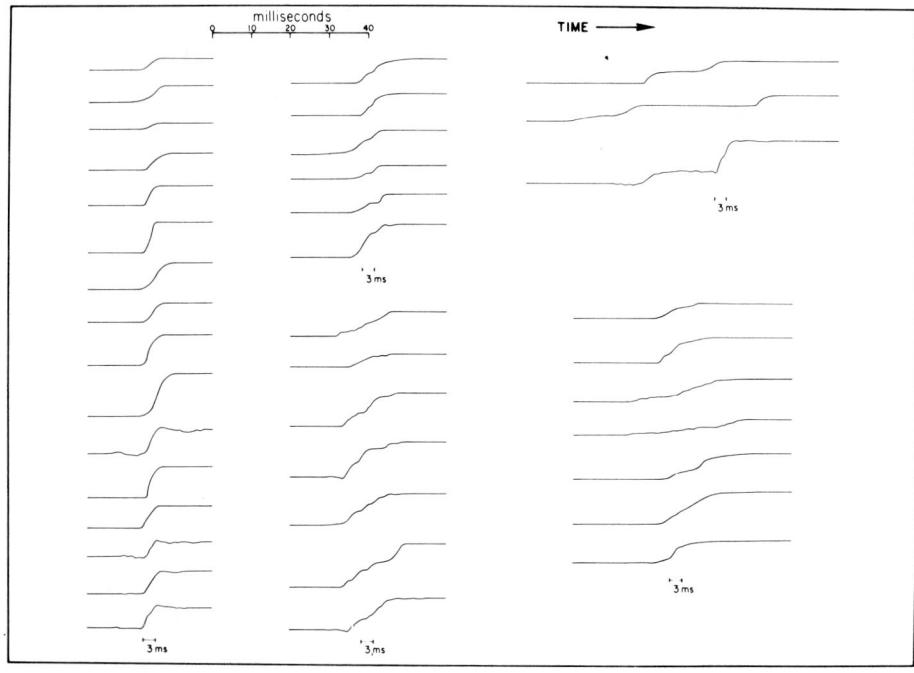

Fig. 5.11. Time functions of fault slip for the elongated fault in the foam-rubber model of Brune (1973).

gives peak particle velocity for these time functions plotted against the total displacement $u$. The displacement is directly proportional to the stress drop, and therefore, this figure can be interpreted as a plot of peak particle velocity, near the center of the fault trace, as a function of stress drop. The results may be compared with the instantaneous stress pulse model of Brune (1970) described above. Assuming the Orowan model, i.e., effective accelerating stress equal to stress drop, the peak particle velocities would be expected to be given by the upper curve, i.e., $\dot{u} = u/1.5$ msec. As can be seen in this figure, all of the events have peak particle velocities below that given by this curve, with the average value for the simple events approximately one-half, i.e., $\dot{u} = u/3.0$ msec. The variations were interpreted as resulting from differences in the point of initiation and direction of propagation of the rupture, an interpretation verified in a later study. The conclusion that the peak particle velocities near the center of the fault should average about one-half the value predicted by the instantaneous stress pulse model is in agreement with results from the propagating circular fault (Kostrov, 1964; Kanamori, 1972; Abe, 1974).

A rather surprising result of the experiment was the occasional occurrence

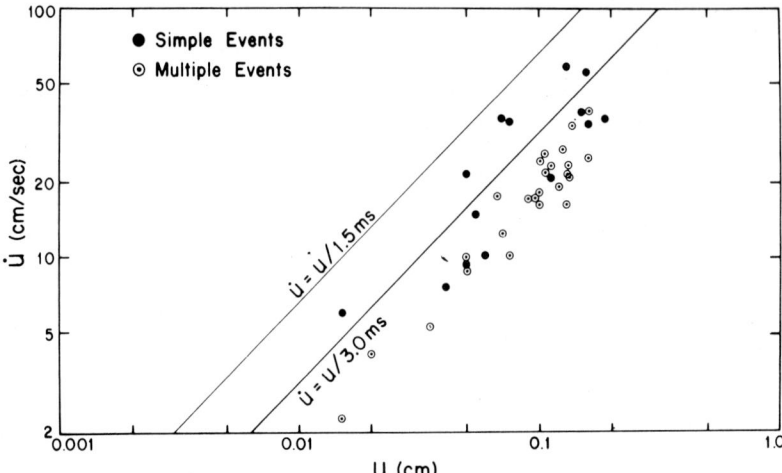

Fig. 5.12. Peak particle velocities plotted versus displacement $u$, for the time functions in Fig. 5.11.

of multiple events, apparently a result of repeated triggering by waves reflecting in the fault plane. The occurrence of multiple events adds an additional complication to attempts to relate peak particle velocities to stress drop, both in the case of laboratory models and in naturally occurring earthquakes. The simplest consequence is that peak particle velocities may be lower than estimated by simple prediction from the stress drop. The cumulative effect of multiple events on the same fault plane causes the stress drop to be larger yet the peak particle velocity is approximately the same in each of the individual events.

Archuleta and Brune (1975) extended the foam-rubber experiment using high-speed photography. The surface of the foam-rubber model was marked by small black beads and an event was filmed on a Fastax camera at a rate of 4400 frames per second. The motion of the beads was measured relative to a stationary grid suspended slightly above the model. This allowed measurement of the dynamic displacement field over the surface near the fault. The positions of each of the beads was obtained by digitizing successive frames of the film. The displacement-time functions were then compared with theoretical predictions based on the observed value of stress drop and the measured velocity of rupture propagation along the surface (approximately $0.72\beta$).

Displacement time functions for twelve points adjacent to the fault trace are shown in Figs. 5.13 and 5.14. In both figures, the left ordinates measure the displacement in units of 0.1 mm, the right ordinate gives the location of the particle along the fault trace (in mm) and the abscissa gives the time in milliseconds. Theoretical curves are taken from the analytical

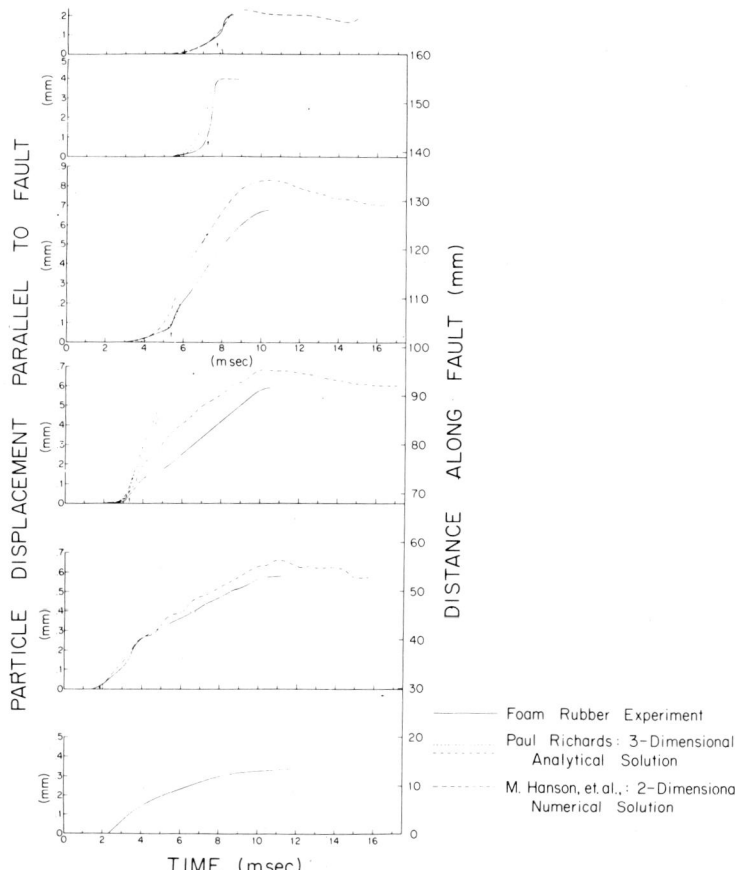

Fig. 5.13. Particle displacement time functions for various points along a circular fault cut during a stick-slip event. Theoretical curves are shown for the growing circular fault model of Richards (1973) and the two-dimensional numerical solution of Hanson et al. (1974).

solution of Richards (1973) for a growing elliptical crack and the numerical results of Hanson et al. (1974) for the case of a one-dimensional unilaterally propagating and stopping shear crack with a rupture velocity of $0.69\beta$. The results of Hanson et al. (1974) were normalized so that the static values of displacement were equal to those observed in the model.

Examination of the data shows that the surface rupture originated about 30 mm from one end of the fault and then propagated bilaterally in both directions, but the length of rupture was much longer in one direction than in the other. Perhaps the most important feature in these diagrams is the increase in peak particle velocity as the rupture propagates down the fault. A plot of peak particle velocity, normalized by $\beta\Delta\sigma/\mu$, as a function of

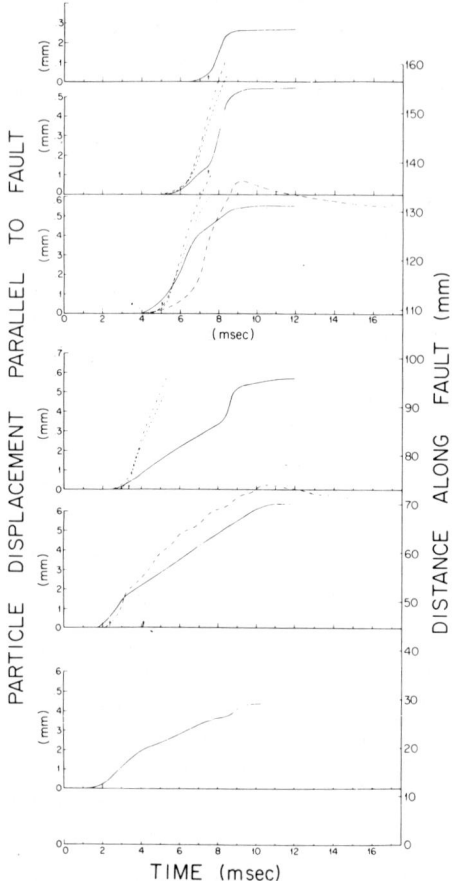

Fig. 5.14. Caption the same as for Fig. 5.13.

distance along the fault is shown in Fig. 5.15. Although there is considerable scatter, the results clearly demonstrate the focussing of energy by rupture propagation. The displacement-time function near the center of the fault is long and relatively low in velocity compared to the time function near the end of the fault, where the particle velocity is higher, but the total displacement smaller because the duration of motion is smaller. This, in part, is due to the displacement being stopped by the rupture reflecting from the end of the fault.

As can be seen from Fig. 15 the peak particle velocities are of the same order of magnitude as predicted by the simple formula $\Delta\sigma\beta/\mu$; however, near the center of the fault and near the point of rupture initiation, the peak particle velocities are only approximately half this value, whereas in the direction of rupture propagation, the particle velocities may exceed $\Delta\sigma\beta/\mu$.

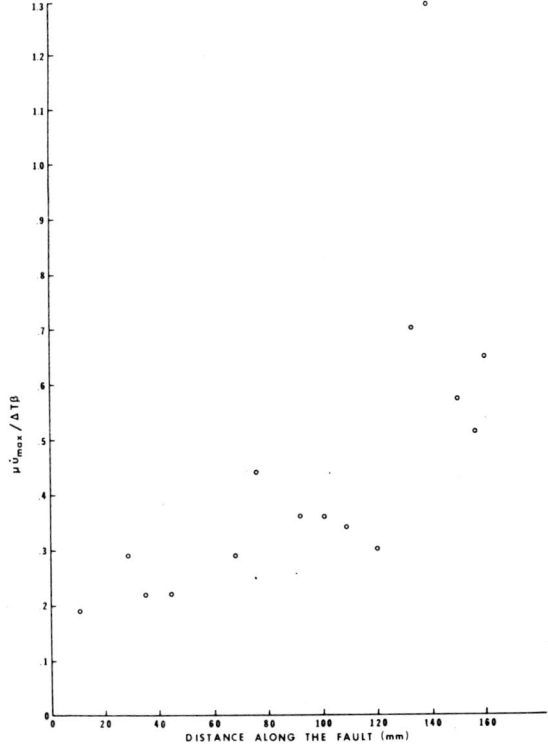

Fig. 5.15. Estimated peak particle velocities for time functions in Fig. 5.14, plotted versus distance along the fault.

Transverse particle displacement (not shown here) is much smaller than the parallel displacement. However, transverse particle velocities were comparable to the parallel particle velocities near the terminal end of the fault. This is probably due to a combination of focussing of high-frequency energy in the direction of propagating and the effect of the rupture stoppage at the end of the fault. No large transverse displacement pulses like that observed for the Parkfield earthquake (Aki, 1968; Housner and Trifunac, 1967) occurred.

The observed displacement-time functions are roughly in agreement with what would be expected on the basis of the theoretical curves presented, but there are important differences. The particle motion diagrams for the growing elliptical crack would not be expected to correspond to the model results except for the very beginning of the time function (because the growing elliptical crack does not contain the effect of fault stoppage). As can be seen in Figs. 5.13 and 5.14, the initial steep part of the displacement-time function curve has a slope approximately comparable to that predicted by the growing elliptical fault model. The two-dimensional, numerical results of Hanson et al. (1974) compare surprisingly well with the results from the foam-rubber model.

The results from the foam-rubber model are encouraging in that they show that our basic ideas about the physical mechanisms involved in growing and stopping ruptures are approximately correct. Future work on such models should lead to more insight on the spontaneous rupture phenomena and give us important clues as to how to proceed both with theoretical and numerical models.

### 5.3.6 Geometrical and boundary-condition effects

In this section, we consider the effects on strong motion of the geometrical parameters of the fault, e.g., depth below the free surface, orientation and slip direction (i.e., strike-slip, normal, or thrust faulting), free-surface amplification, fault breakout at the surface, edge effects, and topographic amplification.

### 5.3.6.1 Depth

Since engineering structures are near the surface, the most important effect of depth is to increase the distance of the energy source from the closest engineering structure and thus reduce the intensity. Very deep earthquakes do essentially no damage even though they may be felt over a large area. On the other hand, where an active fault zone intersects the surface, the effective depth of energy release may be near zero. Small earthquakes usually occur at depths large compared to their dimensions and thus the intensity of shaking at the surface is greatly reduced even though near the hypocenters high accelerations are expected. Occasionally small earthquakes at very shallow depth may generate anomalously high accelerations over a limited area. This may have been the case for the small ($M = 4.7$) Bear Valley earthquake of 4 September 1972 (Morrill and Mathiesen, 1972) which gave an acceleration of $0.69g$ at the Melindy Ranch station. The purely geometrical effect of depth for small earthquakes can be easily estimated. Amplitudes will decrease as $R^{-1}$ where $R$ is the distance from the source to the point of energy release (not the projected surface distance).

Another effect of depth is to change the partitioning of energy between surface waves and body waves. Body waves diminish with distance as $R^{-1}$ whereas surface waves diminish as $R^{-1/2}$. Whenever the distance to the source is comparable to or greater than the source depth, surface waves may be the predominant waves generating strong motion. The effect of the general increase in elastic-wave velocities in the earth is to trap energy in surface waves and decrease the attenuation with distance away from the fault. For large shallow earthquakes that break the surface, most of the energy in the near field may be due to waves trapped in near-surface layers. At present the only methods capable of adequately representing surface waves are dislocation and numerical modeling which incorporate vertical variation in rock properties.

*5.3.6.2 Fault orientation*

The effects of variation in the orientation of faulting are quite complex and for a given case need to be computed using techniques such as dislocation modelling or numerical modelling. Geologists classify faults into three basic categories related to the stress systems they generate or are caused by. Unfortunately, there are not enough strong-motion records for the various types of faulting to make firm conclusions about the effect of fault type on strong motion.

Strike-slip faulting is associated with horizontal motions along the fault plane (usually near vertical) and may be caused by a stress system where the maximum and minimum principal stresses are horizontal and the intermediate principal stress is vertical. Strike-slip motion is associated with transcurrent or transform faults such as the San Andreas fault near Parkfield, and the Imperial fault near El Centro.

Normal faulting is associated with dip-slip motion (with a large vertical component, hanging wall downthrown) along a steeply dipping fault plane where the minimum principal stress is horizontal, the maximum principal stress vertical, and the intermediate principal stress horizontal (along the strike of the fault). Normal faulting is common in areas of crustal tension such as the Basin and Range Province of Nevada, the region of the Fairview Peak—Dixie Valley earthquakes of 1954. Unfortunately, no strong motion records have been obtained in the near field of large normal faulting earthquakes.

Thrust faulting is associated with dip-slip motion (with a large vertical component, hanging wall upthrown) along a shallow dipping fault plane where the maximum principal stress (axis of compression) is horizontal, the minimum principal stress vertical, and the intermediate principal stress horizontal. Thrust faulting is common in areas of crustal compression (e.g., the great Alaska earthquake of 1964, the Tehachapi earthquake of 1952, and the San Fernando earthquake of 1971). Relatively high effective accelerating stresses may be associated with thrust faulting (compared to strike-slip faulting and normal faulting) although present evidence is too limited to prove this. The San Fernando earthquake which generated very high particle velocities and high accelerations (approximately $1.25g$) had a large component of thrusting. The great Indian earthquake of 12 June 1897 for which Oldham (see Richter, 1958) cires evidence for acceleration greater than $1g$ was associated with thrust faulting.

*5.3.6.3 Free-surface amplification*

Many earthquake models are for the case of an infinite unbounded medium. In applying these results, one must make a correction for the free-surface amplification. SH waves and nearly vertically incident SV and P waves are amplified by a factor of approximately two at the surface. At non-vertical angles of incidence, SV waves and P waves may be amplified and phase shifted by varying amounts which can be calculated from well known formulas (see, for example, Ewing et al., 1957), but it is rare that such corrections can be confidently applied except in a most general way. Numerical or dislocation methods may be needed to examine the effect of the free surface in particular cases.

### 5.3.6.4 Amplification by topography and complex earth structure

Topography can cause amplification and attenuation by focussing and defocussing energy. Topographic amplifications may, in part, be responsible for the high accelerations observed on the Pacoima Dam record of the San Fernando earthquake (Reimer et al., 1974; Wong and Jennings, in press). Topographic focussing has been calculated for a symmetric triangular hill by Boore (1973) using a finite-difference technique and Smith (1974) using a finite-element technique. Amplifications of about 2 can occur for wavelengths near the dimensions of the hill or at a sharp corner in the topography.

Complex near-surface structure can cause resonances and amplifications strongly dependent on wavelength. Trifunac (1971, 1973) has treated analytically the two-dimensional problem of plane SH waves incident on a semicircular valley. Smith (1974) compared the results with results from a finite-element calculation. Narrow frequency band resonances occur with amplifications of 5—10.

### 5.3.6.5 Fault breakout

Related to the surface amplification is the phenomenon of fault breakout at the surface. Breakout removes a constraint on the fault motion and can lead to increased slip near the surface, with consequent high amplitudes of strong motion. Several studies (Savage, 1965; Wyss and Brune, 1967) have found far-field evidence of high energy release when a fault intersects the surface (the breakout phase). Some of the high accelerations in the later parts of the Pacoima Dam accelerogram may have been caused by fault breakout (Bolt, 1972; Hanks, 1974; Boore and Zoback, 1974b).

### 5.3.6.6 Transverse motion

Very near a fault the predominant motion will be parallel to the fault slip. However, at high frequencies transverse motion may be comparable, especially near the rupture front and at the ends of the fault, or near asperities which arrest the fault motion. For the case of the continuously growing elliptical crack, the transverse accelerations are comparable or greater than the parallel accelerations (Richards, 1973). Archuleta and Brune (1975), studying a foam-rubber model, found transverse velocities at the end of a semicircular crack to be comparable to the parallel velocities near the center of the fault, although the displacements were considerably less. The Parkfield earthquake of 1966 apparently caused very large transverse motions at one station very near the subsequent fault creep line. The interpretation of the record has been the subject of several papers, mutually controversial (Aki, 1968; Trifunac and Udwadia, 1974; Anderson, 1973).

There are also small permanent transverse motions, but these are small compared to the parallel permanent motions. They are cancelled by following earthquakes so that there are no net cumulative transverse displacements averaged over geologic time.

## 5.3.7. Complexity, scattering and attenuation effects

### 5.3.7.1 Source complexity

The models and concepts considered above are generally much simplified over conditions likely to occur in nature. Natural earthquakes occur along faults with non-uniform friction and strength, and which bend, branch into other faults, die out, and jump in an en-échelon pattern, etc. Furthermore, rocks in the fault zone have usually been crushed and altered into fault gouge with properties much different than the surrounding rock. Under these conditions one might expect that ruptures would not propagate smoothly in the plane. A more accurate representation might be an erratic motion superimposed on a generally smooth slip (Haskell, 1966; Aki, 1967), or in the extreme case, a superposition of multiple events relatively uncorrelated in time and space. Many large earthquakes show evidence of being complex multiple events, e.g., the Alaskan earthquake (Wyss and Brune, 1967), the El Centro earthquake (Trifunac and Brune, 1970), and others (Chandra, 1970; Wu, 1968; Niazi, 1969). In the case of the Alaskan earthquake the rupture started with a 30 km deep event of magnitude about 6.5 near the edge of the subsequent rupture zone, and was succeeded by several larger events, or bursts of energy release, identifiable for more than a minute and extending for a distance of 250 km from the initial event. Energy release continued to a distance of about 500 km from the initial event. The largest identifiable event in the series occurred at a distance of 165 km from the initial event and 44 sec later. This event could have been associated with the rupture intersecting the surface near the oceanic trench (breakout phase). The pattern of damage is consistent with the hypothesis that the largest source of energy release was associated with this event rather than the initial event (Hudson and Cloud, 1973).

In the case of the El Centro earthquake the strong-motion record shows four events in the initial period of high energy release as well as a number of aftershocks occurring in the first few minutes after the main event. The last clear event in the initial period of high energy release was recorded 25 sec after the triggering of the recorder and appears to have originated some 60 km down the fault (from the initial event which was relatively near the strong-motion accelerograph). When distance attenuation is taken into account, it is seen that this event was probably considerably larger than the other events in the series and was probably associated with near-source accelerations near 1g (Trifunac and Brune, 1970; Trifunac, 1972b). The effects of complexity include: (1) a complex pattern of intensity and acceleration in the epicentral region; and (2) distortion of the radiation pattern from that which would be expected from a smoothly propagating rupture, e.g., spreading and degrading of the shock front.

*5.3.7.2 Scattering and incoherency*

At shallow depths the earth is quite inhomogeneous on a length scale comparable to wavelengths which commonly constitute the maximum accelerations (i.e., frequencies of the order of 3—10 Hz). Combined with irregularities in the source mentioned above, this causes the wave field at 5 Hz to be relatively incoherent. As a result, the mathematical focussing of energy in the direction of rupture propagation is degraded. Strong-motion records seldom show the simple pulses predicted by theoretical calculations. In an incoherent wave field there may be regions of anomalously high amplitude (constructive interference) and regions of anomalously low amplitudes (destructive interference).

To a first approximation, scattering, considered independent of physical attenuation described below, does not alter the energy content or spectrum of the signal but only makes it phase incoherent. The same amount of energy flows out of the source region. Only if energy is trapped within the source region for times comparable to the travel times across the region will the effect of additional physical attenuation (over that suffered by direct propagation) significantly reduce the amount of energy passing from the source region.

*5.3.7.3 Attenuation*

Since the peak accelerations in the epicentral region of earthquakes are associated with relatively high frequency seismic waves (3—10 Hz), physical attenuation, in association with scattering, play an important role in determining the pattern of strong motion. Physical attenuation (non-infinite $Q$) absorbs high-frequency seismic energy and converts it into heat. Thus, high-frequency energy will be attenuated at relatively short distances from the fault plane, independent of the fault size. This effect differs from that of geometrical spreading, which is controlled by the dimensions of the source. The effect of physical attenuation is accentuated by the fact that for very high amplitudes the rocks may act nonlinearly and drastically lower the effective $Q$. A reasonable guess at the effective $Q$ at high amplitudes may be about 50 in solid materials. For a $Q$ of 50 at 5 Hz, the $e^{-1}$ distance is 10 km, thus the accelerations beyond 10 km from the fault trace would be severely reduced, independent of the fault size. Less competent rocks, e.g., sedimentary rocks, may have even higher attenuation. Even in the direction of rupture propagation we would not expect very high accelerations to persist more than a few kilometers from the end of the fault break because focussing by rupture propagation is probably balanced by increased nonlinear damping. Uncertainty about attenuation is one of the greatest sources of error in estimating upper limits for particle velocity and acceleration.

5.4 CONCLUSION — ESTIMATES OF MAXIMUM PROBABLE NEAR-SOURCE GROUND MOTION

From the physical considerations outlined above, we conclude that in the near field of earthquakes, accelerations of greater then $1g$ and velocities greater than 100 cm/sec are possible on solid rock. The presently available strong-motion data, although limited, support this conclusion.

It is also clear that the variability of stresses, source mechanism, local structure, etc. is such that one cannot predict from one situation to the next with uncertainties less than a factor of about two in velocity and acceleration. It is easily conceivable that certain combinations of circumstances could lead to accelerations grater than $2g$ and velocities greater than 200 cm/sec in solid rocks. On the other hand, the upper limits for velocities and accelerations in less solid rocks might be overestimated because of uncertainties in the response of materials to high transient stresses.

Further uncertainty in predicting strong ground motion is associated with uncertainties in the stress model. Stress drop is relatively reliably observed because it can be related to permanent displacements observed in the field after the earthquake, but it is not known whether or not the effective accelerating stress can be assumed to be approximately equal to the stress drop. The foregoing estimates of maximum near-source acceleration were based on a stress drop (and effective accelerating stress) of 100 bars. Most small earthquakes have considerably lower stress drops. For large earthquakes, the stress drop is usually not uniform over the whole fault zone. Thus, most earthquakes will show considerably lower near-source accelerations and velocities than estimated from the above maximizing assumptions. On the other hand, some special circumstances might combine to yield considerably higher stress drops, of the order of 1 kbar, and consequently, much higher accelerations and velocities. There are no well documented cases of stress drops as high as a kilobar, but the statistical sampling of both stress drop and strong motion is inadequate to be very confident that this cannot happen.

The question of probability is of paramount importance when applying our present understanding of earthquake-source physics to the design of buildings (Brune and Lomnitz, 1974). Perhaps we should conclude for certain areas that the "possible" accelerations and velocities are very high, but the probabilities of occurrence are so low, because of the special circumstances required, that such high velocities and accelerations can be ignored. On the other hand, we may conclude that in other areas design requirements need to be considerably increased because of high probabilities of large velocities and accelerations. The final assessment of risk from high accelerations and velocities will require consideration of all the factors outlined above. What is most needed is more strong-motion data in the near field of large earthquakes. In the meantime, we also need to obtain more information on fault-zone inhomogeneity, complexity, and behavior of rocks under high transient stresses.

ACKNOWLEDGEMENTS

The author is indebted to Dr. Gerald Frazier, Mr. Ralph Archuleta and Dr. Paul Richards for their critical reading of the manuscript. Ralph Archuleta helped the author in some of the calculations and participated in numerous helpful discussions. The author is especially indebted to Mrs. Neenah Rohner for her patience and skill in typing the manuscript and Mr. Don Betts for drafting the figures.

REFERENCES

Abe, K., 1974. Fault parameters determined by near- and far-field data: The Wakasa Bay earthquake of March 26, 1963. *Bull. Seismol. Soc. Am.*, 64: 1369.
Aki, K., 1966. Generation and propagation of G waves from the Niigata earthquake of June, 16, 1964, 2. Estimation of earthquake moment, released energy, and stress—strain drop from G wave spectrum. *Bull. Earthquake Res. Inst., Tokyo Univ.*, 44: 73—88.
Aki, K., 1967. Scaling law of seismic spectrums. *J. Geophys. Res.*, 72: 1217—1231.
Aki, K., 1968. Seismic displacements near a fault. *J. Geophys. Res.*, 73: 5359—5376.
Ambraseys, N.N., 1969. Maximum intensity of ground movements caused by faulting. *Proc. 4th World Conf. Earthquake Eng., Santiago.*
Anderson, T., 1973. A model source-time function for the Parkfield, California, earthquake. *68th Ann. Natl. Meet. Seismol. Soc. Am., Golden, Colo.* (abstract).
Andrews, D.J., 1975. From antimoment to moment: Plane-strain models of earthquakes that stop. *Bull. Seismol. Soc. Am.*, 65: 163—182.
Archuleta, R. and Brune, J.N., 1975. Surface strong motion associated with a stick-slip event in a foam rubber model of earthquakes. *Bull. Seismol. Soc. Am.*, 65: 1059—1071.
Barenblatt, G.I., 1959. The formation of equilibrium cracks during brittle fracture. General ideas and hypotheses. Axially-symmetric cracks. *Appl. Math. Mech.*, 23: 622—636.
Benioff, H., 1955. Mechanism and strain characteristics of the White Wolf fault as indicated by the aftershock sequence. In: *Earthquakes in Kern County, California Dur-*

ing 1952. *Div. Mines, San Francisco, Bull.*, 171, Pt. II.
Ben-Menahem, A., 1961. Radiation patterns of seismic surface waves from finite moving sources. *Bull. Seismol. Soc. Am.*, 51 (3): 401—435.
Bolt, B.A., 1972. San Fernando rupture mechanism and the Pacoima strong-motion record. *Bull. Seismol. Soc. Am.*, 62: 1053—1061.
Boore, D., 1973. The effect of simple topography on seismic waves: Implications for the recorded accelerations at Pacoima Dam. *Bull. Seismol. Soc. Am.*, 63: 1603—1609.
Boore, D. and Zoback, M.D., 1974a. Near-field motions from kinematic models of propagating faults. *Bull. Seismol. Soc. Am.*, 64: 321—342.
Boore, D. and Zoback, M.D., 1974b. Two-dimensional kinematic fault modeling of the Pacoima Dam strong-motion recordings of the February 9, 1971, San Fernando earthquake. *Bull. Seismol. Soc. Am.*, 64: 555—570.
Brune, J.N., 1970. Tectonic stress and the spectra of seismic shear waves from earthquakes. *J. Geophys. Res.*, 75: 4997—5009.
Brune, J.N., 1971 (Correction). Tectonic stress and the spectra of seismic shear waves from earthquakes. *J. Geophys. Res.*, 76: 5002.
Brune, J.N., 1973. Earthquake modelling by stick-slip along pre-cut surfaces in stressed foam rubber. *Bull. Seismol. Soc. Am.*, 63: 2105—2119.
Brune, J.N. and Allen, C.R., 1967. A low-stress drop, low-magnitude earthquake with surface faulting: The Imperial, California earthquake of March 4, 1966. *Bull. Seismol. Soc. Am.*, 57: 501— 514.
Brune, J.N. and Lomnitz, C., 1974. Recent seismological developments relating to earthquake hazard. *Geofis. Int.*, 14: 49—63.
Brune, J.N., Henyey, T.L. and Roy, R.F., 1969. Heat flow, stress, and the rate of slip along the San Andreas fault, California. *J. Geophys. Res.*, 74: 3821—3827.
Burridge, R., 1973. Admissible speeds for plane-strain self-similar shear cracks with friction but lacking cohesion. *Geophys. J. R. Astron. Soc.*, 35: 439.
Burridge, R. and Halliday, G.S., 1971. Dynamic shear cracks with friction as models for shallow focus earthquakes. *Geophys. J.*, 25: 261.
Burridge, R. and Willis, J., 1969. The self-similar problem of the expanding elliptical crack in an anisotropic solid. *Proc. Cambridge Philos. Soc.*, 66: 443.
Canitez, N. and Toksöz, M.N., 1972. Static and dynamic study of earthquake source mechanism: San Fernando earthquake. *J. Geophys. Res.*, 77: 2583—2594.
Chandra, U., 1970. The Peru—Bolivia border earthquake of August 15, 1963. *Bull. Seismol. Soc. Am.*, 60: 639—646.
Cherry, J.T., Halda, E.J. and Hamilton, K.G., 1976. A deterministic approach to the prediction of free field ground motion and response spectra from stick-slip earthquakes. *Earthquake Eng. Struct. Dyn.*
Chinnery, M.A., 1964. The strength of the earth's crust under horizontal shear stress. *J. Geophys. Res.*, 69: 2085—2089.
Dieterich, J.H., 1973. A deterministic near-field source model. *Proc. 5th World Conf. Earthquake Eng., Rome.*
Dieterich, J.H., 1974. Earthquake mechanisms and modelling. *Ann. Rev. Earth Planet. Sci.*, 2: 275—301.
Dziewonski, A. and Gilbert, J.F., 1974. Temporal variation of the seismic moment tensor and the evidence of precursive compression for two deep earthquakes. *Nature*, 247: 185—188.
Ewing, M., Jardetzky, W. and Press, F., 1957. *Elastic Waves in Layered Media.* McGraw Hill, New York.
Hanks, T., 1974. The faulting mechanism of the San Fernando earthquake, *J. Geophys. Res.*, 79: 1215.
Hanson, M.E., Sanford, A.R. and Shaffer, R.J., 1974. A source function for a dynamic brittle unilateral shear fracture. *Geophys. J. R. Astron. Soc.*, 38: 365—376.

Haskell, N.A., 1964. Total energy and energy spectral density of elastic wave radiation from propagating faults. *Bull. Seismol. Soc. Am.*, 54: 1811—1841.
Haskell, N.A., 1966. Total energy and energy spectral density of elastic wave radiation from propagating faults, II. *Bull. Seismol. Soc. Am.*, 56: 125—140.
Haskell, N.A., 1969. Elastic displacements in the near-field of a propagating fault. *Bull. Seismol. Soc. Am.*, 59: 865—908.
Housner, G.W., 1955. Properties of strong-motion earthquakes. *Bull. Seismol. Soc. Am.*, 45: 197—218.
Housner, A.W. and Trifunac, M.D., 1967. Analysis of accelerograms — Parkfield earthquake. *Bull. Seismol. Soc. Am.*, 57: 1193—1220.
Hudson, D.E. and Cloud, W., 1973. Seismological background for engineering studies of the earthquake. In: *The Great Alaska Earthquake of 1964: Engineering*. National Academy of Sciences.
Ida, Y., 1973. The maximum acceleration of seismic ground motion. *Bull. Seismol. Soc. Am.*, 63: 959—968.
Jeffreys, H., 1962. *The Earth*. Cambridge University Press, London.
Johnson, T., Wu, F.T. and Scholz, C.H., 1973. Source parameters for stick slip and for earthquakes. *Science*, 179: 278—280.
Kanamori, H., 1972. Determination of effective tectonic stress associated with earthquake faulting: The Tottori earthquake of 1943. *Phys. Earth Planet. Inter.*, 5: 426—434.
Keylis-Borok, V.I., 1959. On estimation of the displacement in an earthquake source and of source dimensions. *Ann. Geofis.*, 12: 205—214.
Keylis-Borok, V.I., 1960. Investigation of the mechanism of earthquakes. *Sov. Res. Geophys.*, 4 (transl., *Tr. Geofiz. Inst.*, 40, 1957) 201 pp. American Geophysical Union, Consultants Bureau, New York.
Kikuchi, M. and Takeuchi, H., 1970. Unsteady propagation of longitudinal-shear cracks. *Zisin*, 23: 304—312 (in Japanese).
Knopoff, L., 1958. Energy release in earthquakes. *Geophys. J.*, 1: 44—52.
Kostrov, B.V., 1964. Self-similar problems of propagation of shear cracks. *J. Appl. Math. Mech.*, 28: 1077.
Love, A.E.H., 1914. Mathematical Research (Presidential Address). *Proc. Lond. Math. Soc.*, Ser. 2, Vol. 4.
Love, A.E.H., 1927. *The Mathematical Theory of Elasticity*. Cambridge University Press, 1934.
Maruyama, T., 1963. On the force equivalent of dynamic elastic dislocations with reference to the earthquake mechanism. *Bull. Earthquake Res. Inst., Tokyo Univ.*, 41: 467—486.
McKenzie, D. and Brune, J.N., 1972. Melting on fault planes during large earthquakes. *Geophys. J. R. Astron. Soc.*, 29: 65—78.
Morrill, B.J. and Mathiesen, R.B., 1972. Strong-motion accelerograph records from 4 September 1972, Stone Canyon earthquake. Eastern Section, Seismol. Soc. Am. *Earthquake Notes*, XLIII (3): 17—20.
Neuber, H., 1958. Theory of Notch Stresses: Principles for exact calculation of strength with reference to structural form and material. Transl. from publ. of Springer-Verlag, Berlin by the Office of Tech. Info, AEC-tr-4547 (2nd ed.).
Niazi, M., 1969. Source dynamics of Dasht-e-Bayaz earthquake of August 31, 1968. *Bull. Seismol. Soc. Am.*, 59: 1843—1861.
Orowan, E., 1960. Mechanism of seismic faulting in rock deformation: A symposium. *Geol. Soc. Am. Mem.*, 79: 323—345.
Randall, M., 1968. Relative sizes of multipolar components in deep earthquakes. *J. Geophys. Res.*, 73: 6140—6142.

Randall, M. and Knopoff, L., 1970. The mechanism at the focus of deep earthquakes. *J. Geophys. Res.*, 75: 4965—4976.

Reid, H.F., 1910. Mechanics of the earthquake. In: *The California Earthquake of April 18, 1906*, 2. Carnegie Inst. of Washington, D.C. (updated in 1969).

Reimer, R.B., Clough, R.W. and Raphael, J.M., 1974. Seismic response of Pacoima Dam in the San Fernando earthquake. *Proc. 5th World Conf. Earthquake Eng.*, 2: 2328—2337.

Richards, P.G., 1973. The dynamic field of a growing plane elliptical shear crack. *J. Solids Struct.*, 9: 843—861.

Richter, C.F., 1958. *Elementary Seismology*. Freeman, San Francisco, 768 pp.

Savage, J.C., 1965. The stopping phase on seismograms. *Bull. Seismol. Soc. Am.*, 55: 47.

Savage, J.C. and Wood, M.D., 1971. The relation between apparent stress and stress drop. *Bull. Seismol. Soc. Am.*, 61: 1381—1388.

Smith, W.D., 1974. The application of finite element analysis to body wave propagation problems. *Geophys. J. R. Astron. Soc.*, 42: 747—768.

Starr, A.T., 1928. Slip in a crystal and rupture in a solid due to shear. *Proc. Cambridge Philos. Soc.*, 24: 489—500.

Trifunac, M., 1971. Surface motion of a semi-cylindrical alluvial valley for incident plane SH waves. *Bull. Seismol. Soc. Am.*, 61: 1755—1770.

Trifunac, M., 1972a. Stress estimates for the San Fernando, California earthquake of February 9, 1971: Main event and thirteen aftershocks. *Bull. Seismol. Soc. Am.*, 62: 721—750.

Trifunac, M., 1972b. Tectonic stress and the source mechanism of the Imperial Valley, California earthquake of 1940. *Bull. Seismol. Soc. Am.*, 62: 1283—1302.

Trifunac, M., 1973. Scattering of plane SH waves by a semi-cylindrical canyon. *Earthquake Eng. Struct. Dyn.*, 1: 267—281.

Trifunac, M. and Brune, J.N., 1970. Complexity of energy release during the Imperial Valley, California earthquake of 1940. *Bull. Seismol. Soc. Am.*, 60: 137—160.

Trifunac, M. and Udwadia, F., 1974. Parkfield, California earthquake of June 27, 1966: A three-dimensional moving dislocation. *Bull. Seismol. Soc. Am.*, 64: 511—534.

Tsai, Y. and Patton, H., 1973. Interpretation of the strong-motion earthquake accelerograms using a moving dislocation model: The Parkfield, California earthquake of June 28, 1966. *68th Ann. Natl. Meet. Seismol. Soc. Am., Golden, Colo.* (abstract).

Tucker, B. and Brune, J.N., 1973. Seismograms, S-wave spectra and source parameters for aftershocks of the San Fernando earthquake of February 9, 1971. *NOAA Spec. Rep., Geological and Geophysical Studies*, Vol. III, Washington, D.C.

White, J.E., 1965. *Seismic Waves Radiation, Transmission, and Attenuation*. McGraw-Hill, New York.

Wong, H.L. and Jennings, P.C., 1975. Effects of canyon topography on strong ground motion. *Bull. Seismol. Soc. Am.*, 65: 1239—1257.

Wu, F., 1968. Parkfield earthquake of June 28, 1966; Magnitude and source mechanism. *Bull. Seismol. Soc. Am.*, 58: 689—709.

Wu, F.T., Thomson, K.C. and Kuenzler, H., 1972. Stick-slip propagation velocity and seismic source mechanism. *Bull. Seismol. Soc. Am.*, 62: 1621.

Wyss, M., 1970. Stress estimates for South American shallow and deep earthquakes. *J. Geophys. Res.*, 75: 1529—1544.

Wyss, M. and Brune, J.N., 1967. The Alaska earthquake of 28 March 1964: A complex multiple rupture. *Bull. Seismol. Soc. Am.*, 57: 1017—1023.

Wyss, M. and Molnar, P., 1972. Efficiency, stress drop, apparent stress, effective stress and frictional stress of Denver, Colorado earthquakes. *J. Geophys. Res.*, 77: 1433—1438.

*Chapter 6*

SEISMICITY

LUIS ESTEVA

*Instituto de Ingeniería, Universidad Nacional Autónoma de México, Mexico*

6.1 ON SEISMICITY MODELS

Rational formulation of engineering decisions in seismic areas requires quantitative descriptions of seismicity. These descriptions should conform with their intended applications: in some instances, simultaneous intensities during each earthquake have to be predicted at several locations, while in others it suffices to make independent evaluations of the probable effects of earthquakes at each of those locations.

The second model is adequate for the selection of design parameters of individual components of a regional system (the structures in a region or country) when no significant interaction exists between response or damage of several such individual components, or between any of them and the system as a whole. In other words, it applies when the damage — or negative utility — inflicted upon the system by an earthquake can be taken simply as the addition of the losses in the individual components.

The linearity between monetary values and utilities implied in the second model is not always applicable. Such is the case, for instance, when a significant portion of the national wealth or of the production system is concentrated in a relatively narrow area, or when failure of life-line components may disrupt emergency and relief actions just after an earthquake. Evaluation of risk for the whole regional system has then to be based on seismicity models of the first type, that is, models that predict simultaneous intensities at several locations during each event; for the purpose of decision making, nonlinearity between monetary values and utilities can be accounted for by means of adequate scale transformations. These models are also of interest to insurance companies, when the probability distribution of the maximum loss in a given region during a given time interval is to be estimated.

Whatever the category to which a seismic risk problem belongs, it requires the prediction of probability distributions of certain ground motion characteristics (such as peak ground acceleration or velocity, spectral density, response or Fourier spectra, duration) at a given site during a single shock or of maximum values of some of those characteristics in earthquakes occurring during given time intervals. When the reference interval tends to infinity, the probability distribution of the maximum value of a given characteristic approaches that of its maximum *possible* value. Because different systems or subsystems are sensitive to different ground motion characteristics, the term *intensity characteristic* will be used throughout this chapter to mean a particular parameter or set of parameters of an earthquake motion, in terms of which the response is to be predicted. Thus, when dealing with the failure probability of a structure, intensity can be alternatively measured — with different degrees of correlation with structural response — by the ordinate of the response spectrum for the corresponding period and damping, the peak ground acceleration, or the peak ground velocity.

In general, local instrumental information does not suffice for estimating the probability distributions of maximum intensity characteristics, and use has to be made of data on subjective measures of intensities of past earthquakes, of models of *local seismicity*, and of expressions relating characteristics with magnitude and site-to-source distance. Models of local seismicity consist, at least, of expressions relating magnitudes of earthquakes generated in given volumes of the earth's crust with their return periods. More often than not, a more detailed description of local seismicity is required, including estimates of the maximum magnitude that can be generated in these volumes, as well as probabilistic (stochastic process) models of the possible histories of seismic events (defined by magnitudes and coordinates).

This chapter deals with the various steps to be followed in the evaluation of seismic risk at sites where information other than direct instrumental records of intensities has to be used: identifying potential sources of activity near the site, formulating mathematical models of local seismicity for each source, obtaining the contribution of each source to seismic risk at the site and adding up contributions of the various sources and combining information obtained from local seismicity of sources near the site with data on instrumental or subjective intensities observed at the site.

The foregoing steps consider use of information stemming from sources of different nature. Quantitative values derived therefrom are ordinarily tied to wide uncertainty margins. Hence they demand probabilistic evaluation, even though they cannot always be interpreted in terms of relative frequencies of outcomes of given experiments. Thus, geologists talk of the maximum magnitude that can be generated in a given area, assessed by looking at the dimensions of the geological accidents and by extrapolating the observations of other regions which available evidence allows to brand as similar to the one of interest; the estimates produced are obviously uncertain, and the degree of uncertainty should be expressed together with the most probable value. Following nearly parallel lines, some geophysicists estimate the energy that can be liberated by a single shock in a given area by making quantitative assumptions about source dimensions, dislocation amplitude and stress drop, consistent with tectonic models of the region and, again, with comparisons with areas of similar tectonic characteristics.

Uncertainties attached to estimates of the type just described are in general extremely large: some studies relating fault rupture area, stress drop, and magnitude (Brune, 1968) show that, considering not unusually high stress drops, it does not take very large source dimensions to get magnitudes 8.0 and greater, and those studies are practically restricted to the simplest types of fault displacement. It is not clear, therefore, that realistic bounds can always be assigned to potential magnitudes in given areas or that, when this is feasible, those bounds are sufficiently low, so that designing structures to withstand the corresponding intensities is economically sound, particularly when occurrence of those intensities is not very likely in the near future. Because uncertainties in maximum feasible magnitudes and in other parameters defining magnitude-recurrence laws can be as significant as their mean values when trying to make rational seismic design decisions, those uncertainties have to be explicitly recognized and accounted for by means of adequate probabilistic criteria. A corollary is that geophysically based estimates of seismicity parameters should be accompanied with corresponding uncertainty measures.

Seismic risk estimates are often based only on statistical information (observed magnitudes and hypocentral coordinates). When this is done, a wealth of relevant geophysical information is neglected, while the probabilistic prediction of the future is made to rely on a sample that is often small and of little value, particularly if the sampling period is short as compared with the desirable return period of the events capable of severely damaging a given system.

The criterion advocated here intends to unify the foregoing approaches and rationally to assimilate the corresponding pieces of information. Its philosophy consists in using the geological, geophysical, and all other available non-statistical evidence for producing a set of alternate assumptions concerning a mathematical (stochastic process) model of seismicity in a given source area. An initial probability distribution is assigned to the set of hypotheses, and the statistical information is then used to improve that probability assignment. The criterion is based on application of *Bayes theorem*, also called the *theorem of the probabilities of hypotheses*. Since estimates of risk depend largely on conceptual models of the geophysical processes involved, and these are known with different degrees of uncertainty in different zones of the earth's crust, those estimates will be derived from stochastic process models with uncertain forms or parameters. The degree to which these uncertainties can be reduced depends on the limitations of the state of the art of geophysical sciences and on the effort that can be put into compilation and interpretation of geophysical and statistical information. This is an economical problem that should be handled, formally or informally, by the criteria of decision making under uncertainty.

## 6.2 INTENSITY ATTENUATION

Available criteria for the evaluation of the contribution of potential seismic sources to the risk at a site make use of *intensity attenuation* expressions that relate intensity characteristics with magnitude and distance from site to source. Depending on the application envisaged, the intensity characteristic to be predicted can be expressed in a number of manners, ranging from a subjective index, such as the *Modified Mercalli intensity*, to a combination of one or more quantitative measures of ground shaking (see Chapter 1).

A number of expressions for attenuation of various intensity characteristics with distance have been developed, but there is little agreement among most of them (Ambraseys, 1973). This is due in part to discrepancies in the definitions of some parameters, in the ranges of values analyzed, in the actual wave propagation properties of the geological formations lying between source and site, in the dominating shock mechanisms, and in the forms of the analytical expressions adopted a priori.

Most intensity-attenuation studies concern the prediction of earthquake characteristics on rock or firm ground, and assume that these characteristics, properly modified in terms of frequency-dependent soil amplification factors, should constitute the basis for estimating their counterparts on soft ground. Observations about the influence of soil properties on earthquake damage support the assumption of a strong correlation between type of local ground and intensity in a given shock. Attempts to analytically predict the characteristics of motions on soil given those on firm ground or on bedrock have not been too successful, however (Crouse, 1973; Hudson and Udwadia, 1973; Salt, 1974), with the exception of some peculiar cases, like Mexico City (Herrera et al., 1965), where local conditions favor the fulfillment of the assumptions implied by usual analytical models. The following paragraphs concentrate on prediction on intensities on firm ground; the influence of local soil is discussed in Chapter 4.

### 6.2.1 Intensity attenuation on firm ground

When isoseismals (lines joining sites showing equal intensity) of a given shock are based only on intensities observed on homogeneous ground conditions, such as *firm ground* (compact soils) or bedrock, they are roughly elliptical and the orientations of the corresponding axes are often correlated with local or regional geological trends (Figs. 6.1—6.3). In some regions — for instance near major faults in the western United States — those trends are well defined and the correlations are clear enough as to permit prediction of intensity in the near and far fields in terms of magnitude and distance to the generating fault or to the centroid of the energy liberating volume. In other regions, such as the eastern United States and most of Mexico, isoseismals seem to elongate systematically in a direction that is a function of the epi-

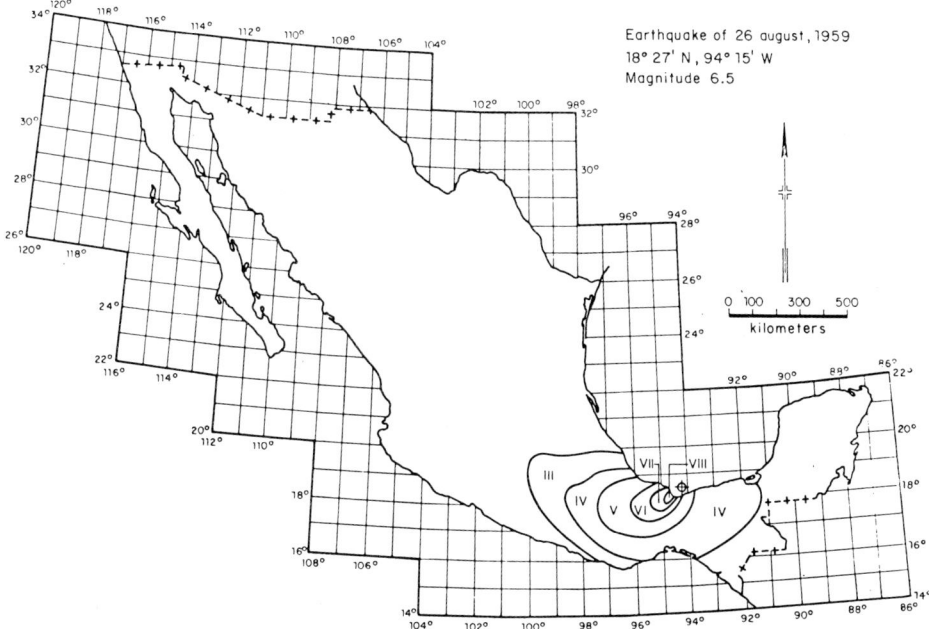

Fig. 6.1. Isoseismals of an earthquake in Mexico. (After Figueroa, 1963.)

central coordinates (Bollinger, 1973; Figueroa, 1963). In that case, intensity should be expressed as a function of magnitude and coordinates of source and site. For most areas in the world, intensity has to be predicted in terms of simple — and cruder — expressions that depend only on magnitude and distance from site to instrumental hypocenter. This stems from inadequate knowledge of geotectonic conditions and from limited information concerning the volume where energy is liberated in each shock.

A comparison of the rates of attenuation of intensities on firm ground for shocks on western and eastern North America has disclosed systematic differences between those rates (Milne and Davenport, 1969). This is the source of a basic, but often unavoidable, weakness of most intensity-attenuation expressions, because they are based on heterogeneous data, recorded in different zones, and the very nature of their applications implies that the less is known about possible systematic deviations in a given zone, as a consequence of the meagerness of local information, the greater weight is given to predictions with respect to observations.

*6.2.1.1 Modified Mercalli intensities*

An analysis of the Modified Mercalli intensities on firm ground reported for earthquakes occurring in Mexico in the last few decades leads to the fol-

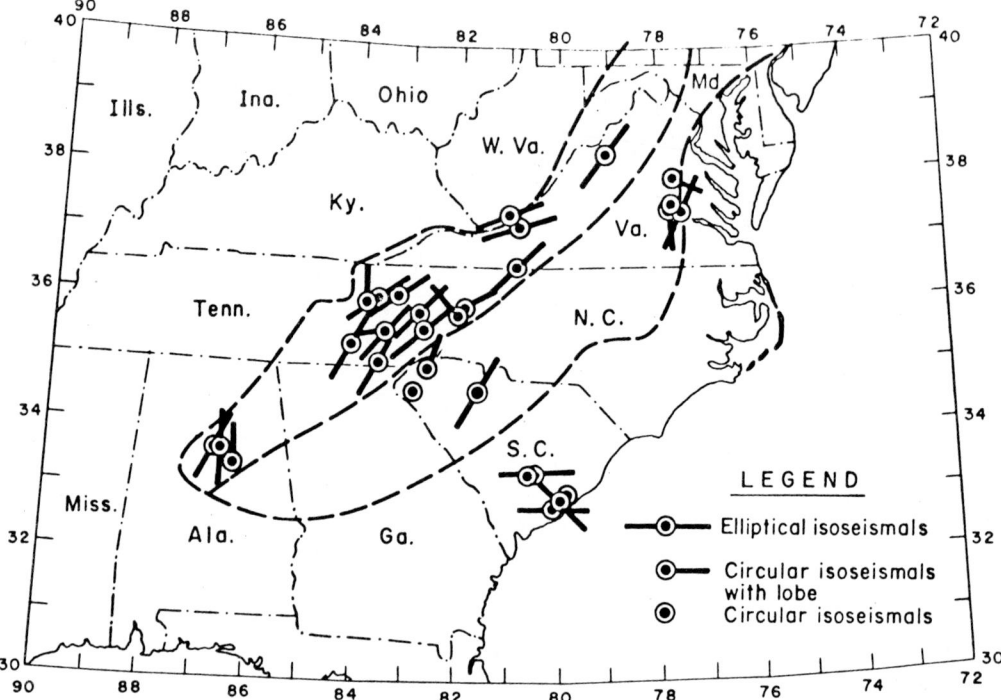

Fig. 6.2. Elongation of isoseismals in the southeastern United States. (After Bollinger, 1973.)

lowing expression relating magnitude $M$, hypocentral distance $R$ (in kilometers) and intensity $I$ (Esteva, 1968):

$$I = 1.45\,M - 5.7\,\log_{10} R + 7.9 \tag{6.1}$$

The prediction error, defined as the difference between observed and computed intensity, is roughly normally distributed, with a standard deviation of 2.04, which means that there is a probability of 60% that an observed intensity is more than one degree greater or smaller than its predicted value.

*6.2.1.2 Peak ground accelerations and velocities*

A few of the available expressions will be described. Their comparison will show how cautiously a designer intending to use them should proceed.

Housner studied the attenuation of peak ground accelerations in several regions of the United States and presented his results graphically (1969) in terms of fault length (in turn a function of magnitude), shapes of isoseismals and areas experiencing intensities greater than given values (Fig. 6.4 and 6.5).

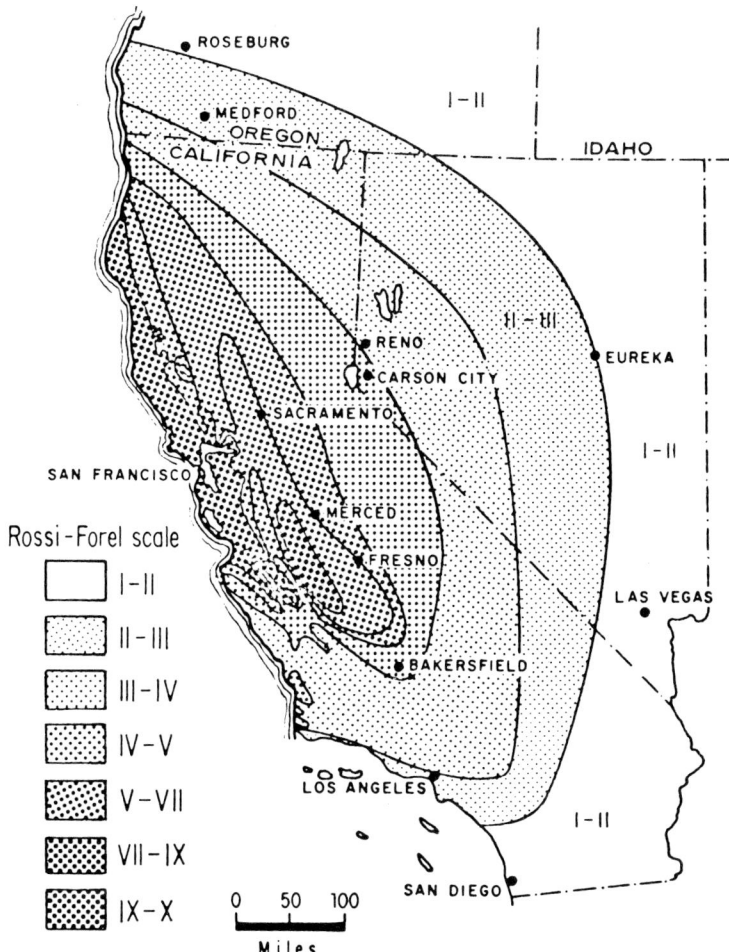

Fig. 6.3. Isoseismals in California. (After Bolt, 1970.)

He showed that intensities attenuate faster with distance on the west coast than in the rest of the country. This comparison is in agreement with Milne and Davenport (1969), who performed a similar analysis for Canada. From observations of strong earthquakes in California and in British Columbia, they developed the following expression for $a$, the peak ground acceleration, as a fraction of gravity:

$$a/g = 0.0069 \; e^{1.6M}/(1.1 \; e^{1.1M} + R^2) \tag{6.2}$$

Here, $R$ is epicentral distance in kilometers. The acceleration varies roughly as $e^{1.64M} R^{-2}$ for large $R$, and as $e^{0.54M}$ where $R$ approaches zero. This reflects to some extent the fact that energy is released not at a single point but from a finite volume. A later study by Davenport (1972) led him

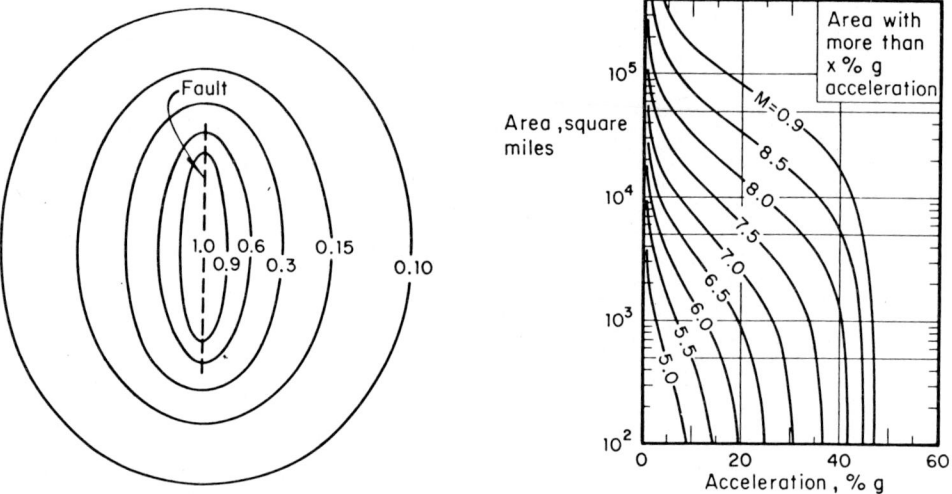

Fig. 6.4. Idealized contour lines of intensity of ground shaking. (After Housner, 1969.)

Fig. 6.5. Area in square miles experiencing shaking of $x$ %g or greater for shocks of different magnitudes. (After Housner, 1969.)

to propose the expression:

$$a/g = 0.279\ e^{0.8M}/R^{1.64} \tag{6.3}$$

The statistical error of this equation was studied by fitting a lognormal probability distribution to the ratios of observed to computed accelerations. A standard deviation of 0.74 was found in the natural logarithms of those ratios.

Esteva and Villaverde (1973), on the basis of accelerations reported by Hudson (1971, 1972a,b), derived expressions for peak ground accelerations and velocities, as follows:

$$a/g = 5.7\ e^{0.8M}/(R + 40)^2 \tag{6.4}$$

$$v = 32\ e^M/(R + 25)^{1.7} \tag{6.5}$$

Here $v$ is peak ground velocity in cm/sec and the other symbols mean the same as above. The standard deviation of the natural logarithm of the ratio of observed to predicted intensity is 0.64 for accelerations and 0.74 for velocities. If judged by this parameter, eqs. 6.3 and 6.4 seem equally reliable. However, as shown by Fig. 6.6, their mean values differ significantly in some ranges.

With the exception of eq. 6.2, all the foregoing attenuation expressions are products of a function of $R$ and a function of $M$. This form, which is acceptable when the dimensions of the energy-liberating source are small com-

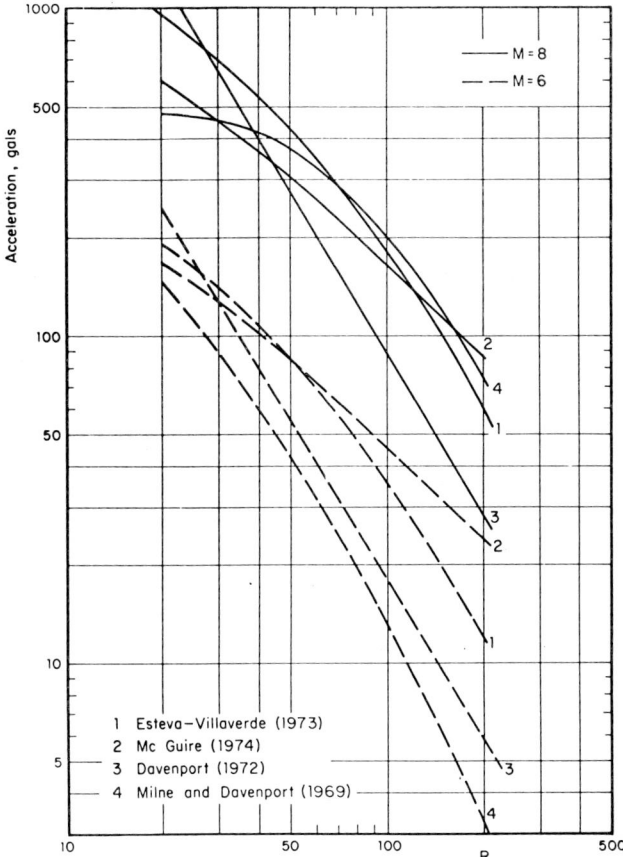

Fig. 6.6. Comparison of several attenuation expressions.

pared with $R$, is inadequate when dealing with earthquake sources whose dimensions are of the order of moderate hypocentral distances, and often greater than them. Although equation errors (probability distributions of the ratio of observed to predicted intensities) have been evaluated by Davenport (1972) and Esteva and Villaverde (1973), their dependence on $M$ and $R$ has not been analyzed. Because seismic risk estimates are very sensitive to the attenuation expressions in the range of large magnitudes and short distances, more detailed studies should be undertaken, aiming at improving those expressions in the mentioned range, and at evaluating the influence of $M$ and $R$ on equation error. Information on strong-motion records will probably be scanty for those studies, and hence they will have to be largely based on analytical or physical models of the generation and propagation of seismic waves. Although significant progress has been lately attained in this direction (Trifunac, 1973) the results from such models have hardly influenced the practice of seismic risk estimation because they have remained either unknown to or imperfectly appreciated by engineers in charge of the corresponding decisions.

### 6.2.1.3 Response spectra

Peak ground acceleration and displacement are fairly good indicators of the response of structures possessing respectively very high and very small natural frequencies. Peak velocity is correlated with the response of intermediate-period systems, but the correlation is less precise than that tying the former parameters; hence, it is natural to formulate seismic risk evaluation and engineering design criteria in terms of spectral ordinates.

Response spectrum prediction for given magnitude and hypocentral or site-to-fault distance usually entails a two-step process, according to which peak ground acceleration, velocity and displacement are initially estimated and then used as reference values for prediction of the ordinates of the response spectrum. Let the second step in the process be represented by the operation $y_s = \alpha y_g$, where $y_s$ is an ordinate of the response spectrum for a given natural period and damping ratio, and $y_g$ is a parameter (such as peak ground acceleration or velocity) that can be directly obtained from the time-history record of a given shock regardless of the dynamic properties of the systems whose response is to be predicted. For given $M$ and $R$, $y_g$ is random and so is $y_s/y_g = \alpha$; the mean and standard deviation of $y_s$ depend on those of $y_g$ and $\alpha$ and on the coefficient of correlation of the latter variables. As shown above, $y_g$ can only be predicted within wide uncertainty limits, often wider than those tied to $y_s$ (Esteva and Villaverde, 1973). The coefficient of variation of $y_s$ given $M$ and $R$ can be smaller than that of $y_g$ only if $\alpha$ and $y_g$ are negatively correlated, which is often the case: the greater the deviation of an observed value of $y_g$ with respect to its expectation for given $M$ and $R$, the lower is likely to be $\alpha$. In other words, it seems that in the intermediate range of natural periods the expected values of spectral ordinates for given damping ratios can be predicted directly in terms of magnitude and focal distance with narrower (or at most equal) margins of uncertainty than those tied to predicted peak ground velocities. For the ranges of very short or very long natural periods, peak amplitudes of ground motion and spectral ordinates approach each other and their standard errors are therefore nearly equal.

McGuire (1974) has derived attenuation expressions for the conditional values (given $M$ and $R$) of the mean and of various percentiles of the probability distributions of the ordinates of the response spectra for given natural periods and damping ratios. Those expressions have the same form as eqs. 6.4 and 6.5, but their parameters show that the rates of attenuation of spectral ordinates differ significantly from those of peak ground accelerations or velocities. For instance, McGuire finds that peak ground velocity attenuates in proportion to $(R + 25)^{-1.20}$, while the mean of the pseudovelocity for a

TABLE 6.I

McGuire's attenuation expressions $y = b_1 \, 10^{b_2 M}(R + 25)^{-b_3}$

| $y$ | $b_1$ | $b_2$ | $b_3$ | $V(y)$ = coeff. of var. of $y$ |
|---|---|---|---|---|
| $a$ gals | 472.3 | 0.278 | 1.301 | 0.548 |
| $v$ cm/sec | 5.64 | 0.401 | 1.202 | 0.696 |
| $d$ cm | 0.393 | 0.434 | 0.885 | 0.883 |
| Undamped spectral pseudovelocities | | | | |
| $T$ = 0.1 sec | 11.0 | 0.278 | 1.346 | 0.941 |
| 0.5 | 3.05 | 0.391 | 1.001 | 0.636 |
| 1.0 | 0.631 | 0.378 | 0.549 | 0.768 |
| 2.0 | 0.0768 | 0.469 | 0.419 | 0.989 |
| 5.0 | 0.0834 | 0.564 | 0.897 | 1.344 |
| 5% damped spectral pseudovelocities | | | | |
| $T$ = 0.1 sec | 10.09 | 0.233 | 1.341 | 0.651 |
| 0.5 | 5.74 | 0.356 | 1.197 | 0.591 |
| 1.0 | 0.432 | 0.399 | 0.704 | 0.703 |
| 2.0 | 0.122 | 0.466 | 0.675 | 0.941 |
| 5.0 | 0.0706 | 0.557 | 0.938 | 1.193 |

natural period of 1 sec and a damping ratio of 2% attenuates in proportion to $(R + 25)^{-0.59}$. These results stem from the way that frequency content changes with $R$ and lead to the conclusion that the ratio of spectral velocity should be taken as a function of $M$ and $R$.

Table 6.I summarizes McGuire's attenuation expressions and their coefficients of variation for ordinates of the pseudovelocity spectra and for peak ground acceleration, velocity and displacement. Similar expressions were derived by Esteva and Villaverde (1973), but they are intended to predict only the maxima of the expected acceleration and velocity spectra, regardless of the periods associated with those maxima. No analysis has been performed of the relative validity of McGuire's and Esteva and Villaverde's expressions for various ranges of $M$ and $R$.

6.3 LOCAL SEISMICITY

The term *local seismicity* will be used here to designate the degree of seismic activity in a given volume of the earth's crust; it can be quantitatively described according to various criteria, each providing a different amount of information. Most usual criteria are based on upper bounds to the magnitudes of earthquakes that can originate in a given seismic source, on the amount of energy liberated by shocks per unit volume and per unit time or on more detailed statistical descriptions of the process.

### 6.3.1 Magnitude-recurrence expressions

Gutenberg and Richter (1954) obtained expressions relating earthquake magnitudes with their rates of occurrence for several zones of the earth. Their results can be put in the form:

$$\lambda = \alpha e^{-\beta M} \tag{6.6}$$

where $\lambda$ is the mean number of earthquakes per unit volume and per unit time having magnitude greater than $M$ and $\alpha$ and $\beta$ are zone-dependent constants; $\alpha$ varies widely from point to point, as evidenced by the map of epicenters shown in Fig. 6.7, while $\beta$ remains within a relatively narrow range, as shown in Fig. 6.8. Equation 6.6 implies a distribution of the energy liberated per shock which is very similar to that observed in the process of microfracturing of laboratory specimens of several types of rock subjected to gradually increasing compressive or bending strain (Mogi, 1962; Scholz, 1968). The values of $\beta$ determined in the laboratory are of the same order as those obtained from seismic events, and have been shown to depend on the heterogeneity of the specimens and on their ability to yield locally. Thus, in heterogeneous specimens made of brittle materials many small shocks precede a major fracture, while in homogeneous or plastic materials the number of small shocks is relatively small. These cases correspond to large and small $\beta$-values, respectively. No general relationship is known to the writer between $\beta$ and geotectonic features of seismic provinces: complexity of crustal structure and of stress gradients precludes extrapolation of laboratory results; and statistical records for relatively small zones of the earth are not, as a rule, adequate for establishing local values of $\beta$. Figure 6.8 shows that for very high magnitudes the observed frequency of events is lower than predicted by eq. 6.6. In addition, Rosenblueth (1969) has shown that $\beta$ cannot be smaller than 3.46, since that would imply an infinite amount of energy liberated per unit time. However, Fig. 6.8 shows that the values of $\beta$ which result from fitting expressions of the form 6.6 to observed data are smaller than 3.46; hence, for very high values of $M$ (above 7, approximately) the curve should bend down, in accordance with statistical evidence.

Expressions alternative to eq. 6.6 have been proposed, attempting to represent more adequately the observed magnitude-recurrence data (Rosenblueth, 1964; Merz and Cornell, 1973). Most of these expressions also fail to recognize the existence of an upper bound to the magnitude that can be generated in a given source. Although no precise estimates of this upper bound can yet be obtained, recognition of its existence and of its dependence on the geotectonic characteristics of the source is inescapable. Indeed, the prac-

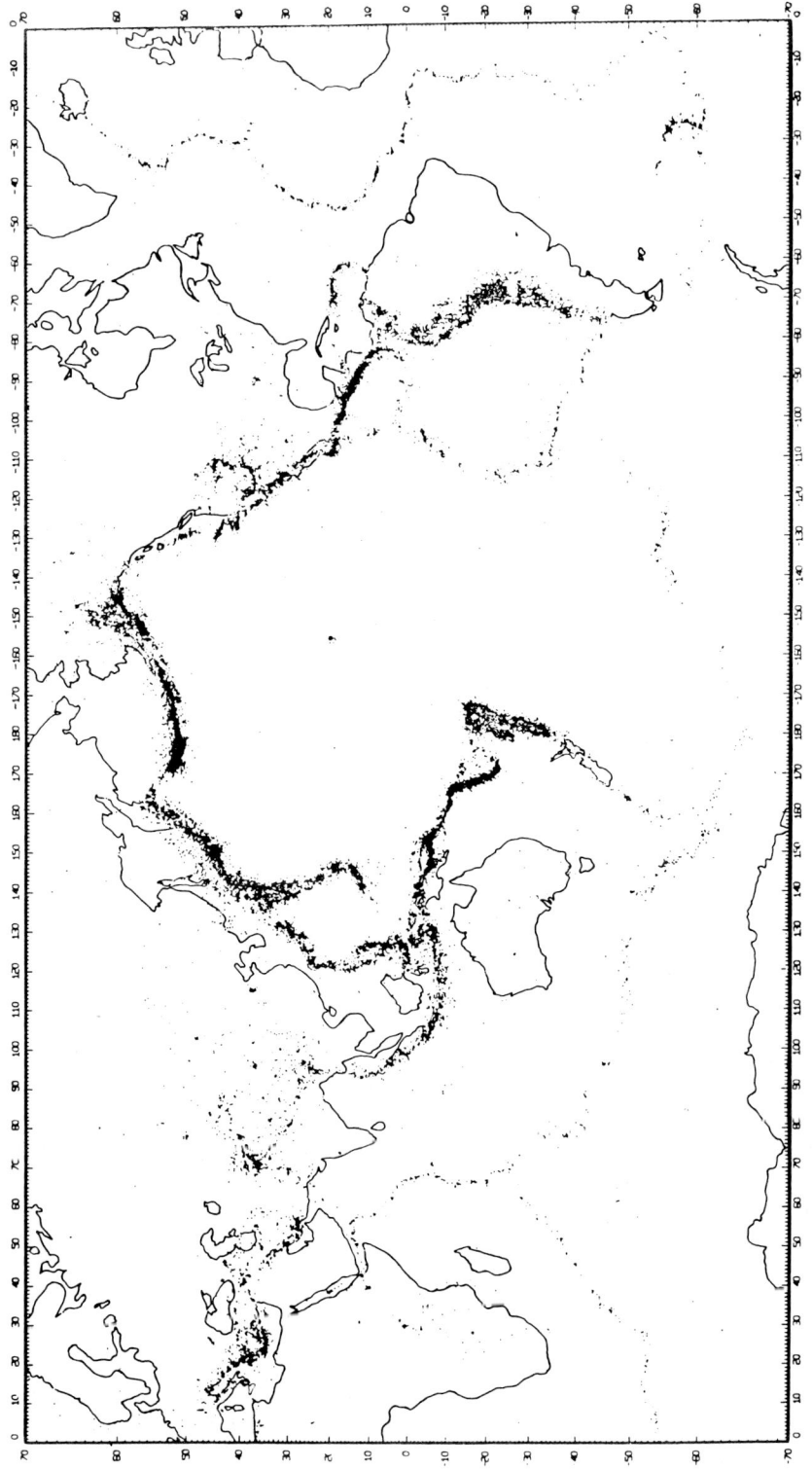

Fig. 6.7. Map showing epicenters for the interval 1961–1967. (After Newmark and Rosenblueth, 1971.)

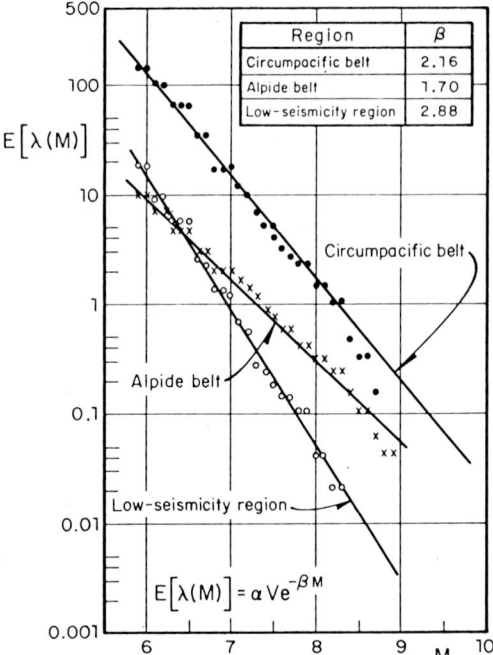

Fig. 6.8. Seismicity of macrozones. (After Esteva, 1968.)

tice of seismic zoning in the Soviet Union has been based on this concept (Gzovsky, 1962; Ananiin et al., 1968) and in many countries design spectra for very important structures, such as nuclear reactors or large dams, are usually derived from the assumption of a maximum credible intensity at a site; that intensity is ordinarily obtained by taking the maximum of the intensities that result at the site when at each of the potential sources an earthquake with magnitude equal to the maximum feasible value for that source is generated at the most unfavourable location within the same source. When this criterion is applied no attention is usually paid to the uncertainty in the maximum feasible magnitude nor to the probability that an earthquake with that magnitude will occur during a given time period. The need to formulate seismic-risk-related decisions that account both for upper bounds to magnitudes and for their probabilities of occurrence suggests adoption of magnitude recurrence expressions of the form:

$$\lambda = \lambda_L G^*(M) \quad \text{for } M_L \leqslant M \leqslant M_U$$
$$= \lambda_L \quad \text{for } M < M_L$$
$$= 0 \quad \text{for } M > M_U \tag{6.7}$$

where $M_L$ = lowest magnitude whose contribution to risk is significant, $M_U$

= maximum feasible magnitude, and $G^*(M)$ = complementary cumulative probability distribution of magnitudes every time that an event $(M \geqslant M_L)$ occurs. A particular form of $G^*(M)$ that lends itself to analytical derivations is:

$$G^*(M) = A_0 + A_1 \exp(-\beta M) - A_2 \exp[-(\beta - \beta_1)M] \tag{6.8}$$

where:

$A_0 = A\beta_1 \exp[-\beta(M_U - M_L)]$

$A_1 = A(\beta - \beta_1) \exp(\beta M_L)$

$A_2 = A\beta \exp(-\beta_1 M_U + \beta M_L)$

$A = [\beta\{1 - \exp[-\beta_1(M_U - M_L)]\} - \beta_1 \{1 - \exp[-\beta(M_U - M_L)]\}]^{-1}$

As $M$ tends to $M_L$ from above, eq. 6.7 approaches eq. 6.6. Adoption of adequate values of $M_U$ and $\beta_1$ permits satisfying two additional conditions: the maximum feasible magnitude and the rate of variation of $\lambda$ in its vicinity. When $\beta_1 \to \infty$, eq. 6.8 tends to an expression proposed by Cornell and Vanmarcke (1969).

Yegulalp and Kuo (1974) have applied the theory of extreme values to estimating the probabilities that given magnitudes are exceeded in given time intervals. They assume those probabilities to fit an extreme type-III distribution given by:

$$F_{M_{max}}(M|t) = \exp[-C(M_U - M)^k t] \quad \text{for } M \leqslant M_U$$
$$= 0 \quad \text{for } M > M_U \tag{6.9}$$

Here $F_{M_{max}}(M|t)$ indicates the probability that the maximum magnitude observed in $t$ years is smaller than $M$, $M_U$ has the same meaning as above, and $C$ and $K$ are zone-dependent parameters. This distribution is consistent with the assumption that earthquakes with magnitudes greater than $M$ take place in accordance with a Poisson process with mean rate $\lambda$ equal to $C(M_U - M)^k$. Equation 6.9 produces magnitude recurrence curves that fit closely the statistical data on which they are based for magnitudes above 5.2 and return periods from 1 to 50 years, even though the values of $M_U$ that result from pure statistical analysis are not reliable measures of the upper bound to magnitudes, since in many cases they turn out inadmissibly high.

For low magnitudes, only a fraction of the number of shocks that take place is detected. As a consequence, $\lambda$-values based on statistical information lie below those computed according to eqs. 6.6 and 6.8 for $M$ smaller than about 5.5. In addition, Fig. 6.9, taken from Yegulalp and Kuo (1974), shows that the numbers of detected shocks fit the extreme type III in eq. 6.9 better than the extreme type-I distribution implied by eq. 6.6., coupled with the assumption of Poisson distribution of the number of events. It is not

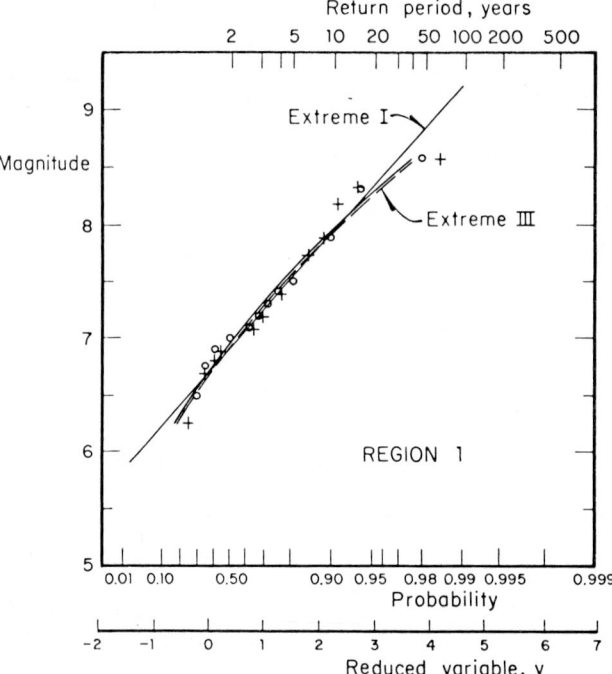

Fig. 6.9. Magnitude statistics in the Aleutian Islands region. (After Yegulalp and Kuo, 1974.)

clear what portion of the deviation from the extreme type-I distribution is due to the low values of the detectability levels and what portion comes from differences between the actual form of variation of $\lambda$ with $M$ and that given by eq. 6.6. The problem deserves attention because estimates of expected losses due to nonstructural damage may be sensitive to the values of $\lambda$ for small magnitudes (say below 5.5) and because the evaluation of the level of seismic activity in a region is often made to depend on the recorded numbers of small magnitude shocks and on assumed detectability levels, i.e. of ratios of numbers of detected and occurred earthquakes (Kaila and Narain 1971; Kaila et al., 1972, 1974).

None of the expressions for $\lambda$ presented in this chapter possess the desirable property that its applicability over a number of non-overlapping regions of the earth's crust implies the validity of an expression of the same form over the addition of those regions, unless some restrictions are imposed on the parameters of each $\lambda$. For instance, the addition of expressions like 6.6 gives place to an expression of the same form only if $\beta$ is the same for all terms in the sum. Similar objections can be made to eq. 6.8. In what follows these forms will be preserved, however, as their accuracy is consistent with the amount of available information and their adoption offers significant advantages in the evaluation of regional seismicity, as shown later.

### 6.3.2 Variation with depth

Depth of prevailing seismic activity in a region depends on its tectonic structure. For instance, most of the activity in the western coast of the United States and Canada consists of shocks with hypocentral depths in the range of 20—30 km. In other areas, such as the southern coast of Mexico, seismic events can be grouped into two ensembles: one of small shallow shocks and one of earthquakes with magnitudes comprised in a wide range, and with depths whose mean value increases with distance from the shoreline (Fig. 6.10). Figure 6.11 shows the depth distribution of earthquakes with magnitude above 5.9 for the whole circum-Pacific belt.

### 6.3.3 Stochastic models of earthquake occurrence

Mean exceedance rates of given magnitudes are expected averages during long time intervals. For decision-making purposes the times of earthquake occurrence are also significant. At present those times can only be predicted within a probabilistic context.

Let $t_i$ (i = 1, ..., n) be the unknown times of occurrence of earthquakes generated in a given volume of the earth's crust during a given time interval, and let $M_i$ be the corresponding magnitudes. For the moment it will be assumed that the risk is uniformly distributed throughout the given volume, and hence no attention will be paid to the focal coordinates of each shock.

Classical methods of time-series analysis have been applied by different researchers attempting to devise analytical models for random earthquake sequences. The following approaches are often found in the literature:

(a) Plotting of histograms of waiting times between shocks (Knopoff, 1964; Aki, 1963).

(b) Evaluation of Poisson's index of dispersion, that is of the ratio of the sample variance of the number of shocks to its expected value (Vere-Jones, 1970; Shlien and Toksöz, 1970). This index equals unity for Poisson processes, is smaller for nearly periodic sequences, and is greater than one when events tend to cluster.

(c) Determination of autocovariance functions, that is, of functions representing the covariance of the numbers of events observed in given time intervals, expressed in terms of the time elapsed between those intervals (Vere-Jones, 1970; Shlien and Toksöz, 1970). The autocovariance function of a Poisson process is a Dirac delta function. This feature is characteristic for the Poisson model since it does not hold for any other stochastic process.

(d) The hazard function $h(t)$, defined so that $h(t)\, dt$ is the conditional probability that an event will take place in the interval $(t, t + dt)$ given that

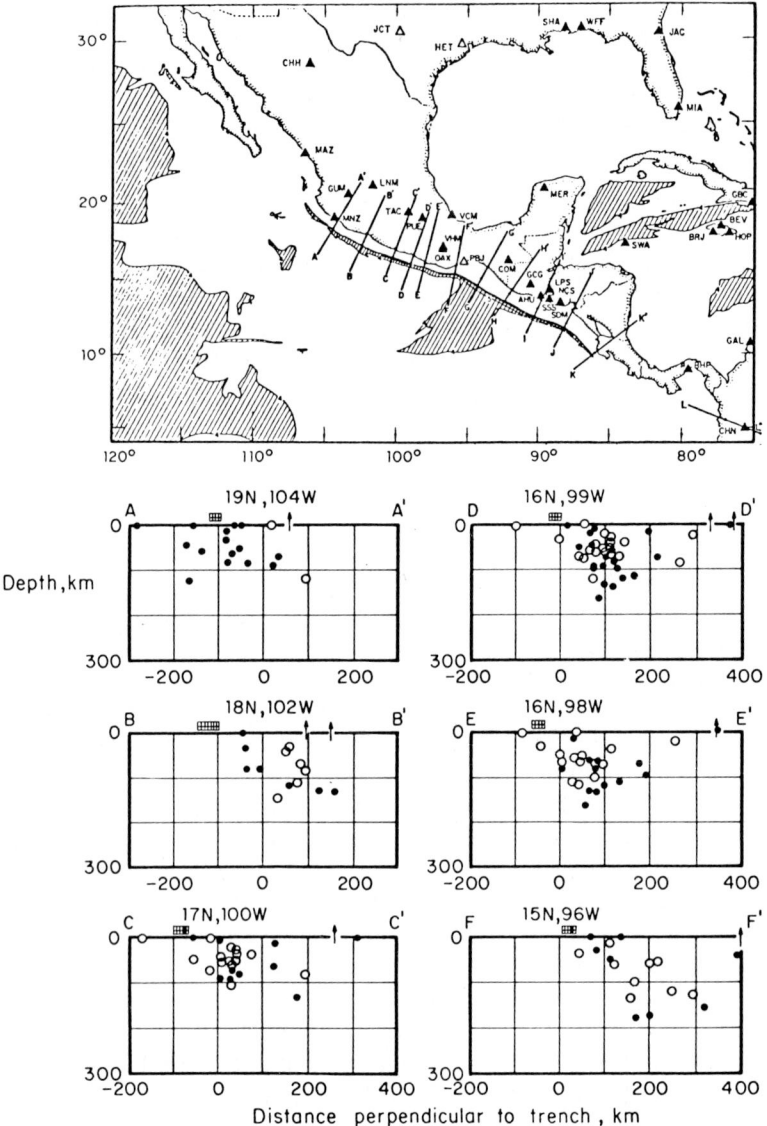

Fig. 6.10. Earthquake hypocenters projected onto a series of vertical sections through Mexico (After Molnar and Sykes, 1969.)

no events have occurred before $t$. If $F(t)$ is the cumulative probability distribution of the time between events:

$$h(t) = f(t)/[1 - F(t)] \tag{6.10}$$

where $f(t) = \partial F(t)/\partial t$.

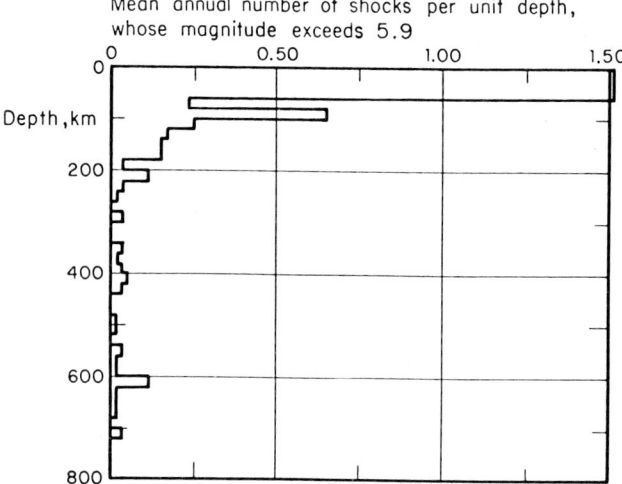

Fig. 6.11. Variation of seismicity with depth. Circum-Pacific Belt, (After Newmark and Rosenblueth, 1971.)

For the Poisson model, $h(t)$ is a constant equal to the mean rate of the process.

### 6.3.3.1 Poisson model

Most commonly applied stochastic models of seismicity assume that the events of earthquake occurrence constitute a Poisson process and that the $M_i$'s are independent and identically distributed. This assumption implies that the probability of having $N$ earthquakes with magnitude exceeding $M$ during time interval $(0, t)$ equals:

$$p_N = [\exp(-\nu_M t)(\nu_M t)^N]/N! \tag{6.11}$$

where $\nu_M$ is the mean rate of exceedance of magnitude $M$ in the given volume. If $N$ is taken equal to zero in eq. 6.11, one obtains that the probability distribution of the maximum magnitude during time interval $t$ is equal to $\exp(-\nu_M t)$. If $\nu_M$ is given by eq. 6.6, the extreme type-I distribution is obtained.

Some weaknesses of this model become evident in the light of statistical information and of an analysis of the physical processes involved: the Poisson assumption implies that the distribution of the waiting time to the next event is not modified by the knowledge of the time elapsed since the last one, while physical models of gradually accumulated and suddenly released energy call for a more general renewal process such that, unlike what happens in the Poisson process, the expected time to the next event decreases as time goes on (Esteva, 1974). Statistical data show that the Poisson assumption may be acceptable when dealing with large shocks throughout the world (Ben-Menahem, 1960), implying lack of correlation between seismicities of different regions; however, when considering small volumes of the earth, of the order of those that can significantly contribute to seismic risk at a site, data often contradict Poisson's model, usually because of clustering of earthquakes in time: the observed numbers of short intervals between events are significantly higher than predicted by the exponential distribution, and values of Poisson's index of dispersion are well above unity (Figs. 6.12 and 6.13). In some instances, however, deviations in the opposite direction have been observed: waiting times tend to be more nearly periodic, Poisson's index of dispersion is smaller than one, and the process can be represented by a renewal model. This condition has been reported, for instance, in the southern coast of Mexico (Esteva, 1974), and in the Kamchatka and Pamir—Hindu Kush regions (Gaisky, 1966 and 1967). The models under discussion also fail to account for clustering in space (Tsuboi, 1958; Gajardo and Lomnitz, 1960), for the evolution of seismicity with time, and for the systematic shifting of active sources along geologic accidents (Allen, Chapter 3 of this book). On account of its simplicity, however, the Poisson process model provides a valuable tool for the formulation of some seismic-risk-related decisions, particularly of those that are sensitive only to magnitudes of events having very long return periods.

*6.3.3.2 Trigger models*

Statistical analysis of waiting times between earthquakes does not favor the adoption of the Poisson model or of other forms of renewal processes, such as those that assume that waiting times are mutually independent with lognormal or gamma distributions (Shlien and Toksöz, 1970). Alternative models have been developed, most of them of the 'trigger type' (Vere-Jones, 1970), i.e. the overall process of earthquake generation is considered as the superposition of a number of time series, each having a different origin, where the origin times are the events of a Poisson process. In general, let $N$ be the number of events that take place during time interval $(0, t)$, $\tau_m$ = origin time of the $m$th series, $W_m(t, \tau_m)$ the corresponding number of events up to instant $t_1$ and $n_t$ the random number of time series initiated in the interval $(0, t)$. The total number of events that occur before instant $t$ is then:

$$N = \sum_{m}^{n_t} W_m(t, \tau_m) \tag{6.12}$$

If origin times are distributed according to a homogeneous Poisson process with mean rate $\nu$, and all $W_m$'s are identically distributed stochastic processes with respect to $(t - \tau_m)$, it can be shown (Parzen, 1962) that the mean and variance of $N$ can be obtained from:

$$E(N) = \nu \int_0^t E[W(t, \tau)] d\tau \tag{6.13}$$

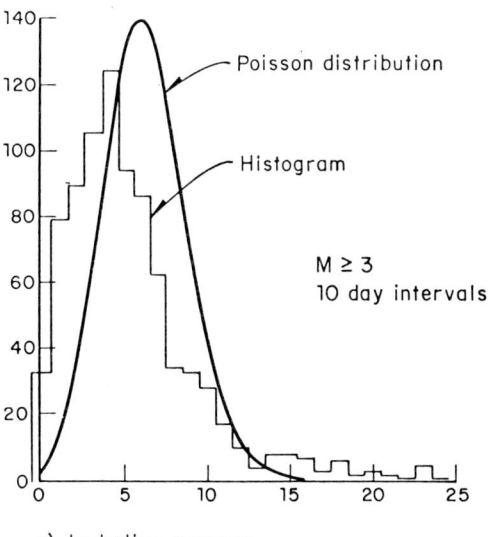

a) Including swarms

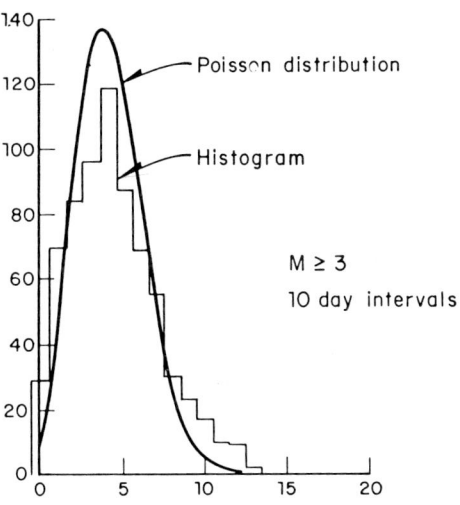

b) Eliminating swarms

Fig. 6.12. Evaluation of Poisson process assumption. (After Knopoff, 1964.)

$$\text{var}(N) = \nu \int_0^t E[W^2(t, \tau)] d\tau \tag{6.14}$$

Parzen (1962) gives also an expression for the probability generating function $\psi_N(Z; t)$ of the distribution of $N$ in terms of $\psi_W(Z; t, \tau)$, the generat-

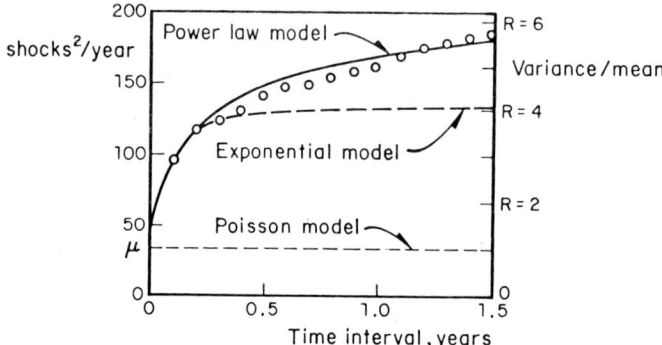

Fig. 6.13. Variance—time curve for New Zealand shallow shocks. (After Vere-Jones, 1966.)

ing function of each of the component processes:

$$\psi_N(Z; t) = \exp\left[-\nu t + \nu \int_0^t \psi_W(Z; t, \tau)d\tau\right] \quad (6.15)$$

where:

$$\psi_W(Z; t, \tau) = \sum_{n=0}^{\infty} Z^n P\{W(t, \tau) = n\} \quad (6.16)$$

and the probability mass function of $N$ can be obtained from $\psi_N(Z; t)$ by recalling that:

$$\psi_N(Z; t) = \sum_{n=0}^{\infty} Z^n P\{N = n\}$$

expanding $\psi_N$ in power series of $Z$, and taking $P\{N = n\}$ equal to the coefficient of $Z^n$ in that expansion. For instance, if it is of interest to compute $P\{N = 0\}$, expansion of $\psi_N(Z; t)$ in a Taylor's series with respect to $Z = 0$ leads to:

$$\psi_N(Z; t) = \psi_N(0; t) + Z\psi_N'(0; t) + \frac{Z^2}{2!}\psi_N''(0; t) + \ldots \quad (6.17)$$

where the prime signifies derivative with respect to $Z$. From the definition of $\psi_N$, $P\{N = 0\} = \psi_N(0; t)$.

Because the component processes of 'trigger'-type time series appear overlapped in sample histories, their analytical representation usually entails study of a number of alternative models, estimation of their parameters, and comparison of model and sample properties — often second-order properties (Cox and Lewis, 1966).

*Vere-Jones models.* Applicability of some general 'trigger' models to represent local seismicity processes was discussed in a comprehensive paper by Vere-Jones (1970), who calibrated them mainly against records of seismic activity in New Zealand. In addition to simple and compound Poisson processes (Parzen, 1962), he considered Neyman-Scott and Bartlett-Lewis models, both of which assume that earthquakes occur in clusters and that the number of events in each cluster is stocastically independent of its origin time. In the Neyman-Scott model, the process of clusters is assumed stationary and Poisson, and each cluster is defined by $p_N$, the probability mass function of its number of events, and $\Lambda(t)$, the cumulative distribution function of the time of an event corresponding to a given cluster, measured from the cluster origin. The Bartlett-Lewis model is a special case of the former, where each cluster is a renewal process that ends after a finite number of renewals. In these models the conditional probability of an event taking place during the interval $(t, t + dt)$, given that the cluster consists of $N$ shocks, is equal to $N\lambda(t)dt$, where $\lambda(t) = \partial \Lambda(t)/\partial t$.

Because clusters overlap in time they cannot easily be identified and separated. Estimation of process parameters is accomplished by assuming different sets of those parameters and evaluating the corresponding goodness of fit with observed data.

Various alternative forms of Neyman-Scott's model were compared by Vere-Jones with observed data on the basis of first- and second-order statistics: hazard functions, interval distributions (in the form of power spectra) and variance time curves. The statistical record comprises about one thousand New Zealand earthquakes with magnitudes greater than 4.5, recorded from 1942 to 1961. Figures 6.13—6.15 show results of the analysis for shallow New Zealand shocks as well as the comparison of observed data with sev-

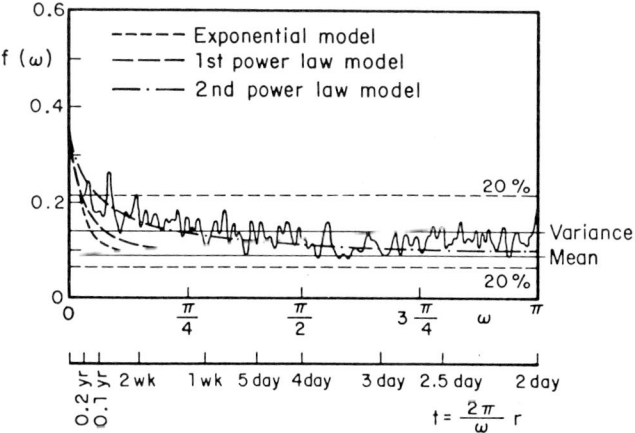

Fig. 6.14. Smoothed periodogram for New Zealand shallow shocks. (After Vere-Jones, 1966.)

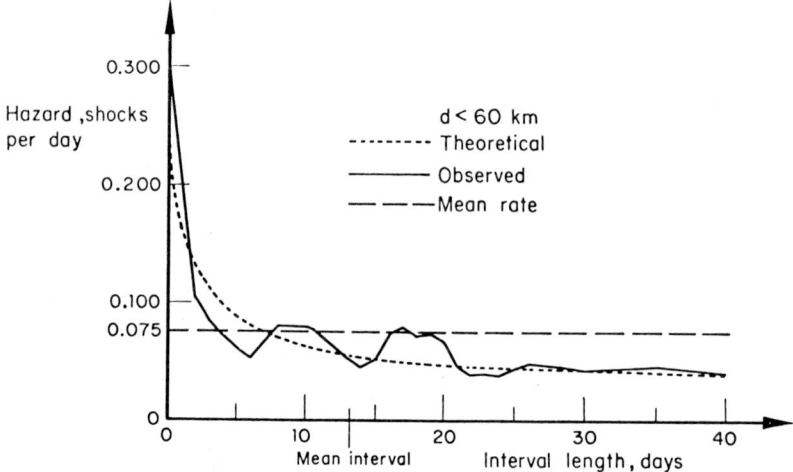

Fig. 6.15. Hazard function for New Zealand shallow shocks. (After Vere-Jones, 1970.)

eral alternative models. The process of cluster origins is Poisson in all cases, but the distributions of cluster sizes ($N$) and of times of events within clusters differ among the various instances: in the Poisson model no clustering takes place (the distribution of $N$ is a Dirac delta function centered at $N = 1$) while in the exponential and in the power-law models the distribution of $N$ is extremely skewed towards $N = 1$, and $\Lambda(t)$ is taken respectively as $1 - e^{-\lambda t}$

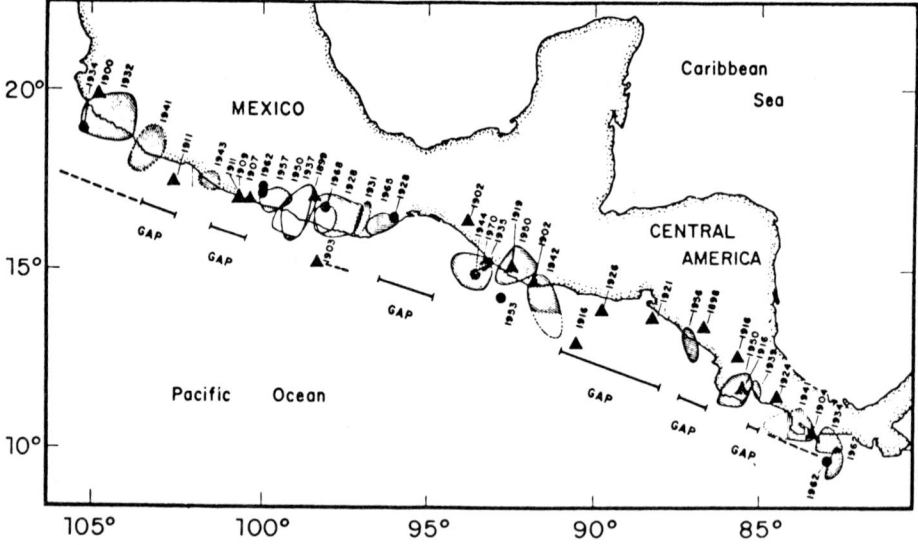

Fig. 6.16. Rupture zones and epicenters of large shallow Middle American earthquakes of this century. (After Kelleher et al., 1973.)

and $1 - [c/(c + t)]^\delta$ for $t \geqslant 0$, and as zero for $t < 0$, where $\lambda$, $c$, and $\delta$ are positive parameters. In Figs. 6.13—6.15, $\delta = 0.25$, $c = 2.3$ days, and $\lambda = 0.061$ shocks/day. The significance of clustering is evidenced by the high value of Poisson's dispersion index in Fig. 6.13, while no significant periodicity can be inferred from Fig. 6.14. Both figures show that the power-law model provides the best fit to the statistics of the samples. A similar analysis for New Zealand's deep shocks shows much less clustering: Poisson's dispersion index equals 2, and the hazard function is nearly constant with time.

Still, data reported by Gaisky (1967) have hazard functions that suggest models where the cluster origins as well as the clusters themselves may be represented by renewal processes. Mean return periods are of the order of several months, and hence these processes do not correspond, at least in the time scale, to the process of alternate periods of activity and quiescense of some geological structures cited by Kelleher et al. (1973), which have led to the concept of 'temporal seismic gaps', discussed below.

*Simplified trigger models.* Shlien and Toksöz (1970) proposed a simple particular case of the Neyman-Scott process; they lumped together all earthquakes taking place during non-overlapping time intervals of a given length and defined them as clusters for which $\lambda(t)$ was a Dirac delta function. Working with one-day intervals, they assumed the number of events per cluster to be distributed in accordance with the discrete Pareto law and applied a maximum-likelihood criterion to the information consisting of 35 000 earthquakes reported by the USCGS from January 1971 to August 1968. The model proposed represents reasonably well both the distribution of the number of earthquakes in one-day intervals and the dispersion index. However, owing to the assumption that no cluster lasts more than one day, the model fails to represent the autocorrelation function of the daily numbers of shocks for small time lags. The degree of clustering is shown to be a regional function, and to diminish with the magnitude threshold value and with the focal depth.

*Aftershock sequences.* The trigger processes described have been branded as reasonable representations of regional seismic activity, even when aftershock sequences and earthquake swarms are suppressed from statistical records, however arbitrary that suppression may be. The most significant instances of clustering are related, however, to aftershock sequences which often follow shallow shocks and only rarely intermediate and deep events. Persistence of large numbers of aftershocks for a few days or weeks has propitiated the detailed statistical analysis of those sequences since last century. Omori (1894) pointed out the decay in the mean rate of aftershock occurrence with $t$, the time elapsed since the main shock; he expressed that rate as inversely proportional to $t + q$, where $q$ is an empirical constant. Utsu (1961) proposed a more general expression, proportional to $(t + c)^{-\zeta}$ where $\zeta$ is a constant; Utsu's proposal is consistent with the power-law expression for $\Lambda(t)$ presented above.

Lomnitz and Hax (1966) proposed a clustering model to represent aftershock sequences; it is a modified version of Neyman and Scott's model, where the process of cluster origins is non-homogeneous Poisson with mean rate decaying in accordance with Omori's law, the number of events in each cluster has a Poisson distribution, and $\Lambda(t)$ is exponential. All the results and methods of analysis described by Vere-Jones (1970) for the stationary process of cluster origins can be applied to the nonstationary case through a transformation of the time scale. Fitting of parameters to four aftershock sequences was accomplished through use of the second-order information of the sample defined on a transformed time scale. By applying this criterion to earthquake sets having magnitudes above different threshold values it was noticed that the degree of clustering decreases as the threshold value increases.

The magnitude of the main shock influences the number of aftershocks and the distribution of their magnitudes and, although the rate of activity decreases with time, the distribution of magnitudes remains stable throughout each sequence (Lomnitz, 1966; Utsu, 1962; Drakopoulos, 1971). Equation 6.6 represents fairly well the distribution of magnitudes observed in most aftershock sequences. Values of $\beta$ range from 0.9 to 3.9 and decrease as the depth increases. Since values of $\beta$ for regular (main) earthquakes are usually estimated from relatively small numbers of shocks generated throughout crust volumes much wider than those active during aftershock sequences, no relation has been established among $\beta$-values for series of both types of events. The parameters of Utsu's expression for the decay of aftershock activity with time have been estimated for several sequences, for instance those following the Aleutian earthquake of March 9, 1957, the Central Alaska earthquake of April 7, 1958, and the Southeastern Alaska earthquake of July 10, 1958 (Utsu, 1962), with magnitudes equal to 8.3, 7.3, and 7.9, respectively; $c$ (in days) was 0.37, 0.40, and 0.01, while $\zeta$ was 1.05, 1.05 and 1.13, respectively. The relationship of the total number of aftershocks whose magnitude exceeds a given value with the magnitude of the main shock was studied by Drakopoulos (1971) for 140 aftershock sequences in Greece from 1912 to 1968. His results can be expressed by $N(M) = A \exp(-\beta M)$, where $N(M)$ is the total number of aftershocks with magnitude greater than $M$, and $A$ is a function of $M_0$, the magnitude of the main shock:

$$A = \exp(3.62 \beta + 1.1 M_0 - 3.46) \tag{6.18}$$

Formulation of stochastic process models for given earthquake sequences is feasible once this relationship and the activity decay law are available for the source of interest. For seismic-risk estimation at a given site the spatial distribution of aftershocks may be as significant as the distribution of magnitudes and the time variation of activity, particularly for sources of relatively large dimensions.

*6.3.3.3 Renewal process models*

The trigger models described are based on information about earthquakes with magnitudes above relatively low thresholds recorded during time intervals of at most ten years. The degrees of clustering observed and the distributions of times between clusters cannot be extrapolated to higher magnitude thresholds and longer time intervals without further study.

Available information shows beyond doubt that significant clustering is the rule, at least when dealing with shallow shocks. However, there is considerable ground for discussion on the nature of the process of cluster origins during intervals of the order of one century or longer. While lack of statistical data hinders the formulation of seismicity models valid over long time intervals, qualitative consideration of the physical processes of earthquake generation may point to models which at least are consistent with the state of knowledge of geophysical sciences. Thus, if strain energy stored in a region grows in a more or less systematic manner, the hazard function should grow with the time elapsed since the last event, and not remain constant as the Poisson assumption implies. The concept of a growing hazard function is consistent with the conclusions of Kelleher et al. (1973) concerning the theory of periodic activation of *seismic gaps*. This theory is partially supported by results of nearly qualitative analysis of the migration of seismic activity along a number of geological structures. An instance is provided by the southern coast of Mexico, one of the most active regions in the world. Large shallow shocks are generated probably by the interaction of the continental mass and the subductive oceanic Cocos plate that underthrusts it and by compressive or flexural failure of the latter (Chapter 2). Seismological data show significant gaps of activity along the coast during the present century and not much is known about previous history (Fig. 6.16). Along these gaps, seismic-risk estimates based solely on observed intensities are quite low, although no significant difference is evident in the geological structure of these regions with respect to the rest of the coast, save some transverse faults which divide the continental formation into several blocks. Without looking at the statistical records a geophysicist would assign equal risk throughout the area. On the basis of seismicity data, Kelleher et al. have concluded that activity migrates along the region, in such a manner that large earthquakes tend to occur at seismic gaps, thus implying that the hazard function grows with time since the last earthquake. Similar phenomena have been observed in other regions; of particular interest is the North Anatolian fault where activity has shifted systematically along it from east to west during the last forty years (Allen, 1969).

Conclusions relative to activation of seismic gaps are controversial because the observation periods have not exceeded one cycle of each process. Nevertheless, those conclusions point to the formulation of stochastic models of seismicity that reflect plausible features of the geophysical processes.

These considerations suggest the use of renewal-process models to represent sequences of individual shocks or of clusters. Such models are characterized because times between events are independent and identically distributed. The Poisson process is a particular renewal model for which the distribution of the waiting time is exponential. Wider generality is achieved, without much loss of mathematical tractability, if inter-event times are supposed to be distributed in accordance with a gamma function:

$$f_T(t) = \frac{\nu}{(k-1)!} (\nu t)^{k-1} e^{-\nu t} \tag{6.19}$$

which becomes the exponential distribution when $k = 1$. If $k < 1$, short intervals are more frequent and the coefficient of variation is greater than in the Poisson model; if $k > 1$, the reverse is true. Shlien and Toksöz (1970) found that gamma models were unable to represent the sequences of individual shocks they analyzed; but these authors handled time intervals at least an order of magnitude shorter than those referred to in this section.

On the basis of hazard function estimated from sequences of small shocks in the Hindu-Kush, Vere-Jones (1970) deduces the validity of 'branching renewal process' models, in which the intervals between cluster centers, as well as those between cluster members, constitute renewal processes.

Owing to the scarcity of statistical information, reliable comparisons between alternate models will have to rest partially on simulation of the process of storage and liberation of strain energy (Burridge and Knopoff, 1967; Veneziano and Cornell, 1973).

### 6.3.4 Influence of the seismicity model on seismic risk

Nominal values of investments made at a given instant increase with time when placing them at compound interest rates, i.e. when capitalizing them. Their real value — and not only the nominal one — will also grow, provided the interest rate overshadows inflation. Conversely, for the purpose of making design decisions, nominal values of expected utilities and costs inflicted upon in the future have to be converted into present or actualized values, which can be directly compared with initial expenditures. Descriptions of seismic risk at a site are insufficient for that purpose unless the probability distributions of the times of occurrence of different intensities — or magnitudes at neighbouring sources — are stipulated; this entails more than simple magnitude-recurrence graphs or even than maximum feasible magnitude estimates.

Immediately after the occurrence of a large earthquake, seismic risk is abnormally high due to aftershock activity and to the probability that damage inflicted by the main shock may have weakened natural or man-made structures if emergency measures are not taken in time. When aftershock activity has ceased and damaged systems have been repaired, a normal risk level is attained, which depends on the probability-density functions of the waiting times to the ensuing damaging earthquakes.

For the purpose of illustration, let it be assumed that a fixed and deterministically known damage $D_0$ occurs whenever a magnitude above a given value is generated at a given source. If $f(t)$ is the probability-density function of the waiting time to the occurrence of the damaging event, and if the risk level is sufficiently low that only the first failure is of concern, the expected value of the actualized cost of damage is (see Chapter 9):

$$\bar{D} = D_0 \int_0^\infty e^{-\gamma t} f(t) \, dt \qquad (6.20)$$

where $\gamma$ is the discount (or compound interest) coefficient and the overbar denotes expectation. If the process is Poisson with mean rate $\nu$, then $f(t)$ is exponential and $\bar{D} \cong D_0 \, \nu/\gamma$; however, if damaging events take place in clusters and most of the damage produced by each cluster corresponds to its first event, the computation of $\bar{D}$ should make use of the mean rate $\nu$ corresponding to the clusters, instead of that applicable to individual events. Table 6.II shows a comparison of seismic risk determined under the alternative assumptions of a Poisson and a gamma model ($k = 2$), both with the same mean return period, $k/\nu$ (Esteva, 1974). Three descriptions of risk are presented as functions of the time $t_0$ elapsed since the last damaging event: $T_1$, the expected time to the next event, measured from instant $t_0$; the expected value of the present cost of failure computed from eq. 6.20, and the hazard function (or mean failure rate). Since clustering is neglected, risk of aftershock occurrence must be either included in $D_0$ or superimposed on that displayed in the table.

This table shows very significant differences among risk levels for both processes. At small values of $t_0$, risk is lower for the gamma process, but it

TABLE 6.II

Comparison of Poisson and gamma processes

| $t_0 \nu/k$ | $\bar{T}_1 \nu/k$ | Poisson process, $k = 1$ $D/D_0$ | | $hk/\nu$ | $T_1 \nu/k$ | Gamma process, $k = 2$ $D/D_0$ | | $hk/\nu$ |
|---|---|---|---|---|---|---|---|---|
| | | $\gamma k/\nu = 10$ | $\gamma k/\nu = 100$ | | | $\gamma k/\nu = 10$ | $bk/\nu = 100$ | |
| 0   |     |        |        |     | 1.0  | 0.0278 | 0.0004 | 0     |
| 0.1 |     |        |        |     | 0.92 | 0.0511 | 0.0036 | 0.367 |
| 0.2 |     |        |        |     | 0.86 | 0.0675 | 0.0059 | 0.667 |
| 0.5 |     |        |        |     | 0.75 | 0.0973 | 0.0100 | 1.333 |
| 1   | 1.0 | 0.0909 | 0.0099 | 1.0 | 0.67 | 0.120  | 0.0132 | 2.000 |
| 2   |     |        |        |     | 0.60 | 0.139  | 0.0158 | 2.667 |
| 5   |     |        |        |     | 0.54 | 0.154  | 0.0179 | 3.333 |
| 10  |     |        |        |     | 0.52 | 0.160  | 0.0187 | 3.633 |
|     |     |        |        |     | 0.50 | 0.167  | 0.0196 | 4.000 |

grows with time, until it outrides that for the Poisson process, which remains constant. The differences shown clearly affect engineering decisions.

6.4 ASSESSMENT OF LOCAL SEISMICITY

Only exceptionally can magnitude-recurrence relations for small volumes of the earth's crust and statistical correlation functions of the process of earthquake generation be derived exclusively from statistical analysis of recorded shocks. In most cases this information is too limited for that purpose and it does not always reflect geological evidence. Since the latter, as well as its connection with seismicity, is beset with wide uncertainty margins, information of different nature has to be evaluated, its uncertainty analyzed, and conclusions reached consistent with all pieces of information. A probabilistic criterion that accomplishes this is presented here: on the basis of geotectonic data and of conceptual models of the physical processes involved, a set of alternate assumptions can be made concerning the functions in question (magnitude recurrence, time, and space correlation) and an initial probability distribution assigned thereto; statistical information is used to judge the likelihood of each assumption, and a posterior probability distribution is obtained. How statistical information contibutes to the posterior probabilities of the alternate assumptions depends on the extent of that information and on the degree of uncertainty implied by the initial probabilities. Thus, if geological evidence supports confidence in a particular assumption or range of assumptions, statistical information should not greatly modify the initial probabilities. If, on the other hand, a long and reliable statistical record is available, it practically determines the form and parameters of the mathematical model selected to represent local seismicity.

*6.4.1 Bayesian estimation of seismicity*

Bayesian statistics provide a framework for probabilistic inference that accounts for prior probabilities assigned to a set of alternate hypothetical models of a given phenomenon as well as for statistical samples of events related to that phenomenon. Unlike conventional methods of statistical inference, Bayesian methods give weight to probability measures obtained from samples or from other sources; numbers, coordinates and magnitudes of earthquakes observed in given time intervals serve to ascertain the probable validity of each of the alternative models of local seismicity that can be postulated on the grounds of geological evidence. Any criterion intended to weigh information of different nature and different degrees of uncertainty should lead to probabilistic conclusions consistent with the degree of confidence attached to each source of information. This is accomplished by Bayesian methods.

Let $H_i$ ($i = 1, ..., n$) be a comprehensive set of mutually exclusive assumptions concerning a given, imperfectly known phenomenon and let $A$ be the observed outcome of such a phenomenon. Before observing outcome $A$ we assign an initial probability $P(H_i)$ to each hypothesis. If $P(A|H_i)$ is the probability of $A$ in case hypothesis $H_i$ is true, then Bayes' theorem (Raiffa and Schlaifer, 1968) states that:

$$P(H_i|A) = P(H_i) \frac{P(A|H_i)}{\Sigma_j P(H_j) P(A|H_j)} \tag{6.21}$$

The first member in this equation is the (posterior) probability that assumption $H_i$ is true, given the observed outcome $A$.

In the evaluation of seismic risk, Bayes' theorem can be used to improve initial estimates of $\lambda(M)$ and its variation with depth in a given area as well as those of the parameters that define the shape of $\lambda(M)$ or, equivalently, the conditional distribution of magnitudes given the occurrence of an earthquake. For that purpose, take $\lambda(M)$ as the product of a rate function $\lambda_L = \lambda(M_L)$ by a shape function $G^*(M,B)$, equal to the conditional complementary distribution of magnitudes given the occurrence of an earthquake with $M \geq M_L$, where $M_L$ is the magnitude threshold of the set of statistical data used in the estimation, and $B$ is the vector of (uncertain) parameters $B_1, ..., B_r$ that define the shape of $\lambda(M)$. For instance, if $\lambda(M)$ is taken as given by eq. 6.8, $B$ is a vector of three elements equal respectively to $\beta, \beta_1$, and $M_U$; if eq. 6.9 is adopted, $B$ is defined by $k$ and $M_U$.

The initial distribution of seismicity is in this case expressed by the initial joint probability density function of $\lambda_L$ and $B$: $f'(\lambda_L, B)$. The observed outcome $A$ can be expressed by the magnitudes of all earthquakes generated in a given source during a given time interval. For instance, suppose that $N$ earthquakes were observed during time interval $t$ and that their magnitudes were $m_1, m_2, ..., m_N$. Bayes' expression takes the form:

$$f''(\lambda_L, B|m_1, ..., m_N; t) = f'(\lambda_L, B) \frac{P[m_1, m_2, ..., m_N; t|\lambda_L, B]}{\int\int P[m_1, m_2, ..., m_N; t|l, b] f'(l,b) dl db} \tag{6.22}$$

where $f''(.)$ is the posterior probability density function, and $l$ and $b$ are dummy variables that stand for all values that may be taken by $\lambda_L$ and $B$, respectively. Estimation of $\lambda_L$ can usually be formulated independently of that of the other parameters. The observed fact is then expressed by $N_L$, the number of earthquakes with magnitude above $M_L$ during time $t$, and the following expression is obtained, as a first step in the estimation of $\lambda(M)$:

$$f''(\lambda_L|N_L; t) = f'(\lambda_L) \frac{P(N_L; t|\lambda_L)}{\int P(N_L; t|l) f'(l) dl} \tag{6.23}$$

### 6.4.1.1 Initial probabilities of hypothetical models

Where statistical information is scarce, seismicity estimates will be very sensitive to initial probabilities assigned to alternative hypothetical models; the opinions of geologists and geophysicists about probable models, about the parameters of these models, and the corresponding margins of uncertainty should be adequately interpreted and expressed in terms of a function $f'$, as required by equations similar to 6.22 and 6.23. Ideally, these opinions should be based on the formulation of potential earthquake sources and on their comparison with possibly similar geotectonic structures. This is usually done by geologists, more qualitatively than quantitatively, when they estimate $M_U$. Initial estimates of $\lambda_L$ are seldom made, despite the significance of this parameter for the design of moderately important structures (see Chapter 9).

Analysis of geological information must consider local details as well as general structure and evolution. In some areas it is clear that all potential earthquake sources can be identified by surface faults, and their displacements in recent geological times measured. When mean displacements per unit time can be estimated, the order of magnitude of creep and of energy liberated by shocks and hence of the recurrence intervals of given magnitudes can be established (Wallace, 1970; Davies and Brune, 1971), the corresponding uncertainty evaluated, and an initial probability distribution assigned. The fact that magnitude-recurrence relations are only weakly correlated with the size of recent displacements is reflected in large uncertainties (Petrushevsky, 1966).

Application of the criterion described in the foregoing paragraph can be unfeasible or inadequate in many problems, as in areas where the abundance of faults of different sizes, ages, and activity, and the insufficient accuracy with which focal coordinates are determined preclude a differentiation of all sources. Regional seismicity may then be evaluated under the assumption that at least part of the seismic activity is distributed in a given volume rather than concentrated in faults of different importance. The same situation would be faced when dealing with active zones where there is no surface evidence of motions. Hence, consideration of the overall behavior of complex geological structures is often more significant than the study of local details.

Not much work has been done in the analysis of the overall behavior of large geological structures with respect to the energy that can be expected to be liberated per unit volume and per unit time in given portions of those structures. Important research and applications should be expected, however, since, as a result of the contribution of plate-tectonics theory to the understanding of large-scale tectonic processes, the numerical values of some of the variables correlated with energy liberation are being determined, and can be used at least to obtain orders of magnitude of expected activity along plate boundaries. Far less well understood are the occurrence of shocks in apparently inactive regions of continental shields and the behavior of complex continental blocks or regions of intense folding, but even there some progress is expected in the study of accumulation of stresses in the crust.

Knowledge of the geological structure can serve to formulate initial probability distributions of seismicity even when quantitative use of geophysical information seems beyond reach. Initial probability distributions of local seismicity parameters $\lambda_L$, $B$ in the small volumes of the earth's crust that contribute significantly to seismic risk at a site, can be assigned by comparison with the average seismicity observed in wider areas of similar tectonic characteristics, or where the extent and completeness of statistical information warrant reliable estimates of magnitude-recurrence curves (Esteva, 1969). In this manner we can, for instance, use the information about the average distribution of the depths of earthquakes of different magnitudes throughout a seismic province to estimate the corresponding distribution in an area of that province, where activity has been low during the observation interval, even though there might be no apparent geophysical reason to account for the difference. Similarly, the expected value and coefficient of variation of $\lambda_L$ in a given area of moderate or low seismicity (as a continental shield) can be obtained from the statistics of the motions originated at all the supposedly stable or aseismic regions in the world.

The significance of initial probabilities in seismic risk estimates, against the weight given to purely statistical information, becomes evident in the example of Fig. 6.16: if Kelleher's theory about activation of seismic gaps is true, risk is greater at the gaps than anywhere else along the coast; if Poisson models are deemed representative of the process of energy liberation, the extent of statistical information is enough to substantiate the hypothesis of reduced risk at gaps. Because both models are still controversial, and represent at most two extreme positions concerning the properties of the actual process, risk estimates will necessarily reflect subjective opinions.

### 6.4.1.2 Significance of statistical information

*Estimation of $\lambda_L$*. Application of eq. 6.23 to estimate $\lambda_L$ independently of other parameters will be first discussed, because it is a relatively simple problem and because $\lambda_L$ is usually more uncertain than $M_U$ and much more so than $\beta$.

A model as defined by eq. 6.19 will be assumed to apply. If the possible assumptions concerning the values of $\lambda_L$ constitute a continuous interval, the initial probabilities of the alternative hypotheses can be expressed in terms of a probability-density function of $\lambda_L$. If, in addition, a certain assumption is made concerning the form of this probability-density function, only the initial values of $E(\lambda_L)$ and $V(\lambda_L)$ have to be assumed. It is advantageous to assign to $\nu = k/E(T)$ a gamma distribution. Then, if $\rho$ and $\mu$ are the parameters of this initial distribution of $\nu$, if $k$ is assumed to be known, and if the observed outcome is expressed as the time $t_n$ elapsed during $n + 1$ consecutive events (earthquakes with magnitude $\geqslant M_L$), application of eq. 6.23 leads to the conclusion that the posterior probability function of $\nu$ is also gamma, now with parameters $\rho + nk$ and $\mu + t_n$. The initial and the posterior expected values of $\nu$ are respectively equal to $\rho/\mu$, and to $(\rho + nk)/(\mu + t_n)$. When initial uncertainty about $\nu$ is small, $\rho$ and $\mu$ will be large and the initial and the posterior expected values of $\nu$ will not differ greatly. On the other hand, if only statistical information were deemed significant, $\rho$ and $\mu$ should be given very small values in the initial distribution, and $E(\nu)$, and hence $\lambda_L$, will be practically defined by $n$, $k$, and $t_n$. This means that the initial estimates of geologists should not only include expected or most probable values of the different parameters, but also statements about ranges of possible values and degrees of confidence attached to each.

In the case studied above only a portion of the statistical information was used. In most cases, especially if seismic activity has been low during the observation interval, significant information is provided by the durations of the intervals elapsed from the initiation of observations to the first of the $n + 1$ events considered, and from the last of these events until the end of the observation period. Here, application of eq. 6.23 leads to expressions slightly more complicated than those obtained when only information about $t_n$ is used.

The particular case when the statistical record reports no events during *at least* an interval $(0, t_0)$ comes up frequently in practical problems. The probability-density function of the time $T_1$ from $t_0$ to the occurrence of the first event must account for the corresponding shifting of the time axis. Furthermore, if the time of occurrence of the last event before the origin is unknown, the distribution of the waiting time from $t = 0$ to the first event coincides with that of the *excess life* in a renewal process at an arbitrary value of $t$ that approaches infinity (Parzen, 1962). For the particular case when the waiting times constitute a gamma process, $T_1$ is measured from $t = 0$, $T$ is the waiting time between consecutive events, and it is known that $T_1 \geq t_0$, the conditional density function of $\tau_1 = (T_1 - t_0)/E(T)$ is given by eq. 6.24 (Esteva, 1974), where $u_0 = t_0/E(T)$:

$$f_{\tau_1}(u|T_1 \geq t_0) = \frac{\sum_{m=1}^{k} \frac{k}{(m-1)!} [k(u + u_0)]^{m-1}}{\sum_{m=1}^{k} \sum_{n=1}^{m} \frac{1}{(n-1)!} (ku_0)^{n-1}} e^{-ku} \qquad (6.24)$$

Consider now the implications of Bayesian analysis when applied to one of the seismic gaps in Fig. 6.16, under the conditions implicit in eq. 6.24. An initial set of assumptions and corresponding probabilities was adopted as described in the following. From previous studies referring to all the southern coast of Mexico, local seismicity in the gap area (measured in terms of $\lambda$ for $M \geq 6.5$) was represented by a gamma process with $k = 2$. An initial probability density function for $\nu$ was adopted such that the expected value of $\lambda(6.5)$ for the region coincided with its average throughout the complete seismic province. Two values of $\rho$ were considered: 2 and 10, which correspond to coefficients of variation of 0.71 and 0.32, respectively. Values in Table 6.III were obtained for the ratio of the final to the initial expected values of $\nu$, in terms of $u_0$.

The last two columns in the table contain the ratios of the computed values of $E''(T_1)$ and $E'(T)$ when $\nu$ is taken as equal respectively to its initial or to its posterior expected value. This table shows that, for $\rho = 10$, that is, when uncertainty attached to the geologically based assumptions is low, the expected value of the time to the next event keeps decreasing, in accordance with the conclusions of Kelleher et al. (1973). However, as time goes on and no events occur, the statistical evidence leads to a reduction in the estimated risk, which shows in the increased conditional expected values of $T_1$. For $\rho = 2$, the geological evidence is less significant and risk estimates decrease at a faster rate.

### 6.4.1.3 Bayesian estimation of jointly distributed parameters

In the general case, estimation of $B$ will consist in the determination of the posterior Bayesian joint probability function of its components, taking as statistical evidence the relative frequencies of observed magnitudes. Thus, if event $A$ is described as the occurrence of $N$ shocks, with magnitudes $m_1, ..., m_N$, and $b_i$ ($i = 1, ..., r$) are values that may be adopted by the components of vector $B$ being estimated, eq. 6.21 becomes:

$$f_B''(b_1, ..., b_r | A) = \frac{f_B'(b_1, ..., b_r) P(A | b_1, ..., b_r)}{\int ... \int f_B'(u_1, ..., u_r) P(A | u_1, ..., u_r) du_1, ..., du_r} \quad (6.25)$$

where $P(A | u_1, ..., u_r)$ is proportional to:

$$\prod_{i=1}^{N} g(m_i | u_1, ..., u_r)$$

and $g(m) = -\partial G^*(m)/\partial m$.

Closed-form solutions for $f''$ as given by eq. 6.25 are not feasible in general. For the purpose of evaluating risk, however, estimates of the posterior first and second moments of $f''$ can be obtained from eq. 6.25, making use of available first-order approximations (Benjamin and Cornell, 1970; Rosenblueth, 1975). Thus, the posterior expected value of $B_i$ is given by $\int f_{B_i}''(u) \, u \, du$, where $f_{B_i}''(u_i) = \int ... \int f_B''(u_1, ..., u_r) \, du_1, ..., du_n$ and the multiple integral is of order $r - 1$, because it is not extended to the dominion of $B_i$. Hence:

$$E''(B_i) = \frac{E_B'[B_i P(A | B_1, ..., B_r)]}{E_B'[P(A | B_1, ..., B_r)]} \quad (6.26)$$

TABLE 6.III

Bayesian estimates of seismicity in one seismic gap

| $u_0 = t_0/E'(T)$ | $E''(\nu)/E'(\nu)$ | | $E''(T_1\|T_1 \geq t_0)/E'(T)$ | |
|---|---|---|---|---|
| | $\rho = 2$ | $\rho = 10$ | $\rho = 2$ | $\rho = 10$ |
| 0 | 1.0 | 1.0 | 0.75 | 0.75 |
| 0.1 | 0.95 | 0.99 | 0.76 | 0.74 |
| 0.5 | 0.75 | 0.94 | 0.91 | 0.71 |
| 1 | 0.58 | 0.87 | 1.14 | 0.73 |
| 5 | 0.20 | 0.54 | 3.11 | 1.05 |
| 10 | 0.11 | 0.36 | 5.47 | 1.55 |
| 20 | 0.06 | 0.22 | 10.50 | 2.48 |

where $E'$ and $E''$ stand for initial and posterior expectation, and subscript $B$ means that expectation is taken with respect to all the components of $B$. Likewise, the following *posterior moments* can be obtained:

Covariance of $B_i$ and $B_j$

$$\text{Cov}''(B_i, B_j) = \frac{E'_B[B_i B_j P(A|B_1, ..., B_r)]}{E'_B[P(A|B_1, ..., B_r)]} - E''(B_i)E''(B_j) \qquad (6.27)$$

Expected value of $\lambda(M)$

$$E''[\lambda(M)] = E''(\lambda_1)E''[G^*(M; B)]$$

$$= E''(\lambda_1) \frac{E'_B[G^*(M; B)P(A|B_1, ..., B_r)]}{E'_B[P(A|B_1, ..., B_r)]} \qquad (6.28)$$

*Marginal distributions.* The posterior expectation of $\lambda(M)$ is in some cases all that is required to describe seismicity for decision-making purposes. Often, however, uncertainty in $\lambda(M)$ must also be acounted for. For instance, the probability of exceedance of a given magnitude during a given time interval has to be obtained as the expectation of the corresponding probabilities over all alternative hypotheses concerning $\lambda(M)$. In this manner it can be shown that, if the occurrence of earthquakes is a Poisson process and the Bayesian distribution of $\lambda_L$ is gamma with mean $\bar{\lambda}_L$ and coefficient of variation $V_L$, the marginal distribution of the number of earthquakes is negative binomial with mean $\bar{\lambda}_L$. In particular, the marginal probability of zero events during time interval $t$ — equivalently, the complementary distribution function of the waiting time between events — is equal to $(1 + t/t'')^{-r''}$, where $r'' = V_L^{-2}$ and $t'' = r''/\bar{\lambda}_L$. The marginal probability-density function of the waiting time, that should be substituted in eq. 6.20, is $\bar{\lambda}_L(1 + t/t'')^{-r''-1}$, which tends to the exponential probability function as $r''$ and $t''$ tend to infinity (and $V_L \to 0$) while their ratio remains equal to $\bar{\lambda}_L$.

Bayesian uncertainty tied to the joint distribution of all seismicity parameters ($\lambda_L, B_1, ..., B_r$) can be included in the computation of the probability of occurrence of a given event $Z$ by taking the expectation of that probability with respect to all parameters:

$$P(Z) = E_{\lambda_L, B}[P(Z); \lambda_L, B_1, ..., B_r)] \tag{6.29}$$

When the joint distribution of $\lambda_L, B$ stems from Bayesian analysis of an initial distribution and an observed event, $A$, this equation adopts the form:

$$P''(Z) = \frac{E'_{\lambda_L, B}[P(Z|\lambda_L, B)P(A|\lambda_L, B)]}{E'_{\lambda_L, B}[P(A|\lambda_L, B)]} \tag{6.30}$$

where ' and " stand for initial and posterior, respectively.

*Spatial variability.* Figure 6.17 shows a map of geotectonic provinces of Mexico, according to F. Mooser. Each province is characterized by the large-scale features of its tectonic structure, but significant local perturbations to the overall patterns can be identified. Take for instance zone 1, whose seismotectonic features were described above, and are schematically shown in Fig. 6.18 (Singh, 1975): the Pacific plate underthrusts the continental block and is thought to break into several blocks, separated by faults transverse to the coast, that dip at different angles. The continental mass is also

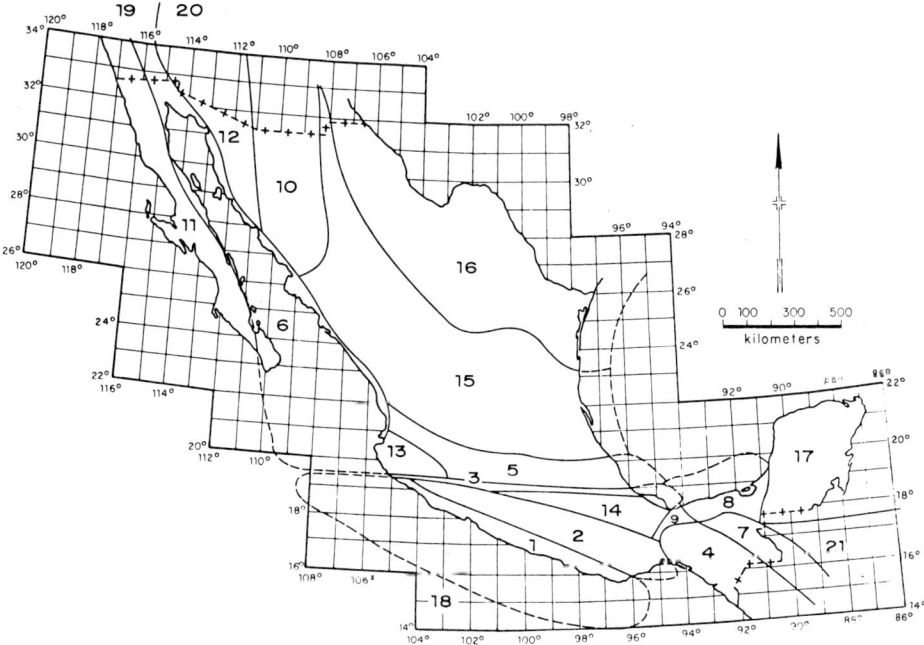

Fig. 6.17. Seismotectonic provinces of Mexico. (After F. Mooser.)

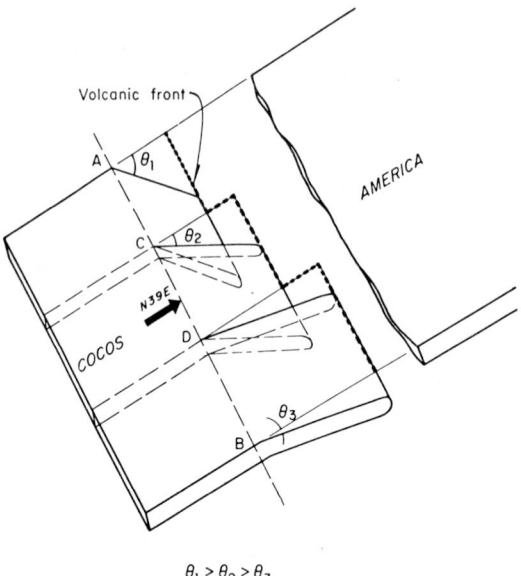

$\theta_1 > \theta_2 > \theta_3$

Fig. 6.18. Schematic drawing of the segmenting of Cocos plate as it subducts below American plate. (After Singh, 1974.)

made up of several large blocks. Seismic activity at the underthrusting plate or at its interface with the continental mass is characterized by magnitudes that may reach very high values and by the increase of mean hypocentral depth with distance from the coast; small and moderate shallow shocks are generated at the blocks themselves. Variability of statistical data along the whole tectonic system was discussed above and is apparent in Fig. 6.10. Bayesian estimation of local seismicity averaged throughout the system is a matter of applying eq. 6.21 or any of its special forms (eqs. 6.22 and 6.23), taking as statistical evidence the information corresponding to the whole system. However, seismic risk estimates are sensitive to values of local seismicity averaged over much smaller volumes of the earth's crust; hence the need to develop criteria for probabilistic inference of possible patterns of space variability of seismicity along tectonically homogeneous zones.

On the basis of seismotectonic information, the system under consideration can first be subdivided into the underthrusting plate and the subsystem of shallow sources; each subsystem can then be separately analyzed. Take for instance the underthrusting plate and subdivide it into $s$ sufficiently small equal-volume subzones. Let $\nu_L$ be the rate of exceedance of magnitude $M_L$ throughout the main system, $\nu_{L_i}$ the corresponding rate at each subzone, and define $p_i$ as $\nu_{L_i}/\nu_L$, with $p_i$ independent of $\nu_L$ ($p_i$ is equal to the probability that an earthquake known to have been generated in the overall system originated at subzone $i$). Initial information about possible space variability of

$\nu_{L_i}$ can be expressed in terms of an initial probability distribution of $p_i$ and of the correlation among $p_i$ and $p_j$ for any $i$ and $j$. Because $\Sigma \nu_{L_i} = \nu_L$, one obtains $\Sigma p_i = 1$. This imposes two restrictions on the initial joint probability distribution of the $p_i'$s: $E'(p_i) = 1$, var$' \Sigma p_i = 0$. If all $p_i'$s are assigned equal expectations and all pairs $p_i, p_j, i \neq j$ are assumed to possess the same correlation coefficient $\rho_{ij} = \rho'$, the restrictions mentioned lead to $E'(p_i) = 1/s$ and $\rho' = -1/(s-1)$. Posterior values of $E(p_i)$ and $\rho_{ij}$ are obtained according to the same principles that led to eqs. 6.25—6.28. Statistical evidence is in this case described by $N$, the total number of earthquakes generated in the system, and $n_i$ ($i = 1, ..., s$) the corresponding numbers for the subzones. Given the $p_i'$s, the probability of this event is the multinomial distribution:

$$P[A|p_1, ..., p_s] = \frac{N!}{n_1!, ..., n_s!} p_1^{n_1} ... p_s^{n_s} \tag{6.31}$$

If the correlation coefficients among seismicities of the various subzones can be neglected, each $p_i$ can be separately estimated. Because $p_i$ has to be comprised between 0 and 1, it is natural to assign it a beta initial probability distribution, defined by its parameters $n_i'$ and $N_i'$, such that $E'(p_i) = n_i'/N_i'$ and var$'(p_i) = n_i'(N_i' - n_i')/[N_i'^2(N_i' + 1)]$ (Raiffa and Schlaifer, 1968). The parameters of the posterior distribution will be:

$$n_i'' = n_i' + n_i, N_i'' = N_i' + N$$

Take for instance a zone whose prior distribution of $\lambda_L$ is assumed gamma with expected value $\lambda_L'$ and coefficient of variation $V_L'$. Suppose that, on the basis of geological evidence and of the dimensions involved, it is decided to subdivide the zone into four subzones of equal dimensions; a-priori considerations lead to the assignment of expected values and coefficients of variation of $p_i$ for those subzones, say $E'(p_i) = 0.25$, $V'(p_i) = 0.25$ ($i = 1, ..., 4$). From previous considerations for $s = 4$ take $\rho_{ij}' = -1/3$ for $i \neq j$. Suppose now that, during a given time interval $t$, ten earthquakes were observed in the zone, of which 0, 1, 3, and 6 occurred respectively in each subzone. If the Poisson process model is adopted, $\lambda_L'$ and $V_L'$ can be expressed in terms of a fictitious number of events $n' = V_L'^{-2}$ occurred during a fictitious time interval $t' = n'/\lambda_L'$; after observing $n$ earthquakes during an interval $t$, the Bayesian mean and coefficient of variation of $\lambda_L$ will be $\lambda_L'' = (n' + n)/(t' + t)$, $V_L'' = (n' + n)^{-1/2}$ (Esteva, 1968). Hence:

$$\lambda_L'' = (V_L'^{-2} + 10)/(V_L'^{-2} \lambda_L'^{-1} + t), \quad V_L'' = (V_L'^{-2} + 10)^{-1/2}$$

Local deviations of seismicity in each subzone with respect to the average $\lambda_L$ can be analyzed in terms of $p_i$ ($i = 1, ..., 4$); Bayesian analysis of the proportion in which the ten earthquakes were distributed among the subzones proceeds according to:

$$E''(p_i|A) = \frac{E'[p_i P(A|p_i, ..., p_4)]}{E'[P(A|p_i, ..., p_4)]} \tag{6.32}$$

The expectations that appear in this equation have to be computed with respect to the initial joint distribution of the $p_i$'s. In practice, adequate approximations are required. For instance, Benjamin and Cornells' (1970) first-order approximation leads to $E''(p_1) = 0.226$, $E''(p_4) = 0.294$.

If correlation among subzone seismicities is neglected, and statistical information of each subzone is independently analyzed, when the $p_i$'s are assigned beta probability-density functions with means and coefficients of variation as defined above, one obtains $E''(p_1) = 0.206$, $E''(p_4) = 0.311$, which are not very different from those formerly obtained; however, when $E'(p_i) = 0.25$ and $V'(p_i) = 0.5$, the first criterion leads to $E''(p_i) = 0.206$, $E''(p_4) = 0.314$, while the second produces 0.131 and 0.416, respectively. Part of the difference may be due to neglect of $\rho'_{ij}$, but probably a significant part stems from inaccuracies of the first-order approximation to the expectations that appear in eq. 6.32; alternate approximations are therefore desirable.

*Incomplete data.* Statistical information is known to be fairly reliable only for magnitudes above threshold values that depend on the region considered, its level of activity, and the quality of local and nearby seismic instrumentation. Even incomplete statistical records may be significant when evaluating some seismicity parameters; their use has to be accompanied by estimates of detectability values, that is, of ratios of the numbers of events recorded to total numbers of events in given ranges (Esteva, 1970; Kaila and Narain, 1971).

## 6.5 REGIONAL SEISMICITY

The final goal of local seismicity assessment is the estimation of regional seismicity, that is, of probability distributions of intensities at given sites, and of probabilistic correlations among them. These functions are obtained by integrating the contributions of local seismicities of nearby sources, and hence their estimates reflect Bayesian uncertainties tied to those seismicities. In the following, regional seismicity will be expressed in terms of mean rates of exceedance of given intensities; more detailed probabilistic descriptions would entail adoption of specific hypotheses concerning space and time correlations of earthquake generation.

### 6.5.1 Intensity-recurrence curves

The case when uncertainty in seismicity parameters is neglected will be discussed first. Consider an elementary seismic source with volume $dV$ and local seismicity $\lambda(M)$ per unit volume, distant $R$ from a site $S$, where intensity-recurrence functions are to be estimated. Every time that a magnitude $M$ shock is generated at that source, the intensity at $S$ equals:

$$Y = \epsilon Y_p = \epsilon b_1 \exp(b_2 M) g(R) \tag{6.33}$$

(see eqs. 6.4 and 6.5), where $\epsilon$ is a random factor and $Y$ and $Y_p$ stand for actual and predicted intensities, $b_1$ and $b_2$ are given constants, and $g(R)$ is a function of hypocentral distance. The probability that an earthquake originating at the source will have an intensity greater than $y$ is equal to the probability that $\epsilon Y_p > y$. If $Y_p$ is expressed in terms of $M$ and randomness in $\epsilon$ is accounted for, one obtains:

$$\nu(y) = \int_{\alpha_u}^{\alpha_L} \nu_p(y/u) f_\epsilon(u) du \tag{6.34}$$

where $\nu$ and $\nu_p$ are respectively mean rates at which actual and predicted intensities exceed given values, $\alpha_U = y/y_U$, $\alpha_L = y/y_L$, $y_U$, and $y_L$ are the predicted intensities that correspond to $M_U$ and $M_L$, and $f_\epsilon$ the probability-density function of $\epsilon$. If eq. 6.33 is assumed to hold:

$$\nu_p(y) = K_0 + K_1 y^{-r_1} - K_2 y^{-r_2} \tag{6.35}$$

where:

$$K_i = [b_1 g(R)]^{r_i} A_i \lambda_L dV \quad (i = 0, 1, 2) \tag{6.36}$$

$$r_0 = 0, \quad r_1 = \beta/b_2, \quad r_2 = (\beta - \beta_1)/b_2 \tag{6.37}$$

Substitution of eq. 6.35 into 6.34, coupled with the assumption that $\ln \epsilon$ is normally distributed with mean $m$ and standard deviation $\sigma$ leads to:

$$\nu(y) = c_0 K_0 + c_1 K_1 y^{-r_1} - c_2 K_2 y^{-r_2} \tag{6.38}$$

where:

$$c_i = \exp(Q_i) \left[ \phi\left(\frac{\ln \alpha_L - u_i}{\sigma}\right) - \phi\left(\frac{\ln \alpha_U - u_i}{\sigma}\right) \right] \tag{6.39}$$

$\phi$ is the standard normal cumulative distribution function, $Q_i = 1/2 \, \sigma^2 r_i^2 + m r_i$, and $u_i = m + \sigma^2 r_i$. Similar expressions have been presented by Merz and Cornell (1973) for the special case of eq. 6.8 when $\beta_1 \to \infty$ and for a quadratic form of the relation between magnitude and logarithm of exceedance rate. Closed-form solutions in terms of incomplete gamma functions are obtained when magnitudes are assumed to possess extreme type-III distributions (eq. 6.9).

Intensity-recurrence curves at given sites are obtained by integration of the contributions of all significant sources. Uncertainties in local seismicities can be handled by describing regional seismicity in terms of means and variances of $\nu(y)$ and estimating these moments from eq. 6.34 and suitable first- and second-moment approximations. Influence of these uncertainties in design decisions has been discussed by Rosenblueth (in preparation).

## 6.5.2 Seismic probability maps

When intensity-recurrence functions are determined for a number of sites with uniform local ground conditions the results are conveniently represented by sets of seismic probability maps, each map showing contours of intensities that correspond to a given return period. For instance, Figs. 6.19 and 6.20 show peak ground velocities and accelerations that correspond to 100 years return period on firm ground in Mexico. These maps form part of a set that was obtained through application of the criteria described in this chapter. Because the ratio of peak ground accelerations and velocities does not remain constant throughout a region, the corresponding design spectra will not only vary in scale but also in shape (frequency content); in other words, seismic risk will usually have to be expressed in terms of at least the values of two parameters (for instance, as in this case, peak ground accelerations and velocities that correspond to various risk levels (return periods)).

## 6.5.3 Microzoning

Implicit in the above criteria for evaluation of regional seismicity is the adoption of intensity attenuation expressions valid on firm ground. Scatter of actual intensities with respect to predicted values was ascribed to differences in source mechanisms, propagation paths, and local site conditions; at least the latter group of variables can introduce systematic deviations in the

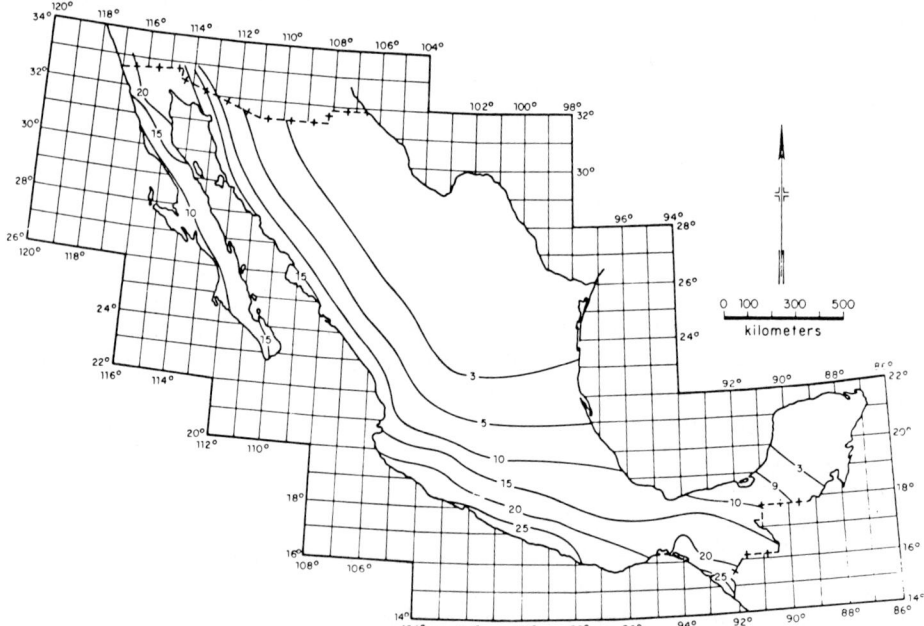

Fig. 6.19. Peak ground velocities with return period of 100 years (cm/sec).

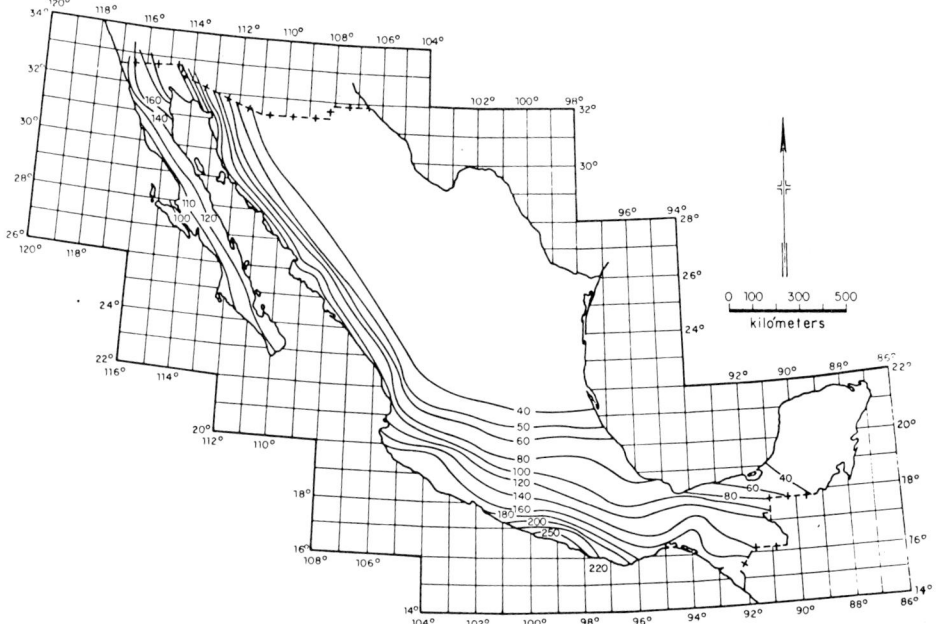

Fig. 6.20. Peak ground accelerations with return period of 100 years (cm/sec$^2$).

ratio of actual to predicted intensities; and geological details may significantly alter local seismicity in a small region, as well as energy radiation patterns, and hence regional seismicity in the neighbourhood. These systematic deviations are the matter of microzoning, that is, of local modification of risk maps similar to Figs. 6.19 and 6.20.

Most of the effort invested in microzoning has been devoted to study of the influence of local soil stratigraphy on the intensity and frequency content of earthquakes (see Chapter 4). Analytical models have been practically limited to response analysis of stratified formations of linear or nonlinear soils to vertically traveling shear waves. The results of comparing observed and predicted behavior have ranged from satisfactory (Herrera et al., 1965) to poor (Hudson and Udwadia, 1972). Topographic irregularities, as hills or slopes of firm ground formations underlying sediments, may introduce significant systematic perturbations in the surface motion, as a consequence of wave focusing or dynamic amplification. The latter effect was probably responsible for the exceptionally high accelerations recorded at the abutment of Pacoima dam during the 1971 San Fernando earthquake.

Present practice of microzoning determines seismic intensities or design parameters in two steps. First the values of those parameters on firm ground are estimated by means of suitable attenuation expressions and then they are amplified according to the properties of local soil; but this implies an arbitrary decision to which seismic risk is very sensitive: selecting the boundary between soil and firm ground. A specially difficult problem stems when trying to fix that boundary for the purpose of predicting the motion at the top of a hill or the slope stability of a high cliff (Rukos, 1974).

It can be concluded that rational formulation of microzoning for seismic risk is still in its infancy and that new criteria will appear that will probably require intensity attenuation models which include the influence of local systematic perturbations. Whether these models are available or the two-step process described above is acceptable, intensity-recurrence expressions can be obtained as for the unperturbated case, after multiplying the second member of eq. 6.34 by an adequate intensity-dependent corrective factor.

REFERENCES

Aki, K., 1963. *Some Problems in Statistical Seismology.* University of Tokyo, Geophysical Institute.

Allen, C.R., 1969. Active faulting in northern Turkey. *Calif. Inst. Tech., Div. Geol. Sci., Contrib.* 1577.

Allen, C.R., St. Amand, P., Richter, C.F. and Nordquist, J.M., 1965. Relationship between seismicity and geologic structure in the southern California region. *Bull. Seismol. Soc. Am.,* 55 (4): 753—797.

Ambraseys, N.N., 1973. Dynamics and response of foundation materials in epicentral regions of strong earthquakes. *Proc. 5th World Conf. Earthquake Eng., Rome.*

Ananiin, I.V., Bune, V.I., Vvedenskaia, N.A., Kirillova, I.V., Reisner, G.I. and Sholpo, V.N., 1968. *Methods of Compiling a Map of Seismic Regionalization on the Example of the Caucasus.* C. Yu. Schmidt Institute of the Physics of the Earth, Academy of Sciences of the USSR, Moscow.

Benjamin, J.R. and Cornell, C.A., 1970. *Probability, Statistics and Decision for Civil Engineers.* McGraw-Hill, New York.

Ben-Menahem, A., 1960. Some consequences of earthquake statistics for the years 1918—1955. *Gerlands Beitr. Geophys.,* 69: 68—72

Bollinger, G.A., 1973. Seismicity of the southeastern United States. *Bull. Seismol. Soc. Am.,* 63: 1785—1808.

Bolt, B.A., 1970. Causes of earthquakes. In: R.L. Wiegel (editor), *Earthquake Engineering.* Prentice-Hall, Englewood Cliffs.

Brune, J.N., 1968. Seismic moment, seismicity and rate of slip along major fault zones. *J. Geophys. Res.,* 73: 777—784.

Burridge, R. and Knopoff, L., 1967. Model and theoretical seismicity. *Bull. Seismol. Soc. Am.,* 57: 341—371.

Cornell, C.A. and Vanmarcke, E.H., 1969. The major influences on seismic risk. *Proc. 4th World Conf. Earthquake Eng. Santiago.*

Crouse, C.B., 1973. Engineering studies of the San Fernando earthquake. *Calif. Inst. Technol., Earthquake Eng. Res. Lab.* Rep. 73-04.

Cox, D.F. and Lewis, P.A.W., 1966. *The Statistical Analysis of Series of Events.* Methuen, London.

Davenport, A.G., 1972. A statistical relationship between shock amplitude, magnitude and epicentral distance and its application to seismic zoning. *Univ. Western Ontario, Faculty Eng. Sci.,* BLWT-4-72.

Davies, G.F. and Brune, J.N., 1971. Regional and global fault slip rates from seismicity. *Nature,* 229: 101—107.

Drakopoulos, J.C., 1971. A statistical model on the occurrence of aftershocks in the area of Greece. *Bull. Int. Inst. Seismol. Earthquake Eng.,* 8: 17—39.

Esteva, L., 1968. Bases para la formulación de decisiones de diseño sísmico. *Natl. Univ. Mexico, Inst. Eng.* Rep. 182.

Esteva, L., 1969, Seismicity prediction: a bayesian approach. *Proc. 4th World Conf. Earthquake Eng. Santiago.*

Esteva, L., 1970. Consideraciones prácticas en la estimación bayesiana de riesgo sísmico. *Natl. Univ. Mexico, Inst. Eng., Rep.* 248.

Esteva, L., 1974. Geology and probability in the assessment of seismic risk. *Proc. 2nd Int. Congr. Int. Assoc. Eng. Geol., Sao Paulo.*

Esteva, L. and Villaverde, R., 1973. Seismic risk, design spectra and structural reliability. *Proc. 5th World Conf. Earthquake Eng., Rome,* pp. 2586—2597.

Figueroa, J., 1963. Isosistas de macrosismos mexicanos. *Ingeniería,* 33 (1): 45—68.

Gaisky, V.N., 1966. The time distribution of large, deep earthquakes from the Pamir—Hindu-Kush. *Dokl. Akad. Nauk Tadjik S.S.R.,* 9 (8): 18—21.

Gaisky, V.N., 1967. On similarity between collections of earthquakes, the connections between them, and their tendency to periodicity. *Fiz. Zemli,* 7: 20—28 (English transl., pp. 432—437).

Gajardo, E. and Lomnitz, C., 1960. Seismic provinces of Chile. *Proc. 2nd World Conf. Earthquake Eng., Tokyo,* pp. 1529—1540.

Gutenberg, B. and Richter, C.F., 1954. *Seismicity of the Earth.* Princeton University Press, Princeton.

Gzovsky, M.G., 1962. Tectonophysics and earthquake forecasting. *Bull. Seismol. Soc. Am.,* 52 (3): 485—505.

Herrera, I., Rosenblueth, E. and Rascón, O.A., 1965. Earthquake spectrum prediction for the Valley of Mexico. *Proc. 3rd Int. Conf. Earthquake Eng., Auckland and Wellington,* 1: 61—74.

Housner, G.W., 1969. Engineering estimates of ground shaking and maximum earthquake magnitude. *Proc. 4th World Conf. Earthquake Eng., Santiago.*

Hudson, D.E., 1971. *Strong Motion Instrumental Data on the San Fernando Earthquake of February 9, 1971.* California Institute of Technology, Earthquake Engineering Research Laboratory.

Hudson, D.E., 1972a. Local distributions of strong earthquake ground shaking. *Bull. Seismol. Soc. Am.,* 62 (6).

Hudson, D.E., 1972b. *Analysis of Strong Motion Earthquake Accelerograms, III, Response Spectra, Part A.* California Institute of Technology, Earthquake Engineering Research Laboratory.

Hudson, D.E. and Vdwadia, F.E., 1973. Local distribution of strong earthquake ground motions. *Proc. 5th World Conf. Earthquake Eng., Rome,* pp. 691—700.

Kaila, K.L. and Narain, H., 1971. A new approach for preparation of quantitative seismicity maps as applied to Alpide Belt—Sunda Arc and adjoining areas. *Bull. Seismol. Soc. Am.,* 61 (5): 1275—1291.

Kaila, K.L., Gaur, V.K. and Narain, H., 1972. Quantitative seismicity maps of India. *Bull. Seismol. Soc. Am.,* 62 (5): 1119—1132.

Kaila, K.L., Rao, N.M. and Narain, H., 1974. Seismotectonic maps of southwest Asia region comprising eastern Turkey, Caucasus, Persian Plateau, Afghanistan and Hindukush. *Bull. Seismol. Soc. Am.,* 64 (3): 657—669.

Kelleher, J., Sykes, L. and Oliver, J., 1973. Possible criteria for predicting earthquake locations and their application to major plate boundaries of the Pacific and the Caribbean. *J. Geophys. Res.,* 78 (14): 2547—2585.

Knopoff, L., 1964. The statistics of earthquakes in southern California. *Bull. Seismol. Soc. Am.,* 54: 1871—1873.

Lomnitz, C., 1966. Magnitude stability in earthquake sequences. *Bull Seismol. Soc. Am.,* 56: 247—249.

Lomnitz, C. and Hax, A., 1966. Clustering in aftershock sequences. In: J.S. Steinhart and T. Jefferson Smith (editors), *The Earth Beneath the Continents.* Am. Geophys. Union, pp. 502-508.

McGuire, R.K., 1974. Seismic structural response risk analysis incorporating peak response regressions on earthquake magnitude and distance. *Mass. Inst. Technol., Dep. Civ. Eng.*, R74-51.

Merz, H.A. and Cornell, C.A., 1973. Seismic risk analysis based on a quadratic magnitude—frequency law. *Bull. Seismol. Soc. Am.*, 63 (6): 1999—2006.

Milne, W.G. and Davenport, A.G., 1969. Earthquake probability. *Proc. 4th World Conf. Earthquake Eng., Santiago*.

Mogi, K., 1962. Study of elastic shocks caused by the fracture of heterogeneous materials and its relations to earthquake phenomena. *Bull. Earthquake Res. Inst. Tokyo*, 40: 125—173.

Molnar, P. and Sykes, L.R., 1969. Tectonics of the Caribbean and Middle America regions from focal mechanisms and seismicity. *Geol. Soc. Am. Bull.*, 80: 1639.

Newark, N.M. and Rosenblueth, E., 1971. *Fundamentals of Earthquake Engineering*. Prentice-Hall, Englewood Cliffs.

Omori, F., 1894. On the aftershocks of earthquakes. *J. Coll. Sci. Imp. Univ. Tokyo*, 7: 111—200.

Parzen, E., 1962. *Stochastic Processes*. Holden Day, San Francisco.

Petrushevsky, B.A., 1966. *The Geological Fundamentals of Seismic Zoning*. Scientific Translation Service, order 5032, Ann Arbor, USA.

Raiffa, H. and Schlaifer, R., 1968. *Applied Statistical Decision Theory*, MIT Press.

Rosenblueth, E., 1964. Probabilistic design to resist earthquakes. *Am. Soc. Civ. Eng., J. Eng. Mech. Div.*, 90 (EM5): 189—249.

Rosenblueth, E., 1969. Seismicity and earthquake simulation. *Rep. NSF-UCEER Conf. Earthquake Eng. Res., Pasadena*, pp. 47-64.

Rosenblueth, E., 1975. *Point Estimates for Probability Moments*. National University of Mexico, Institute of Engineering, Mexico City.

Rosenblueth, E., in preparation. Optimum design for infrequent disturbances.

Rukos, E., 1974. *Análysis dinamico de la margen izquierda de Chicoasén*. National University of Mexico, Institute of Engineering, Mexico City.

Salt, P.E., 1974. Seismic site response. *Bull. N. Z. Natl. Soc. Earthquake Eng.*, 7 (2): 63—77.

Scholz, C.H., 1968. The frequency—magnitude relation of microfracturing and its relation to earthquakes. *Bull. Seismol. Soc. Am.*, 58: 399—417.

Shlien, S. and Toksöz, M.N., 1970. A clustering model for earthquake occurrences. *Bull. Seismol. Soc. Am.*, 60 (6): 1765—1787.

Singh, S.K., 1975. *Mexican Volcanic Belt: Some Comments on a Model Proposed by F. Mooser*. National University of Mexico, Institute of Engineering, Mexico City.

Trifunac, M.D., 1973. Characterization of response spectra by parameters governing the gross nature of earthquake source mechanisms. *Proc. 5th World Conf. Earthquake Eng., Rome*, pp. 701—704.

Tsuboi, C., 1958. Earthquake province. Domain of sympathetic seismic activities. *J. Phys. Earth.*, 6 (1): 35—49.

Utsu, T., 1961. A statistical study on the occurrence of aftershocks. *Geophys. Mag., Tokyo*, 30: 521—605.

Utsu, T., 1962. On the nature of three Alaska aftershock sequences of 1957 and 1958. *Bull. Seismol. Soc. Am.*, 52: 179—297.

Veneziano, D. and Cornell, C.A., 1973. Earthquake models with spatial and temporal memory for engineering seismic risk analysis. *Mass. Inst. Technol., Dep. Civ. Eng.*

Vere-Jones, D., 1970. Stochastic models for earthquake occurrence. *J. R. Stat. Soc.*, 32 (1): 1—45.

Wallace, R.E., 1970. Earthquake recurrence intervals on the San Andreas Fault. *Geol. Soc. Am. Bull.*, 81: 2875—2890.

Yegulalp, T.M. and Kuo, J.T., 1974. Statistical prediction of the occurrences of maximum magnitude earthquakes. *Bull. Seismol. Soc. Am.*, 64 (2): 393—414.

*Chapter 7*

# TSUNAMIS

ROBERT L. WIEGEL

*Professor of Civil Engineering, University of California, Berkeley, Calif., U.S.A.*

## 7.1 INTRODUCTION

### 7.1.1 Some data

Tsunamis are the long water waves (with wave 'periods' in approximately the 5—60 minute range) generated impulsively by mechanisms such as underwater tectonic displacements associated with earthquakes, high-speed subaqueous slides, rock-slides into reservoirs, bays or the ocean, and exploding islands. They may be caused by the tectonic displacement of an entire body of water such as a lake (Wiegel and Camotim, 1962).

The horizontal component of velocity $V_h$ at which the water is displaced from the source by one of the mechanisms mentioned above is important, with the speed being measured relative to $\sqrt{gd}$ (where $g$ is the acceleration of gravity and $d$ is the water depth). However, as long as the Froude number ($V_h/\sqrt{gd}$) is high, theory and hydraulic experiments show it is not as important as the amount of water that is displaced (see, for example, Wiegel et al., 1970; Hatori, 1970).

It is likely that the major cause of large-scale catastrophic tsunamis is a rapidly occurring tectonic displacement of the ocean bottom, with the displacement having a substantial vertical component (dip-slip), as shown in Fig. 7.1 (Iida, 1970; see also, Balakina, 1970 and Watanabe, 1970). One would expect that strike-slips would have to occur through a seamount or submarine cliff to generate a tsunami, and then, owing to the rapid decrease of the ground displacement with distance from the fault (Bonilla, 1970), it is unlikely that major tsunamis would be generated by this mechanism as the waves would be rather short (Garcia, 1972). However, earthquakes associated with strike-slip faults (as well as with other types of faulting) may trigger a submarine earthquake, which in turn may generate a tsunami.

Tsunamis are important because of the loss of life and great property damage that result from large ones. More than 27,000 people were killed and 10,000 houses destroyed in Japan by the tsunami of June 15, 1896 (Leet, 1948). A great tsunami which struck Chile, Hawaii, California, Japan and all other coastal areas bordering the Pacific Ocean, occurred in conjunction with the Chilean earthquake of May 23, 1960 (Committee for Field Investigation

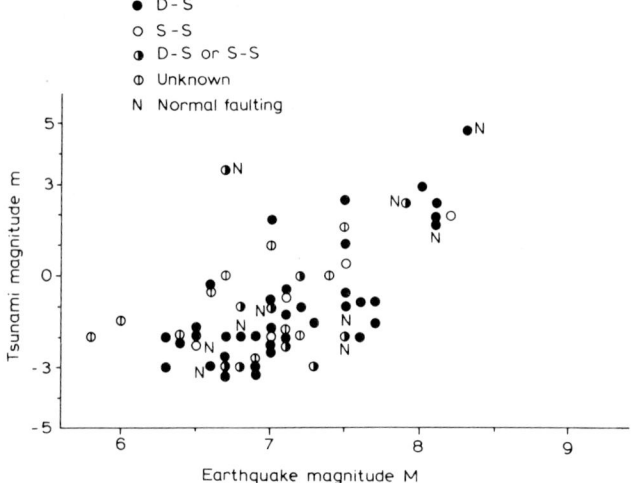

Fig. 7.1. Relationship between earthquake magnitude and tsunami magnitude, classified according to the type of faulting. (From Iida, 1970.)

of the Chilean Tsunami of 1960, 1961). Another catastrophic tsunami inundated many coastal regions of the Pacific, the origin of it being the large tectonic displacement in the Gulf of Alaska that occurred during the Alaska earthquake of March 27, 1964 (National Academy of Sciences, 1972).

*7.1.2 Relationships among earthquake magnitudes, aftershock areas, tectonic displacements, and tsunami damage*

Most earthquakes that occur offshore are of small enough magnitude and/or are of such deep focal depth that no noticeable tsunamis accompany them (Iida, 1963a, 1970). It is probable that the reason for this is that large tectonic displacements are unlikely to be associated with these types of earthquakes, as can be seen in Figs. 7.2 and 7.3.

Iida (1963a, 1970) found a relationship among the magnitude (Richter, $M$) and focal depth of submarine earthquakes, and the magnitude of associated tsunamis, if any occurred. His results for tsunamis generated near Japan are shown in Fig. 7.4. For seismic conditions to the left of line $A$, no tsunamis of appreciable height have been observed. For seismic conditions of the type defined by the region to the right of line $B$, great tsunamis have been generated. Information on tsunami magnitude, $m$, tsunami energy and maximum run-up elevation of tsunamis in Japan, is given in Table 7.I.

The close relationship between earthquake aftershock area and the region of tectonic displacement has been shown for the Alaska earthquake of March 27, 1964 by Berg et al. (1972), as can be seen in Fig. 7.5. In the same figure

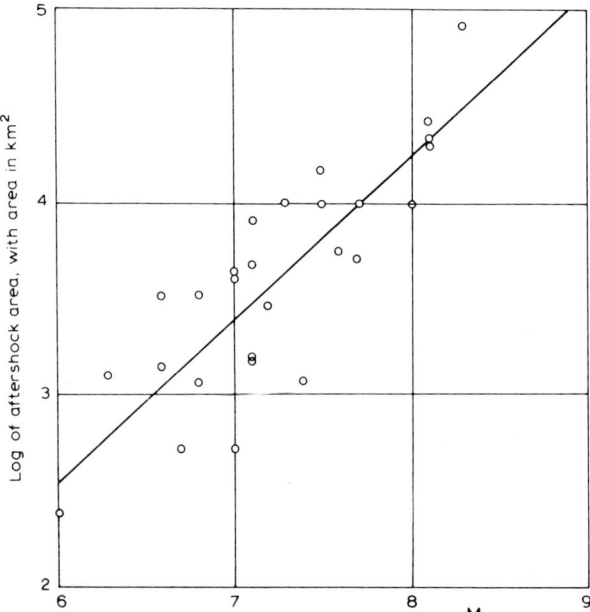

Fig. 7.2. Relationship between aftershock area and magnitude $M$ of tsunamigenic earthquakes. (From Iida, 1963a.)

it has been shown that the underwater portion of the aftershock area is also the area of tsunami generation as determined by other methods. Using the method of inverse water wave refraction drawings, Fig. 7.6, Hatori (1970) made similar analyses for a number of earthquakes off the coast of Japan which were accompanied by tsunamis. His findings, together with those of Utsu and Seki (1955) are shown in Fig. 7.7, which are similar to the results of study by Kajiura (1967) shown in Fig. 7.8.

There are many papers in the technical literature on tsunamis generated in the ocean (see, for example, Wiegel, 1970), but few papers on the subject of the smaller, but very important tsunamis generated in bays and reservoirs. Most of these have been generated by high-speed subaqueous landslides (see Wiegel, 1955, for a model study of this phenomenon) and rockslides into the body of water. The highest wave that has been documented, to the author's knowledge, was created in Lituya Bay, Alaska, by a large high-speed rockslide, triggered by an earthquake, that crashed into the bay (Miller, 1960). The slide was roughly triangular in cross-section, with dimensions of 2400 ft and 3000 ft along the slope, a maximum thickness of about 300 ft normal to the slope, with the center of gravity of the mass of rock being about 2000 ft above the water surface. The volume and weight of the rock mass were about 40 million cubic yards and 90 million tons, respectively. The potential en-

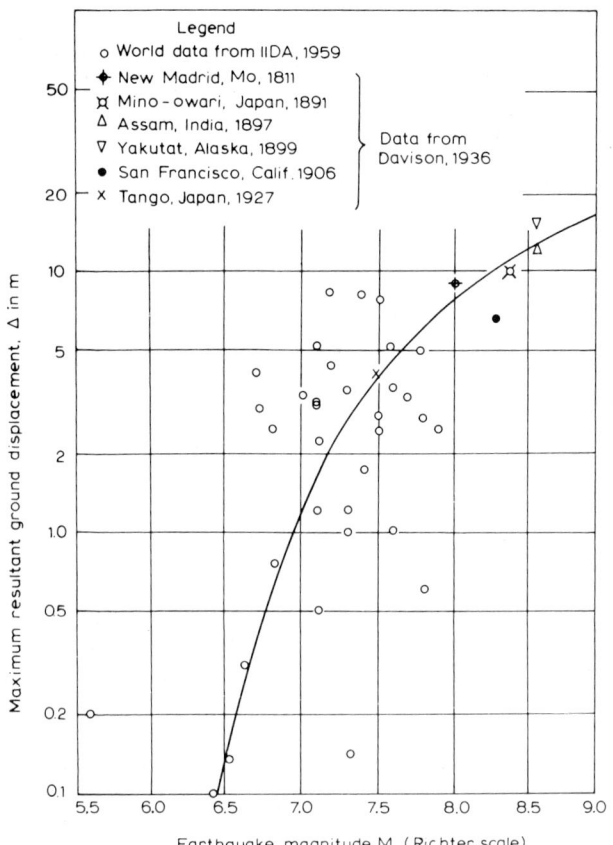

Fig. 7.3. Trend of dependence of maximum resultant ground displacement on earthquake magnitude. (From Wilson et al., 1962.)

ergy of the slide was about $3.5 \cdot 10^{14}$ ft-lb, with the potential energy of the resulting main wave being about 2% of the slide energy. The slide occurred along one side of the head of a T-shaped bay, causing water to surge up the opposite side to a maximum elevation of more than 1700 ft (520 m) above the water level, and a wave nearly solitary in form about 200 ft high moved down the main part of the bay and out to sea, Fig. 7.9.

The generation of large waves by subaqueous slides were a prominent feature of the Alaska earthquake of March 27, 1964 (Coulter and Migliaccio, 1966; Grantz et al. 1964; McCulloch, 1966; Plafker and Mayo, 1965). Grantz et al. (1964) state:

"A number of highly destructive waves were experienced locally in coastal areas during and almost immediately after the March 27 earthquake. Several of these early waves, notably the first ones that hit Seward and Valdez, were definitely generated by sub-

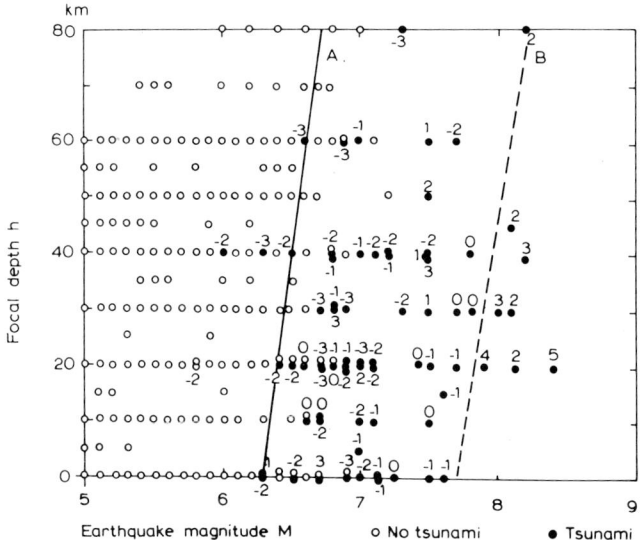

Fig. 7.4. Relationship between earthquake magnitude and focal depth of submarine earthquakes during the period 1926 to 1968. The numeral outside of the circle is the tsunami magnitude in round numbers. (From Iida, 1970.)

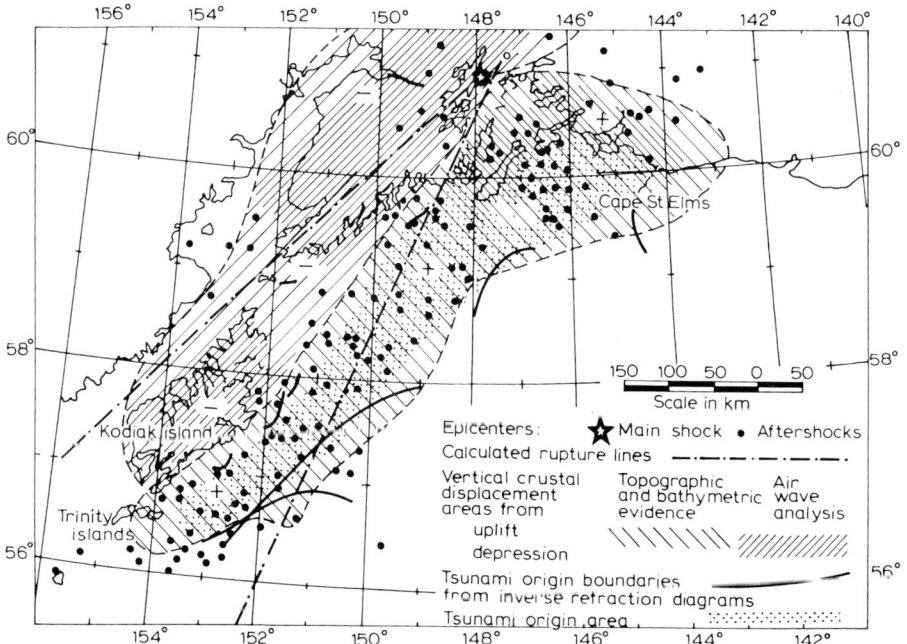

Fig. 7.5. Source area of the major Alaska tsunami, as indicated by various kinds of evidence. (From Berg et al., 1972.)

TABLE 7.I

Magnitude, energy, and run-up elevation of tsunamis in Japan

(From Iida, 1963a)

| Tsunami magnitude classification ($m$) | Tsunami energy | | Maximum run-up elevation | |
|---|---|---|---|---|
| | (ergs) | (ft-lb) | (m) | (ft) |
| 5 | $25.6 \cdot 10^{23}$ | $18.9 \cdot 10^{16}$ | >32 | >105 |
| 4.5 | 12.8 | 9.4 | 24—32 | 79—105 |
| 4 | 6.4 | 4.7 | 16—24 | 52.5—79 |
| 3.5 | 3.2 | 2.4 | 12—16 | 39.2—52.5 |
| 3 | 1.6 | 1.2 | 8—12 | 26.2—39.2 |
| 2.5 | 0.8 | 0.59 | 6—8 | 19.7—26.2 |
| 2 | 0.4 | 0.29 | 4—6 | 13.1—19.7 |
| 1.5 | 0.2 | 0.15 | 3—4 | 9.9—13.1 |
| 1 | 0.1 | 0.074 | 2—3 | 6.6—9.9 |
| 0.5 | 0.05 | 0.037 | 1.5—2 | 4.9—6.6 |
| 0 | 0.025 | 0.018 | 1—1.5 | 3.2—4.9 |
| —0.5 | 0.0125 | 0.0092 | 0.75—1 | 2.5—3.2 |
| —1 | 0.006 | 0.0044 | 0.50—0.75 | 1.6—2.5 |
| —1.5 | 0.003 | 0.0022 | 0.30—0.50 | 1.0—1.6 |
| —2 | 0.0015 | 0.0011 | <0.30 | <1.0 |

marine landslides; they are described on earlier pages. The origin of other waves that were responsible for the loss of at least 43 lives and extensive property damage is not certain; their general distribution and direction are shown on figure 7. Because Prince William Sound is sparsely populated, the loss in lives and property there from earthquake-generated waves is proportionally extremely high. This tragic loss occurred even though the earthquake and attendant waves came at low tide and probably after the entire eastern part of the Sound was tectonically elevated 4 to 7.5 feet."

They also state:

"At an unknown time after the earthquake, a surge of water shot into Valdez Arm and Port Valdez (loc. 14 on fig. 7; fig. 11). The wave overtopped and destroyed a navigation light in Valdez Passage, struck the north shore of Port Valdez near the Cliff mine, was reflected to the southeast shore, which it struck at Jackson Point, and dissipated toward the head of Port Valdez. The wave destroyed wooden buildings at the Cliff mine and deposited debris 170 feet above sea level. Runup reached 220 feet, as indicated by broken and freshly scarred shrubs. At Jackson Point the Dayville cannery was floated away and the wave heaped driftwood to a height of 17 feet above the highest tide level."

"Mr. Jackson, who was at sheltered Sawmill Cove outside Valdez Passage (fig. 11), reports that during and immediately after the earthquake the sea withdrew briefly and then rose rapidly to about the highest tide level but that no large waves were observed. The villages of Tatitlek and Ellamar, on the mainland east of Valdez Passage, were sheltered by Bligh Island from destructive waves from the south. Residents there report a 15-foot withdrawal of water immediately after the earthquake; following withdrawal, the

231

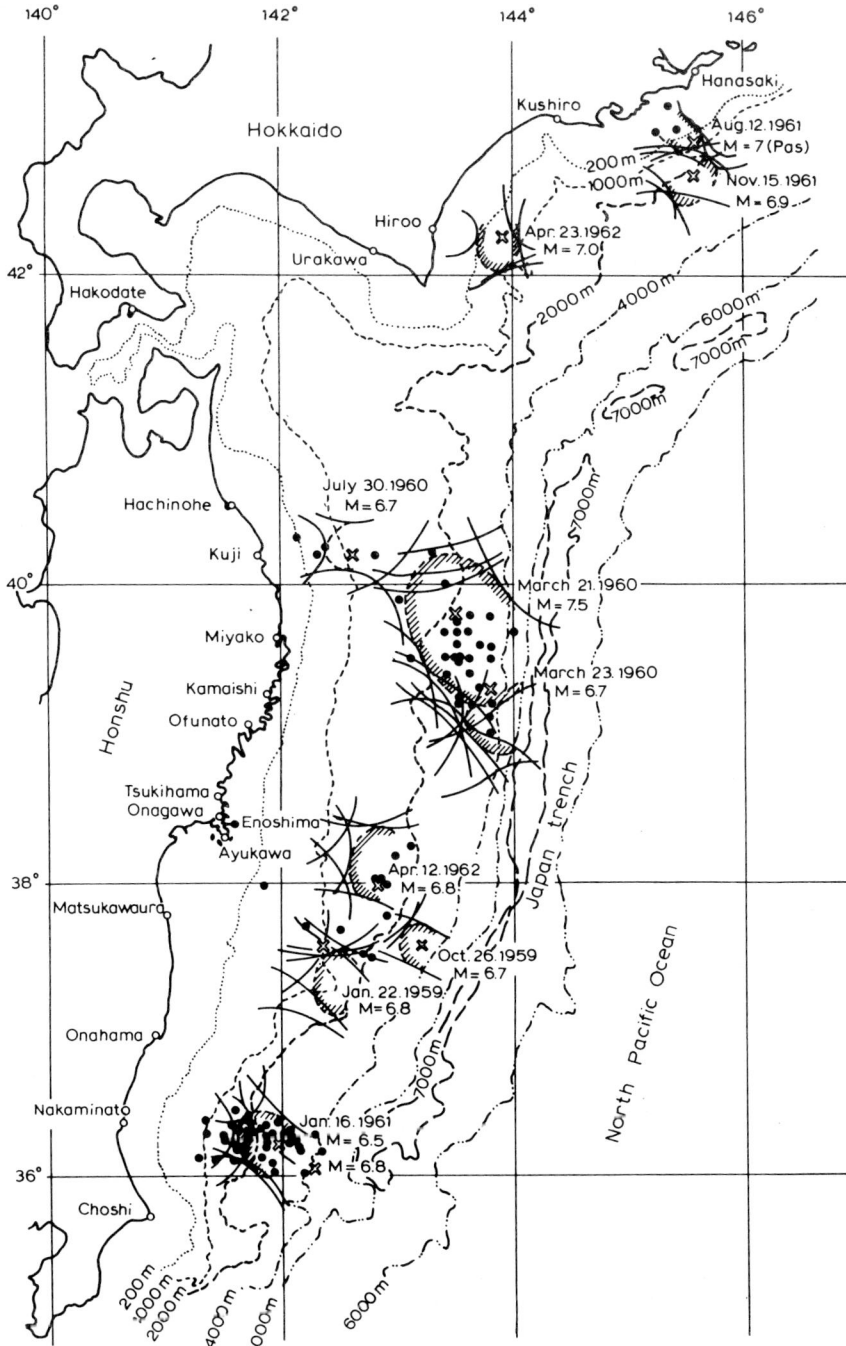

Fig. 7.6. Geographic distribution of estimated source areas of tsunamis in northeast Japan during the period 1959–1962 and aftershocks. (After Hatori, 1970.)

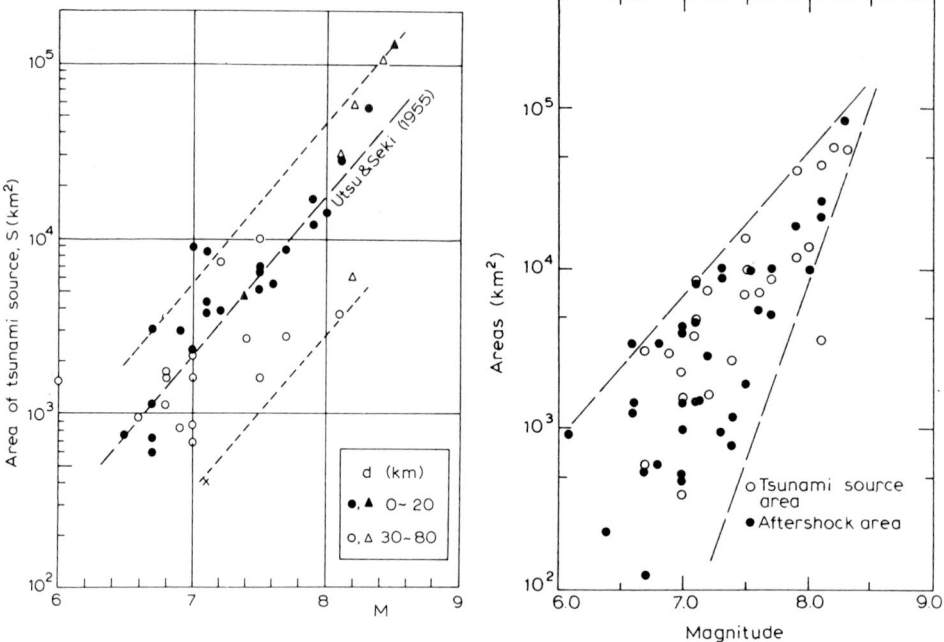

Fig. 7.7. Relation between the area of tsunami source and earthquake magnitude. (From Hatori, 1970.)

Fig. 7.8. Relation of aftershock and tsunami origin areas to magnitude for earthquake in Japan. (After Kajiura, 1967.)

water level rose rapidly about 18 feet, approximately to the highest tide level despite a 4- to 6-foot tectonic uplift of the land during the earthquake. A generally similar sequence of events is reported by eyewitnesses residing on Peak Island, which is in a sheltered location behind the Naked Islands; by Ed Brenner, who was in Port Wells at the time of the earthquake; and by residents of Whittier."

The figures referred to in the above quotations are not shown in the present chapter. However, much of the information is shown in Fig. 7.10, which is from Plafker and Mayo (1965). It is important to note that wave run-up elevations are being referred to, and that these occurred at a time when the tide level was very nearly at MLLW.

## 7.1.3 Landslide and subaqueous slide-generated tsunamis

There are a number of cases of landslides and rockfalls into reservoirs, lakes, bays and fjords which have created large waves, many of which have caused loss of life and considerable damage (Wiegel et al., 1970, 1972).

Fig. 7.9. Lituya Bay, Alaska. September, 1954 (top) and August, 1958 (bottom). (After Miller, 1960.) Courtesy U.S. Geological Survey.

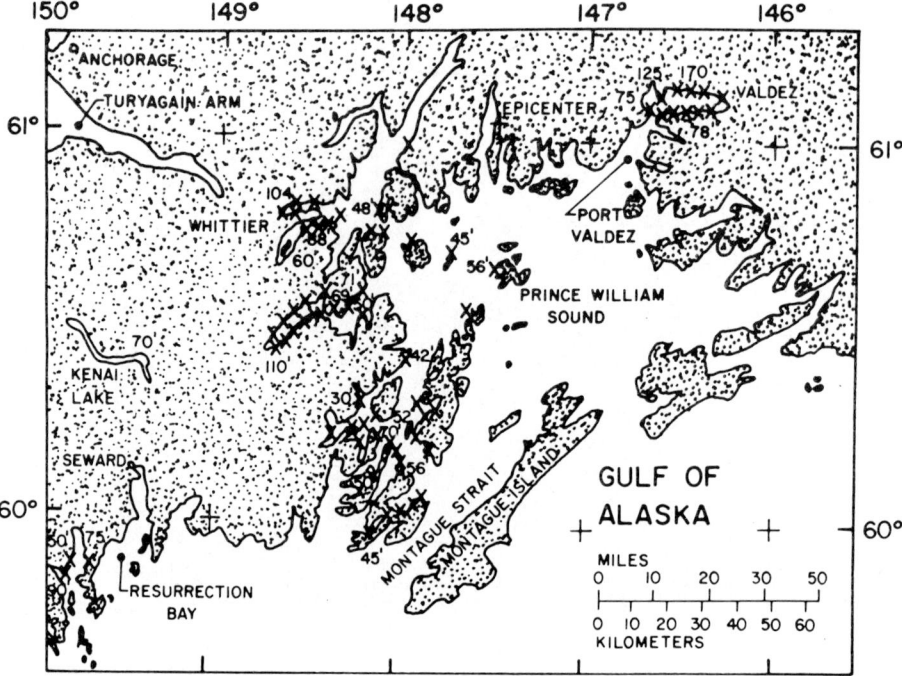

Fig. 7.10. Generalized distribution of larger destructive local waves and known subaqueous slides in Prince William Sound and part of the Kenai Peninsula. Shorelines damaged by the waves with run-up heights in excess of 40 ft. above lower low water are indicated by an "X". Numerals are the measured maximum run-up heights. Solid triangle indicates known subaqueous slide. (From Plafker and Mayo, 1965.)

These slides have been rapidly moving ones. It has been estimated from seismograph records and the area from which the slide occurred that the Vaiont slide moved at from 50 to 100 ft/sec (about 12—30 m/sec) *, and apparently other slides have reached speeds as great as 352 ft/sec (107 m/sec) (see Wiegel et al., 1970, 1972). Shreve (1966) estimated the minimum speeds of six hugh landslides to be from 70 to 270 ft/sec (about 20 to 80 m/sec). These estimates were made from the elevations reached by the landslide on the opposite side of the canyon using only the conservation of energy, neglecting friction. Mitchell (1910) states that the great Frank slide in Alberta, Canada (40,000,000 cubic yards; about 30,000,000 m$^3$) moved about 2.5 miles in 1—1.5 min. (145—220 ft/sec, about 44—67 m/sec). This

---

* The figures given in the following few paragraphs were in the English system in the original references. The conversion to the metric system is given in approximate terms so as not to infer greater accuracy than existed.

slide was from 400 to 500 ft (about 120—150 m) thick at its center. Mitchell also described the Elm slide in Switzerland (12,000,000 cubic yards, about 9,000,000 m³), falling about 1500 ft (about 460 m) and climbing to an elevation of over 300 ft (about 90 m) on the opposite side of the valley, then deflected down the level valley floor. This slide moved so fast that eyewitnesses described the wind it generated to be so great that trees were blown about like matches and houses were lifted through the air like feathers by its force alone. In January, 1970, a slide of several hundred thousand cubic yards on the Eel River, California moved 1700 ft (about 520 m) to the bottom of a 30°-slope in a few seconds and then continued 650 ft (about 200 m) across the river channel (Dukleth, 1970). About a week prior to this, a larger slide (about 1,500,000—2,000,000 cubic yards; about 1,100,000—1,500,000 m³) moved downward 3200 ft (about 975 m) to the base of a 30°-slope, in about 20—30 sec, and then across the South Fork of the Smith River (Dwyer, 1970), the speed of this slide was from 110 to 160 ft/sec (34—49 m/sec).

Consider a range of speeds from 50 to 250 ft/sec (about 15—75 m/sec), and a range of water depths from 150 to 600 ft (about 45—180 m). Then $V_{avg}/\sqrt{gd}$ can range from about 0.36 to 3.6 values of $\lambda/d$ (where $\lambda$ is the thickness of the slide) that is from less than one-half to more than one.

The two-dimensional laboratory studies of Wiegel et al. (1970, 1972), together with the laboratory and theoretical work of Noda (1970) and of Garcia (1972) show that the phenomenon of water wave generation by such high-speed slides is understood. Owing to this, it is desirable to make three-dimensional hydraulic model studies of possible high-speed rockslides into reservoirs, to determine the characteristics and motions of the resulting waves. Such a study was made for Mica Dam on the Columbia River (Western Canada Hydraulic Laboratories, Ltd., 1970).

## 7.2 THEORY OF THE GENERATION OF TSUNAMIS

### 7.2.1 Initial elevation or depression of the water surface

Before investigating the problem of waves generated by moving sources such as landslides or tectonic displacements, it is useful to study a simpler source such as an initial elevation or depression of a finite area of the surface of a body of water.

Consider a two-dimensional case, with an initial surface elevation (or depression) of $E(x)$ (Kranzer and Keller, 1959). The record of the water surface $y_s$ as a function of $x$ and $t$ is:

$$y_s(x, t) = \frac{1}{\sqrt{x}} \left[ \frac{-C_G(k)}{C'_G(k)} \right]^{1/2} \overline{E}(k) \sin 2\pi \left( \frac{t}{T} - \frac{x}{L} - \frac{1}{8} \right) \quad (7.1)$$

for $x < t\sqrt{gd}$

Here $x$ is the horizontal coordinate, $t$ is time, $d$ is the water depth, $k$ is the wave number ($2\pi/L$), $L$ is the component wave length, $C_G$ is the speed at which wave (or wave number $k$) energy is propagated, $C'_G(k)$ is the first derivative of $C_G(k)$, $T$ is the component wave period, and $\overline{E}(k)$ is the Fourier transform of $E(x)$. This equation is valid only when linear water wave theory is valid, and even then it is not valid for the leading part of the first wave (to the first wave crest). $C_G$ is defined by the equation:

$$C_G = \tfrac{1}{2} C \left(1 + \frac{2kd}{\sinh 2kd}\right) \tag{7.2}$$

and $C$, the wave phase speed is defined by:

$$C = \sqrt{gd}\,(\tanh kd/kd)^{1/2} \tag{7.3}$$

For the simple case of a rectangular initial elevation (or depression) of the water surface of height $h$ over a length $\lambda$, $\overline{E}(k)$ is given by:

$$\overline{E}(k) = \sqrt{2/\pi}\,\frac{h \sin k\lambda}{k} \tag{7.4}$$

The initial source and its Fourier transform for two values of $\lambda$ are shown in Fig. 7.11, and some records of waves generated in the laboratory from such a source are shown in Fig. 7.12. There are several useful things that can be learned from these four equations:

(1) Groups of waves are created.

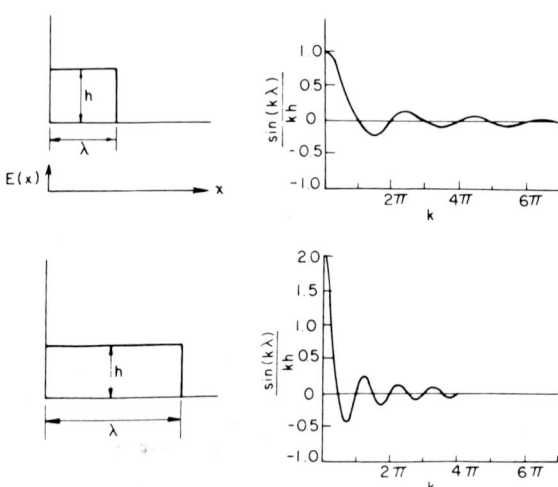

Fig. 7.11. Fourier cosine transform: $E(k)/\sqrt{2/\pi} = \int_0^\infty E(x) \cos kx\,dx = (h \sin k\lambda)/k$.

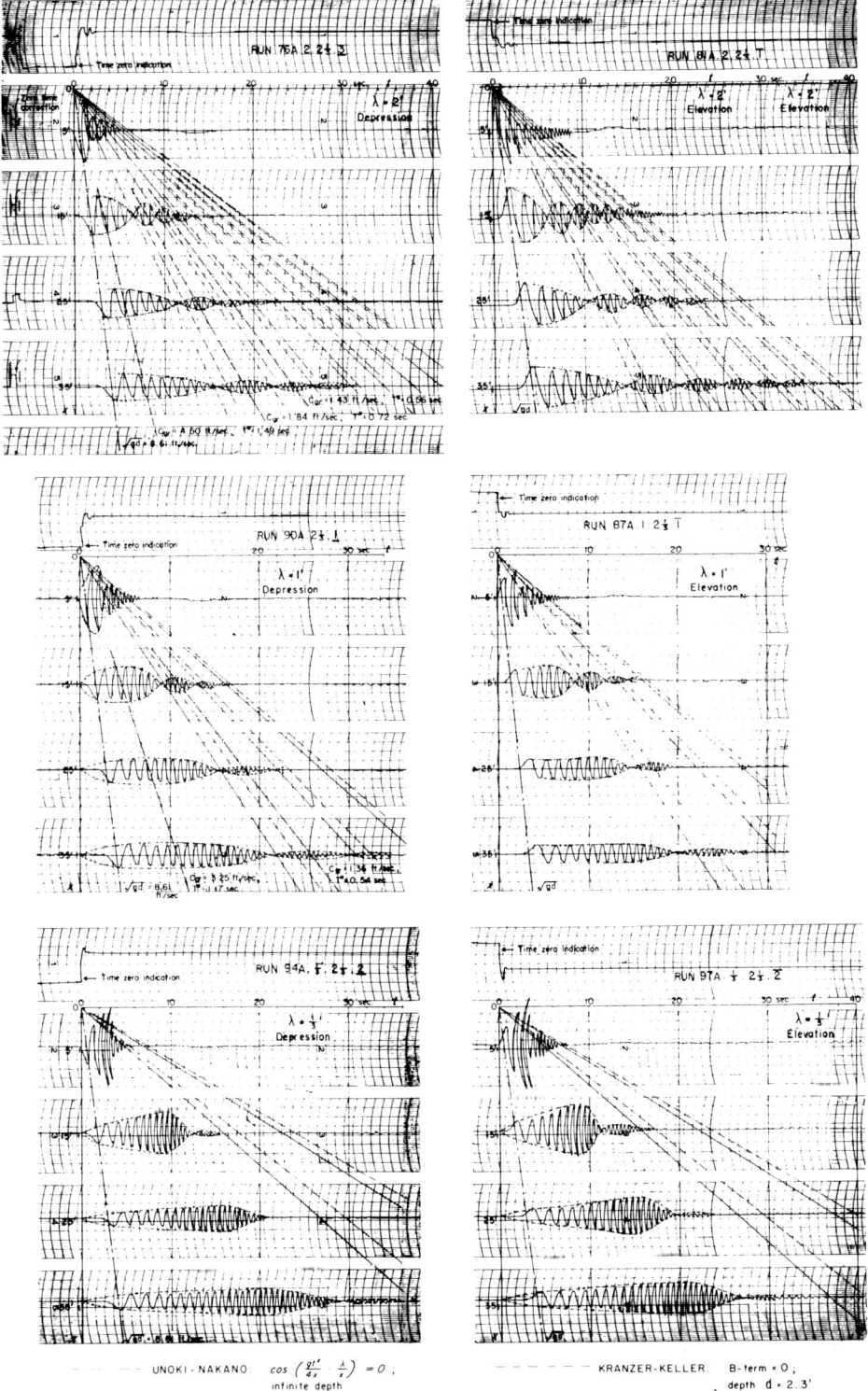

Fig. 7.12. Impulsively generated waves, for $d = 2.3'$, and $\lambda = 1/3'$, $1'$, $2'$. (From Prins, 1958.)

(2) The larger the value of $\lambda/d$, the closer to the leading edge is the location of the highest wave.

(3) The larger the value of $\lambda/d$, the greater is the portion of the total energy that is concentrated in the smaller wave numbers (the longer waves).

(4) The length (or period) of the 'individual' wave that passes a location $x$ at a given time $t$ is determined by $C_G(k)$.

(5) The larger $\lambda/d$, the less dispersive is the system ($C_G(k) \to C(k)$), and the less likely is it that linear theory will be valid.

(6) The greater $x/d$, the smaller is $y_{s,\max}/h$.

For values of $\lambda/d$ greater than about $1/2$, the first wave becomes the highest. This height can be approximated by the equation:

$$y_s(x, t) \approx \frac{6^{5/6}\Gamma(4/3)\overline{E}(0)}{4\pi^{1/2}d^{2/3}x^{1/3}} \approx \frac{0.45\, h\lambda}{d^{2/3}x^{1/3}} \tag{7.5}$$

Equation 7.5 shows that for the initial condition being considered, the greater the initial elevation, the greater the length of the initial elevation, and the more shallow the water, the higher the wave will be. Furthermore, the height of this wave will decrease as it moves away from the source.

The phenomenon is more complicated than the above equations indicate, even when linear theory is appropriate. F. Ursell (personal communication, 1958) solved the problem using the method of steepest descents (Chester et al., 1957). For the case of an initial rectangular elevation (or depression), he found:

$$y_s(x, t) \approx \frac{2h}{\pi} \int_0^\infty \cos kx \, \cos(t\sqrt{gk \tanh kd}) \, \frac{\sin k}{k} dk \tag{7.6}$$

Except for the wave front, Kelvin's principle of stationary phase was used by Ursell, and he obtained the same solution as Kranzer and Keller (1959). The inequality for eq. 7.1, $x < \sqrt{gd}$, must be determined by matching numerical results from eqs. 7.1 and 7.4 with those given subsequently herein. Near the wave front the method of steepest descents was used by Ursell to evaluate the integral in eq. 7.6 for $x/t\sqrt{gd} < 1$ and $0 < \beta < \infty$:

$$y_s(x, t) \approx \frac{h}{(t\sqrt{g/d})^{1/3}} p_0(\beta) A_i[-t\sqrt{g/d})^{2/3} \alpha(\beta)] \tag{7.7}$$

in which $A_i$ is the Airy integral:

$$A_i(X) = \int_0^\infty \cos(Xu + \tfrac{1}{3}u^3)\,du \tag{7.8}$$

which satisfies the differential equation:

$$\left(\frac{d^2}{dX^2} - X\right) A_i(X) = 0 \tag{7.9}$$

This integral has been described and tabulated (Abramowitz and Stegun, 1964, p. 446) and $\beta$ is defined by:

$$\frac{x}{t\sqrt{gd}} = \frac{1}{2}\left(1 + \frac{2\beta}{\sinh 2\beta}\right) \sqrt{\frac{\tanh \beta}{\beta}} \tag{7.10}$$

with $\alpha(\beta)$ and $p_0(\beta)$ given by:

$$\alpha(\beta) = \left(\frac{\beta}{\sinh 2\beta}\right)^{2/3} \left(\frac{\tanh \beta}{\beta}\right)^{1/3} \left(\frac{-2\beta + \sinh 2\beta}{\frac{4}{3}\beta^3}\right)^{2/3} \beta^2 \tag{7.11}$$

$$p_0(\beta) = \frac{\sin(\beta\lambda/d)}{\beta} \left[\frac{2\sqrt{\alpha(\beta)}}{-\frac{d^2}{d\beta^2}(\sqrt{\beta \tanh \beta})}\right]^{1/2} \tag{7.12}$$

in which:

$$\frac{d^2}{d\beta^2}(\sqrt{\beta \tanh \beta}) = -(\beta \tanh \beta)^{1/2} \left[\frac{1}{\cosh^2 \beta} + \frac{1}{4\beta^2}\left(1 - \frac{2\beta}{\sinh 2\beta}\right)^2\right] \tag{7.13}$$

For $x/t\sqrt{gd} > 1$, substitution of $\gamma = i\beta$ (for $0 < \gamma < 1/2\,\pi$) into eq. 7.7 results in:

$$y_s(x, t) = \frac{h}{(t\sqrt{g/d})^{1/3}} p_0(\gamma) A_i [-(t\sqrt{g/d})^{2/3}\alpha(\gamma)] \tag{7.14}$$

Here, $\gamma$ is defined by:

$$\frac{x}{t\sqrt{gd}} = \frac{1}{2}\left(1 + \frac{2\gamma}{\sin 2\gamma}\right) \sqrt{\frac{\tan \gamma}{\gamma}} \tag{7.15}$$

and $\alpha(\gamma)$ and $p_0(\gamma)$ are given by:

$$\alpha(\gamma) = -\left(\frac{\gamma}{\sin 2\gamma}\right)^{2/3} \left(\frac{\tan \gamma}{\gamma}\right)^{1/3} \left(\frac{2\gamma - \sin 2\gamma}{\frac{4}{3}\gamma^3}\right)^{2/3} \gamma^2 \tag{7.16}$$

$$p_0(\gamma) = \frac{\sinh(\gamma\lambda/d)}{\gamma} \left[\frac{2\sqrt{\alpha(\gamma)}}{-\frac{d^2}{d\gamma^2}(\sqrt{-\gamma \tan \gamma})}\right]^{1/2} \tag{7.17}$$

in which:

$$\left[\frac{2\sqrt{\alpha(\gamma)}}{-\dfrac{d^2}{d\beta^2}(\sqrt{-\gamma\tan\gamma})}\right]^{1/2}$$

$$= \left[\frac{2\left(\dfrac{\gamma}{\sin 2\gamma}\right)^{1/3}\left(\dfrac{\tan\gamma}{\gamma}\right)^{1/6}\left(\dfrac{2\gamma-\sin 2\gamma}{\dfrac{4\gamma^3}{3}}\right)^{1/3}}{(\gamma\tan\gamma)^{1/2}\left\{\dfrac{1}{\cos^2\gamma} - \dfrac{1}{4\gamma^2}\left(1-\dfrac{2\gamma}{\sin 2\gamma}\right)^2\right\}}\gamma\right]^{1/2} \quad (7.18)$$

For small values of $\beta$, Ursell found that;

$$\beta^2 \to a\left(1 - \dfrac{1}{t\sqrt{gd}}\right) \tag{7.19a}$$

$$\alpha(\beta) \to \dfrac{\beta^2}{2^{2/3}} \tag{7.19b}$$

$$-\dfrac{d^2}{d\beta^2}(\sqrt{\beta\tanh\beta}) \to \beta \tag{7.19c}$$

with similar approximations occurring for small values of $\gamma$, so that eqs. 7.7 and 7.14 become:

$$y_s(x,t) \approx \dfrac{2^{1/3}\lambda h}{(t\sqrt{g/d})^{1/3}d} A_i\left[\dfrac{2^{1/3}}{(t\sqrt{g/d})^{1/3}d}(x-t\sqrt{gd})\right] \tag{7.20a}$$

$$\approx \dfrac{1.26\,\lambda h}{(t\sqrt{g/d})^{1/3}d} A_i\left[\dfrac{1.26}{(t\sqrt{g/d})^{1/3}}\left(\dfrac{x}{d} - t\sqrt{\dfrac{g}{d}}\right)\right] \tag{7.20b}$$

For the point on the wave front where $x = t\sqrt{gd}$, $A_i(0) = 0.355$, and eq. 7.20a becomes:

$$y_s(x,t) \approx \dfrac{(0.355)\,2^{1/3}\lambda h}{\left(\dfrac{x}{\sqrt{gd}}\sqrt{\dfrac{g}{d}}\right)^{1/3}d} = \dfrac{0.45\,h\lambda}{d^{2/3}x^{1/3}} \tag{7.21}$$

which is the same as eq. 7.5. Equation 7.20b was used to plot the curve in Fig. 7.13. For this case the wave crest travels more slowly than $\sqrt{gd}$, and the wave front elevation where $t = x/\sqrt{gd}$ is lower than the crest. Calculations of the leading wave for relatively large values of $(\gamma/d)(d/x)^{1/3}$ show that the velocity of the first crest exceeds $\sqrt{gd}$. Kajiura's theory also shows this.

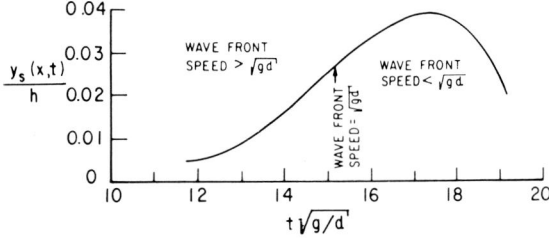

Fig. 7.13. $y_s(x, t)/h$ vs. $t\sqrt{g/d}$ for $\lambda/d = 0.145$ and $x/d = 15.2$. Calculated using Ursell's approximate theory. (From Wiegel et al., 1970.)

Equations 7.7 and 7.20a were used to calculate the time history of a wave front, and they were compared with one set of Prins (1958) data. It was found that the theory is good for these conditions (small $\lambda/d$, small $h/d$, and large $x/d$). Comparison of measurements and theory for relatively large values of $\lambda/d$ and small values of $x/d$, but still for small values of $h/d$, are shown in Fig. 7.14. The results are not as good as for the other conditions. Equation 7.5 predicted the crest height correctly for $x/d = 2.15$, but not the value of $t\sqrt{g/d}$ at which the crest occurred. Equation 7.20a was adequate for prediction purposes for $x/d = 6.45$.

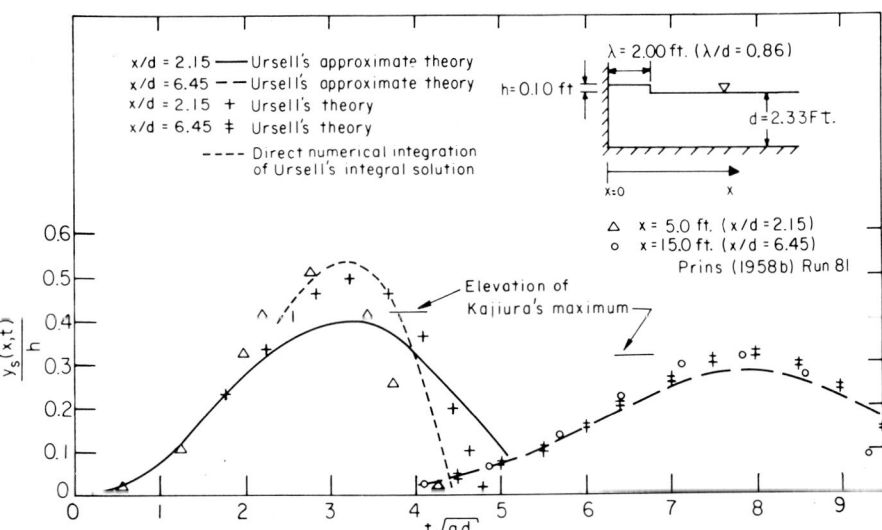

Fig. 7.14. Comparison of measured values of first wave for $x/d = 2.15$ and 6.45 (with $h = 0.10$ ft., $\lambda = 2.00$ ft. and $d = 2.33$ ft.) with theoretical calculations for Ursell's theory and Kajiura's theory. (From Wiegel et al., 1970.)

As $\lambda/d$ and $h/d$ (or $h/\lambda$) increase, the waves become nonlinear. With increasing values of $\lambda/d$ and $h/d$ the waves reduce to one, with a small tail, and it becomes a quasi-solitary wave which is not dispersive. If $\lambda/d$ and/or $h/d$ are/is increased still further, a bore forms. The observed wave characteristics as a function of these two parameters are shown in Fig. 7.15, with examples of wave records obtained from tests in a hydraulic model study shown in Fig. 7.16. It is interesting to note that for the nonlinear case, the hydraulic model tests show the waves generated by an initial depression of the water surface to be quite different from those from an initial elevation.

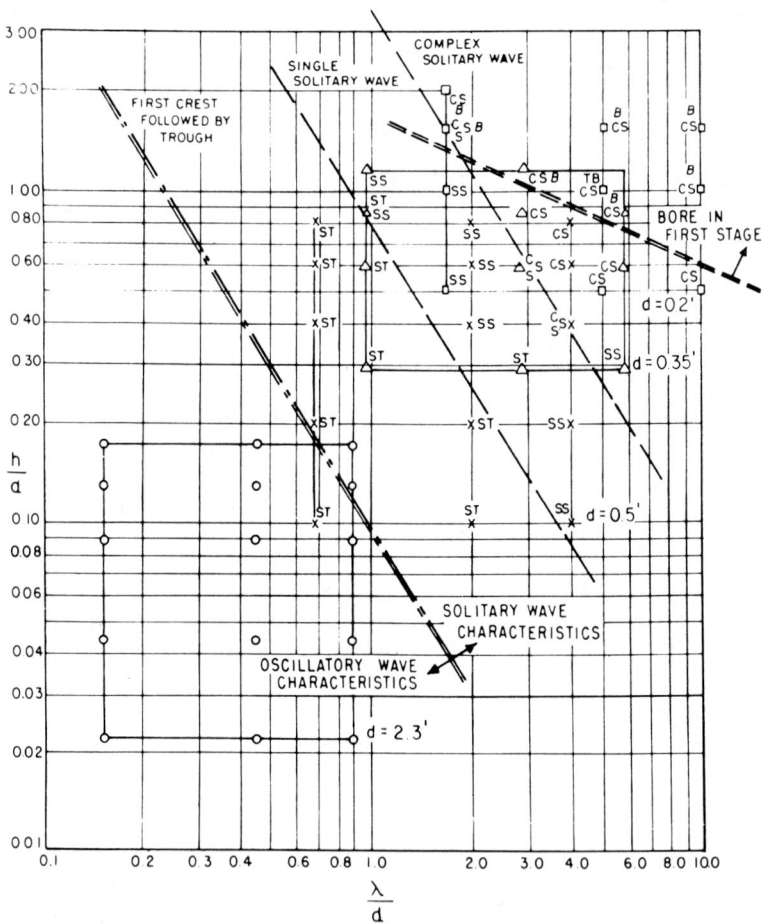

Fig. 7.15. Relation between $\lambda/d$, $h/d$ and characteristics of leading wave for an initial elevation. (From Prins, 1958.)

a.

b.

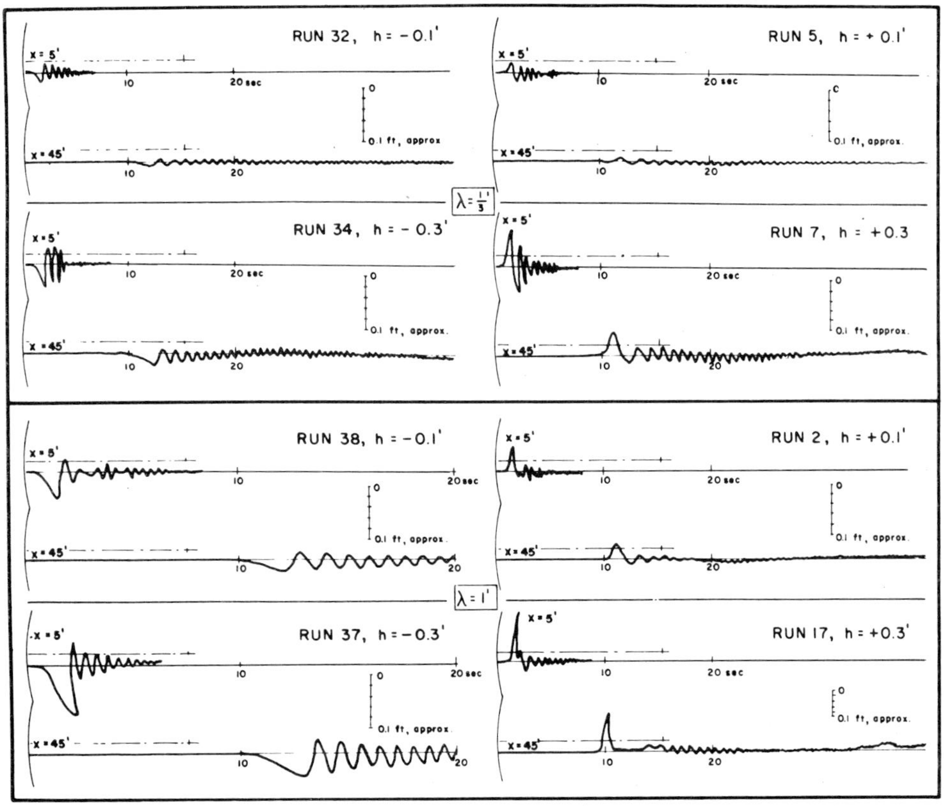

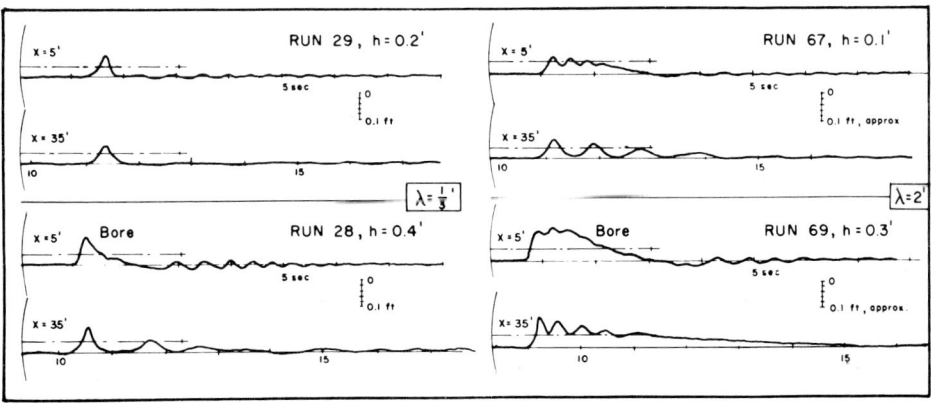

Fig. 7.16. (a) Leading parts of the wave patterns at depth $d = 0.5'$ and (b) at depth $d = 0.2'$. (After Prins, 1958.)

## 7.2.2 Vertical displacement of bottom: linear theory

The importance of the theory presented in the section above is that the same result is obtained if a segment of the ocean bottom is moved vertically, with the motion occurring instantaneously, provided the right-hand side of eq. 7.1 is multiplied by $1/\cosh kd$ (see Takahasi, 1963, for experimental confirmation of this idea). This results in an increase in the relative importance of the longer wave components.

Referring to eqs. 7.1 and 7.4, one can draw the conclusion that a large source is required to generate tsunamis that have most of their energy in long-period waves. This is confirmed by the empirical relationship between the maximum period of a tsunami and the earthquake magnitude found by Iida (1963b), Fig. 7.17, when one considers at the same time the relationship shown in Fig. 7.7 between the earthquake magnitude and the area of the tsunami source.

## 7.2.3 Moving boundary: linear theory and hydraulic-model studies

Wiegel et al. (1970, 1972) made a hydraulic model study of the waves generated by the free fall of a weighted rectangular box into a body of water, which represented a high-speed rock slide. They originally used a Froude number based upon the average speed with which the body dropped through the water to the bottom of the tank, $V_{avg}/\sqrt{gd}$. Later (1972), they concluded that a more representative Froude number should be based upon the average rate at which the water was forced to move horizontally from under the falling body:

$$V_h/\sqrt{gd} = (V_{avg}/\sqrt{gd})(\lambda/h) \tag{7.22}$$

Here, $h$ represents the elevation, above the reservoir (or lake, etc.) bottom, of the bottom of the model rock slide; this would have an upper limit of $d$,

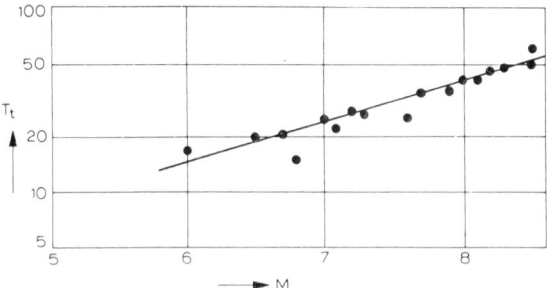

Fig. 7.17. Relationship between the maximum period of a tsunami and earthquake magnitude. (From Iida, 1963b.)

if the bottom of the rock slide, at the start, was above the water surface. The ranges of $V_{avg}/\sqrt{gd}$ and $\lambda/h$ used in the tests were those that have been shown in the Introduction to exist for prototype high-speed slides.

In the work reported by Wiegel et al. (1970, 1972), it was found that although a body dropped from above the water surface made a large splash, this rather spectacular phenomenon had essentially no effect upon the water waves that were generated. Studies were made, using neutrally buoyant particles, to study the motion of the water being forced from under the falling body. For large body speeds, when the body approached the bottom of the tank a pronounced horizontal jet was formed with much mixing. This phenomenon, however, did not appear to have much influence on the generation of the water waves.

Recognizing the importance of the horizontal speed of water flow from under the falling body, Noda developed a theory for water wave generation based upon this concept for his Ph.D. thesis. Owing to the likelihood of knowing the velocity of a rock slide rather than the water velocities, Noda (1969, 1970) started with a given displacement—time history of the fall, then calculated the velocity—space—time history of the water being forced from under the falling body. He then made sensitivity studies using this mathematical model, and found the resulting waves to be rather insensitive to variations from the simple case of a uniform fall speed.

Noda made use of the basic mathematical formulation of Kennard (1949), for two specific types of problems. He developed an asymptotic solution which was based upon the stationary phase method of Kelvin where applicable, and then based upon the method of steepest descents after the work of Ursell (1958; see Chester et al., 1957 for the basic mathematical theory). Noda also developed a numerical method of solving the integral equation, and determined the range of parameters for which the asymptotic solution could be used. This is shown in Fig. 7.18. In this figure, $V$ is the uniform speed of fall (designated by $N = 1$, $C_1 = 0$ in the figure) of the vertically falling body, $x$ is the horizontal coordinate, $x^*$ is the dimensionless horizontal coordinate ($x^* = x/d$), $\eta_{max}$ is defined by a sketch in the figure, and $\lambda$ is the horizontal dimension of the falling body. $V^*$ is the dimensionless velocity (Froude number), $V/\sqrt{gd}$.

Noda also solved the problem of waves generated by the horizontal motion of a vertical wall. The results of his direct integral solution are shown in Fig. 7.19 for the case of the uniform velocity of the horizontally moving vertical end wall (designated by $N = 1$ $V^* = 1$ in the figure). In this figure, $T^*$ is the dimensionless time the wall is in motion ($T^* = T\sqrt{g/d}$, where $T$ is the time the wall is in motion), and $t$ is time (sec).

Noda defined regions of types of waves observed in the laboratory experiments he made with a vertically falling body. His definitions of wave types are shown in Fig. 7.20 (see also Table 7.II), together with the regions he defined. Similar results were obtained by Das and Wiegel (1972) for a

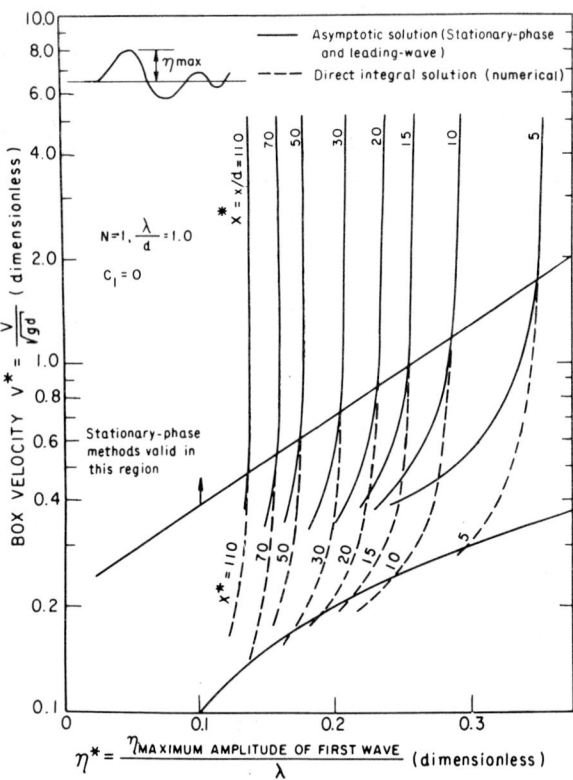

Fig. 7.18. Box-drop problem: Comparison of asymptotic solutions with direct numerical solutions. (From Noda, 1970.)

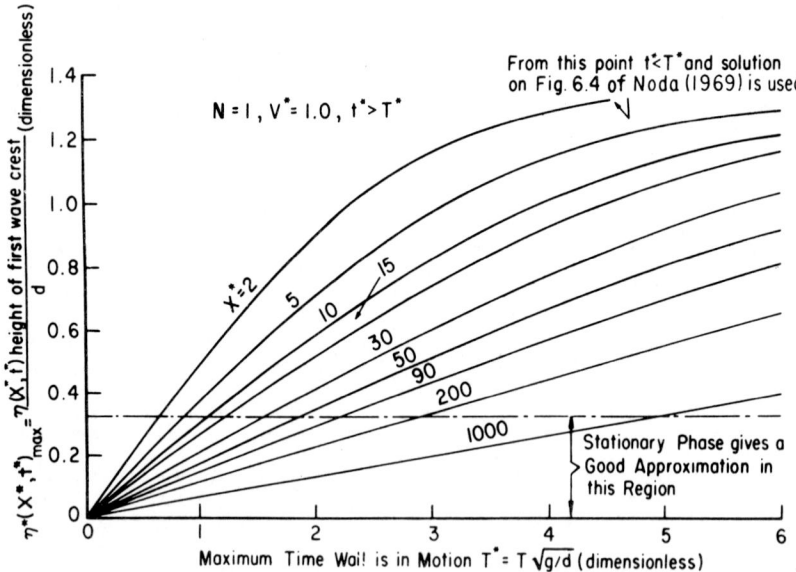

Fig. 7.19. Wall-motion problem: Direct integral solution for $t^* > T^*$. (From Noda, 1969.)

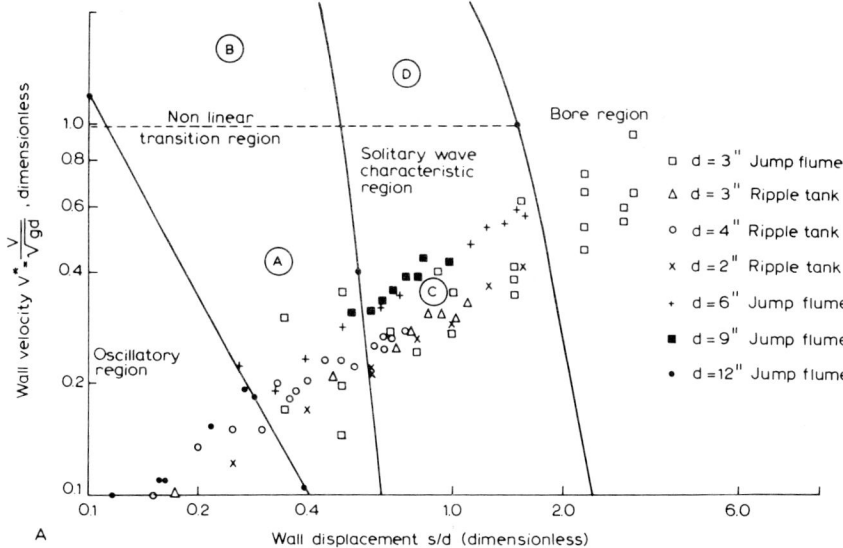

Fig. 7.20a. Regions of wave characteristics. (From Das and Wiegel, 1972.)

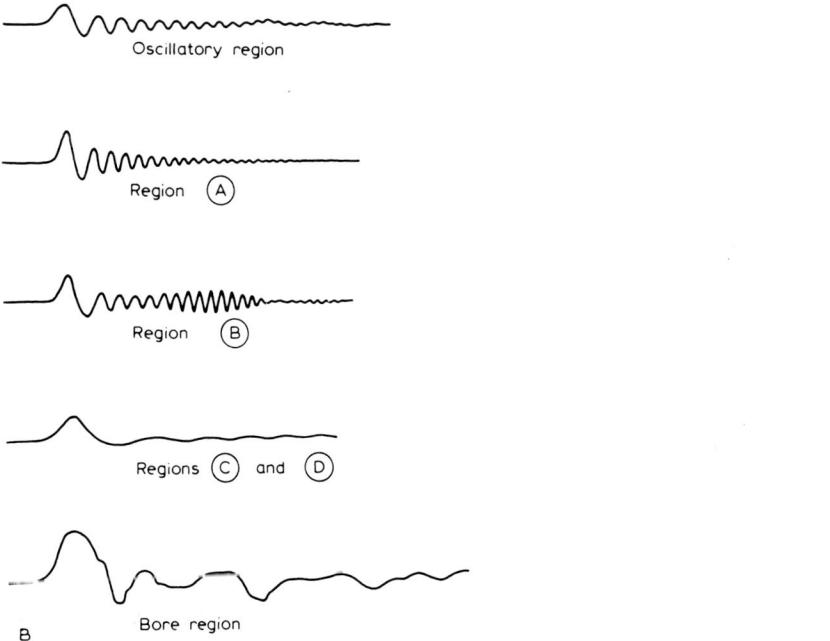

Fig. 7.20b. Wave characteristics for box-drop problem. (From Noda, 1970.)

TABLE 7.II

Range of experimental conditions and associated wave characteristics (from Das and Wiegel, 1972)

| Experiment number | Water depth $d$, (inches) | $S/d$ | $V^* = V/\sqrt{gd}$ | $T^* = T\sqrt{g/d}$ | Wave characteristics | Flume where studied |
|---|---|---|---|---|---|---|
| 1 | 2 | 0.25—1.55 | 0.12—0.41 | 2.06—3.75 | oscillatory-nonlinear-solitary | ripple tank |
| 2 | 3 | 0.17—1.10 | 0.009—0.32 | 1.7—3.4 | oscillatory-nonlinear-solitary | ripple tank |
| 3 | 3 | 0.35—3.16 | 0.17—0.93 | 1.13—5.67 | oscillatory-nonlinear-solitary-bore region | jump flume |
| 4 | 4 | 0.15—0.75 | 0.10—0.28 | 1.47—2.70 | oscillatory-nonlinear-solitary | ripple tank |
| 5 | 6 | 0.26—1.58 | 0.19—0.57 | 1.20—2.80 | oscillatory-nonlinear-solitary | jump flume |
| 6 | 9 | 0.53—0.99 | 0.31—0.43 | 1.75—2.30 | nonlinear-solitary | jump flume |
| 7 | 12 | 0.067—0.29 | 0.08—0.19 | 0.85—1.54 | oscillatory | jump flume |

horizontally moving vertical wall. For the horizontal motion, $S/d$ was used in place of $\lambda/d$. $S$ is the total horizontal displacement of the wall, so that it is equivalent to $\lambda$.

### 7.2.4 Large high-speed horizontal motion of vertical plane boundary

There may be cases where the horizontal motion of a vertical boundary is not small compared with the water depth. Noda (1970) found for the limiting case that the elevation of the crest of the maximum wave generated is given by:

$$\eta_{max}/d = 1.32\,(V/\sqrt{gd}) \qquad (7.23)$$

where $V$ is the uniform speed of the moving boundary. Bakhmeteff (1933) obtained a solution for the formation of a bore by a horizontal moving boundary. He found that $\eta_{max}$ (the elevation of the crest of the bore above the undisturbed water surface) was related to the speed at which the boundary moved, $V$, and the water, depth, $d$, by the equation:

$$V/\sqrt{gd} = \left(\frac{\eta_{\max} + d}{d}\right) \left(\frac{\eta_{\max} + d}{2\eta_{\max}}\right)^{1/2} \qquad (7.24)$$

Miller and White (1966) obtained a number of measurements of the heights of bores generated by the horizontal movement of a vertical wall. Their results are given in Fig. 7.21, together with some results of Das and Wiegel (1972).

### 7.2.5 'Exact' two-dimensional numerical solution for waves generated by a moving boundary

Garcia (1972) developed a two-dimensional numerical model for the generation and initial travel of waves using the Eulerian incompressible hydrodynamic modeling method known as ABMAC (Arbitrary Boundary Marker and Cell). The ABMAC technique solves the complete Navier-Stokes equations for the problem, thus giving an 'exact' solution. The term 'exact' is in quotation marks as the numerical method is approximate. The Navier-Stokes equation and continuity equation are approximated by finite-difference equations which are solved over a finite computational mesh.

Garcia also made a number of hydraulic model experiments in the laboratory to check the validity of his numerical model. These tests were made using a positive horizontal displacement of a vertical wall, a wall with a 1 : 1 slope, and a wall with a 1 : 2 slope, and both a positive and negative displacement of an underwater step. Typical results are shown in Figs. 7.22 and 7.23. It can be seen that the numerical model predicts the hydraulic model waves quite well.

Garcia used numerical calculations, the laboratory measurements of Das and Wiegel (1972) and his own laboratory measurements for the case of the horizontal motion of a vertical wall to show the relationships among $N_{F\text{avg}}$, $\lambda/d$ and $H/d$, together with a description of the characteristic of the wave (nonbreaking, breaking, or bore), Fig. 7.24.

After checking the validity of the numerical model, Garcia used it to study the characteristics of waves generated by a simplified hypothesized motion of an extension of the San Andreas Fault through the eastern end of the Mendocino Escarpment off Cape Mendocino, California. This region is seismically quite active (Cameron, 1961; Bolt et al., 1968). The purpose of this study was to determine whether or not a major strike-slip fault motion through a major sea-floor escarpment could generate a significant tsunami. Owing to limitations of the numerical techniques, or more accurately the storage capability of the largest existing digital computers, a two-dimensional approximation to the three-dimensional source motion was modeled. It was

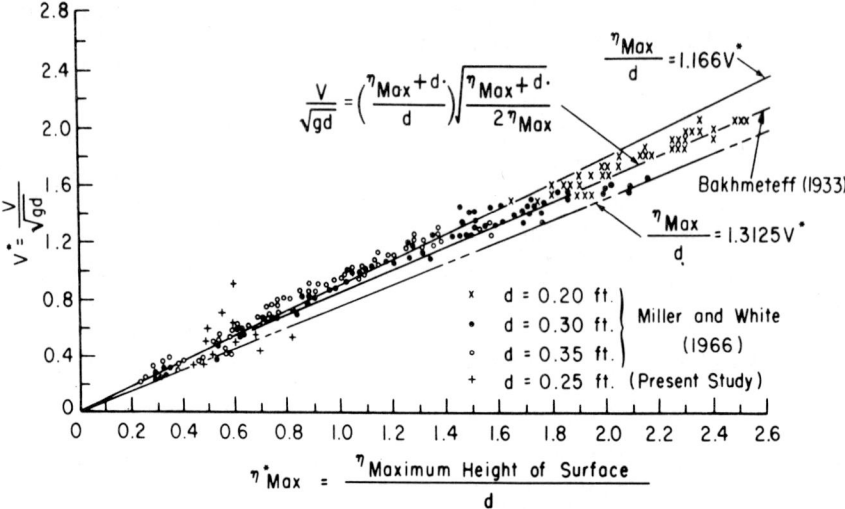

Fig. 7.21. Wall-motion problem: Comparison of analytic solution (Noda, 1970) with experimental data for nonlinear waves. (From Das and Wiegel, 1972.)

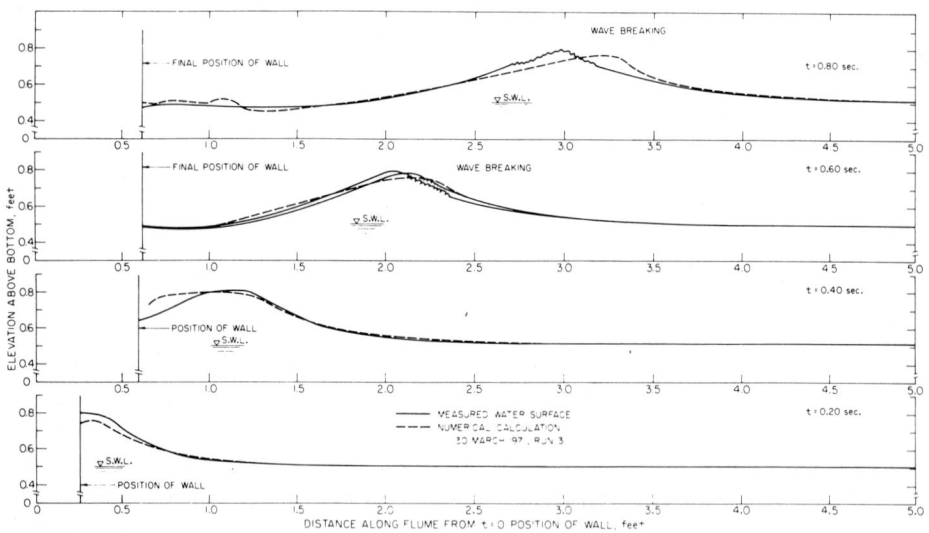

Fig. 7.22. Measured water surface profile for wave generated by horizontal motion of vertical wall, compared with theory. $d = 0.50$ ft., $\lambda = 0.62$ ft, and $N_{Favg} = 0.35$. (From Garcia, 1972.)

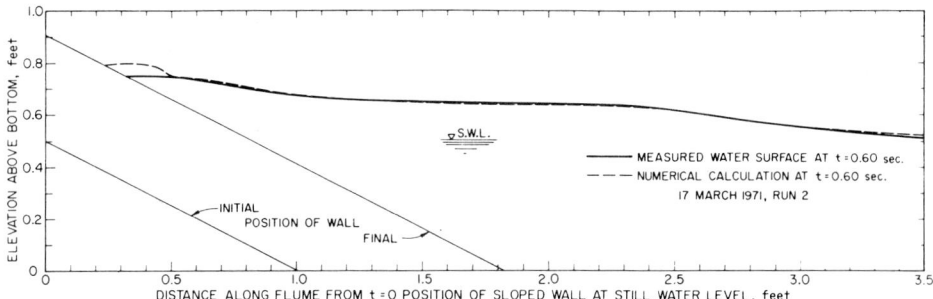

Fig. 7.23. Measured water surface profile for wave generated by horizontal motion of 1 : 2 slope wall, compared with theory. $d = 0.50$ ft., $\lambda = 0.82$ ft., and $N_{Favg} = 0.36$. (From Garcia, 1972.)

recognized that the results of this type of approximation would predict tsunami wave heights that would be higher some distance (even a few wave lengths) from the source than would occur for a real case owing to the rapid decrease in ground movement with distance from the main fault.

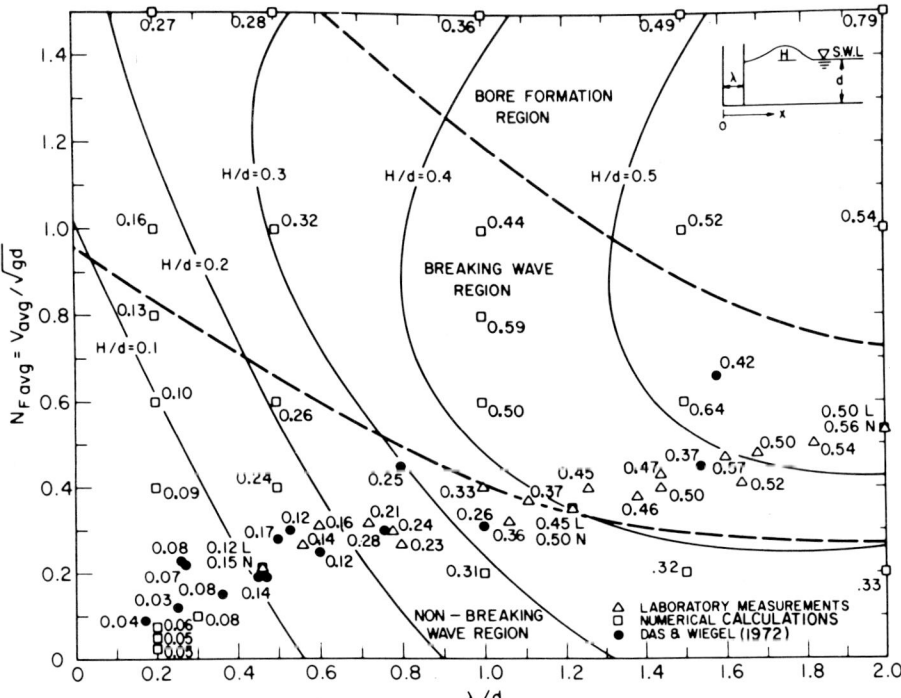

Fig. 7.24. Range of data and wave characteristics, vertical wall $N_{Favg} = V_{avg}/\sqrt{gd}$ vs. $\lambda/d$ for $x/d = 10.0$. Numbers give value of $H/d$. (From Garcia, 1972.)

For example, the fault displacement of branch and secondary faults has been found to decrease to 20% of the main fault displacement at about 5—6 km from the main fault (Bonilla, 1970).

Two locations through the Mendocino Escarpment (one of them through the Gorda Escarpment portion) were modeled. Several displacements were used in the numerical model. There is considerable uncertainty about the interval of time that it takes for such tectonic motions to occur, so calculations were made for several time intervals, from 1 to 60 sec. It was believed that these two durations of motion more than bracketed the probable durations of such occurrences.

A reasonable approximation to one profile through the Mendocino Escarpment is shown in Fig. 7.25 (Profile 2, 128° West). Resulting waves from two types of motion are shown in Figs. 7.26 and Fig. 7.27, both for horizontal motions of 10 m occurring in 5 sec (probably more rapid than could occur). Figure 7.26 shows the waves that should result if only the south face of the escarpment moved horizontally, while Fig. 7.27 shows the waves that would result if the entire 'ridge' moved (i.e., a quasi-dipole source). The two sets of waves are quite different at the source, but in both cases the waves move away from the source both towards the north and towards the south. In the 'dipole' case, the initial disturbance moving northward is a wave trough, while the initial disturbance moving towards the south is a crest. In the case of only one side of the 'ridge' moving, both of the initial disturbances are wave crests.

Garcia found in the seven cases he studied that the 'period' of the first wave, measured from the first crest to the second crest was short, ranging from 1.5 to 1.7 min. This is because the size of the actual source is relatively small even though the hypothesized fault displacement was rather large (10 m). The fact that the tsunami waves generated by this mechanism have short 'periods' (hence, short wave lengths) compared with the tsunamis generated by vertical displacement of large areas of the ocean bottom is of great practical importance. The run-up and flooding of coastal land, even close to the

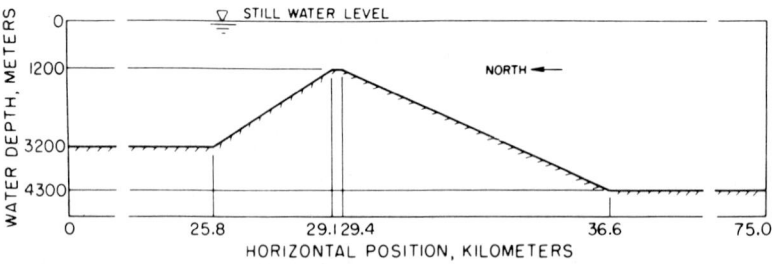

Fig. 7.25. Bottom profile 2. (From Garcia, 1972.)

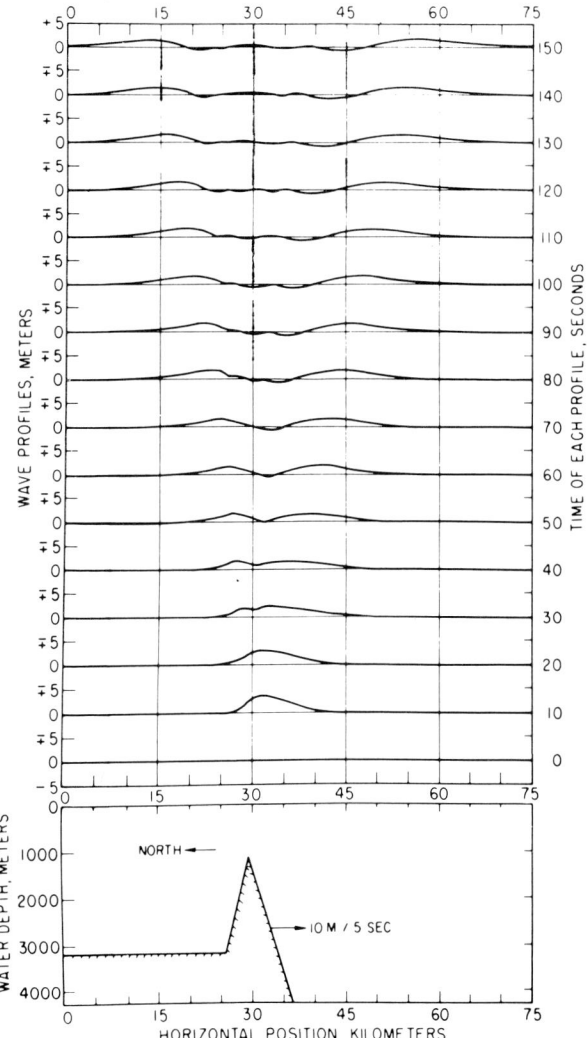

Fig. 7.26. Wave profiles at 10-sec intervals for 10 m in 5 sec displacement of south face of Profile 2. (From Garcia, 1972.)

source, will not be as great as for the much longer tsunamis, all other things being equal. Furthermore, the waves will disperse rather rapidly as they move across the ocean. The elevation of the crest of the first (largest) wave was about 1.25 m at a distance of about 50 km from the source. Furthermore, there was almost no difference in this value for a source movement duration of 1, 5 or 10 sec.

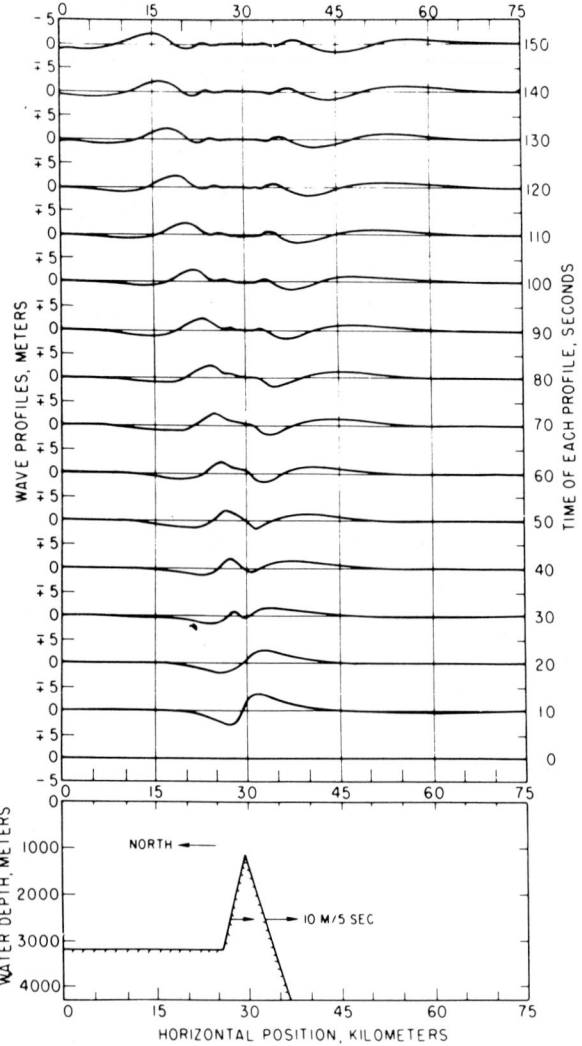

Fig. 7.27. Wave profiles at 10-sec intervals for 10 m in 5 sec displacement of whole ridge of Profile 2. (From Garcia, 1972.)

## 7.3 TSUNAMI SOURCES AND TRAVEL ACROSS THE OCEAN

### 7.3.1 Sources

As stated previously, the tsunamis that affect the entire Pacific Ocean are usually generated by large submarine tectonic displacements such as the one that occurred during the 1964 Alaskan earthquake (Fig. 7.5) and the 1960

Chilean earthquake (Plafker and Savage, 1970). The sources for these two very large events (and also the Rat Islands earthquake in the Aleutians, February 4, 1965; see Jordan et al., 1965) were roughly rectangular in shape, so that they acted as an elongated source for the tsunami waves, rather than a circular or nearly circular source of the type that occurs off the east coast of Japan (see Fig. 7.6). Owing to this, these waves traveled away from the source with strong directional characteristics (see the experimental work of Takahasi, 1963, and the numerical calculations of Hwang et al., 1972).

If the ocean were of uniform depth, the maximum wave heights would be nearly along a great circle route drawn normal to the center of the long axis of the source, but modified slightly by the Coriolis force (Wiegel, 1964a). Refraction modifies this, but it is still a useful approximation to the real solution. Using this concept, one can easily explain the high tsunami waves in the general vicinity of Crescent City, California due to the March 27, 1964 Alaskan earthquake. Approximate sources (aftershock areas, when known) were plotted on a globe for a number of earthquakes, and the great-circle paths were drawn from these source areas in the manner described above. A photograph of the results is shown in Fig. 7.28. It is clear from this figure why the Hawaiian Islands have been subjected to large waves from so many tsunamis.

The concept of plate tectonics is very useful in drawing some conclusions regarding the locations and types of sources of tsunamis. The estimated motions of a source so chosen can be used as an input to a theory of the type developed by Garcia (1972) to calculate numerically the tsunami waves generated by the hypothesized source. Then, using numerical techniques, it is possible to calculate the travel paths and successive wave front positions of the tsunami waves as they travel across the ocean (or, better, the use of a finite-difference numerical method; for a discussion of this and an application using the linearized long-wave equations, see Houston and Garcia, 1974). This permits a reasonable method for estimating the relative importance of different sources of tsunami for any particular site that is of interest. The effects on the tsunami waves at a particular site of the local geometry (refraction, diffraction, wave-trapping, run-up, reflection, seiching, etc.) is quite difficult and not as well understood as are the phenomena of tsunami generation and the travel of these waves across the ocean.

Dr. Gaylord Miller, NOAA Joint Tsunami Research Effort (Honolulu, Hawaii), has developed a series of charts of the Pacific Ocean area which show the location of offshore earthquakes, both those for which tsunamis have occurred and those which were not accompanied by tsunamis. Some very useful information can be obtained from these charts. For example, on February 3, 1923 a $M = 8.3$ earthquake occurred offshore the east coast of Kamchatka. The maximum tsunami wave was only about 0.21 m high at Crescent City, California, about 0.24 m high at San Diego, California, quite low at the Golden Gate (San Francisco, California), although it was quite

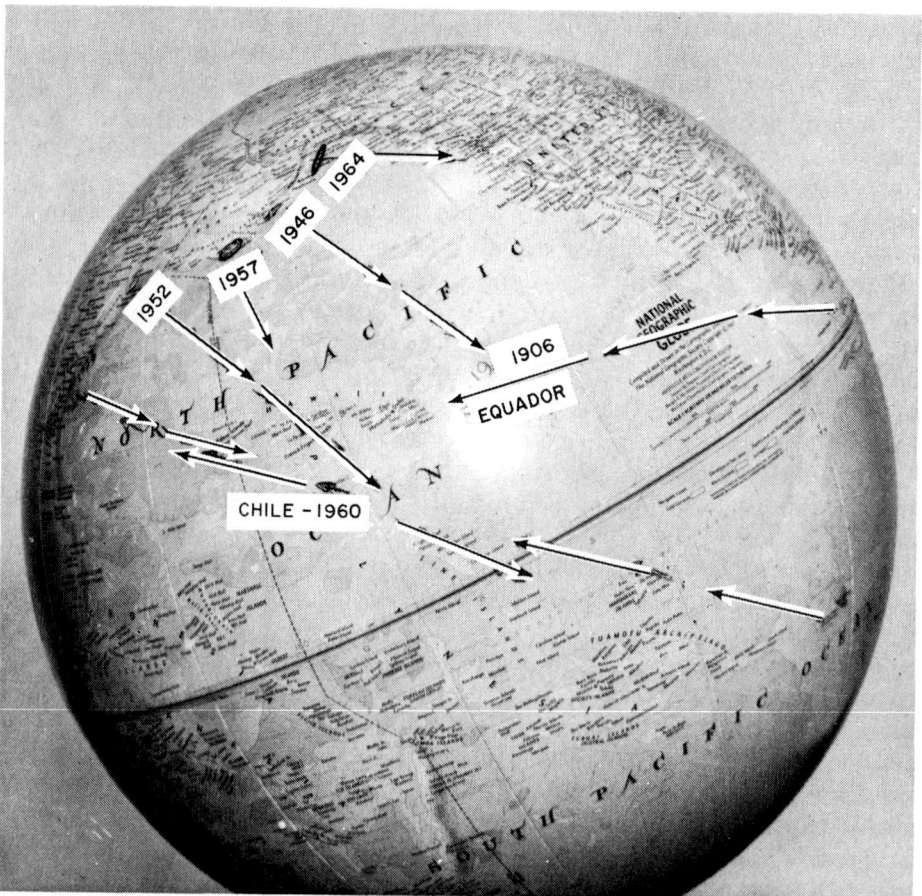

Fig. 7.28. Photograph of a globe showing sources and great-circle paths of several tsunamis.

high at some places in the Hawaiian Islands. From another source (Iida et al., 1967) it appears that the tsunami run-up (or wave height, it is not stated) was from 6 to 8 m at two places along the coast of Kamchatka near the source. The September 7, 1918 earthquake offshore the east coast of the Kuril Islands generated tsunami waves 6—12 m high (or run-up elevation ?) at the Kuril Islands, but these waves were apparently not noticeable along the coast of California.

Dr. Miller's charts are not yet complete, but they are still under development. For example, the June 15, 1896 earthquake offshore the east coast of Japan was not shown. The tsunami waves associated with this earthquake were rather high at Hilo, Hawaii, but were barely noticeable at the Golden Gate. The March 2, 1933 earthquake which occurred offshore the Sanriku coast of Japan generated tsunamis that were quite high along the Sanriku coast, but barely noticeable at the Golden Gate. The November 4, 1952 earthquake offshore Kamchatka was rather high at Hilo, Hawaii, and was a few feet high at the Golden Gate.

Based upon a number of earthquakes and tsunamis generated off Japan, it appears that the geometry of the source areas is such that they do not generate tsunamis that are large along the coast of California. This seems to be true of a number of other source areas. The most dangerous source areas, from the standpoint of the California coast are southern Chile, certain portions of the Aleutian Trench, and some areas of Kamchatka. This is due to a combination of the size, geometry, and orientation of the source area and the fact that the tectonic displacement includes a rather large vertical component.

The effect of the orientation of the source areas can be understood better by studying them on a globe rather than by looking at flat hydrographic charts. As stated previously, the March 27, 1964 Alaskan earthquake and associated tectonic displacement was such that the great circle path from the center of the tsunami source was 'beamed' directly to about the California—Oregon border region. One could realistically assume that the axes of tectonic displacements just north of the Aleutian Trench would be roughly parallel to the trench. It would appear, then, that the location of the source of the 1964 Alaska tsunami would be about the worse case from the standpoint of the coast of northern and central California. Some rather extensive numerical calculations made in connection with the Amchitka Island test also indicated that the orientation and geometry of the 1964 Alaska source area was particularly bad from the standpoint of the coast of central and northern California (Hwang et al. 1970, 1971). Similar 'sensitivity' studies have been made for Japan and perhaps for some other areas.

*7.3.2 Directional characteristics*

Very little work has been done on the directional characteristics of waves generated by three-dimensional tectonic displacements. Hatori (1963) found that prototype data showed the relationship of the tsunami wave height ($H_a$) in the direction of the major axis of a source of length $a$ to the wave height ($H_b$) in the direction of the minor axis of length $b$ is given by the equation developed by Momoi (1962) for an instantaneously and uniformly elevated elliptic source:

$$H_b/H_a = a/b \tag{7.25}$$

Thus, for a tsunami generated by the type of tectonic displacement that occurred during the 1960 Chilean and the 1964 Alaskan earthquakes, $H_b$ can be larger than $H_a$ by a factor of 5 or 6.

Takahasi (1963) made a number of hydraulic model tests using several circular cylinders placed in the bottom of a large basin. The cylinders could be either raised or depressed very rapidly. Tests made with two adjoining cylinders showed that the 'beaming' of wave energy was strongly directional. The results of both cylinders being raised or lowered simultaneously was quite different from the case in which one cylinder was raised while the other was depressed. Additional tests were made using a row of six circular cylinders located side by side on a straight line oriented east—west. These were all raised simultaneously. It was found that the amplitude of the waves decreased as $1/r^{1/3}$ to $1/r^{1/4}$ as they moved toward the south, and that they decreased as $1/r^{3/2}$ toward the east. Here, $r$ is the distance from the source.

These laboratory experiments show that a 'beaming' of the type that occurred for the 1964 Alaska earthquake tsunami results mostly from the shape and orientation of the source.

Liu and Wiegel (1974) made a three-dimensional hydraulic model study of the waves generated by a horizontally moving plunger, with a step in the bottom, similar to the two-dimensional work of Garcia (1972). An example of their results is given in Fig. 7.29, in which the ratio of the crest elevation of the first wave for the three-dimensional case to the two-dimensional case $(H_3/H_2)$ is given as a function of distance from the source $(x/d)$ along the direction of the source motion $(\theta = 0)$. It was found that $H_3/H_2$ decreased in proportion to $(x/d)^{-3/5}$ for $x/d > 1.5$. The effect of the direction from the source can be seen in Figs. 7.30 and 7.31. In these figures $\theta$ is the horizontal angle measured from the centerline along the direction of the source motion. Liu and Wiegel made a few measurements with a wider

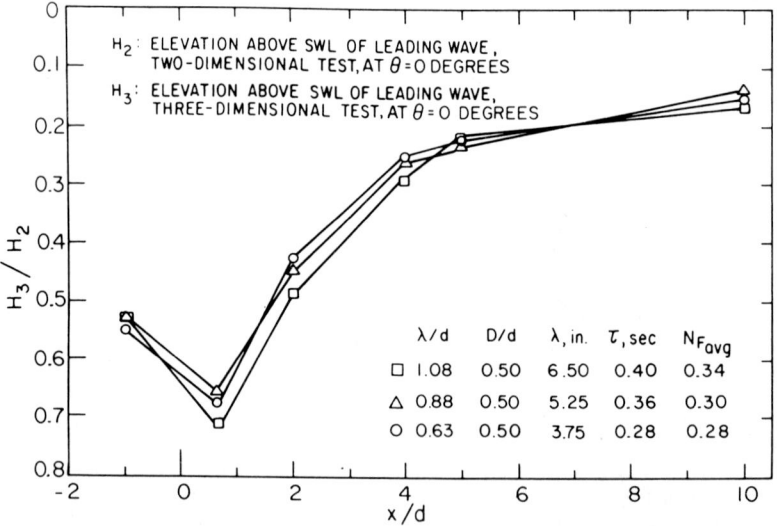

Fig. 7.29. $H_3/H_2$ vs. $x/d$, for $W = 0.50$ ft., $d/W = 1.0$. (From Liu and Wiegel, 1974.)

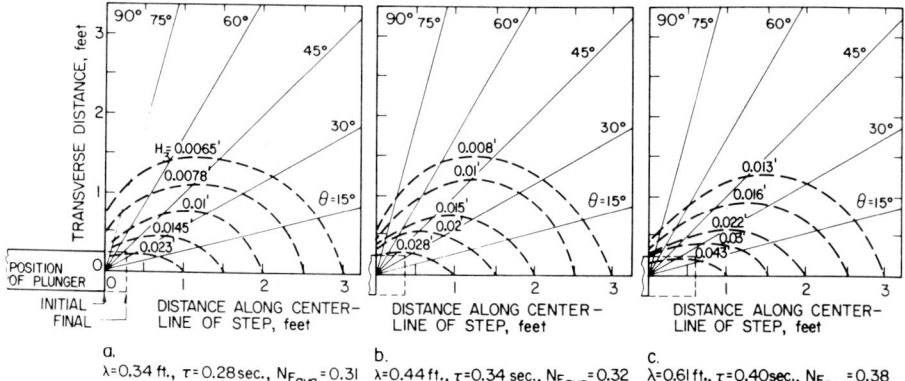

Fig. 7.30. Contours of equal crest elevation above SWL. $D/d = 0.50$, $d/W = 1.0$, $d = 0.50$ ft., $W = 0.50$ ft. (From Liu and Wiegel, 1974.)

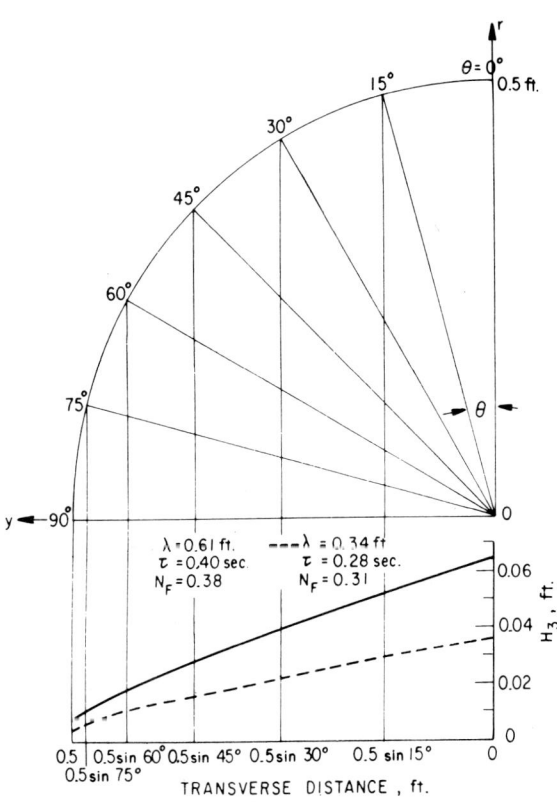

Fig. 31. Elevation of first wave crest above SWL at $r = 0.50$ ft. Radius vs. projection onto the transverse axis, $W = 0.50$ ft. (From Liu and Wiegel, 1974.)

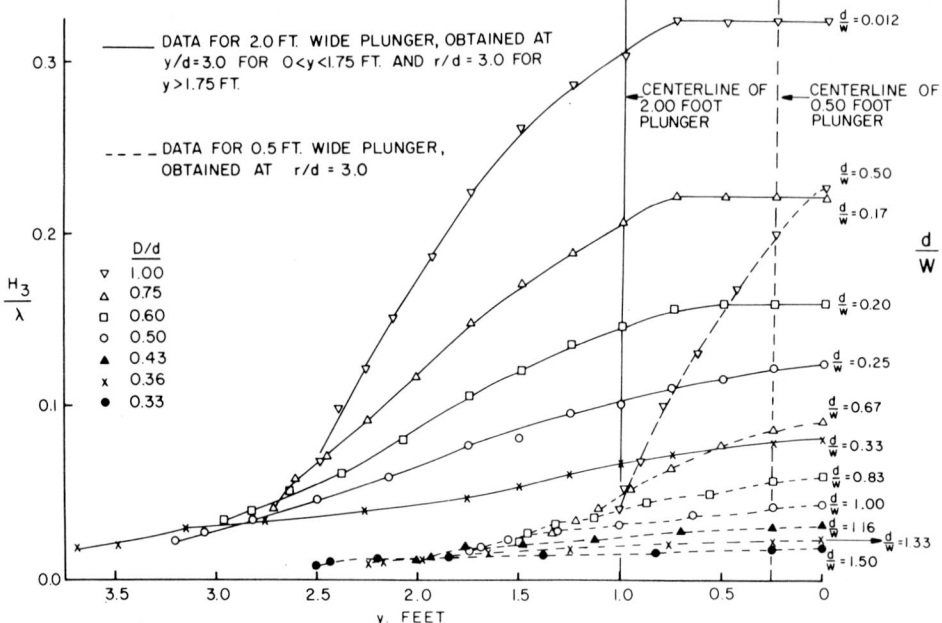

Fig. 7.32. Comparison of $H_3/\lambda$ vs. $y$ for the narrow plungers for several values of $d/W$ and $D/d$, with $N_{F\text{avg}} = 0.44$.

source (four times the width of the first source), and a comparison of some results for the two cases is shown in Fig. 7.32. It can be seen that if the curves for the $W = 0.50$ ft model are shifted so that $y = 0$ for them is made to coincide with $y = 1.75$ ft for the $W = 2.0$ ft model, the two sets of curves are similar in shape.

### 7.3.3 Travel across the ocean

Tsunami waves traveling across the ocean are affected by refraction owing to their great wave length (of the large tsunamis) and the variable water depth. Travel paths and wave front positions were obtained by graphical means for several tsunamis, using hydrographic charts. Although refraction drawings of reasonable accuracy can be constructed for small areas of the ocean by this method, the use of a large terrestrial globe should be used instead of charts.

An example of the refraction drawing for the Chilean tsunami of May 24, 1960 is shown in Fig. 7.33 (Takano et al., 1961). A globe 75 cm in diameter was used, with depths being transferred to the globe from hydrographic charts at the appropriate locations. Owing to the uncertainty of the source location,

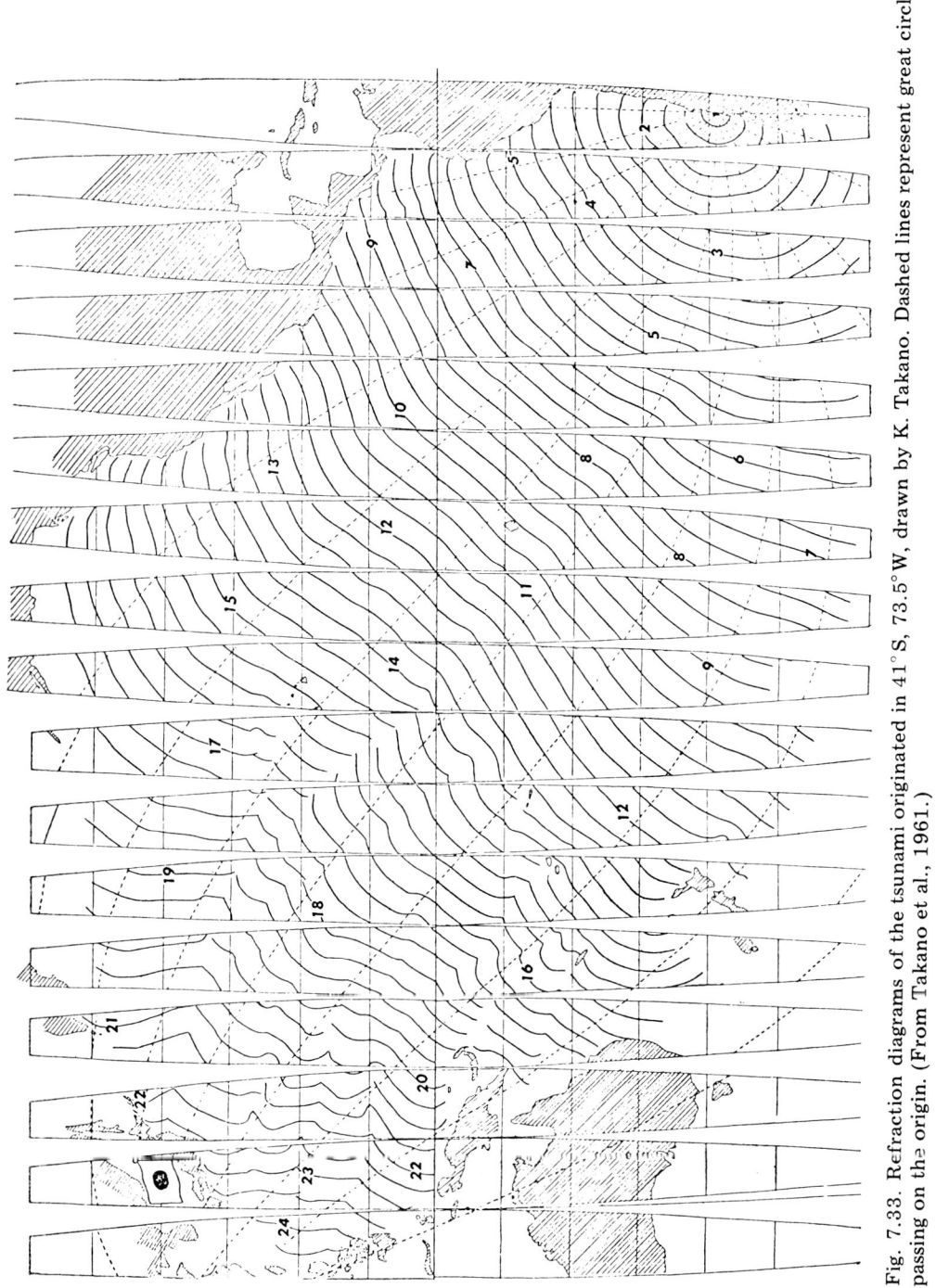

Fig. 7.33. Refraction diagrams of the tsunami originated in 41°S, 73.5°W, drawn by K. Takano. Dashed lines represent great circles passing on the origin. (From Takano et al., 1961.)

three refraction diagrams were constructed, each for a different origin. The dashed lines shown in the figure represent the great-circle paths passing through the origin. It is clear that there is a tendency for the tsunami energy to concentrate in the region of Japan, owing to this factor. Japan is situated nearly at the 'anti-pole' from the tsunami source. If the depth of the ocean were constant, the wave front as it moved across the ocean would always be perpendicular to the great circles. The numbers shown on the figure are wave travel times from the source, in hours.

The refraction diagrams in the paper cited were constructed with the origin being a point source. In reality, it was a very long source. In addition, diffraction and reflection effects are ignored in the construction of refraction diagrams. Numerical methods have been developed by which the travel of tsunami waves across the ocean can be calculated, for such a source, formulated in a finite-difference representation of the equations of motion, for a spherical coordinate system (Hwang et al., 1972; see also Houston and Garcia, 1974). Care must be exercised in choosing the proper grid spacing,

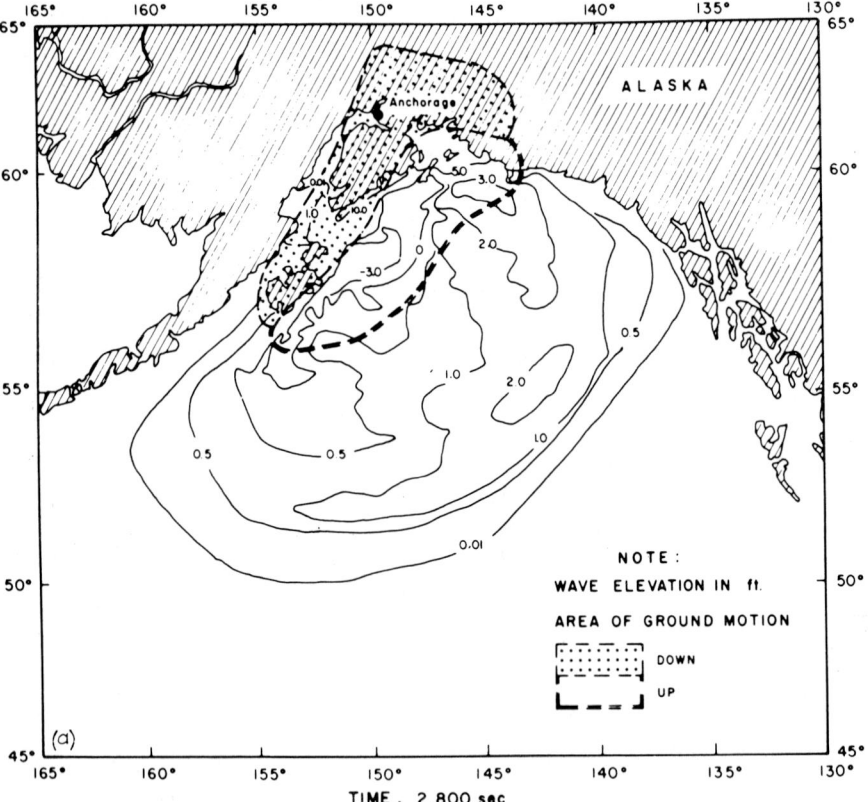

Fig. 7.34a. For caption see next page.

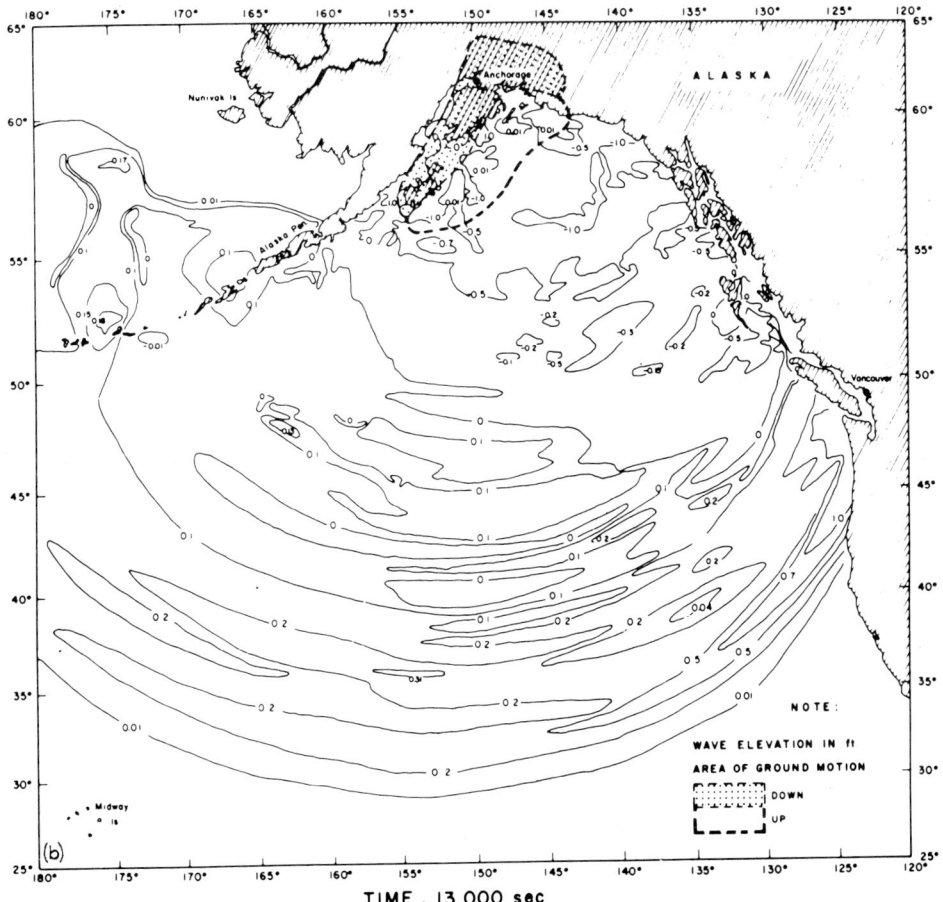

Fig. 7.34. Surface elevation contours. (a) 2,800 sec and (b) 13,000 sec after the 1964 Alaskan earthquake as determined by the revised numerical model. (From Hwang et al., 1972.)

which must be commensurate with both computational needs and the accuracy of the ocean depths and their locations. This has been done for several tsunamis, including the March 27, 1964 Alaskan tsunami (Hwang et al., 1972). The surface elevation contours at two different times after the earthquake are shown in Fig. 7.34.

## 7.4 EFFECTS ALONG THE COAST

The generation and travel of tsunamis are better understood than their behavior along coasts, from a quantitative standpoint. Refraction, diffraction, reflection, energy dissipation, trapped modes, resonance, run-up and draw-down are all important phenomena.

## 7.4.1 Refraction

Tsunamis are long waves compared with the depth of the body of water in which they are generated. Owing to this their speed of advance is approximated quite closely by the formula:

$$C = (gd)^{1/2} \tag{7.26}$$

although the equation for the speed at which the first crest of the wave train travels is much more complicated. Thus, a wave traveling at an angle to the bottom contours will bend, as the portion of the wave moving in deeper water will travel farther in a given time than will the portion of the wave in water that is less deep. This process is known as refraction, and an example on an oceanic scale is shown in Fig. 7.33.

Refraction is a mechanism of primary importance in regard to the height of tsunami waves along a portion of the coast. For example, consider the refraction diagram shown in Fig. 7.34 for the 1960 Chilean tsunami. If orthogonal lines were drawn perpendicular to the wave fronts, and the assumption made that there was no flow of energy across the orthogonals it would be seen that the portion of the tsunami waves traveling along the Pacific Coast of North America would be moving at glancing incidence. Thus, as the waves bent toward the coast their fronts spread over a relatively great distance, so that the energy density of them was not as great at the coast as it would have been for the case without refraction.

## 7.4.2 Wave trapping

A fundamental difficulty is encountered in constructing refraction drawings, when it is necessary to take care of reflection from the shore. At some water depth it is necessary to make a subjective decision that the incident wave reflects, with the angle of reflection being equal to the angle of incidence (unless the reflection is of the Mach type). Once this is done, it is a rather simple manner to construct wave refraction diagrams for the reflected waves. For some offshore conditions these reflected waves refract in such a manner that they become trapped in the nearshore zone or on the continental shelf (Isaacs et al., 1951; Williams and Isaacs, 1952). An example of such a situation was found to exist for tsunami waves at Hilo, Hawaii by Palmer et al. (1966), as is shown in Fig. 7.35.

## 7.4.3 Mach reflection

On occasions, tsunamis in certain coastal regions have exhibited characteristics which do not seem to be accounted for when they are studied

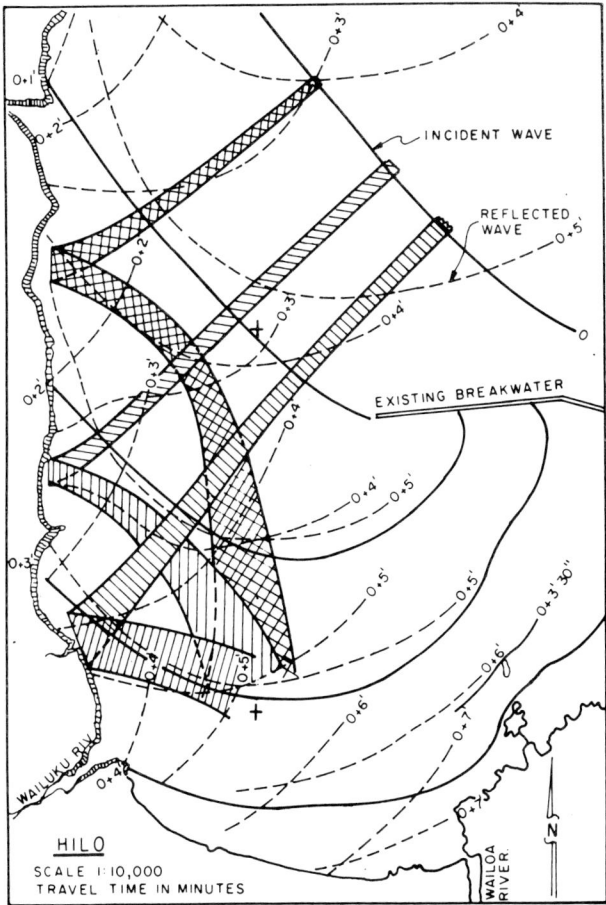

Fig. 7.35. 1960 tsunami refraction diagram for Hilo, Hawaii. (From Palmer et al., 1966.)

by means of the commonly used types of refraction diagrams, diffraction calculations and wave run-up theory and measurements. It was suggested to the author by Professor J.D. Isaacs in 1956 that this might be due to something analogous to the Mach-reflection phenomenon in acoustics. A number of aspects of this phenomenon have been investigated experimentally by several graduate students under the author's direction (Perroud, 1957; Chen, 1961; Sigurdsson and Wiegel, 1962; Nielsen, 1962).

The first experiments were performed with a solitary wave incident to an oblique, vertical, impervious, smooth barrier. It was found that the phenomenon existed. Subsequent tests were made with periodic waves, and it was found that the Mach-reflection occurred for values of $L/d$ (ratio of wave length to water depth) greater than about two. Several experiments were

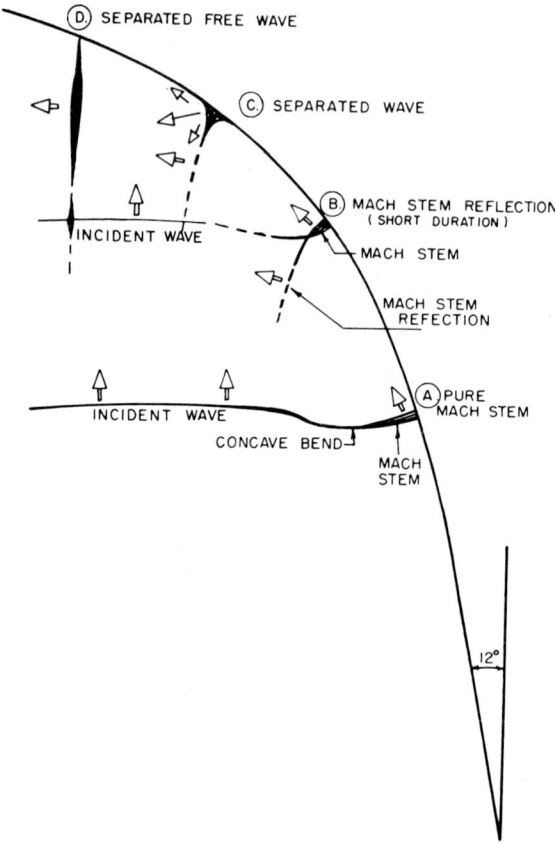

Fig. 7.36. Schematic development of wave crests along vertical curved breakwater. Wave heights are indicated by thickness of crest lines. Arrows show travel directions. (From Nielsen, 1962.)

performed with types of models which were of importance to coastal engineering problems. One of these was an undulating vertical wall, which was found to affect only the small details of the wave motion. A second model consisted of a curved, vertical impervious barrier (similar to vertical wall jetties or breakwaters). Experiments in a hydraulic model showed that once the Mach-stem formed, it became so strong that it ultimately became independent of the incident wave, as is shown in Fig. 7.36. The importance of this phenomenon on the wave height at the boundary can be seen in Fig. 7.37.

The Mach-reflection has been observed for regular wind-generated waves along the East Mole of the Harbor of Eckernförde on the German Baltic Coast and reported on by Fuhrboter et al. (1969). Measurements of wave

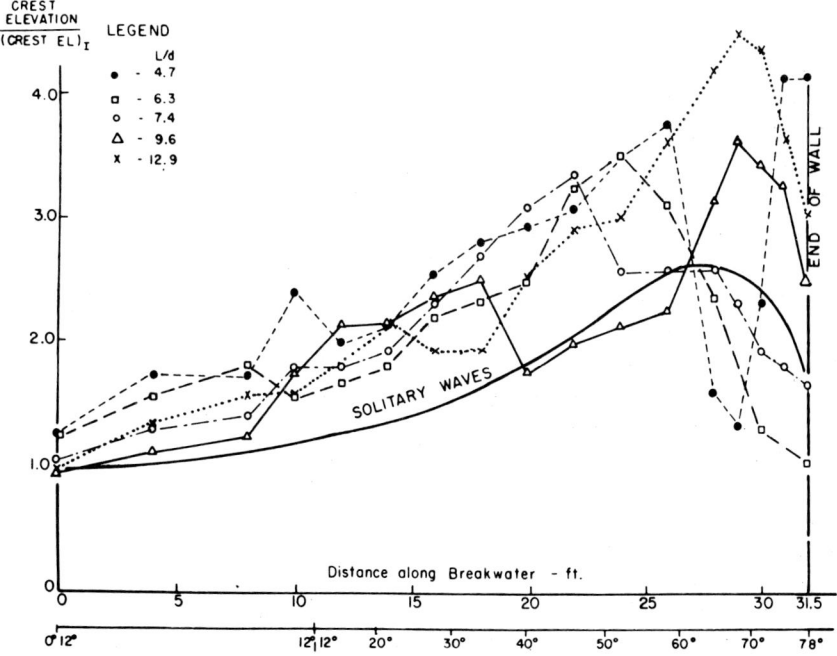

Fig. 7.37. Maximum wave elevation along curved vertical breakwater for different wave lengths (Richmond model). $H_I/d \simeq 0.14$, $d = 1.0$ ft. (From Nielsen, 1962.)

heights were made at a number of places along the mole, together with some measurements of wave pressures acting on the structure.

Hilo, Hawaii, is subject to severe damage from tsunamis which originate at a number of locations. The orientation, topography and hydrography of the region are such that it appeared likely that a Mach-stem might have been associated with the April 1, 1946 tsunami which originated in the Aleutian Islands of Alaska. The height of the tsunami should increase by a factor of 4—5 as it moves onto the shallow portion of the reef off Hilo, due to shoaling effects alone, and it was believed that this, together with the Mach-stem effect, could account for the characteristics of several of the waves of the tsunami as they were observed. It was observed (M.L. Child of Hilo, Hawaii) that the two waves that did the most damage came in as a bore in a southerly direction along the cliff that forms the west border of the bay, swinging easterly and running up through the streets of the town and into the lee area of the breakwater, and waves also came over the top of the breakwater. A photograph taken at the time shows a wave that looks remarkably like a Mach-stem (Wiegel, 1964b).

From some of the observations of the tsunami of 22 May 1960, originating in Chile, it appeared as if something similar must have happened, but the author could not visualize how it could have occurred.

In order to study the gross characteristics of tsunamis at Hilo, a 1 : 15,000 (1 : $\sqrt{15,000}$ = 1 : 122 time and velocity scale) undistorted model was constructed of fiberglass. The model was approximately 8 ft on a side, so that the entire bay could be included, as well as the reef. A portion of the ocean was included, to a depth of 6000 to 7000 ft, prototype. The model was placed at one end of an 8-ft wide by 6-ft deep by 200-ft long tank so that a number of waves could be measured before reflections could be of importance. A series of runs were made with periods ranging from 8 to 24 min. prototype, and with waves from the N, E, and SE directions.

It was found, for waves from the north, that a wave which had the appearance of a Mach-stem was generated along the west cliff and rolled into the town of Hilo in a manner that was similar to observations made of the actual tsunami. It was also found that the shoaling effect was about as theory predicted; this is, the wave height increased by a factor of about 4 over the reef, with respect to the wave height in the deep-water portion of the tank.

After the tests had been run, it was brought to the investigator's attention that due to refraction in the ocean, the tsunamis generated off Chile would most likely approach Hilo Bay from an easterly direction, rather than from a southeasterly direction as was originally supposed. Because of this, the results of the model tests for the waves from the east will be described herein. A remarkable phenomenon was observed in the 12—20 min. (prototype) period range. Referring to Fig. 7.38a, the initial wave refracted to about the position shown as (1)—(1). The northerly portion started to reflect from the coast while the southerly portion continued to move toward shore. This resulted in the pattern (2A)—(2A) as the reflected portion and (2)—(2) as the continuing portion. As the reflected portion (2A)—(2A) moved down the coast, it became independent of (2)—(2). At the same time the southerly tip of (2)—(2) diffracted into the harbor, raising the water level. About at the same time (2A)—(2A) progressed to position (3)—(3) with the portion near the coast being considerably higher than the portion offshore. The portion near the coast ran right along the coast, reaching positions (4)—(4) and (5)—(5) as a high wave running on top of the water which had diffracted into the harbor from (2)—(2). It then ran into the town of Hilo. The author believes that something similar to this must have happened during the actual tsunami.

The transformation of (2A)—(2A) to (3)—(3) was probably caused by a combination of refraction and reflection, together with some nonlinear effects because of the relatively large wave height to water depth ratio along the coast, and the height of the tsunami at Hilo was probably due to this combined with the diffracted wave. At some place between (2A)—(2A) and (3)—(3) a Mach-stem type of phenomenon evolved and because of its strength became independent of the normally reflected portion of the wave.

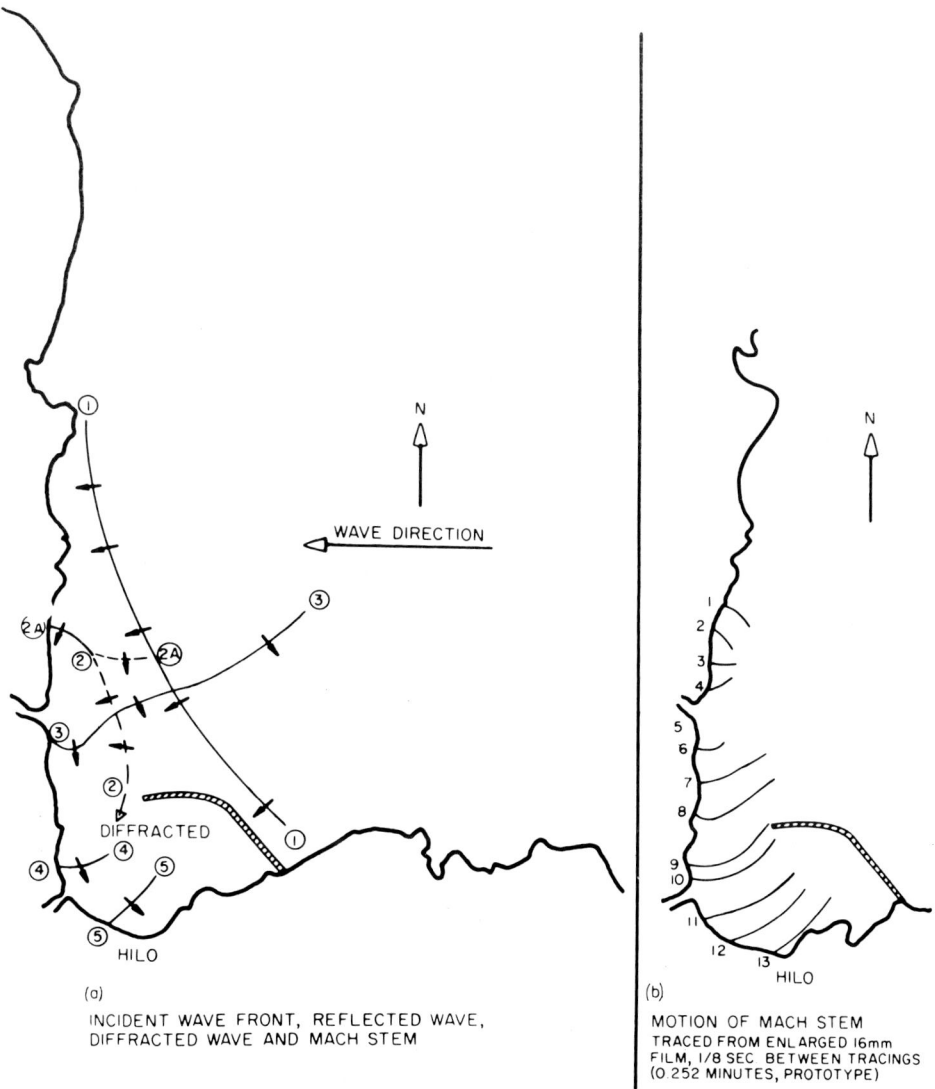

Fig. 7.38. Some results from a 1 : 15,000 scale-model study of tsunamis (16 minute, prototype) at Hilo, Hawaii. Run x, 8 Feb 1963, wave from the east (in 7,000 ft. of water, prototype). (From Wiegel, 1964b.)

In Fig. 7.38b are shown the successive positions of this Mach-stem type of wave as it moves along the coast. These positions were traced from enlargements of a 16-mm motion picture taken during the model study, for a wave of 8-sec period (16-min. period in the prototype).

## 7.4.4 Resonance

A general feature of resonance is that the greater the build-up of an oscillation in a bay or harbor, the larger will be the number of waves necessary to cause the peak response of the bay or harbor. Thus, if the highest waves occur shortly after the arrival of the first wave, then one would expect that damping is of considerable importance. For example, there is considerable evidence that damping is of major importance in San Francisco Bay, California (Wiegel, 1970).

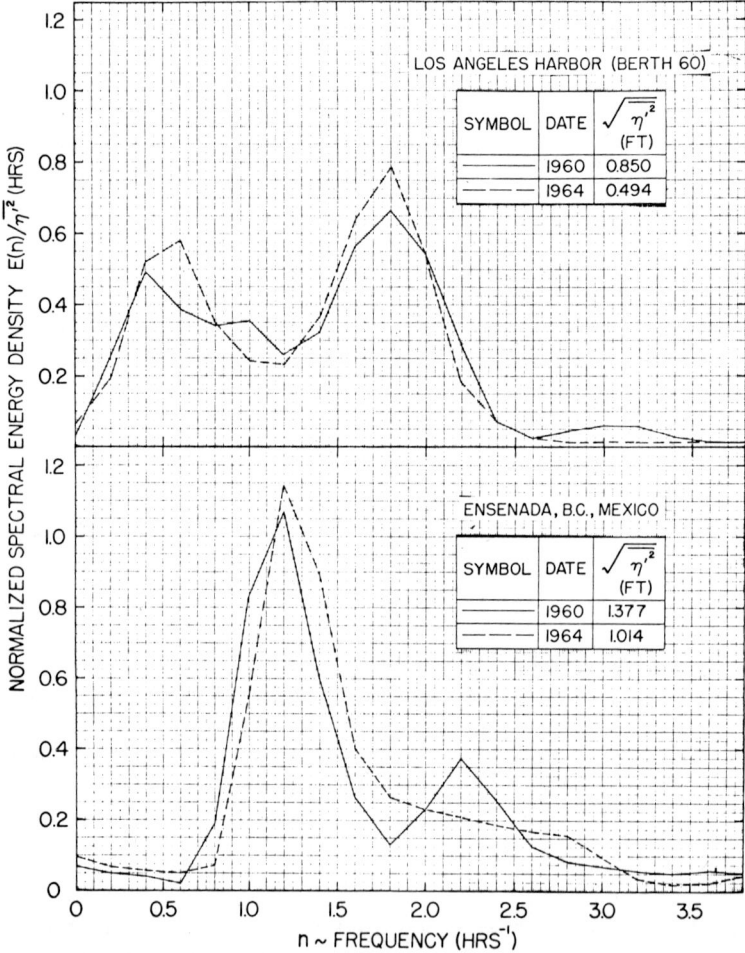

Fig. 7.39. Normalized energy density spectra for tsunamis at Los Angeles Harbor (Berth 60) and Ensenada, B.C., Mexico caused by Chilean (1960) and Alaskan (1964) earthquakes. (From Raichlen, 1972.)

Some bays and harbors (and, perhaps, continental shelfs) exhibit the characteristics of oscillations at natural frequencies with tsunami wave spectra being the forcing function (Wilson, 1971). A good example of this is Los Angeles Harbor (Berth 60), Fig. 7.39. The normalized energy density spectra are similar for both the 1960 Chilean tsunami and the 1964 Alaskan tsunami (Raichlen, 1972). The same is true for another location, Ensenada, Baja California (Mexico), but the energy density spectra for Ensenada are different from those of Los Angeles Harbor. The energy density spectra at Santa Monica, California (Fig. 7.40) is similar in shape to the one at Ensenada, but the frequency of the peak is different for the two places. In these figures the energy densities are normalized by dividing the actual values by the mean square of the marigram (the tide record after the predicted astronomical tide has been subtracted), $\overline{\eta'^2}$ (also called the variance). Values of the standard deviation ($\sqrt{\overline{\eta'^2}}$) are shown in the figures.

Some problems associated with calculating the response of bays to tsunamis are discussed by Houston and Garcia (1974).

## 7.4.5 Run-up and draw-down

The problem of the run-up and draw-down of tsunami waves on a shore is complicated. There is no exact mathematical model covering the general

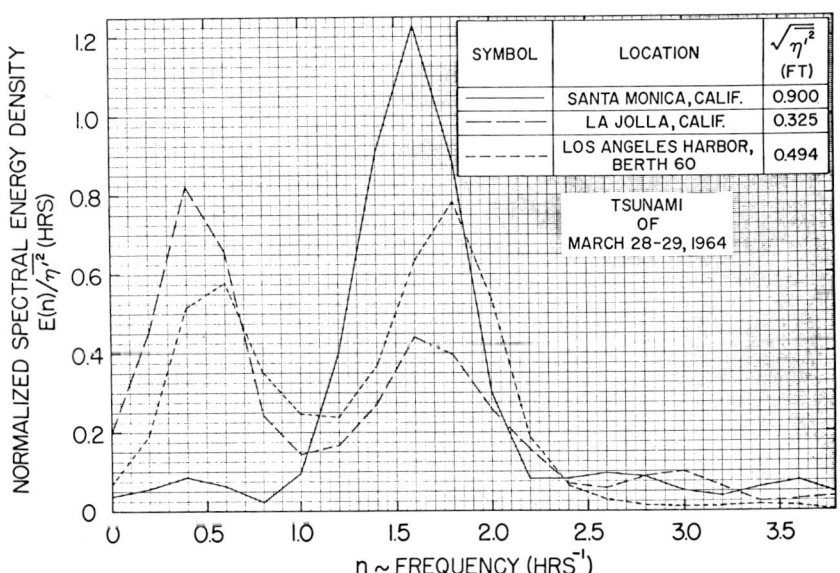

Fig. 7.40. Normalized energy density spectra for tsunamis at three southern California locations for the Alaskan earthquake of March 27, 1964. (From Raichlen, 1972.)

Fig. 7.41. Tsunami were run-up and wave receding over Puumaile Hospital grounds, Hilo, Hawaii, April 1, 1946. (Courtesy, Modern Camera Center, Hilo, Hawaii.)

case, but there are several approximate theories with limited ranges of validities. A discussion of some of these theories is available in Wiegel (1970), and in the report by Houston and Garcia (1974).

There are two essential difficulties involved in obtaining reliable run-up and draw-down information, aside from the mathematical difficulties for even simple boundaries. The first is the lack of precise tsunami wave data offshore to be used as the input to calculating run-up and draw-down, and the second is the irregularity and roughness of the shore (see, for example, the photograph in Fig. 7.41 of a tsunami wave running over land).

Owing to this, it is usually necessary to make use of observations, together with statistical inferences from these observations. Some results of observed wave run-ups along the coast of Japan are shown in Fig. 7.42.

There are combinations of conditions where a bore might form during the run-up of tsunami waves. However, it is most common for the run-up to have the appearance of a rapidly rising tide. The draw-down occurs in a similar manner, and can be quite rapid. During the 1964 Alaskan tsunami, an observer, Mr. Vern Knight at the Aeolian Yacht Club (Alameda) in San Francisco Bay, told the author that he saw fish that were stranded, 'flapping on the mud flats', during at least one draw-down.

There are a number of tide gages in harbors that have recorded tsunami waves (or at least the response of the harbor to tsunami waves). As far as the author knows, there has been no systematic study of the relationship between the tsunami wave crest height as measured by a tide gage and the associated run-up nearby. A study made by the author at Crescent City,

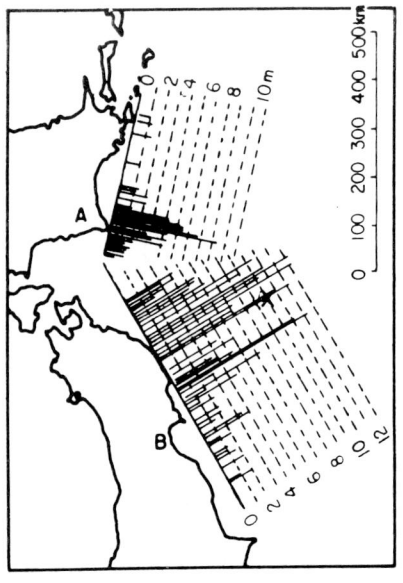
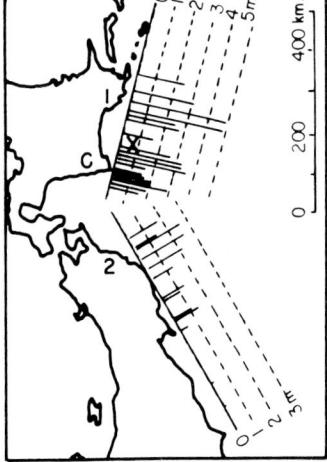
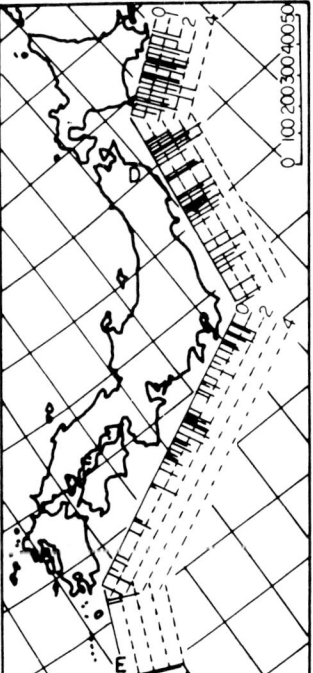

Fig. 7.42. Tsunami height distribution along the coast of Japan. (From Watanabe, 1970.) (a) The distribution of maximum height along the open coast for the Sanriku-Oki tsunami of March 2, 1933. × = origin of tsunami. (b) The distribution of maximum height along the open coast for the Tokachi-Oki tsunami of March 4, 1952. × = origin of tsunami. (c) The distribution of maximum height along the open coast for the Chilean tsunami of May 23, 1960.

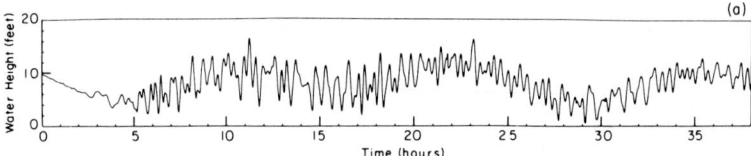

Fig. 7.43a. Water height recorded at Crescent City, California during the tsunami of May 22, 1960. (From Borcherdt and Borgman, 1970.)

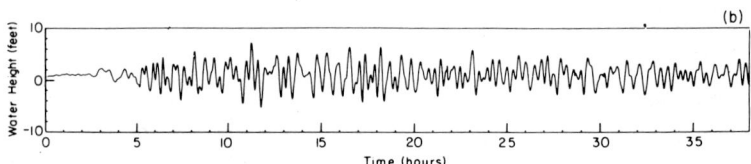

Fig. 7.43b. Water height recorded at Crescent City, California minus predicted astronomical water height during the tsunami of May 22, 1960. The time scale begins at 0:00 a.m. PST. (From Borcherdt and Borgman, 1970.)

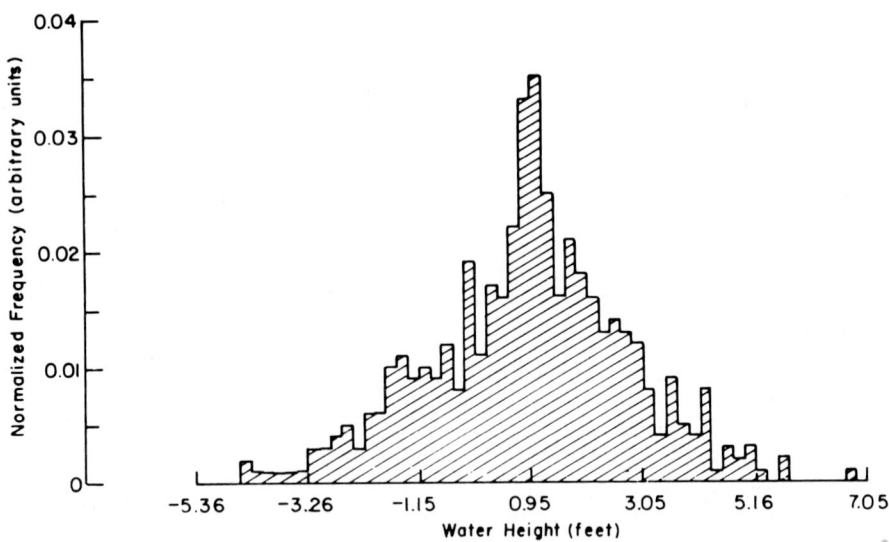

Fig. 7.44. Normalized histogram for estimated tsunami water height above astronomical water height as recorded at Crescent City, California during the tsunami of May 22, 1960. (From Borcherdt and Borgman, 1970.)

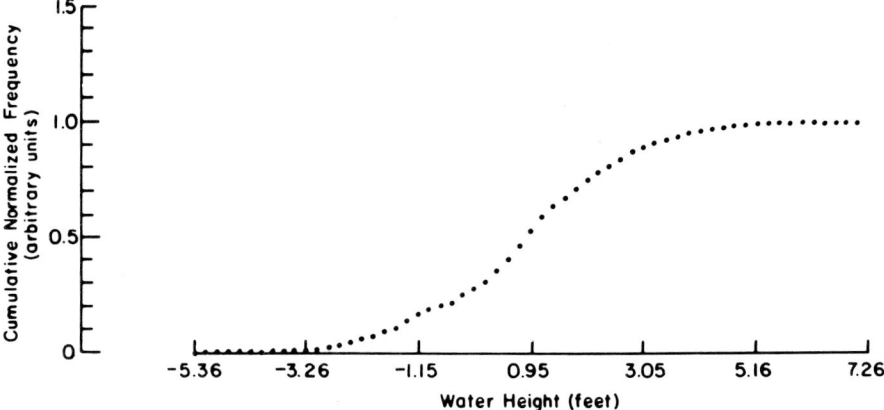

Fig. 7.45. Cumulative histogram for estimated tsunami water height above astronomical water height as recorded at Crescent City, California during the tsunami of May 22, 1960. (From Borcherdt and Borgman, 1970.)

California, indicated that the values were about the same for the tsunami studied. This appears to be the case, also, for Hilo, Hawaii, during the 1960 Chilean tsunami (Houston and Garcia, 1974, pp. A14—A16). It seems reasonable as the tide gages were located a small fraction of a tsunami wave length from the shoreline.

No systematic study has been made, to the author's knowledge, to determine whether or not tsunami wave crests and troughs are equally distributed about the astronomical tide level, on the average. A number of marigrams of tsunamis should be studied in the manner of Borcherdt and Borgman (1970), Figs. 7.43, 7.44 and 7.45. The data they obtained for the 1960 Chilean tsunami at Crescent City showed the run-up to exceed the draw-down, on the average.

7.5 DISTRIBUTION FUNCTIONS

It is necessary to make use of measured tsunamis, or in their absence to make use of observed maximum tsunami wave run-ups at a site to obtain distribution functions for a particular site. Such distribution functions are available for a few places (Soloviev, 1970; Wiegel, 1970; Rascón and Villarreal, 1974). Dates, heights, sources and other information on tsunamis in the Pacific Ocean have been catalogued (U.S. Coast and Geodetic Survey, 1964; Adams, 1967; Iida et al., 1967). These sources can be used to develop distribution functions for some other areas.

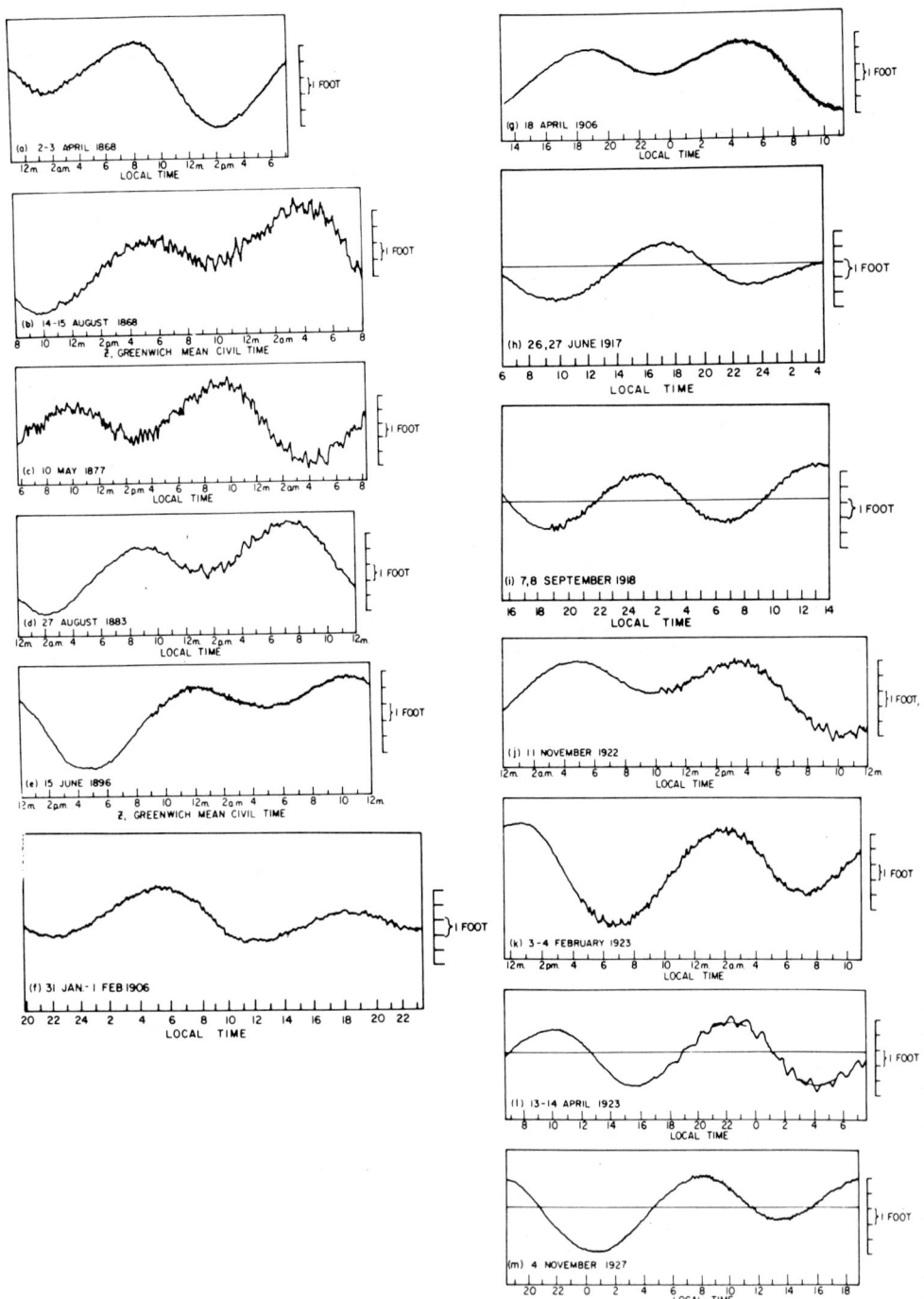

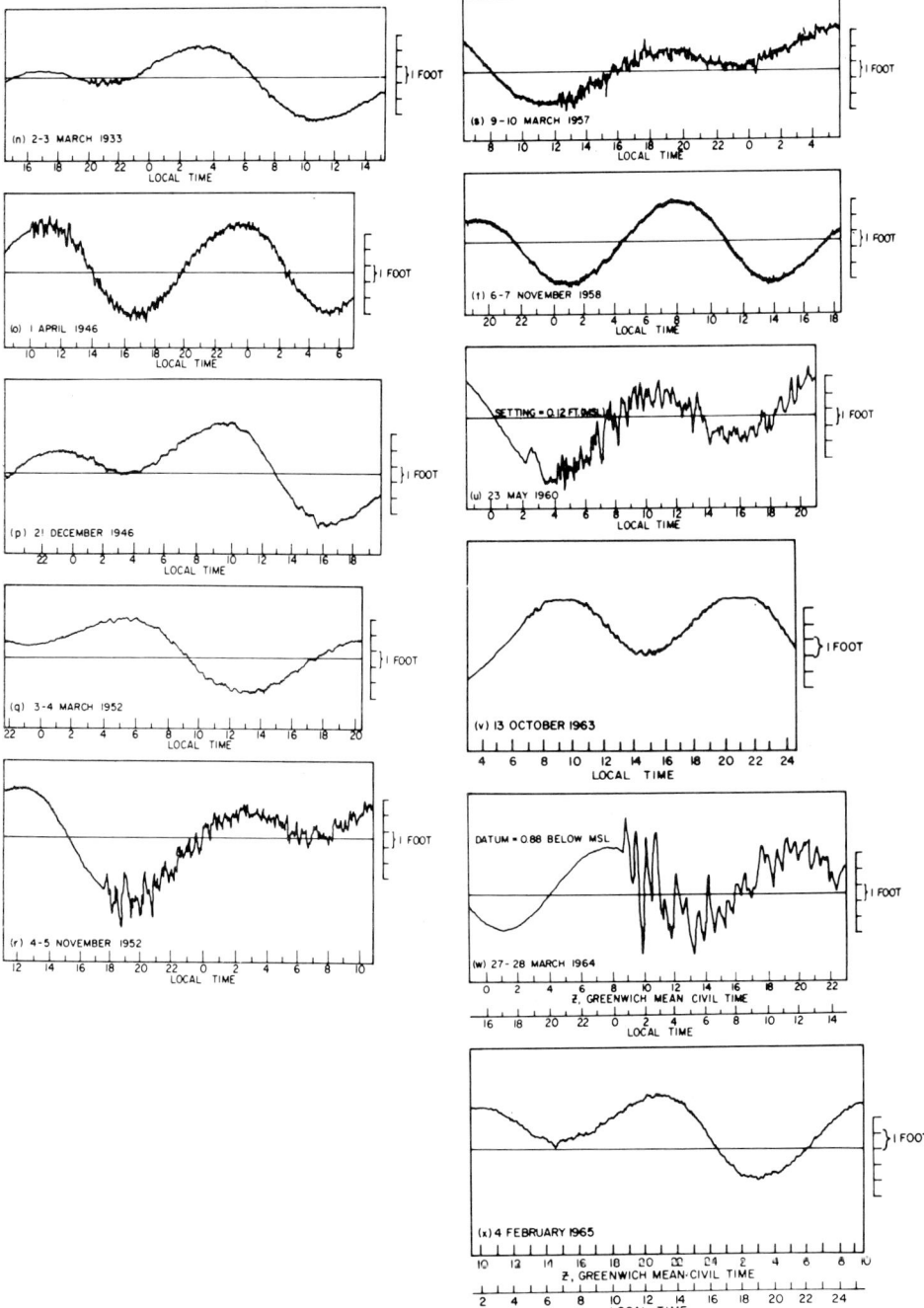

Fig. 7.46. Traces of U.S. Coast and Geodetic Survey tide-gage records for a number of tsunamis, Presidio, San Francisco Bay, California.

TABLE 7.III

Height of highest wave observed on tide gage records showing tsunamis at The Presidio (or Fort Point), entrance to San Francisco, Bay, California, 1868—1974, listed according to rank at the tide gage

| No. | Date | Height of highest tsunami wave (trough to crest) in feet | Source |
|---|---|---|---|
| 1 | 27 Mar 1964 | 7.50 | Gulf of Alaska |
| 2 | 23 May 1960 | 3.75 | South Chile |
| 3 | 4 Nov 1952 | 3.50 | Kamchatka |
| 4 | 9—10 Mar 1957 | 1.50—1.75 | Aleutian Islands |
| 5 | 1 Apr 1946 | 1.25—1.50 | East Aleutian Islands |
| 6 | 13—14 Apr 1923 | 1—1.25 | Kamchatka |
| 7 | 9—10 May 1877 | 1 + | Chile |
| 8 | 27 Aug 1883 | ~1 | Krakatoa Volcano Explosions |
| 9 | 14—15 Aug 1868 | ~1 | Peru—Chile |
| 10 | 11 Nov 1922 | 0.75—1 | North Chile |
| 11 | 3 Feb 1923 | 0.5 + | Kamchatka |
| 12 | 2—3 Mar 1933 | 0.5 | Sanriku, Japan |
| 13 | 26 Jun 1917 | ~0.5 | Samoa Islands |
| 14 | 31 Jan—1 Feb 1906 | ~0.5 | Colombia—Equador |
| 15 | 7 Sep 1918 | 0.25—0.50 | Kuril Islands |
| 16 | 18 Apr 1906 | 0.25—0.50 | San Francisco (San Andreas Fault) |
| 17 | 2 Apr 1868 | 0.25—0.50 | Sandwich (Hawaiian) Islands |
| 18 | 4 Feb 1965 | 0.25—0.50 | Rat Islands (Aleutians) |
| 19 | 13 Oct 1963 | 0.25—0.50 | Kuril Islands |
| 20 | 15 Jun 1896 | 0.25 | Sanriku, Japan |
| 21 | 4 Nov 1927 | ~0.25 | Point Arguello, California |
| 22 | 19—21 Dec 1946 | ~0.25 | Nankaido, Japan |
| 23 | 4 Mar 1952 | ~0.25 | Tokachi, Hokkaido, Japan |
| 24 | 6—7 Nov 1958 | ~0.25 | Kuril Islands |

*7.5.1 Entrance to San Francisco Bay, California, and other locations*

Tide gages have been in operation at the entrance to San Francisco Bay for more than one hundred years. The author has obtained records from the U.S. Coast and Geodetic Survey (now a part of NOAA), and other sources, for 24 tsunamis. These are ranked according to the height of the highest wave in Table 7.III. Copies of the tide gage records are shown in Fig. 7.46. The distribution function for this location, together with several other locations are given in Fig. 7.47.

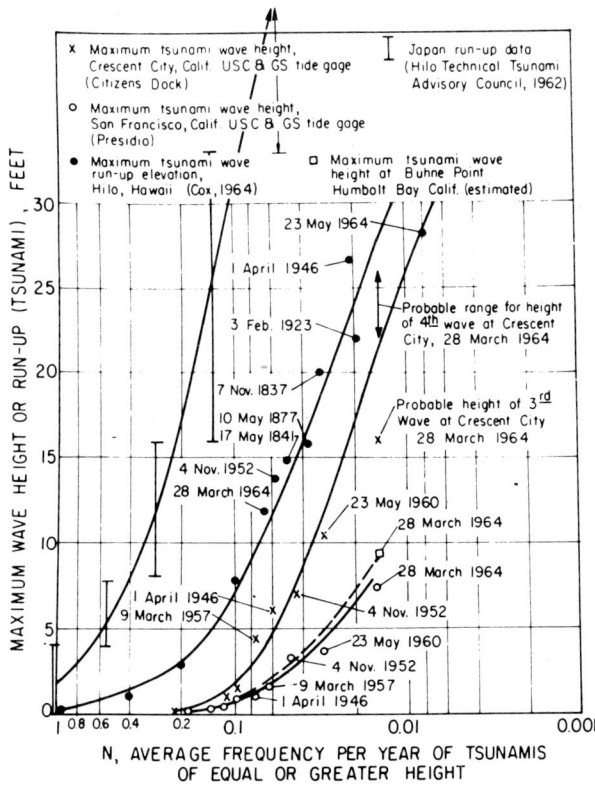

Fig. 7.47. Distribution function for maximum tsunami waves. (From Wiegel, 1964a.)

*7.5.2 Risk*

For areas located an appreciable distance from the origin of a tsunami, it is possible to provide a warning system, thus providing a mechanism for increasing public safety. Such a system exists: the International Tsunami Warning System (TWS). The Honolulu Observatory is the Operational Center of the TWS, and distributions tsunami watches and warnings for the International Tsunami Information Center (ITIC) to warn service participants through the Pacific Basin. The TWS makes use of 31 seismic stations, 51 tide stations and 47 dissemination points scattered throughout the Pacific Basin under the varying control of 15 different nations with general guidance from UNESCO's Intergovernmental Oceanographic Commission (IOC), through the International Coordination Group for the Tsunami Warning System in the Pacific (ICG/ITSU) and thence through ITIC. The present address of the operational center of the TWS is: Honolulu Observatory, NOAA-National Weather Service, 91-270 Fort Weaver Road, Ewa Beach, Hawaii 96706, U.S.A.

Engineers are always faced with a difficult problem in assessing the risk associated with the design of a structure where geophysical phenomena are involved. The main difficulty is an insufficient length of observations. Use is made of distribution functions of the type given in Fig. 7.47, together with the more or less subjective use of general types of formulae (such as the Poisson distribution) which have been developed from studies of time series for which a greater number of data are available. The Poisson distribution is given by:

$$R = 1 - e^{-ND} \qquad (7.27)$$

where $R$ is the chance of a given height being exceeded in $D$ years duration (the 'return period'), and $N$ is the expected number of occurrences per year, as is given in Fig. 7.47. This method has been used by Wiegel (1970) for tsunami risk evaluation, based upon the work of Wemelsfelder (1961) for storm surges.

In order to make better use of the few data that are usually available for a particular location, an extrapolation can be made to predict maximum tsunami wave heights for long recurrence intervals, using Bayes' theorem. For details on the application of Bayes' theorem to the problem of tsunamis the reader is referred to the paper by Rascón and Villarreal (1974).

7.6 COMBINED TIDE AND TSUNAMI PROBABILITIES

For many purposes, it is desirable to be able to calculate the probabilities of occurrence of combined astronomical tide and tsunamis. The time at which a tsunami is generated cannot be predicted, and must be considered to be a phenomenon which occurs randomly in time. The water heights due to astronomical tides can be specified for a given location at any particular time. However, when considering the problem of estimating the water height

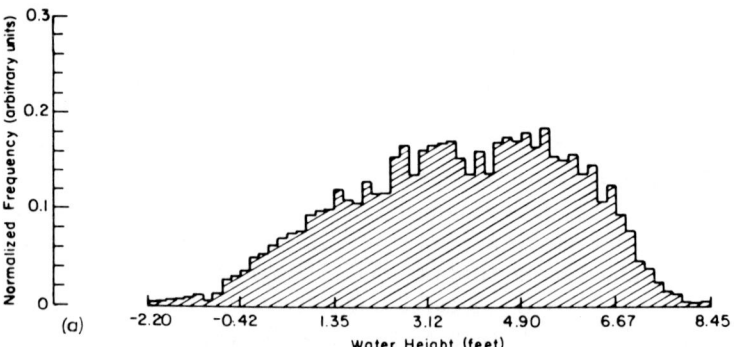

Fig. 7.48a. For caption see next page.

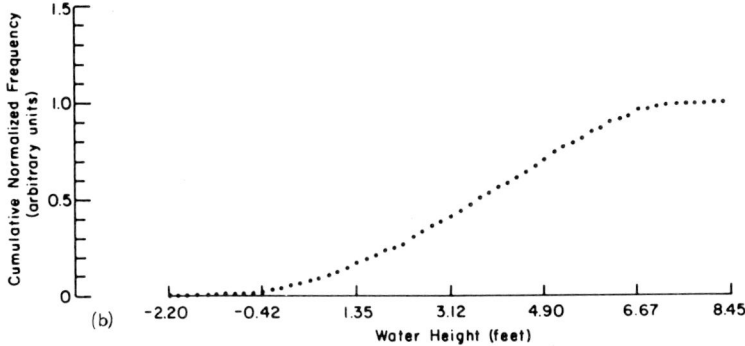

Fig. 7.48. Histograms (a: normalized and b: cumulative) for 369-day series of astronomical water height, beginning January 1, 1960 at Crescent City, California. (From Borcherdt and Borgman, 1970.)

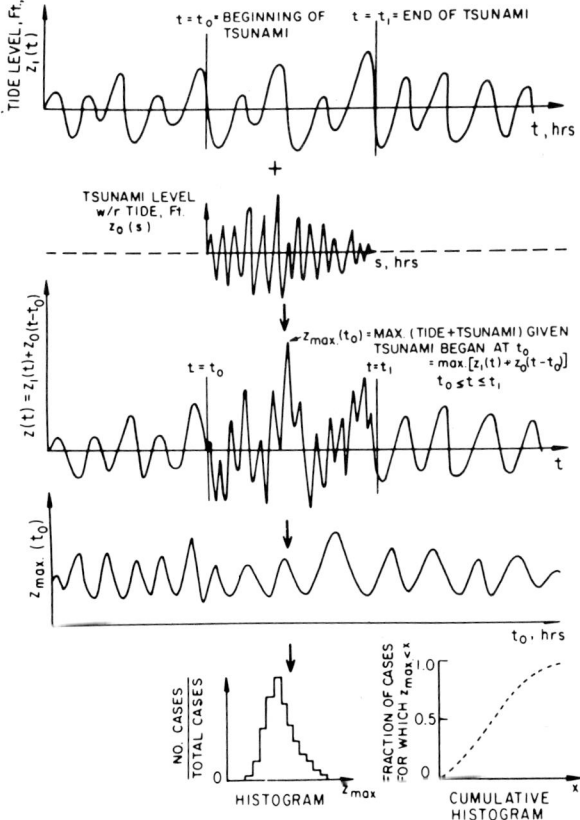

Fig. 7.49. Schematic illustration of method for combining known tsunamis with astronomical tide for random time of occurrence. (From Petrauskas and Borgman, 1971.)

due to both a tsunami and astronomical tides, it is necessary to consider the astronomical water height as a known function of a random variable.

Borcherdt and Borgman (1970) used a well-known (Shureman, 1940) technique for predicting the height of the astronomical tide. They prepared a simple computer program to calculate the tide height for any location for which the tidal constituents are available. The normalized histogram and the cumulative histogram for the 369-day series, for Crescent City, California, beginning January 1, 1960 are shown in Fig. 7.48. They did several similar sets of calculations, and found the shape of the histograms to be similar to that of a beta distribution. They made use of the tide calculations to subtract them from the tide-gage record at Crescent City of the May 22, 1960 tsunami. The resulting normalized histogram and cumulative histogram are shown in Figs. 7.44 and 7.45.

Petrauskas and Borgman (1971) continued the work cited above and developed a computer method for calculating the data necessary to construct a frequency diagram for the fraction of the year the astronomical tides

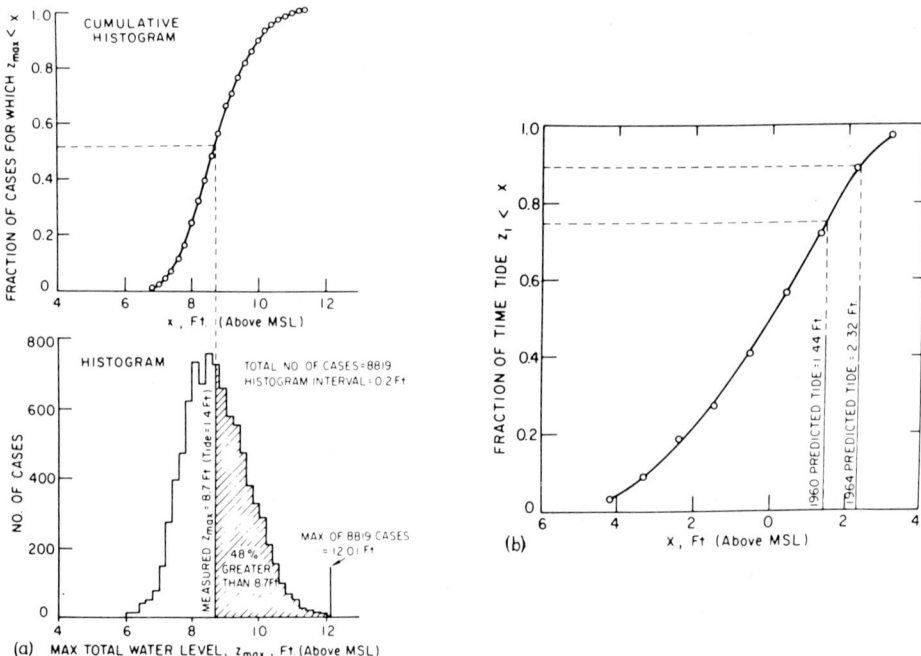

Fig. 7.50a. Histogram and cumulative histogram for $z_{max}$ calculated by combining 1960 hourly predicted tide with 1960 tsunami hourly maxima. (From Petrauskas and Borgman, 1971.)

Fig. 7.50b. Cumulative histogram of the Crescent City predicted tide in feet above mean sea level. (From Petrauskas and Borgman, 1971.)

would combine with a given recorded tsunami to produce a specified combined water-level elevation. The technique is shown schematically in Fig. 7.49. The results for the May 22, 1960 tsunamis at Crescent City are shown in Fig. 7.50. The 8.7 ft. elevation above mean sea level associated with this tsunami could easily have been higher if the tsunami had occurred at another time of the year when the tides were more extreme. In fact, 48% of the time during the year the combined water-level elevations would have been greater than that which occurred. A similar study made for the March 27, 1964 tsunami at Crescent City showed that in only 15% of the year would the combined tide and tsunami have resulted in a maximum water level greater than that which occurred.

REFERENCES

Abramowitz, M. and Stegun, I., 1964. *Handbook of Mathematical Functions with Formulas, Graphs, and Mathematical Tables.* U.S. National Bureau of Standards, Applied Mathematics Series No. 55.

Adams, W.M., 1967. *An Index to Tsunami Literature to 1966.* Data Rep. No. 8, HIG 67-21, Hawaii Institute of Geophysics, University of Hawaii.

Bakhmeteff, B.A., 1933. Discussion of water pressures on dams during earthquakes. *Trans. ASCE,* 98: 460—468.

Balakina, L.M., 1970. Relationship of tsunami generation and earthquake mechanism in the northwestern Pacific. In: W.M. Adams (editor), *Tsunamis in the Pacific Ocean.* East—West Center Press, Univ. of Hawaii, Honolulu, pp. 47—55.

Berg, E., Cox, D.C., Furumoto, A.S., Kajiura, K., Kawasumi, H. and Shima, E., 1972. Source of the major tsunami. In: *The Great Alaska Earthquake of 1964: Oceanography and Coastal Engineering.* Committee on the Alaska Earthquake of the Division of Earth Sciences, National Research Council, National Academy of Sciences, Washington, D.C., pp. 122—139.

Bolt, B.A., Lomnitz, C. and McEvilly T.V., 1968. Seismological evidence on the tectonics of central and northern California and the Mendocino Escarpment. *Bull. Seismol. Soc. Am.,* 58 (6): 1725—1767.

Bonilla, M.G., 1970. Surface faulting and related effects. In: R.L. Wiegel (editor), *Earthquake Engineering.* Prentice-Hall, Englewood Cliffs, N.J., pp. 47—74.

Borcherdt, R.D. and Borgman, L.E., 1970. *Empirical Probability Distributions for Astronomical Water Height.* Tech. Rep. No. HEL 16-6, Hydraulic Engineering Laboratory, University of California, Berkeley, Calif., 32 pp.

Cameron, J.B., 1961. Earthquakes in northern California coastal region, II. *Bull. Seismol. Soc. Am.,* 51 (3): 337—354.

Chen, T.C., 1961. *Experimental Study on the Solitary Wave Reflection Along a Straight Sloped Wall at Oblique Angle of Incidence.* U.S. Army Corps of Engineers, Beach Erosion Board, Tech. Memo. No. 124.

Chester, C., Friedman, B. and Ursell, F., 1957. An extension of the method of steepest descents. *Proc. Cambr. Philos. Soc.,* 53 (3): 599—611.

Committee for Field Investigation of the Chilean Tsunami of 1960, 1961. *The Chilean Tsunami of May 24, 1960.* Maruzen, Tokyo, 397 pp.

Coulter, H.W. and Migliaccio, R.R., 1966. Effects of the earthquake of March 27, 1964 at Valdez, Alaska. *U.S. Geol. Surv. Prof. Pap.,* 542-C, 36 pp.

Das, M.M. and Wiegel, R.L., 1972. Waves generated by horizontal motion of a wall. *J. Waterways, Harbors, Coastal Eng. Div., Proc. ASCE*, 98 (WW1): 49—65.

Davison, C., 1936. *Great Earthquakes*. Murby, London.

Dukleth, G.W., 1970. Letter to John R. Teerink, Resources Agency, State of California.

Dwyer, M.J., 1970. Letter to P.J. Lorens, Resources Agency, State of California.

Fuhrboter, A., Hager, M., Hensen, W. and Tomczak, G., 1969. Methods for deciding limiting design conditions to be allowed for in the satisfactory economic conception of maritime structures with reference to the probability of damage to the destruction of the structure as determined by its characteristics in relation to the duration and direction of the swell and waves. *XXII Int. Navigation Congr., Paris, PIANC, Section II, Subject 5*, Permanent International Association of Navigation Congresses, pp. 5—28.

Garcia, W.J., Jr., 1972. *A study of Water Waves Generated by Tectonic Displacements*. Thesis, Dep. Civ. Eng., Univ. of California, Berkeley, Calif., 114 pages.

Grantz, A., Plafker, G. and Kachadoorian, R., 1964. Alaska's Good Friday earthquake, March 27, 1964: a preliminary geologic evaluation. *U.S. Geol. Surv. Circular*, 491: 35 pp.

Hatori, T., 1963. Directivity of tsunamis. *Bull Earthquake Res. Inst., Univ. of Tokyo*, 41: 61—81.

Hatori, T., 1970. Dimensions and geographic distribution of tsunami sources near Japan. In: W.M. Adams (editor), *Tsunamis in the Pacific Ocean*. East—West Center Press, University of Hawaii, Honolulu, pp. 69—83.

Houston, J.R. and Garcia, A.W., 1974. *Type 16 Flood Insurance Study: Tsunami Predictions for Pacific Coastal Communities*. Tech. Rep. H-74-3, U.S. Army Waterways Experiment Station, Vicksburg, Miss., 128 pages.

Hwang, L.-S., Divorky, D. and Yuen, A., 1971. *Amchitka Tsunami Study*. Tetra Tech, Inc., Rep. NVO-289-7, Nevada Operations Office, U.S. Atomic Energy Commission.

Hwang, L. S., Butler, H.L. and Divorky, D.J., 1972. Tsunami model: Generation and open-sea characteristics. *Bull. Seismol. Soc. Am.*, 62 (6): 1579—1596.

Iida, K., 1963a. Magnitude, energy and generation mechanisms of tsunamis and a catalog of earthquakes associated with tsunamis. *Proc. Tsunami Meetings Associated with the Tenth Pacific Science Congress. IUGG Monogr.*, 24: 7—18.

Iida, K., 1963b. On the estimation of tsunami energy. *Proc. Tsunami Meetings Associated with the Tenth Pacific Congress. IUGG Monogr.*, 24: 167—173.

Iida, K., 1970. The generation of tsunamis and the focal mechanism of earthquakes. In: W.M. Adams (editor), *Tsunamis in the Pacific Ocean*. East—West Center Press, Univ. of Hawaii, Honolulu, pp. 1—18.

Iida, K., Cox, D.C. and Pararas-Caragannis, G., 1967. *Preliminary Catalog of Tsunamis Occurring in the Pacific Ocean*. Data Rep. No. 5, HIG 67-10, Hawaii Institute of Geophysics, University of Hawaii. August 1967.

Isaacs, J.D., Williams, E.A. and Eckart, C., 1951. Reflection of surface waves by deep water. *Trans. Am. Geophys. Union*, 32 (1): 37—40.

Jordan, J.N., Lander, J.F. and Black, R., 1965. Aftershocks of the 4 February 1965 Rat Islands earthquake. *Science*, 148 (3675): 1323—1324.

Kajiura, K., 1963. The leading wave of a tsunami. *Bull Earthquake Res. Inst., Univ. Tokyo*, 41 (3): 531—571.

Kajiura, K., 1967. Tsunami. *Zisin (J. Seismol. Soc. Japan)*, 20 (4): 219—222.

Kennard, E.H., 1949. Generation of surface waves by a moving partition. *Q. Appl. Math.*, 7 (3): 303—312.

Kranzer, HC. and Keller, J.B., 1959. Water waves produced by explosions. *J. Appl. Phys.*, 30 (3): 398—407.

Leet, D.L., 1948. *Causes of Catastrophe: Earthquakes, Volcanos, Tidal Waves, Hurricanes*. McGraw-Hill, New York.

Liu, S.-L. and Wiegel, R.L., 1974. *Three-dimensional Hydraulic Model Study of Water*

*Waves Generated by Horizontal Tectonic Displacements.* Tech. Rep. No. HEL 16-10, Hydraulic Engineering Laboratory, University of California, Berkeley, Calif.

McCulloch, D.S., 1966. Slide-induced waves, seiching and ground fracturing caused by the earthquake of March 27, 1964 at Kenai Lake, Alaska. *U.S. Geol. Surv. Prof. Pap.*, 543-A, 41 pp.

Miller, D.J., 1960. Giant waves in Lituya Bay, Alaska. *U.S. Geol. Surv. Prof. Pap.*, 354-C.

Miller, R.L. and White, R.V., 1966. *A Single-Impulse System for Generating Solitary, Undulating Surge, and Gravity Shock Waves in a Laboratory.* Fluid Dynamics and Sediment Transport Lab. Rep. No. 5, Dep. of Geophysical Science, Univ. of Chicago.

Mitchell, G.E., 1910. Landslides and rock avalanches. *Natl. Geogr. Mag.*, 21 (4): 277—287.

Momoi, T., 1962. Some remarks on generation of waves from ellipitical wave origin. *Bull. Earthquake Res. Inst., Tokyo Univ.*, 40 (2): 297—307.

National Academy of Sciences, 1972. *The Great Alaska Earthquake of 1964: Oceanography and Coastal Engineering.* Committee on the Alaska Earthquake of the Division of Earth Sciences, National Research Council, Washington, D.C., 556 pp.

Nielsen, A.H., 1962. *Diffraction of Periodic Waves Along a Vertical Breakwater for Small Angles of Incidence.* Tech. Rep. No. HEL 1-2, Hydraulic Engineering Laboratory, University of California, Berkeley, Calif.

Noda, E., 1969. *Theory of Water Waves Generated by a Time-Dependent Boundary Displacement.* Tech. Rep. No. HEL 16-5, Hydraulic Engineering Laboratory, University of California, Berkeley, Calif.

Noda, E., 1970. Water waves generated by landslides. *J. Waterways, Harbors and Coastal Eng. Div., Proc. ASCE*, 96 (WW4): 835—855.

Palmer, R.Q., Mulvihill, M.E. and Funasaki, G.T., 1966. Hilo harbor tsunami model-reflected waves superimposed. *Coastal Engineering: Santa Barbara Specialty Conf., 1965, ASCE*, pp. 24—31.

Perroud, P.H., 1957. *The Solitary Wave Reflection Along a Straigth Vertical Wall at Oblique Incidence.* Thesis, Dep. Civil Engineering; also: *IER Tech. Rep.* No. 99-3, Univ. of California, Berkeley, Calif.

Petrauskas, C. and Borgman, L.E., 1971. *Frequencies of Crest Heights for Random Combinations of Astronomical Tides and Tsunamis Recorded at Crescent City, California.* Tech. Rep. No. HEL 16-8, Hydraulic Engineering Laboratory, Univ. of California, Berkeley, Calif., 64 pp.

Plafker, G. and Mayo, L.R., 1965. *Tectonic Deformation, Subaqueous Slides and Destructive Waves Associated with the Alaskan March 27, 1964 Earthquake: An Interim Geologic Evaluation.* U.S. Geological Survey, Openfile Report.

Plafker, G. and Savage, J.C., 1970. Mechanism of the Chilean earthquakes of May 21 and 22, 1960. *Geol. Soc. Am. Bull.*, 81: 1001—1030.

Prins, J.E., 1958. Characteristics of waves generated by a local disturbance. *Trans. Am. Geophys. Union*, 39 (5): 865—874.

Raichlen, F., 1972. Discussion of tsunami-responses of San Pedro Bay and Shelf, Calif. *J. Waterways, Harbors Coastal Eng. Div., Proc. ASCE*, 98 (WWI): 104—110.

Rascón, O.A. and Villarreal, A.G., 1974. Estudio estadístico de los tsunamis observados en la costa mexicana del Pacífico. *Ing. (México)*, 44 (1).

Shreve, R.L., 1966. Sherman landslide, Alaska. *Science*, 154 (3757): 1639—1643.

Shureman, P., 1940. *Manual of Harmonic Analysis and Prediction of Tides.* U.S. Dept. of Commerce, U.S. Coast and Geodetic Survey, Spec. Publ., 98 (revised 1958).

Sigurdsson, G. and Wiegel, R.L., 1962. Solitary wave behavior at concave barriers. *The Port Engineer (Calcutta)*, pp. 4—8.

Soloviev, S.L., 1970. Recurrence of tsunamis in the Pacific. In: W.M. Adams (editor), *Tsunamis in the Pacific Ocean: Proc. Int. Symp. on Tsunamis and Tsunami Research, Honolulu, 1969.* Univ. of Hawaii, East—West Center Press, p. 149—164.

Takahasi, R., 1963. On some model experiments on tsunami generation. *Proc. Tsunami Meetings Associated with the Tenth Pacific Science Congr., IUGG Monogr.* 24, p. 235—248.

Takano, K., Nagata, Y., Sudo, H. and Takeda, A., 1961. Drawing of refraction diagrams and analysis of the Chilean tsunami of May 22, 1960 for a terrestrial globe. In: Committee for Field Investigation of the Chilean Tsunami of 1960 (editor), *Report of the Field Investigation Committee for Chilean Tsunami of May 24, 1960*. Maruzen, Tokyo, p. 46—51.

U.S. Coast and Geodetic Survey, 1964. *Annotated Bibliography on Tsunamis. IUGG Monogr.* 27, Paris, 249 pp.

Utsu, T. and Seki, A., 1955. A relation between the area of aftershock and the energy of main-shock. *Zisin (J. Seismol. Soc. Japan)*, [II], Vol. 7: 233—240 (in Japanese).

Watanabe, H., 1970. Statistical studies of tsunami sources and tsunamigenic earthquakes occurring in and near Japan. In: W.M. Adams (editor), *Tsunamis in the Pacific Ocean*. East—West Center Press, Univ. Hawaii, Honolulu, p. 99—117.

Wemelsfelder, P.J., 1961. On the use of frequency curves of storm floods. *Proc. Seventh Conf. on Coastal Engineering, The Hague, 1960*. Council on Wave Research, The Engineering Foundation, Berkeley, Calif., p. 617—632.

Western Canada Hydraulic Laboratories Ltd. 1970. *Hydraulic Model Studies: Wave Action Generated by Slides into Mica Reservoir*. Report to CASECO Consultants, Ltd., Vancouver, B.C., Canada, for British Columbia Hydro and Power Authority: Columbia River Development, Mica Project, 74 pp.

Wiegel, R.L., 1955. Laboratory studies of gravity waves generated by the movement of a submerged body. *Trans. Am. Geophys. Union*, 36 (5): 759—774.

Wiegel, R.L., 1964a. *Tsunami Information in Regard to Proposed Nuclear Power Plant Site, Pacific Gas and Electric Company at Bodega Head, California*. Consultant Report to Pacific Gas and Electric Company, San Francisco, Calif.

Wiegel, R.L., 1964b. Water wave equivalent of Mach-reflection. *Proc. Ninth Conf. on Coastal Engineering, Lisbon, 1964*, ASCE, pp. 82-102(b).

Wiegel, R.L., 1964c. *Oceanographical Engineering*. Prentice-Hall, Englewood Cliffs, N.J.

Wiegel, R.L., 1970. Tsunamis. In: R.L. Wiegel (editor), *Earthquake Engineering*. Prentice-Hall, Englewood Cliffs, N.J., pp. 253—306.

Wiegel, R.L. and Camotim, D., 1962. Model study of oscillations of Hebgen Lake. *Bull. Seismol. Soc. Am.*, 52 (2): 273—277.

Wiegel, R.L., Noda, E.K., Kuba, E.M., Gee, D.M. and Tornberg, G.F., 1970. Water waves generated by landslides in reservoirs. *J. Waterways Harbors Div., Proc. ASCE*, 96 (WW2): 307—333.

Wiegel, R.L., Noda, E.K., Kuba, E.M., Gee, D.M. and Tornberg, G.F., 1972. Closure to "Water waves generated by landslides in reservoirs." *J. Waterways, Harbors and Coastal Eng. Div. Proc. ASCE*, 98 (WW1): 72—74.

Williams, E.A. and Isaacs, J.D., 1952. The refraction of groups and of the waves they generate in shallow water. *Trans. Am. Geophys. Union*, 33 (4): 523—530.

Wilson, B.W., 1971. Tsunami-responses of San Pedro Bay and Shelf, Calif. *J. Waterways, Harbors and Coastal Eng. Div., Proc. ASCE*, 97 (WW2): 239—258.

Wilson, B.W., Webb, L.M. and Hendrickson, J.A., 1962. *The Nature of Tsunamis, Their Generation and Dispersion in Water of Finite Depth*. National Engineering Science Co., Pasadena, Calif., Tech. Rep. SN 57-2.

Chapter 8

STRUCTURAL RESPONSE TO EARTHQUAKES

ERIK H. VANMARCKE
Massachusetts Institute of Technology, Cambridge, Mass., U.S.A.

8.1 INTRODUCTION

Almost every step in the seismic design process is beset by uncertainty. This chapter deals with one key step in this process: prediction of the response of structures with known dynamic characteristics to ground motions only partially specified, for example through their peak acceleration or through a set of smooth response spectra. Engineers are interested not in the details of the response motion but in the few response parameters that facilitate assessment of the system's performance during an earthquake. The performance of linear systems is most often evaluated in terms of the maximum response which is compared with the allowable (damage or failure) response value. System performance is sometimes measured by the cumulative effect of repeated cycles of response motion as when fatigue or liquefaction can be a problem. In the broader context of overall seismic safety assessment of structures, this chapter seeks to provide the conditional system response *given* a (rough) description of the ground motion. To determine the "unconditional" seismic safety, results derived here must be combined with those of a seismic risk analysis and with information about the structures' dynamic properties and resistance, in ways covered in Chapter 9.

The system types treated here are linear one-degree and multi-degree structures, uncoupled soil-structure and structure-equipment systems, and simple nonlinear structures. For each system type, three different approaches to the response analysis are discussed. Each analysis procedure requires a different representation of the ground motions to be expected at the site and provides a different level of information about system response and performance.

The first and simplest procedure is based directly on the (smooth, expected) response spectra. In analyzing linear elastic multi-degree systems by the *response spectrum approach*, the response spectrum is used to predict the peak response for each mode of the structure. The individual modal maxima are then combined, usually by the square root of the sum of the squares, to provide the *expected* peak response of the complete structure. This method gives no information about the degree to which actual responses might deviate from the predicted value. Similar approximate procedures have been proposed to predict, for example, the response of light equipment in buildings and of simple nonlinear systems directly from the specified response spectra.

The second procedure involves step-by-step *time integration* based on one or more recorded or simulated accelerograms. Recorded ground motions which are reasonably representative of the type of motion to be expected at a site are often not available. Moreover, engineers are aware that, due to the stochastic nature of earthquake ground motion, the information obtained from a structural response analysis using a single record is quite unreliable. It is possible to generate by computer a set of artificial motions which cannot be distinguished, as regards over-all statistical properties, from actual recorded ground motions. Frequently, these simulated motions are "designed" to have computed response spectra which oscillate around the specified expected response spectra for the site. By calculating structural responses for each record in a sample, one can in fact construct relative frequency curves for a response parameter of interest, or at least obtain reasonably stable estimates of its average and perhaps of its standard deviation. The step-by-step integration procedure has the advantage of being *generally* applicable, e.g., in the case of complex nonlinear systems which other procedures cannot handle, but it is expensive and time-consuming.

The third procedure is a random vibration analysis which extends the first procedure, having as its aim the *direct* prediction of the probability distribution, or the first and higher moments of a dynamic response parameter of interest in terms of the dynamic properties of the structure and a statistical description of the earthquake. The most convenient ground-motion representation for this purpose is in terms of the spectral-density function and an equivalent duration of strong-motion shaking. This immediately raises a question about the relationship between the spectral-density function and the prescribed smooth response spectrum. This question is dealt with in some detail in this chapter. A blending of the theories of structural dynamics and probability has been underway for some time, but only recently have the major obstacles preventing its useful application to *seismic* analysis of general linear systems been cleared. One obstacle has been the lack of a workable methodology for dealing, in probabilistic terms, with the *transient* nature of the ground motion and the structural response, and another (obstacle) the difficulty in dealing with *maximum* values as principal response parameters. Random vibration analyses of relatively complex linear systems and of some simple inelastic systems are presented in this chapter. These analyses hold a clear advantage over the other procedures in that they yield information about the *distribution* of structural response, allowing direct assessment of the probability of exceeding intolerable response levels. Moreover, the solutions often take a simple analytical form, which facilitates incorporating these results in an overall seismic safety analysis.

The next section examines in some detail the common characterizations of the earthquake threat at a site, i.e., response spectra, representative (real or artificial) ground motions and spectral-density functions. The question of compatibility between these three representations is explored in Section 8.3, which outlines the basic methodology of random vibration applied to seismic response of simple one-degree systems. From this, a method is derived to generate artificial earthquake motions whose computed response spectra "match" a set of prescribed smooth response spectra.

Throughout the chapter, attention is restricted to investigating the dynamic response under only one component of ground motion. For linear systems, the analyses can relatively easily be extended if the earthquake excitation is represented by three statistically independent translational components of ground motion. Recent evidence (Penzien and Watabe, 1975) supports the use of such a model.

Finally, every effort has been made to present results which are of practical value in earthquake engineering. Especially in the random vibration coverage, much of which is new and based on the research by the writer and his associates, mathematical rigor is frequently relaxed and many proofs and derivations are left in the references.

## 8.2 COMMON GROUND-MOTION REPRESENTATIONS

### 8.2.1 Response spectra

For elastic structures, the response parameter which has most practical value is the maximum response. In particular, for a simple one-degree structure with a natural frequency $\omega_n$ and damping ratio $\zeta$, this quantity is usually represented by the peak relative displacement ($S_D$), the pseudo-velocity ($S_V = \omega_n S_D$) or the pseudo-acceleration ($S_A = \omega_n^2 S_D$). The pseudo-velocity is approximately equal to the maximum relative velocity of the one-degree structure, while the pseudo-acceleration is close to its maximum absolute acceleration. Response spectra are plots of these (peak) response parameters as functions of natural frequency for different values of damping. It is common to characterize the earthquake threat at a site in terms of a set of smooth response spectra for use in design. One procedure to construct these spectra is to start from predictions of the peak values of ground acceleration, velocity and displacement, which are obtained based on an evaluation of the site seismicity and geological setting or on a formal seismic-risk analysis, as described in Chapter 6. According to the method suggested by Newmark and Hall (1969, 1973), a point with coordinates ($\omega_n, \zeta$) on the "expected" response spectra can then be determined by multiplying the peak ground acceleration, velocity of displacement (depending on the period $\omega_n$) by a factor which depends on the damping $\zeta$.

Of course, actual peak responses are unpredictable even when the peak

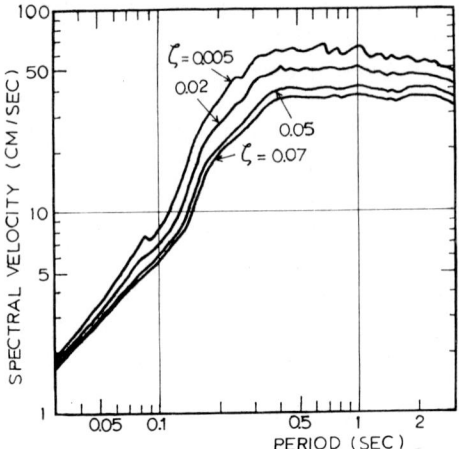

Fig. 8.1. Average response spectra for four damping values based on 39 strong-motion records scaled to a common maximum acceleration, 0.3g.

ground-motion values are known exactly. The response spectra of two apparently similar time histories will have peaks and valleys at different natural frequencies. The maximum responses of a given one-degree system to many different accelerograms, all scaled to the same peak acceleration, will be randomly distributed about an average response value. For example, Fig. 8.1 shows the average response spectra for four damping values based on 39 his-

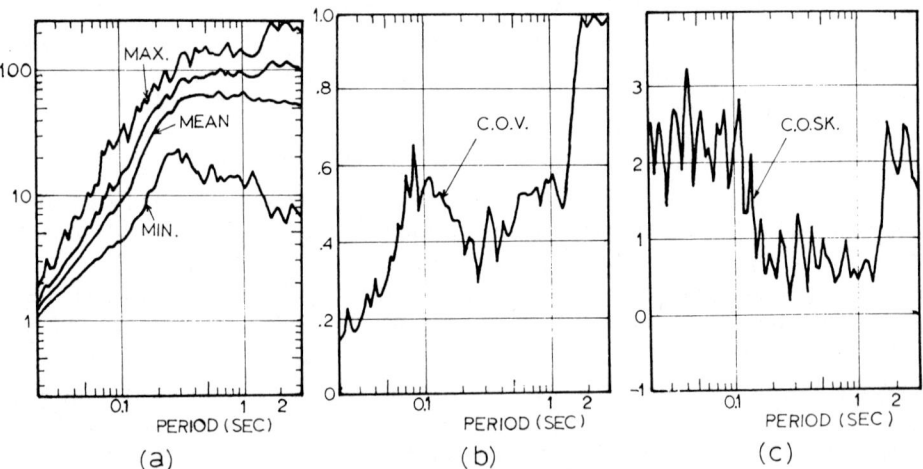

Fig. 8.2. Statistics of the scaled response spectra for 0.5% damping: (a) average, smallest and largest value; (b) coefficient of variation; and (c) coefficient of skewness. Plot (a) also shows the average plus one standard deviation.

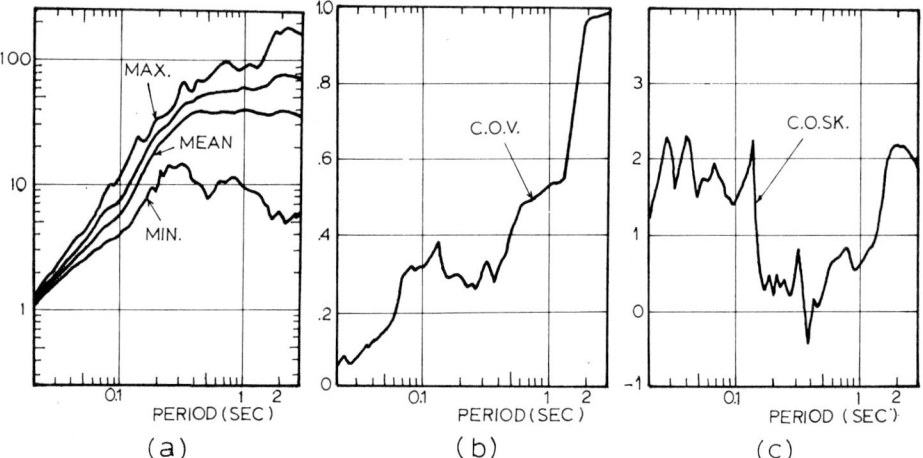

Fig. 8.3. Statistics of the scaled response spectra for 5% damping: (a) average, smallest and largest value; (b) coefficient of variation; and (c) coefficient of skewness. Plot (a) also shows the average plus one standard deviation.

torical strong-motion records, all scaled to a common maximum acceleration. The response spectra are computed at 80 period values equally spaced on a logarithmic scale. Note that segments of the average response spectra curves are parallel to 0° and 45° lines on the log-log plot. Figures 8.2 and 8.3 show several other statistics of the response spectra for 0.5% and 5% damping, respectively: (a) the average, smallest, and largest computed response; (b) the coefficient of variation (= ratio standard deviation/average); and (c) the coefficient of skewness of the response spectrum values. The coefficient of variation (c.o.v.) indicates the degree of uncertainty in the prediction of the peak response at different natural periods. It increases very markedly with period, except in the period range around 0.2—0.4 seconds, where spectral accelerations tend to be constant. In this range, typical c.o.v. values are 0.3 and 0.4, for the 0.5 and 5% damped response spectra, respectively. This variability stems from the fact that the ground motions (in this sample of 39) were recorded at different sites, during different earthquakes, have different durations, etc. The histogram of peak responses of a given one-degree structure is strongly positively skewed. Note in Figs. 8.2 and 8.3, the extreme skewness at periods below 0.15 seconds or above 1.5 seconds.

The information on variability and skewness plays an important role in assessing the degree of conservatism inherent in a set of prescribed smooth response spectra, or *design spectra*, scaled to the design maximum ground acceleration. For example, design spectra based on the average plus one standard deviation may correspond to different exceedance probabilities at different periods; moreover, these probabilities will tend to be unconservative unless an appropriate highly skewed probability model is chosen.

The response spectra statistics can also be evaluated for accelerograms scaled to a common maximum ground velocity, or maximum ground displacement. The resulting computed coefficients of variation will then be smaller at higher periods. Notice, however, that "actual" maximum ground displacements are very sensitive to debatable base-correction criteria applied to accelerograms (Trifunac et al., 1973).

Finally, it is useful to categorize available earthquake records according to site soil conditions, magnitude, focal distance or strong-motion duration (McGuire, 1974; Newmark et al., 1973). Statistical analyses based on a *subset* of ground motions with more narrowly defined properties will tend to yield smaller coefficients of variation than those quoted previously (Figs. 8.2b and 8.3b). Due to the relative scarcity of data, however, all response spectra statistics will be less reliably estimated.

*8.2.2 Simulated earthquakes*

A typical accelerogram has the appearance of a transient, stochastic function (Housner, 1947, 1959), and it is not difficult to generate by computer many accelerograms which cannot be distinguished, as regards over-all statistical properties, from recorded ground motions. Of major interest are the maximum acceleration and the duration of strong shaking. Other important parameters are those governing the relative frequency content and the variation of motion intensity with time. One commonly used method of numerical simulation of earthquakes is based on the fact that any periodic function can be expanded into a series of sinusoidal waves:

$$x(t) = \sum_{i=1}^{n} A_i \sin(\omega_i t + \phi_i) \tag{8.1}$$

$A_i$ is the amplitude and $\phi_i$ is the phase angle of the $i$th contributing sinusoid. By fixing an array of amplitudes and then generating different arrays of phase angles, different motions which are similar in general appearance (i.e., in frequency content) but different in the "details", can be generated. The computer uses a "random number generator" subroutine to produce strings of phase angles with a uniform distribution in the range between 0 and $2\pi$.

The *total power* of the steady-state motion, $x(t)$, is $\sum_{i=1}^{n} (A_i^2/2)$. Assume now that the frequencies $\omega_i$ in eq. 8.1 are chosen to lie at equal intervals $\Delta \omega$. Figure 8.4 shows a function $G(\omega)$ whose value at $\omega_i$ is equal to $A_i^2/2\Delta\omega$ so that $G(\omega_i)\Delta\omega = A_i^2/2$. Allowing the number of sinusoids in the motion to become very large, the total power will become equal to the area under the continuous curve $G(\omega)$, which is in effect the *spectral-density function*. Formal definitions of $G(\omega)$ can be found in many textbooks (Crandall and Mark, 1963; Lin, 1967). $G(\omega)$ expresses the relative importance (i.e., the relative contribution to the total power) of sinusoids with frequencies within some specified band of frequencies. When $G(\omega)$ is nar-

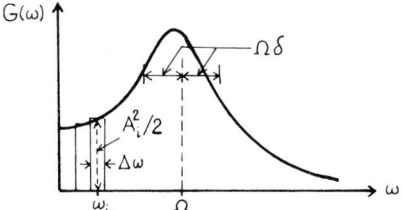

Fig. 8.4. The spectral-density function $G(\omega)$ and the spectral parameters $\Omega$ and $\delta$.

rowly centered around a single frequency, then eq. 8.1 will generate nearly sinusoidal functions as shown in Fig. 8.5a. On the other hand, if the spectral-density function is nearly constant over a wide band of frequencies, components with widely different frequencies will compete to contribute equally to the motion intensity, and the resulting motions will resemble portions of earthquake records, as illustrated in Fig. 8.5b. Of course, the total power and the relative frequency content of the motions produced by using eq. 8.1 do not vary with time. To simulate in part the transient character of real earthquakes, the stationary motions $x(t)$ are usually multiplied by a deterministic intensity function such as the "boxcar", trapezoidal (Hou, 1968), and exponential (Shinozuka, 1973) functions shown in Fig. 6.

Bycroft (1960) and Brady (1966) simulated "white noise" (for which $G(\omega)$ is theoretically constant) for all frequencies to represent earthquake ground motion. Actually, the simplest workable form of $G(\omega)$ is that corresponding to a band-limited white noise. The spectral density is constant in the frequency range from 0 to $\omega_0$, as shown in Fig. 8.7a:

$$G(\omega) \begin{cases} = G_0 & 0 \leq \omega \leq \omega_0 \\ = 0 & \omega > \omega_0 \end{cases} \tag{8.2}$$

Based on Kanai's study (1961) of the frequency content of a limited number

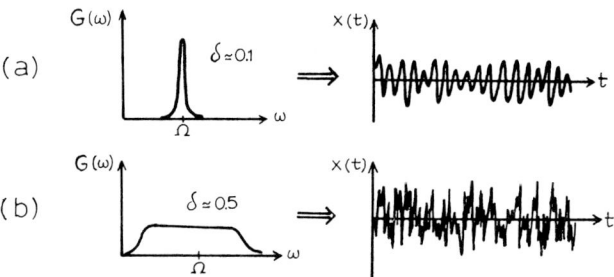

Fig. 8.5. Spectral-density functions corresponding to different bandwidths.

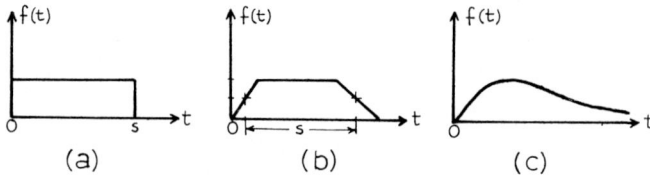

Fig. 8.6. Intensity envelope functions: (a) boxcar; (b) trapezoidal (Hou, 1968); (c) exponential (Shinozuka, 1973).

of recorded strong ground motions, Tajimi (1960) suggested the following widely quoted formula for the spectral-density function (s.d.f.) of ground motion (Fig. 8.7b):

$$G(\omega) = \frac{[1 + 4\zeta_g^2(\omega/\omega_g)^2]G_0}{[1-(\omega/\omega_g)^2]^2 + 4\zeta_g^2(\omega/\omega_g)^2} \qquad (8.3) *$$

Sample functions $x(t)$ with spectral densities corresponding to eq. 8.3 can be obtained by filtering "ideal white noise" (i.e., with $\omega_0 = \infty$ in eq. 8.2) through a simple oscillator with natural frequency $\omega_g$ and viscous damping $\zeta_g$ (Franklin, 1965; Penzien and Liu, 1969). These may be interpreted as the "predominant ground frequency" and the "ground damping", respectively. The values $\omega_g = 4\pi$ and $\zeta_g = 0.60$ have been suggested for firm ground sites. $G_0$ is a measure of ground intensity. Extensions of eq. 8.3 have been proposed for model ground motions whose spectral density shows more than one dominant spectral peak (Liu and Jhaveri, 1969). Theoretical spectral-density shapes such as those described by eqs. 8.2 and 8.3 are obtained by examining, smoothing and/or averaging of the squared Fourier amplitudes $|f(\omega)|^2$ of

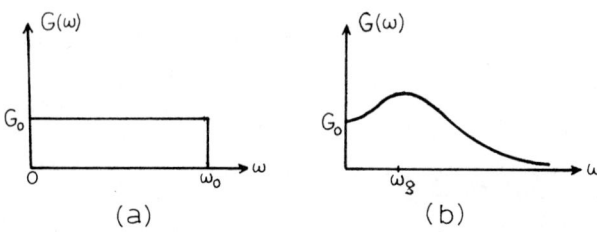

Fig. 8.7. (a) Band-limited white noise spectral density, and (b) Kanai-Tajimi spectral density.

---

* Soviet researchers (Barstein, 1960; Bolotin, 1969) suggested a probabilistic approach to earthquake engineering based on the autocorrelation function of a form corresponding to the s.d.f. in eq. 8.3.

actual strong-motion records. This stems from the basic fact that, for stationary random processes, the expected value of $|f(\omega)|^2$ and the spectral density functions $G(\omega)$ are proportional (Jenkins, 1961; Rosenblueth, 1964).

Ground-motion models which account for the time-varying nature of the relative frequency content have also been proposed, (e.g., Liu, 1970). In reference to eq. 8.1, it suffices to allow the amplitudes $A_i$ to vary slowly with time. The spectral content can then be described by an "evolutionary" spectral-density function $G(\omega, t)$, i.e., $G(\omega_i, t)$ is proportional to $A_i^2$ (Priestley, 1967). As will be seen in Section 8.3.2, these more sophisticated models are particularly useful in describing the frequency content of structural *response*, or of motions at the point of support of secondary systems attached to structures. However, time-invariant models in which $G(\omega)$ reflects the frequency content during the most intense part of the ground motion are believed to be sufficiently accurate for the purpose of seismic-response prediction, for all but certain nonlinear systems.

The properties of spectral-density functions are further discussed in the next section, and additional comments about simulation of earthquakes which are compatible with smooth response spectra are offered in Section 8.3.5.

### 8.2.3 Spectral-density functions

In random vibration analysis of seismic response, an earthquake can be represented as a limited duration segment of a stationary random function with a given spectral-density function $G(\omega)$. The representation is at once powerful and simple. It allows the application of some important results of stationary probabilistic dynamic analysis, while the specification of a limited duration and a sudden start captures the essential transient character of earthquake ground motion. The integral over frequency of $G(\omega)$ equals the average total power, or the variance $\sigma^2$ for motions which fluctuate about a zero mean value, e.g., ground acceleration and linear system response. For the band-limited noise motion (eq. 8.2), the variance is:

$$\sigma^2 = \int_0^\infty G(\omega) d\omega = G_0 \omega_0 \tag{8.4}$$

and for the Kanai-Tajimi spectrum (eq. 8.3), the variance is:

$$\sigma^2 = \int_0^\infty G(\omega) d\omega = \frac{\pi G_0 \omega_g}{4 \zeta_g}(1 + 4\zeta_g^2)^{1/2} \tag{8.5}$$

Actually, a more useful way for dealing with the frequency content of ground motions is through the *normalized* spectral-density function:

$$G^*(\omega) = \frac{1}{\sigma^2} G(\omega) \qquad (8.6a)$$

or its cumulative spectral distribution:

$$F^*(\omega) = \int_0^\omega G^*(\omega)d\omega \qquad (8.6b)$$

which increases from 0 to 1 as $\omega$ goes from 0 to $\infty$. Note the analogy between normalized s.d.f. and the probability density function (p.d.f.) of any random variable: both are nonnegative and have unit area. The moments of the spectral density function $G(\omega)$ are:

$$\lambda_i = \int_0^\infty \omega^i G(\omega)d\omega = \sigma^2 \int_0^\infty \omega^i G^*(\omega)d\omega = \sigma^2 \lambda_i^* \qquad (8.7)$$

in which $\lambda_i^*$ is the $i$th moment of the unit area spectral density. It is clear that $\sigma^2 = \lambda_0$ and $\lambda_0^* = 1$. A measure of where the spectral is concentrated along the frequency axis is (see Fig. 8.4):

$$\Omega = \sqrt{\lambda_2/\lambda_0} = \sqrt{\lambda_2^*} \qquad (8.8)$$

which is analogous to the root mean square (r.m.s.) of a random variable. A convenient measure of the spread or the dispersion of the s.d.f. about its center frequency is (Vanmarcke, 1969; 1972):

$$\delta = \sqrt{1 - \lambda_1^2/\lambda_0\lambda_2} = \sqrt{1 - \lambda_1^{*2}/\lambda_2^*} \qquad (8.9)$$

which is dimensionless, always lies between 0 and 1, and increases with increasing bandwidth. Pursuing the analogy between $G^*(\omega)$ and the p.d.f. of a random variable, $\delta$ is equivalent to the ratio of the standard deviation of the r.m.s. value (see Fig. 8.4). It is clear that $\delta$ will be large if $G(\omega)$ has two or more fairly widely separated peaks. Important time-domain interpretations of the spectral parameters $\Omega$ and $\delta$ are discussed in Section 8.3.3. In the frequency domain, these two parameters provide a summary description of $G^*(\omega)$. In fact, it is possible to develop (Chebychev-type) bounds on $F^*(\omega)$ in terms of $\Omega$ and $\delta$ (Vanmarcke, 1972). * A parameter whose definition is similar to

---

* One such inequality is:

$$1 - F^*(\omega) \leq \delta^2[(\omega/\Omega) - 1]^{-2}$$

For example, taking $\delta = 0.2$ and $\omega = 3\Omega$ yields the inequality $1 - F^*(2\Omega) \leq 0.01$; in words, for processes having a $\delta$-factor equal to 0.2, the fraction of the total power contributed by components with frequencies larger than $3\Omega$ is less than 1%.

δ has been proposed by Longuet-Higgins (1952):

$$\epsilon = \sqrt{1 - \lambda_2^2/\lambda_0\lambda_4} = \sqrt{1 - \lambda_2^{*2}/\lambda_4^*} \qquad (8.10)$$

It also varies between 0 and 1, and it is closely related to the kurtosis parameter for random variables.

In evaluating higher spectral moments, problems due to lack of convergence are sometimes encountered. These are due to the fact that various proposed algebraic expressions (e.g., eq. 8.3), while providing a good fit to computed power spectra in the central frequency region, do not properly represent the ground motion in the upper frequency tail. Owing to recording and processing limitations, strong-motion accelerograms provide little information about motion frequency content beyond a circular frequency $\omega_0 = 2\pi/\Delta T$, where $\Delta T$ is approximately 0.02 seconds. Consequently, it is difficult to evaluate contributions to, say, the second spectral moment, due to frequencies beyond $\omega_0$; one is, in effect, restricted to computing *partial* moments of $G(\omega)$, in a limited frequency range $(0, \omega_0)$. The values of the spectral parameters corresponding to these partial s.d.f.'s given in eqs. 8.2 and 8.3 are listed in Table 8.I. Spectral shape parameters can also be computed from the squared Fourier amplitudes $|f(\omega)|^2$ of actual strong-motion records. (Sixsmith and Roesset, 1970). It suffices to substitute $|f(\omega)|^2$ for $G(\omega)$ in the definitions, eqs. 8.8 and 8.9. Some results are presented in Table 8.II.

The r.m.s. value $\sigma$ is closely tied to the maximum ground acceleration, $A$, both of which are more difficult to predict than the spectral parameters $\Omega$ and $\delta$. The relation between $\sigma$ and $A$ is examined in some detail in Section 8.3.4. In particular, it is shown there that the *median* maximum acceleration, $\hat{A}$, can *theoretically* be calculated as follows:

$$\hat{A} = \sigma \times \sqrt{2 \ln\left(2.8 \frac{\Omega s}{2\pi}\right)} \qquad (8.11)$$

TABLE 8.I

Approximate values for parameters of relative frequency content

|  | $\Omega$ (eq. 8.8) | $\delta$ (eq. 8.9) | $\epsilon$ (eq. 8.10) |
|---|---|---|---|
| Band-limited white noise spectral density | $\sqrt{\frac{1}{3}}\omega_0 = 0.58\,\omega_0$ | 0.5 | 0.66 |
| Kanai-Tajimi spectral density for $\omega \leq \omega_0 = 4\omega_g$ and $\zeta_g = 0.6$ | ~2.1 $\omega_g$ | 0.67 | 0.96 |

TABLE 8.II

Spectral parameters computed from squared Fourier amplitudes $|f(\omega)|^2$ of earthquakes (Sixsmith and Roesset, 1970)

| | $\Omega$ (rad/sec) | $\delta$ | $\epsilon$ |
|---|---|---|---|
| El Centro 1940 N—S | 31.35 | 0.73 | 0.97 |
| El Centro 1940 E—W | 25.51 | 0.64 | 0.96 |
| Olympia N10W | 36.07 | 0.65 | 0.93 |
| Olympia N80E | 30.85 | 0.62 | 0.94 |
| Taft N69W | 27.71 | 0.66 | 0.96 |
| Taft S21W | 27.46 | 0.64 | 0.96 |

in which $s$ = strong-motion duration, and $(\Omega s)/2\pi$ = expected number of cycles of ground motion. The difficulty in the direct assessment of $\sigma^2$ from a ground-motion record stems from the fact that the motion intensity actually varies with time and that the choice of the length of the motion is critical in computations of average power. In modeling ground motions for the purpose of random vibration analysis, a reasonable procedure is to select an appropriate value for $\sigma$, to choose an equivalent duration of strong shaking, and to determine the spectral shape function $G^*(\omega)$, either directly from the response spectrum $S_V$ by the method outlined in Section 8.3.5, or by selecting a theoretical shape (e.g., the Kanai-Tajimi form, eq. 8.3).

## 8.3 RANDOM VIBRATION-BASED PREDICTION OF RESPONSE SPECTRA

This section is devoted to a random vibration analysis of linear one-degree systems with different natural frequencies $\omega_n$ (from 0 to $\infty$) which are undamped or have small viscous damping ratios $\zeta$. Stationary response statistics are derived first, and the required modifications to deal with the transient character of the excitation and the response are discussed subsequently. In the solution developed in this section, the response spectral value $y_{s;p}$ corresponding to an exceedance probability $p$ and a strong-motion duration $s$ is expressed as a multiple of $\sigma_y(s)$, the standard deviation of the one-degree system response (for given $\omega_n$ and $\zeta$) evaluated at $s$, as follows:

$$y_{s;p} = r_{s;p}\, \sigma_y(s) \tag{8.12}$$

The situation is depicted in Fig. 8.8. The problem of determining the peak factor $r_{s;p}$ is formidable, requiring the solution of the so-called first-passage problem. Exact solutions to this problem do not yet exist, but good, practical approximate solutions are now available. When $p = 0.5$, eq. 8.12 yields the *median* response spectra. While it is not uncommon to find the factor

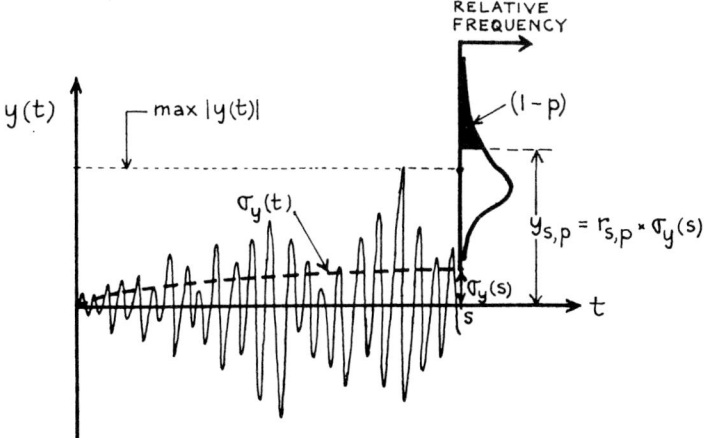

Fig. 8.8. Relationship between the response standard deviation $\sigma_y(s)$, the peak factor $r_{s,p}$ and the maximum response $y_{s,p}$.

$r_{s;0.5}$ treated as a constant, it may actually lie anywhere between 1.25 and 3.50 for typical ground motions, as will be seen in this section. The methodology presented herein can be used for an earthquake-response analysis of rather general linear systems. In this section, complete results are derived for one-degree systems only. More complex linear systems are dealt with in Sections 8.4 and 8.5.

### 8.3.1 Stationary response variance

A basic result of stationary random vibration of linear systems is the following relationship between the spectral-density functions (s.d.f.) of input and output (see, for example, Crandall and Mark, 1963):

$$G_y(\omega) = G(\omega) \, |H(\omega)|^2 \tag{8.13}$$

in which $G_y(\omega)$ = the output s.d.f., $G(\omega)$ = the input s.d.f., and $|H(\omega)|$ = the amplification function of the linear system, i.e., the amplitude of the steady-state response of the system to a sinusoidal input with unit amplitude and frequency $\omega$. Also, the variance of the response, $\sigma_y^2$, is equal to the area under $G_y(\omega)$:

$$\sigma_y^2 = \int_0^\infty G_y(\omega) d\omega = \int_0^\infty G(\omega) \, |H(\omega)|^2 \, d\omega \tag{8.14}$$

For a linear one-degree system with natural frequency $\omega_n$ and damping ratio $\zeta$, whose input and output are support acceleration and relative displacement response, respectively, the squared amplification function is:

$$|H(\omega)|^2 = [(\omega_n^2 - \omega^2)^2 + 4\zeta^2 \omega_n^2 \omega^2]^{-1} \qquad (8.15)$$

The relative displacement response variance $\sigma_y^2$ is obtained by inserting eq. 8.15 into eq. 8.14. Also, the standard deviation of the pseudo-acceleration response is $\sigma_a = \omega_n^2 \sigma_y$.

Note that, at the extremes of the frequency scale, this formulation leads to the following desirable, not commonly recognized, results. First, $\sigma_a$ approaches the standard deviation of the ground acceleration ($\sigma$) when $\omega_n \to \infty$ (and $|H(\omega)|^2 \to \omega_n^{-4}$):

$$\sigma_a = \omega_n^2 \sigma_y \to \left[ \int_0^\infty G(\omega) d\omega \right]^{1/2} = \sigma \qquad (8.16)$$

At the other extreme, when $\omega_n \to 0$ and $|H(\omega)| \to \omega^{-2}$ (see eq. 8.15), we obtain:

$$\sigma_y^2 = \int_0^\infty \omega^{-4} G(\omega) d\omega = \sigma_{gr.\,displ.}^2 \qquad (8.17)$$

The integrand $\omega^{-4} G(\omega)$ is actually the spectral-density function of the *ground displacement* \*, and therefore $\sigma_y^2$ is equal to the variance of the ground displacement when $\omega_n = 0$.

As shown in Fig. 8.9, the earthquake excitation spectral-density function often varies relatively smoothly in the immediate vicinity of the system's natural frequency $\omega_n$, while $|H(\omega)|^2$ exhibits a sharp peak at $\omega_n$. This effect is more pronounced as the system damping decreases. It leads to the following useful approximation for the pseudo-acceleration response variance, $\sigma_a^2 = \omega_n^4 \sigma_y^2$:

$$\sigma_a^2 = \omega_n^4 \int_0^\infty G(\omega)|H(\omega)|^2 d\omega \simeq \omega_n^4 G(\omega_n) \int_0^\infty |H(\omega)|^2 d\omega - \omega_n G(\omega_n)$$

$$+ \int_0^{\omega_n} G(\omega) d\omega = G(\omega_n) \omega_n \left( \frac{\pi}{4\zeta} - 1 \right) + \int_0^{\omega_n} G(\omega) d\omega \qquad (8.18)$$

Figure 8.9 illustrates the meaning of the two terms on the right-hand side of eq. 8.18. The first term accounts for the contribution in a narrow frequency

---

\* This follows from eq. 8.13 and from the fact that the amplification function of a hypothetical system whose input is the second derivative of the output, equals $\omega^{-2}$. Of course, in this section both ground acceleration and ground displacement are assumed to be stationary random processes.

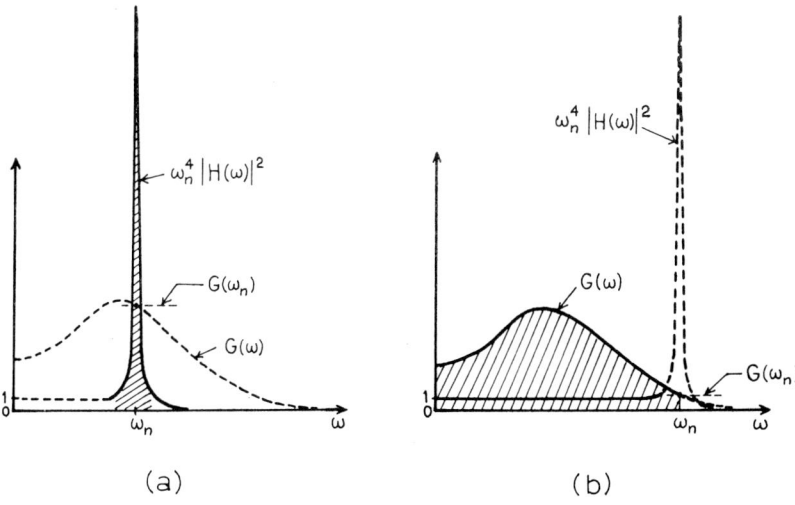

Fig. 8.9. Computation of the variance of one-degree system response to wide-band input. The first term (in eq. 8.18) accounts for the contribution in a narrow frequency range around $\omega_n$, the second term for the contribution in the frequency range $(0,\omega_n)$.

range around $\omega_n$, the second the contribution in frequency range below $(0, \omega_n)$. The term $\omega_n G(\omega_n)$ is subtracted because the area it represents would otherwise be counted twice. The relative importance of the second term in eq. 8.18, the partial area under $G(\omega)$, increases for higher natural frequencies. For lightly damped systems with intermediate natural frequencies, the first term will strongly dominate. Equation 8.18 correctly predicts $\sigma_a^2$ when the excitation is an ideal white noise, i.e., $G(\omega) = G_0$, for all $\omega$. We have then:

$$\sigma^2 = G_0 \omega_n [(\pi/4\zeta) - 1] + G_0 \omega_n = (\pi G_0 \omega_n)/4\zeta \tag{8.19}$$

A widely used approximation for the variance of the response to wide band excitation is obtained by substituting $G_0$ in eq. 8.19 by $G(\omega_n)$. It is evident that this result is of little use in predicting responses at higher frequencies.

It is convenient to express the response variance $\sigma_a^2$ in terms of the ground-motion variance $\sigma^2$, i.e., to evaluate the ratio *:

$$F_a^2 = \frac{\sigma_a^2}{\sigma^2} \simeq G^*(\omega_n) \omega_n \left(\frac{\pi}{4\zeta} - 1\right) + F^*(\omega_n) \tag{8.20}$$

where $G^*(\omega_n)$ is the unit-area ground-motion s.d.f., and $F^*(\omega_n)$ is the normalized cumulative spectrum. It is not important that the expression for

---

* For light damping, the factor $[(\pi/4\zeta) - 1]$ may be replaced by $\pi/4\zeta$.

$F^*(\omega_n)$ in eq. 8.20 be accurate in the range where the first term predominates. But $F^*(\omega)$ should approach 1 at high values of $\omega_n$, when the contribution due to the first term vanishes. If $F^*(\omega)$ is not easily obtainable, a convenient *approximation* which satisfies this criterion is the Rayleigh cumulative function:

$$F^*(\omega_n) = 1 - \exp(-\omega_n^2/\Omega^2) \tag{8.21}$$

implying a s.d.f. shape which has the correct mean-square frequency $\Omega^2$ and an exponentially decaying upper tail. The fact that spectra such as the Kanai-Tajimi form poorly represents the very high frequency content of actual ground motions, further justifies the use of approximations as eq. 8.21.

The approximation solution for the ratio $\sigma_a^2/\sigma^2$ is now evaluated for the two types of s.d.f.'s introduced. For the band-limited noise, the solution is:

$$F_a^2 = \frac{\sigma_a^2}{\sigma^2} = \begin{cases} (\omega_n/\omega_0)(\pi/4\zeta) & \omega_n \leq \omega_0 \\ 1 & \omega_n > \omega_0 \end{cases} \tag{8.22}$$

When $G(\omega)$ has a Kanai-Tajimi form (eq. 8.2), then the variance $\sigma^2$ is given by eq. 8.5, and the approximation for $F_a^2$ takes the form *:

$$F_a^2 = \frac{\sigma_a^2}{\sigma^2} \simeq \frac{\omega_n}{\omega_g} \frac{\zeta_g}{\zeta} (1 + 4\zeta_g^2)^{-1/2} \frac{(\omega_g^4 + 4\zeta_g^2 \omega_n^2 \omega_g^2)}{(\omega_g^2 - \omega_n^2)^2 + 4\zeta_g^2 \omega_n^2 \omega_g^2} + F^*(\omega_n) \tag{8.23}$$

The factor $F_a$ given by eqs. 8.22 and 8.23 is plotted in Figs. 8.10 and 8.11, respectively, for $\zeta = 0.05$. For this damping and for the period values considered, the response variance approaches the steady state very quickly. For *very* lightly damped systems, the transient character of the response must be accounted for.

*8.3.2 Transient response variance*

The steady-state value given by eq. 8.18 may never be closely approached when the one-degree system's natural frequency or damping is very small, i.e., when the product $\zeta\omega_n$ is small. As was first shown by Caughey and Stumpf (1961), the response variance will build up from zero (at the time when the earthquake strikes) to a maximum value, near the end of the (equivalent stationary) motion duration s. The frequency content of the one-degree system response will evolve in a way which can perhaps most conveniently be described by the *time-dependent* spectral density function

---

* Exact analytical expressions for $F(\omega_n)$ are available (Pulgrano and Ablowitz, 1969) thought they are rather lengthly. The approximation, eq. 8.21, has been used in the computations leading to Fig. 8.10.

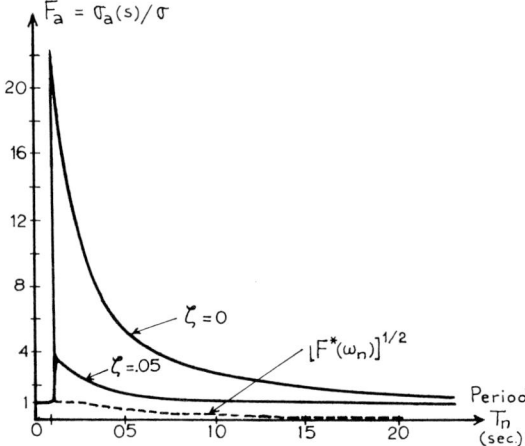

Fig. 8.10. The amplification factor $F_a$ for a band-limited white-noise excitation.

$G_y(\omega,t)$ (Corotis and Vanmarcke, 1975). For a broad class of functions $G_y(\omega,t)$, the time-dependent variance of the transient response can be obtained by integration over all frequencies:

$$\sigma_y^2(t) = \int_0^\infty G_y(\omega,t)\,d\omega \tag{8.24}$$

The function $G_y(\omega,t)$ will depend on the input s.d.f. $G(\omega)$ and on the system properties. For any linear system with impulse-response function $h(t)$, it is possible to define the truncated Fourier transformation, or the "time-

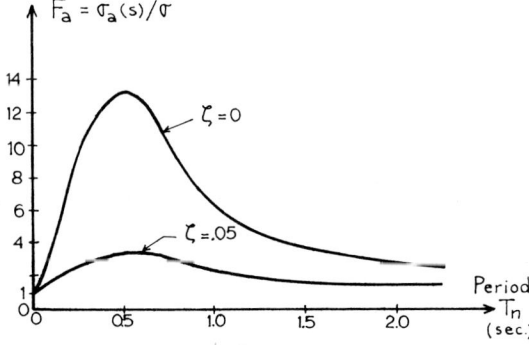

Fig. 8.11. The amplification factor $F_a$ for excitation with Kanai-Tajimi spectral-density function.

dependent transfer function", as follows:

$$H(\omega,t) = \int_0^t h(t-\tau)\exp(i\omega\tau)d\tau \tag{8.25}$$

which converges to the transfer function $H(\omega)$ when $t \to \infty$. When the system is suddenly exposed to a steady excitation with s.d.f. $G(\omega)$, the time-dependent response s.d.f. will be given by:

$$G_y(\omega,t) = G(\omega)|H(\omega,t)|^2 \tag{8.26}$$

In the case at hand, $G(\omega)$ is wide-band and smoothly varying and the system is a simple oscillator with impulse-response function:

$$h(t) = \begin{cases} \dfrac{1}{\omega_1}\exp(-\zeta\omega_n t)\sin\omega_1 t & t \geq 0 \\ 0 & t < 0 \end{cases} \tag{8.27}$$

where $\omega_1 = \omega_n\sqrt{1-\zeta^2}$ is the damped natural frequency of the system. The transient squared amplification function $|H(\omega,t)|^2$ has the following form (Caughey and Stumpf, 1961; Hammond, 1968; Corotis et al., 1972):

$$|H(\omega,t)|^2 = |H(\omega)|^2 \left[1 - 2\exp(-\omega_n\zeta t)\left\{\left(\cos\omega_1 t + \frac{\omega_n\zeta}{\omega_1}\sin\omega_1 t\right)\cos\omega t \right.\right.$$
$$+ \frac{\omega}{\omega_1}\sin\omega_1 t \sin\omega t - \exp(-\omega_n\zeta t)\left(\frac{1}{2} + \frac{\omega_n\zeta}{\omega_1}\sin\omega_1 t \cos\omega_1 t\right.$$
$$\left.\left.+ \frac{(\omega_n\zeta)^2 - \omega_1 + \omega^2}{2\omega_1^2}\sin^2\omega_1 t\right)\right\}\right] \tag{8.28}$$

Integrating this expression over all frequencies yields approximately:

$$\int_0^\infty |H(\omega,t)|^2 d\omega \simeq \frac{\pi}{4\zeta\omega_n^3}[1 - \exp(-2\zeta\omega_n t)] \tag{8.29}$$

The foregoing integral increases from 0 to the stationary value $\pi/(4\zeta\omega_n^3)$, which will be achieved when $t \gg 1/\zeta\omega_n$. Comparison between eq. 8.29 and the stationary value motivates the definition of a fictitious *time-dependent damping*:

$$\zeta_t = \frac{\zeta}{1 - \exp(-2\zeta\omega_n t)} \tag{8.30}$$

so that the right-hand side of eq. 8.29 can be written as $\pi/4\zeta_t\omega_n^3$. Of course, $\zeta_t = \zeta$ when $t \to \infty$. Actually, the parameter $\zeta_t$ is particularly useful in that it allows the entire set of spectral shapes $|H(\omega, t)|^2$ (eq. 8.28) to be crudely approximated by:

$$|H(\omega,t)|^2 \simeq [(\omega_n^2 - \omega^2)^2 + 4\zeta_t \omega_n^2 \omega^2]^{-1} \tag{8.31}$$

which has the familiar form of the squared amplification function $|H(\omega)|^2$ given by eq. 8.15. The damping parameter $\zeta_t$ decays from a very high value down to the actual system damping $\zeta$: the rate of decay is governed by the product $\zeta \omega_n$. The approximate form yields, for all values of $t$, not only about the same total area, but also the same central frequency ($\omega_n$) and about the same "bandwidth" as the exact form (Corotis and Vanmarcke, 1975). The main advantage resulting from the use of eq. 8.31 is that all the stationary results obtained in the previous section can now be applied to the transient response situation, simply by substituting $\zeta$ by $\zeta_t$.

Note that the use of eqs. 8.30 and 8.31 also provides a convenient way for treating the response of the *undamped* system, for which the stationary condition is, of course, never closely approached. In this case:

$$\zeta_t = \frac{\zeta}{1 - \exp(-2\zeta\omega_n t)} \xrightarrow[\zeta=0]{} \frac{1}{2\omega_n t} \tag{8.32}$$

A result of particular importance is the pseudo-acceleration response variance evaluated at the end of the motion duration, $s$:

$$\sigma_a^2(s) = \omega_n^4 \sigma_y^2(s) = \omega_n^4 \int_0^\infty G(\omega)|H(\omega,s)|^2 d\omega \simeq G(\omega_n)\omega_n \left(\frac{\pi}{4\zeta_s} - 1\right)$$

$$+ \int_0^{\omega_n} G(\omega) d\omega \tag{8.33}$$

Also, the ratio of the transient acceleration response variance to the ground acceleration variance is:

$$F_a^2 = \frac{\sigma_a^2(s)}{\sigma^2} \simeq G^*(\omega)\omega_n \left(\frac{\pi}{4\zeta_s} - 1\right) + F^*(\omega_n) \tag{8.34}$$

These results are analogous in form to eqs. 8.18 and 8.20 and, in fact, converge to them when the product ($\zeta\omega_n s$) grows large. Again, if the damping is light, little accuracy is lost by replacing the factor $[(\pi/4\zeta_s) - 1]$ by $\pi/4\zeta_s$ in eqs. 8.33 and 8.34.

The ratio $F_a$ is evaluated for the undamped case ($\zeta = 0$) in Figs. 8.10 and 8.11 which correspond to a ground motion with white noise and Kanai-Tajimi spectral density, respectively.

### 8.3.3 Other pertinent response statistics

The distribution of the maximum response, as well as some other useful response "level crossing" statistics, depends importantly on the higher moments of the *response* spectral-density function. If the response is *stationary*, its spectral moments can be obtained in the same way as those of the ground motion (see eq. 8.7).

$$\lambda_{i,y} = \int_0^\infty \omega^i G_y(\omega) d\omega \qquad (8.35)$$

Throughout this section, the subscript $y$ is used to characterize response parameters. When $i = 0$, eq. 8.35 defines the variance, i.e., $\lambda_{0,y} = \sigma_y^2$. It is well-known that $\lambda_{2,y} = \sigma_{\dot{y}}^2 =$ the variance of the derivative of $y(t)$. The center frequency is:

$$\Omega_y = \sqrt{\lambda_{2,y}/\lambda_{0,y}} = \sigma_{\dot{y}}/\sigma_y \qquad (8.36)$$

For Gaussian processes, $\Omega_y$ is closely related to $\nu_a$, the average number of times per second the response $y(t)$ exceeds the response level $a$, or the average rate of *a-upcrossings* (Rice, 1945):

$$\nu_a = (\Omega_y/2\pi) \exp(-a^2/2\sigma_y^2) = \nu_0 \exp(-r^2/2) \qquad (8.37)$$

in which $r = (a/\sigma_y)$ and $\nu_0 = \Omega_y/2\pi$. Note that for $a = 0$, we obtain $\nu_0 = \Omega_y/2\pi$. Also, the average number of times per second that $y(t)$ moves outside the range $(-a, a)$ is $2\nu_a$. Another useful spectral parameter is:

$$\delta_y = \sqrt{1 - \lambda_{1,y}^2/\lambda_{0,y}\lambda_{2,y}} \qquad (8.38)$$

which measures the spread or the variability in the frequency content of the response motion; it is dimensionless and lies between 0 and 1. The value of $\delta_y$ is small for narrow-band processes (it equals zero for a pure sinusoid with random phase angle) and relatively large for wide-band processes. In the time domain, $\delta_y$ is equal to the ratio $\sigma_{\dot{r}}/\sigma_{\dot{y}}$, in which $\sigma_{\dot{r}}$ is the r.m.s. value of the slope of the *envelope* $r(t)$ of the function $y(t)$, and $\sigma_{\dot{y}}$ is the r.m.s. of the slope $y(t)$ (see Fig. 8.12). The envelope definition used here is that of

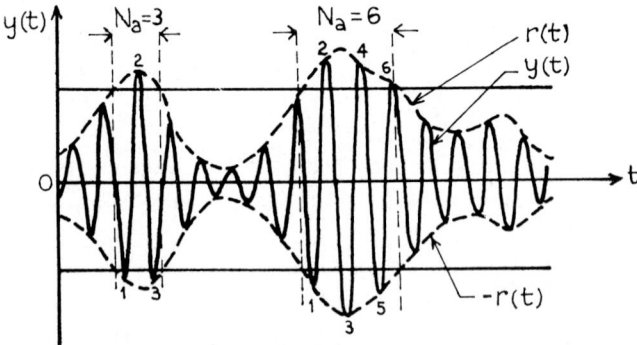

Fig. 8.12. The envelope $r(t)$ of the stationary random process $y(t)$.

Cramer and Leadbetter (1967) which is essentially equivalent to that of Rice (1945). It exists for any stationary random process regardless of band width. Of course, the concept of an envelope is most useful for narrow-band random functions. The mean number of times per second the envelope of $y(t)$ exceeds the level $a$, or the mean rate of envelope crossings, is given by (Cramer and Leadbetter, 1967; Vanmarcke, 1972):

$$n_a = \sqrt{2\pi}\delta_y \left(\frac{a}{\sigma_y}\right) \nu_a \qquad (8.39)$$

The relationship between the mean crossing rates $2\nu_a$ and $n_a$ is interesting. At low threshold levels, $n_a$ will be less than $2\nu_a$: the crossings of $y(t)$ outside the range $(-a, a)$ tend to occur in clumps which immediately follow envelope crossings. Lyon (1961) has argued that the quotient $2\nu_a/n_a$ can be interpreted as the mean clump size. This concept is particularly useful when the ratio is well above 1. But for high thresholds, the ratio $2\nu_a/n_a$ may become much smaller than one, as many envelope crossings are not followed by $a$-crossings. An estimate of the mean clump size which accounts for this effect is (Vanmarcke, 1969, 1975):

$$E[N_a] = \frac{1}{1 - \exp(-n_a/2\nu_a)} = \frac{1}{1 - \exp[-\sqrt{\pi/2}\delta_y(a/\sigma_y)]} \qquad (8.40)$$

which is plotted in Fig. 8.13 for several values of $\delta_y$. Note that $E[N_a] \to 1$ for high threshold levels. When the mean clump size is large, i.e., when the product $\delta_y(a/\sigma_y)$ is small, $E[N_a] \simeq 2\nu_a/n_a = [\sqrt{\pi/2}\delta_y(a/\sigma_y)]^{-1}$.

These definitions can easily be extended to the *non-stationary response* situation when the frequency content of $y(t)$ can be described in terms of a time-dependent (or evolutionary) spectral-density function $G_y(\omega, t)$. The

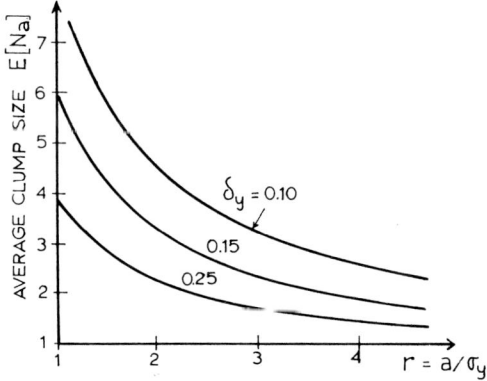

Fig. 8.13. The average clump size as a function of the reduced threshold level $r = a/\sigma_y$ and the bandwidth factor $\delta_y$.

time-dependent moments $\lambda_{i,y}(t)$ can then be obtained by integration of $\omega^i G(\omega, t)$ over all frequencies, i.e.,

$$\lambda_{i,y}(t) = \int_0^\infty \omega^i G(\omega) \, |H(\omega,t)|^2 \, d\omega \tag{8.41}$$

Furthermore, time-dependent response statistics, e.g., $\Omega_y(t)$ and $\delta_y(t)$, can be obtained by substituting the appropriate time-dependent moments in the "stationary case" definitions in much the same way as the variance $\sigma_y^2(t) = \lambda_{0,y}(t)$:

$$\lambda_{i,y}(t) \simeq \omega_n^{-4} \left[ G(\omega_n) \left\{ \omega_n^4 \int_0^\infty \omega^i \, |H(\omega,t)|^2 \, d\omega - \frac{\omega_n^{i+1}}{i+1} \right\} \right.$$
$$\left. + \int_0^{\omega_n} \omega^i G(\omega) d\omega \right] \tag{8.42}$$

where $|H(\omega,t)|^2$ may be substituted by the right-hand side of eq. 8.31. This equation becomes identical to eq. 8.33 when $i = 0$. At intermediate and low natural frequencies, the second term in eq. 8.42 will be unimportant. In this case, one obtains the central frequency $\Omega_y(t) \simeq \omega_n$ and the dispersion parameter $\delta_y(t) \simeq [(4/\pi)\zeta_t]^{1/2}$, where $\zeta_t$ is given by eq. 8.30. At very high naural frequencies, the second term will be predominante in eq. 8,42, and the spectral parameters of the response approach those of the ground motion, i.e., $\Omega_y(t) \to \Omega$ and $\delta_y(t) \to \delta$.

In general, the spectral shape parameters $\Omega_y(t)$ and $\delta_y(t)$ will lie in between the values obtained for these limiting cases. The following formulas can then be used. They express the parameters of the spectral density function, say $G_T(\omega)$, which is the sum of a number of positive functions $G_k(\omega)$, each contributing a fraction $p_k$ to the total area under $G_T(\omega)$. The spectral parameters of $G_T(\omega)$ are $\Omega_T$ and $\delta_T$, those of $G_k(\omega)$ are $\Omega_k$ and $\delta_k$. The weights $p_j$ sum to one. We have:

$$\Omega_T = \left( \sum_k p_k \Omega_k^2 \right)^{1/2} \tag{8.43}$$

$$\delta_T = \left[ 1 - \left\{ \sum_k p_k (\Omega_k/\Omega_T) \sqrt{1 - \delta_k^2} \right\}^2 \right]^{1/2} \tag{8.44}$$

In the case at hand, there are two contributions with weights $p_1 = 1 - p_2$ and $p_2 = \sigma^2 F^*(\omega_n)/\sigma_y^2(s)$, and the parameters of interest are $\Omega_T = \Omega_y(t)$, $\delta_T = \delta_y(t)$, $\delta_1 = (4/\pi\zeta_t)^{1/2}$, $\delta_2 = \delta$, $\Omega_1 = \omega_n$ and $\Omega_2 = \Omega$. The two limiting cases referred to earlier correspond to $p_1 = 0$ and $p_1 = 1$, respectively.

### 8.3.4 Prediction of maximum response

The purpose of this section is to evaluate the factor $r_{s,p}$ by which the response standard deviation $\sigma_y(s)$ must be multiplied to predict the level

$y_{s,p}$ below which the absolute value of the response $y(t)$ will remain, with probability $p$, during the time interval $(0, s)$. The task at hand is equivalent to finding the probability $L_a(s)$ that the system response fails to make a "passage" across a specified response level $a$ during the time interval $s$. The *first-passage problem* has for decades been the subject of considerable research, and an exact solution does not yet exist. Moreover, few of the proposed approaches are of practical value to earthquake engineers. *

Most of the literature on the first-passage problem deals with the stationary response to Gaussian white noise of a lightly damped linear one-degree system. The reader is referred to Crandall (1970) for an excellent state-of-the-art review. Below, a relatively simple approximate procedure is presented to predict the maximum responses of a rather general linear system exposed, suddenly and for a limited time, to steady-state Gaussian excitation. The proposed solution is based on research by the writer and his associates (Vanmarcke, 1969 and 1975; Corotis et al., 1972). The case considered first is when the random motion at hand is *stationary* and Gaussian and the starting condition is random.

### 8.3.4.1 Stationary response

Theoretical and simulation studies (Crandall et al., 1966; Ditlevsen, 1971) have confirmed that the probability $L_a(t)$ decays approximately exponentially with time, as follows **:

$$L_a(s) = A \exp(-\alpha s) \qquad (8.45)$$

in which $L_a(s)$ is the probability that $|y(t)|$ remains below the level $a$ during the interval $(0, s)$, $A = L_a(0)$ = the probability of starting below the threshold, and $\alpha$ = the decay rate. At high levels, $A \simeq 1$, and $\alpha \simeq 2\nu_a$ (see eq. 8.37):

$$L_a(t) = \exp(-2\nu_a t) = \exp[-(\Omega_y/\pi)\exp(-a^2/2\sigma_y^2)t] \qquad (8.46)$$

Cramer (1966) has shown that this result is asymptotically exact (when the level $a$ increases to infinity). Davenport (1964) independently derived this result. It is consistent with the assumption that high-level crossings occur according to a Poisson process. For $a$-levels of practical interest, however, the use of eq. 8.46 results in an error whose magnitude strongly depends on the *bandwidth* of the process. Numerical simulation studies indicate that the error tends to be on the unsafe side for wide-band processes and low threshold levels (Ditlevsen, 1971) and on the safe side for narrow-band processes (Crandall et al., 1966). For wide-band processes, the main effect is that the

---

* In fact, the most common approach to this problem has been to avoid it, by adopting a constant peak factor, e.g., by predicting maximum response based on the "$3\sigma$-rule".
** The validity of the exponential approximation is doubtful at very small values of $s$, e.g., when only a few cycles of motion have elapsed. This is of little practical concern here.

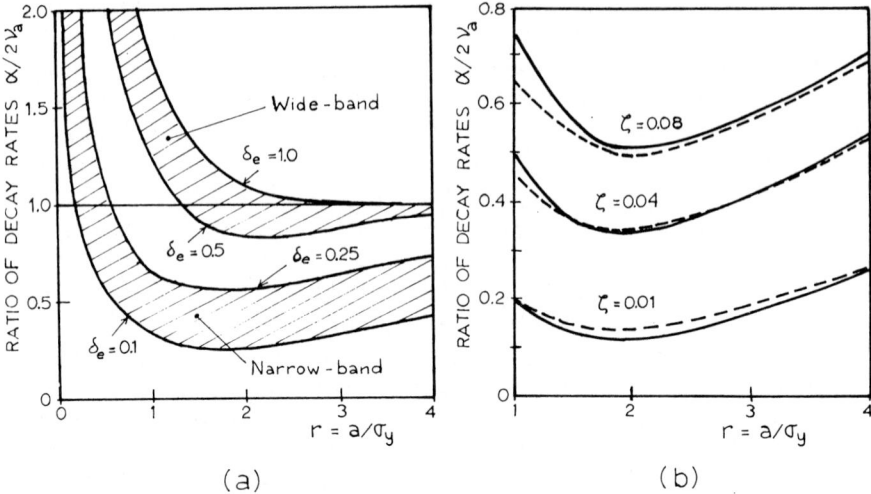

Fig. 8.14. The ratio of decay rates $\alpha/2\nu_a$: (a) predicted by eq. 8.47 for different threshold levels $r$ and bandwidth factors $\delta_e$; and (b) specifically, for the stationary response of a linear one-degree-of-freedom system under Gaussian white-noise excitation. (Both the theoretical value (eq. 8.47) and the empirical value obtained by simulation (Crandall et al., 1966) are shown.)

Poisson crossing assumption makes no allowance for the time motion actually spends above the level $a$. For narrow-band processes, it is important to account for the fact that level crossings tend to occur not independently in accordance with the Poisson model, but in clusters or clumps.

The use in eq. 8.45 of the following approximation for $\alpha$ yields much improved estimates of $L_a(t)$ which are in close agreement with those obtained by simulation (Crandall et al., 1966):

$$\alpha = 2\nu_a \frac{1 - \exp(-\sqrt{\pi/2}\delta_e r)}{1 - \exp(-r^2/2)} \qquad (8.47)$$

in which $r = a/\sigma_y$ is the reduced level and $\delta_e = \delta_y^{1+b}$ is a bandwidth measure; $b$ is a semi-empirical nonnegative constant estimated at about 0.2. Choosing $b = 0$ (or $\delta_e = \delta_y$) will result in slightly more conservative probability estimates (i.e., $r_{s,p}$ will be slightly higher). The ratio $\alpha/2\nu_a$ given by eq. 8.47 is plotted in Fig. 8.14 for several values of $\delta_e$ as a function of the reduced level $r$. Note that it converges to one at high threshold levels in accordance with Cramer's result. An improved estimate of the probability $A$ (in eq. 8.45) *

---

* $A$ is interpreted here as the probability that the first peak of $|y(t)|$ (immediately after the start) will be below the level $a = r\sigma_y$.

is $[1 - \exp(-r^2/2)]$, which also approaches one at high levels. For the derivation of eq. 8.47, see Vanmarcke (1975).

The approximation for $L_a(s)$ based on eqs. 8.45 and 8.47 depends on the motion parameters $\sigma_y$, $\Omega_y$ and $\delta_y$, all of which are defined in terms of the first few spectral moments of $y(t)$. The reduced level $r_{s,p}$ corresponding to reliability $p$ and duration $s$ is obtained by inverting $p = \exp(-\alpha s)$. The result can be expressed in terms of the factors $n = (\Omega_y s/2\pi)(-\log p)^{-1}$ and $\delta_e$. When $p = e^{-1} = 0.368$, $(-\log p)^{-1} = 1$, and $n$ equals the average number of cycles of response motion $(\Omega_y s/2\pi)$. The values of $(-\log p)^{-1}$ corresponding to $p = 0.5, 0.9$, and $0.99$ are about $1.4, 10, 100$, respectively. Note that the factor $(-\log p)^{-1}$ may be substituted by $(1-p)^{-1}$ when the reliability $p$ is very close to one.

The reduced level $r_{s,p}$ is plotted in Fig. 8.15 as a function of $n$ for several values of $\delta_e$. For large values of $\delta_e$, the solution approaches the upper-bound curve which constitutes the solution to eq. 8.46. The exact equation for the upper-bound curve is:

$$r_{s,p} = \sqrt{2 \log 2n} \tag{8.48}$$

in which:

$$n = (\Omega_y s/2\pi)(-\log p)^{-1} \tag{8.49}$$

An approximate expression for the other curves in Fig. 8.15 is:

$$r_{s,p} = [2 \log\{2n[1 - \exp(-\delta_e\sqrt{\pi \log 2n})]\}]^{1/2} \tag{8.50}$$

As an illustration, consider the seismic response of a linear one-degree system with a damping value not less than 2% and an intermediate natural

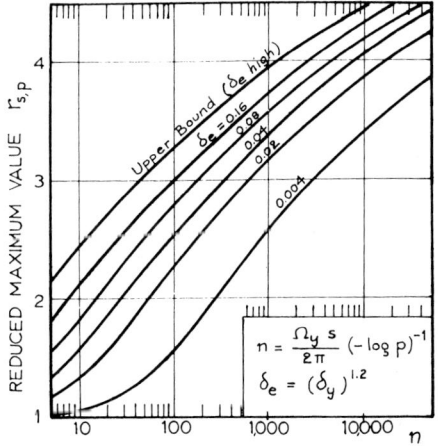

Fig. 8.15. The reduced maximum value $r_{s,p}$ as a function of $n = (\Omega_y s/2\pi)(-\log p)^{-1}$ for several values of $\delta_e = (\delta_y)^{1.2}$.

period. The response will quickly approach a steady-state condition and it is reasonable to use the stationary solution for $r_{s,p}$. Our interest is in the dependence of the maximum response on *damping*, an intriguing topic of earthquake engineering research (Housner and Jennings, 1964; Arias and Husid, 1962; Bycroft, 1960; Brady, 1966). It has been noted that the ratio of response spectra at intermediate periods corresponding to two different damping factors $\zeta_1$ and $\zeta_2$ (both not less than 0.02) is approximately $(\zeta_2/\zeta_1)^{0.4}$, in apparent disagreement with the result that the ratio of the stationary standard deviations is $(\zeta_2/\zeta_1)^{0.5}$. It appears that the peak factor ratio (heretofore thought not to depend upon damping) can account for the discrepancy, i.e., a factor $(\zeta_2/\zeta_1)^{0.1}$. Take $s = 10$ sec, $(\Omega_y/2\pi) = 1$ cps and $p = 0.50$, so that $n = 1 \times 10 \times 1.4 = 14$. The upper-bound value (eq. 8.48) of $r_{s,p}$ is $(2 \log 28)^{1/2} = 2.58$. Table 8.III lists the peak factors corresponding to the damping values 2%, 5% and 10%. Also shown are some peak factor ratios predicted by theory which agree very well with the empirical result $(\zeta_1/\zeta_2)^{0.1}$. Actually, Bycroft (1960) has shown that the ratio of response spectra is closer to the ratio of standard deviations, $(\zeta_2/\zeta_1)^{0.5}$ at low periods (say, 0.2—0.5 seconds) or when both $\zeta_1$ and $\zeta_2$ are high (say, 10% or more). Both of these effects are correctly predicted by eq. 8.50

*8.3.4.2 Transient response*

A major advantage of the solution just presented is that it can easily be extended to obtain first-passage probability estimates for transient response whose frequency content is described in terms of the time-dependent spectral density function $G_y(\omega, t)$. It is possible to evaluate the time-dependent decay rate $\alpha(t)$ in the expression (Amin et al., 1969):

$$L_a(s) = \exp\left[-\int_0^s \alpha(t)\,\mathrm{d}t\right] \tag{8.51}$$

which is a direct extension of eq. 8.45. In this case, $A = L_a(0) = 1$, since the response $y(t)$ builds up from the rest. Recall that our aim is to determine the factor $r_{s,p}$ which must be multiplied by the standard deviation $\sigma_y(s)$ to pre-

TABLE 8.III

Peak factors corresponding to the damping values 2, 5 and 10%

| $\zeta$ | $\delta_e$ | $(r_{s,p})_\zeta$ | $\dfrac{(r_{s,p})_\zeta}{(r_{s,p})_{0.02}}$ | $\left(\dfrac{\zeta}{0.02}\right)^{0.1}$ |
|---|---|---|---|---|
| 0.02 | 0.11 | 2.06 | 1.0 | 1.0 |
| 0.05 | 0.19 | 2.26 | 1.10 | 1.10 |
| 0.10 | 0.29 | 2.38 | 1.16 | 1.17 |

dict maximum-response fractiles. A direct but impractical approach is to substitute the parameters $\sigma_y$, $\Omega_y$ and $\delta_y$ in the expression for $\alpha$ (eq. 8.47) by their time-dependent equivalents, and to solve eq. 8.51 *numerically* for $r_{s,p} = a/\sigma_y(s)$. A much simpler approximate procedure is outlined below. As the response variance $\sigma_y^2(t)$ increases with time, from 0 to $\sigma_y^2(s)$, the failure rate of $\alpha(t)$ will increase much more rapidly, and the integral $\int_0^s \alpha(t)\,dt$ in eq. 8.51 will be dominated by contributions corresponding to values of $t$ close to $s$. This motivates the introduction of an "equivalent stationary response" duration $s_0$ such that:

$$L_a(s) = \exp\left[-\int_0^s \alpha(t)\,dt\right] = \exp[-s_0 \alpha(s)] \tag{8.52}$$

where, clearly, $s_0 \leqslant s$. The ratio $(s_0/s)$ can be roughly estimated from the ratio:

$$m = \sigma^2(s)/\sigma^2(s/2) \tag{8.53}$$

by the following equation *:

$$s_0/s \simeq \exp[-2(m-1)] = \exp[-2\{\sigma_y^2(s)/\sigma_y^2(s/2) - 1\}] \tag{8.54}$$

For lightly damped one-degree systems, $1 \leqslant m \leqslant 2$, with the upper bound applicable when $\omega_n \beta s \to 0$. When $m = 1$, $s_0 = s$. The attractive feature of eq. 8.52 is, of course, that the stationary first-passage solution can again be used to derive $r_{s,p}$. It suffices to substitute the quantities $s\Omega_y$ and $\delta_y$ in eqs. 8.48—8.50 by $s_0 \Omega_y(s)$ and $\delta_y(s)$, respectively.

It is now possible to evaluate peak factor *spectra* (see Fig. 8.16) which show $r_{s,p}$ as a function of natural period for various damping factors. Figure 8.17 shows the *median* ($p = 0.50$) peak factor spectra for a typical earthquake ground motion, say, with Kanai-Tajimi frequency content and a strong-motion duration of 10 seconds. At very low periods, the peak factor $r_{s,p} = r_{10,0.50}$ essentially equals that of the ground motion (the input parameters are $\Omega_y(s) = \Omega$, $\delta_y(s) = \delta$, and $m = 1$). At moderate and high periods, the peak factor is determined using the parameters $\Omega_y(s) = \omega_n$, $\delta_y(s) = [(4/\pi)\zeta_s]^{1/2}$ and:

$$m = \sigma_y^2(s)/\sigma_y^2(s/2) = [1 - \exp(-2\zeta\omega_n s)]/[1 - \exp(\zeta\omega_n s)] \tag{8.55}$$

Note that for very long periods or very light damping values, i.e., when the product $\zeta \omega_n s \to 0$, the parameters become $\delta_y(s) \to (\pi \omega_n s/2)^{-1/2}$ and $m \to 2$. In the transition period range (see Fig. 8.17), the input parameters are combinations of those just given (see Section 8.3.3).

---

* Equation 8.54 is obtained by approximating the area under $\alpha(t)$ (from 0 to $s$) in terms of the values $\alpha(s)$ and $\alpha(0.5\,s)$, which in turn depend principally on the respective transient variances. Note that $m$ is the ratio of the transient variances at $s$ and $0.5\,s$, respectively.

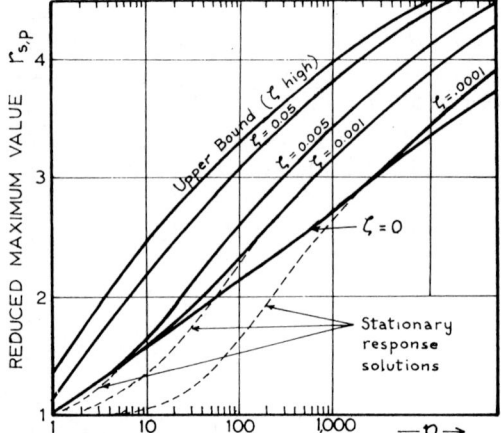

Fig. 8.16. The reduced maximum value $r_{s,p}$ as a function of $n = (\omega_n s/2\pi)(-\log p)^{-1}$ for the response to suddenly applied ideal white noise. For a given probability $p$ and duration $s$, $r_{s,p}$ depends on natural frequency $\omega_n$ and damping $\zeta$.

Multiplying the median peak factor spectra (median "$r$-spectra") by the "$\sigma$-spectra" (see Fig. 8.11) yields predictions of the *median* (50% exceedance probability) *response spectra*. It is evident that the peak factors significantly affect response-spectra predictions. For example, it is easy to verify that under white-noise excitation, the damped pseudovelocity "$\sigma$-spectrum" is constant while the "$r$-spectrum" decreases with period (see Fig. 8.18). This leads to a predicted median undamped response spectrum which decreases

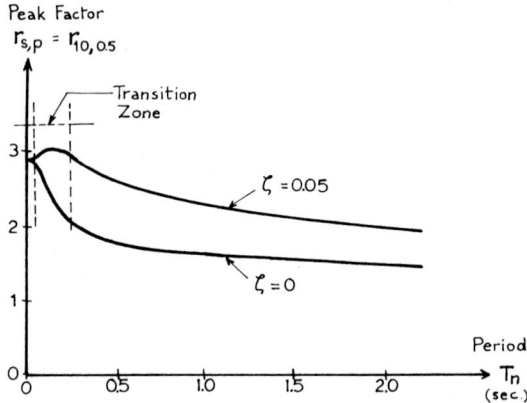

Fig. 8.17. Median peak factor spectra (or "$r$-spectra"): the reduced median maximum values as natural period for several damping values ($p = 0.5$; $s = 15$ sec; $G(\omega)$ = Kanai-Tajimi spectral density).

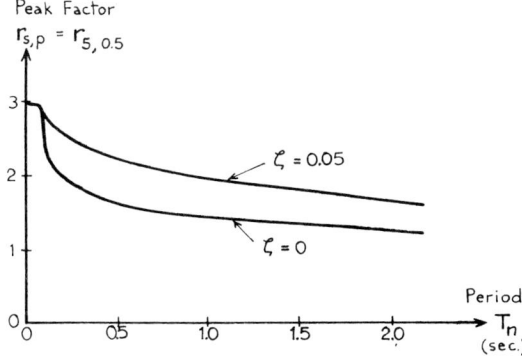

Fig. 8.18. Median peak factor spectra (or "$r$-spectra") for band-limited white-noise excitation.

with increasing periods, an effect which has been observed in simulation studies (Bycroft, 1960; Brady, 1966). The peak factor theory also explains satisfactorily troublesome discrepancies noted by many researchers (e.g., Jennings, 1964; Arias and Husid, 1962) between predictions of the ratio of damped and undamped response spectra based on real or simulated earthquakes versus those based on random vibration theory (neglecting peak factor effects).

*8.3.5 Compatibility of ground-motion representations*

The preceding sections have established a relationship between the response spectrum and the spectral-density function of ground motions at a site. The relationship is not unique, though; it depends on the chosen strong-motion duration $s$, on the exceedance probability level $p$ assigned, and on the damping level $\zeta$ involved. The first step in the procedure to obtain $G(\omega)$ from a given damped response spectrum $S_V$, is to divide the latter by the peak factor $r_{s,p}$ at a set of target period values $\omega_n^{(i)}$. This yields the "$\sigma$-spectrum" (i.e., $\sigma_a(s)$ as a function of period) from which $G(\omega)$ can be obtained by a simple iterative procedure based on eq. 8.33. One starts at the lowest natural frequency $\omega_n^{(1)}$, when the contribution of the integral term on the right-hand side of eq. 8.33 is negligible. At an arbitrary frequency $\omega_n^{(i)}$, the integral of $G(\omega)$ up to $\omega_n^{(i)}$ is evaluated numerically and $G(\omega_n^{(i+1)})$ can be found from eq. 8.33. Having obtained $G(\omega)$, one can proceed to *simulate* a set of ground motions whose computed response spectra will "match" the specified smooth response spectra (see Section 8.2.2).

Note that the $S_V$ to $G(\omega)$ conversion procedure is reversible: start from the s.d.f. $G(\omega)$, one can evaluate the response spectra for various damping values corresponding to different specified probability levels. This dual capability helps resolve important questions relating to: (1) the compatibility of the different methods of representing the ground motion at a site: (2) the "internal consistency" of a set of specified damped response spectra: and (3) the level of conservatism built into a set of scaled smooth response spectra. In relation to the last point, it must be emphasized that there need not necessarily be a close relationship between the *design* response spectrum at a site and the spectral-density function of strong earthquakes that might occur at the site (McGuire, 1974; Vanmarcke and Cornell, 1972). A design response spectrum is developed to cover different possible ground motions (e.g., caused by large, distant earthquakes or close, moderate earthquakes) each with its own frequency content and expected duration. The common practice of generating simulated motions for use in design based on a single spectral density function appears to be justified only if one earthquake source accounts for nearly all of the total site seismic risk, or if the *local* geology is principally responsible for shaping the frequency content of the ground motion. Otherwise, it is necessary to re-examine the data on seismicity and attenuation to determine the relative likelihood of occurrence of the various types of site ground motion and frequency content. These issues need considerable further study.

The last point, not least in importance, concerns the *variability* of the response spectra. Statistical analyses based on a set of recorded accelerograms from different real earthquakes and sites (see Section 8.2.1), will yield higher coefficients of variation than those obtained based on random vibration analysis or on a set of simulated ground motions with common expected frequency content and duration. The variability in responses obtained by random vibration (or by time integration of simulated motions) reflects only inherent uncertainty due to random phasing, not the uncertainty in such ground-motion parameters as duration and dominant ground frequency. Agreement among coefficients of variation can only be achieved by combining the (joint probability) distribution of the ground-motion parameters with the "conditional" response distribution derived in the preceding sections.

8.4 MULTI-DEGREE-OF-FREEDOM SYSTEMS

A basic step in the seismic analysis of a structure or a soil-structure system is to construct a dynamic model, frequently a lumped-parameter model whose parameters are the elements of the mass and stiffness matrices. These may be determined by any one of several conventional procedures (see, for example, Biggs, 1964). The natural frequencies and shapes of the normal modes can then be determined by solving the eigenvalue problem. In the normal-mode method, the $n$-degree system response at a predetermined point $B$ on the structure (see Fig. 8.20) is expressed in terms of the modal coefficients $c_k$ and the generalized modal coordinates $y_k(t)$, $k = 1, 2, ..., n$. In particular, the displacement at point $B$ relative to the ground is:

$$y(t) = \sum_{k=1}^{n} \phi_{kB}\Gamma_k y_k(t) = \Sigma c_k y_k(t) \tag{8.56}$$

Each component, $y_k(t)$, is the response of a one-degree system characterized by the (undamped) natural frequency $\omega_k$ and an assigned percentage of critical damping $\zeta_k$. Also, $c_k = \phi_{kB}\Gamma_k$, where $\phi_{kB}$ is the characteristic shape ordinate for mode $k$ at point $B$, and $\Gamma_k$ is the participation factor in the $k$th mode. For more details about the normal-mode technique, the reader is referred to a textbook on structural dynamics (e.g., Biggs, 1964; Hurty and Rubinstein, 1964; Clough and Penzien, 1975).

A response quantity closely related to $y(t)$ is the absolute acceleration response:

$$z(t) = x(t) + \ddot{y}(t) = x(t) + \sum_{k=1}^{n} c_k \ddot{y}_k(t) \tag{8.57}$$

where $\ddot{y}(t)$ is the second derivative of $y(t)$, $\ddot{y}_k(t)$ is the second derivative of the $k$th modal coordinate, and $x(t)$ is the ground acceleration.

In complex systems where damping varies substantially in nature (hysteretic vs. viscous) or magnitude throughout the system, classical modal analysis is not strictly applicable. The difficulty with normal-mode approximations lies in the assignment of modal damping (Caughey, 1960b). For example, in soil-structure interaction analysis, the damping in part of the system is viscous while in other parts it is more nearly hysteretic. Roesset et al. (1973) have suggested a method for determining weighted modal damping. An alternative to the modal-analysis approach is a solution by frequency-domain analysis (e.g., Sarrazin et al., 1972).

### 8.4.1 Response-spectrum approach

The response-spectrum approach is the simplest and most common way to estimate the maximum system response, say, the maximum relative displacement, due to a ground motion. The response spectrum is utilized to produce the maximum modal displacements $S_{Dk}$. These are read from the response spectrum at the frequency $\omega_k$ and for the damping coefficient $\zeta_k$. The maximum displacement at $B$ (relative to the ground) is often predicted by the "root-sum-square" rule (Rosenblueth, 1951):

$$S = \sqrt{\sum_{k=1}^{n} c_k^2 S_{Dk}^2} \tag{8.58}$$

Similarly, the modal maximum *accelerations* at mass point $B$ on the structure are $A_{kB} = |\omega_k^{-2} S_{Dk} c_k|$. When multiplied by the mass (at $B$), these accelerations are the equivalent static forces which may be used to compute the maximum stresses in the structure due to mode $k$. The modal stresses at a point may then be combined by the root-sum-square procedure to predict total maximum stress. Other rules have been suggested which attempt to account for the interaction between modal contributions (which becomes more pronounced when modal frequencies are close or when the damping is high). Rosenblueth and Elorduy (1969) have suggested the expression:

$$S^2 = \sum_{k=1}^{n} c_k^2 S_{Dk}^2 + \sum\sum_{k \neq j} \frac{c_j c_k S_{Dj} S_{Dk}}{1 + \epsilon_{jks}^2} \tag{8.59}$$

where:

$$\epsilon_{jks} = \frac{|\omega_k' - \omega_j'|}{\zeta_k' \omega_k + \zeta_j' \omega_j}$$

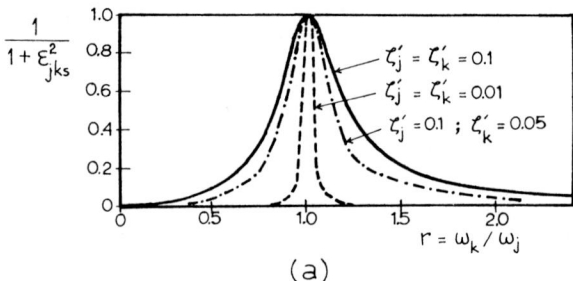

(a)

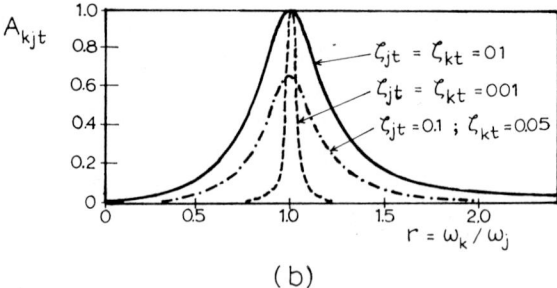

(b)

Fig. 8.19. The factors $(1 + \epsilon_{jks}^2)^{-1}$ and $A_{jks}$ which account for interaction between modes $j$ and $k$ of a multi-degree-of-freedom system.

with $\omega'_k = \omega_k(1 - \zeta'^2_k)^{1/2}$ and $\zeta'_k = \zeta_k + (2/\omega_k s)$.* The "cross-coefficient" is plotted as a function of the ratio $r = \omega_j/\omega_k$ for several combinations of damping factors in Fig. 8.19a. Rascón and Villareal (1974) have evaluated eqs. 8.58 and 8.59 for two-degree systems by comparing exact and predicted responses. They found that eq. 8.59 gives much better results than eq. 8.58 when the two natural frequencies are close. Also, both approximations tend to underestimate the exact maximum response for undamped systems. The ratio of exact to predicted undamped response is about 1.2—1.25.

*8.4.2 Random vibration approach*

The random vibration method of analysis seeks to determine the distribution of the maximum response in a direct way. The basic procedure is identical to that followed for one-degree systems. The ground motion is characterized by its spectral-density function $G(\omega)$ and duration $s$. The system relating input acceleration $x(t)$ and output relative displacement $y(t)$ has the impulse-response function:

$$h(t) = \sum_{k=1}^{n} c_k h_k(t) \qquad (8.60)$$

where $h_k(t)$ is the impulse-response function of a one-degree system (see eq. 8.27) with parameters $\omega_k$ and $\zeta_k$. The truncated Fourier transform of $h(t)$, or the time-dependent transfer function, is:

$$H(\omega,t) = \int_0^t h(t-\tau) \exp(-i\omega\tau)\, d\tau = \sum_{k=1}^{n} c_k H_k(\omega,t) \qquad (8.61)$$

and the time-dependent spectral-density function of $y(t)$ equals:

$$G_y(\omega,t) = G(\omega)|H(\omega,t)|^2 = G(\omega) \sum_{k=1}^{n}\sum_{j=1}^{n} c_k c_j H_k(\omega,t) H_j^*(\omega,t) \qquad (8.62)$$

in which $H_j^*(\omega, t)$ = the complex conjugate of $H_j(\omega, t)$. Although the expression $G_y(\omega, t)$ has the form of a double summation, significant contributions to its moments with respect to frequency (spectral moments) usually come from the terms for which $j = k$. This is particularly true when modal frequencies are well-separated and when damping values are low. Algebraic manipulation allows one to express eq. 8.62 approximately as follows (Vanmarcke,

---

* Note that the damping factor $\zeta'$ is duration-dependent much in the same way as $\zeta_t$ (see eq. 8.32). The two definitions are equivalent for $t = s = \infty$ and for $\zeta = 0$.

1972) *:

$$G_y(\omega,t) \simeq G(\omega) \sum_{k=1}^{n} |H_k(\omega,t)|^2 \left(c_k^2 + \sum_{j \neq k} c_j c_k A_{kjt}\right) \quad (8.63)$$

in which $A_{kjt}$ is a factor which depends on the ratio of modal frequencies $r = \omega_j/\omega_k$ and on the equivalent damping values $\zeta_{kt} = \zeta_k[1 - \exp(-2\zeta_k\omega_k t)]^{-1}$ and $\zeta_{jt} = \zeta_j(1 - \exp(-2\zeta_j\omega_j t))^{-1}$

$$A_{kjt} = \frac{8r\zeta_{kt}(\zeta_{jt} + \zeta_{kt}r)[(1-r^2)^2 - 4r(\zeta_{kt} - \zeta_{jt}r)(\zeta_{jt} - \zeta_{kt}r)]}{8r^2[(\zeta_{kt}^2 + \zeta_{jt}^2)(1-r^2)^2 - 2(\zeta_{jt}^2 - \zeta_{kt}^2 r^2)(\zeta_{kt}^2 - \zeta_{jt}^2 r^2)] + (1-r^2)^4} \quad (8.64)$$

The factor $A_{kjt}$ is plotted in Fig. 8.19b as a function of $r$ for different pairs of values of $\zeta_{jt}$ and $\zeta_{kt}$. At $r = 1$, $A_{kjt} = 2\zeta_{kt}/(\zeta_{kt} + \zeta_{jt})$, which is equal to one if $\zeta_{jt} = \zeta_{kt}$.** $A_{kjt}$ vanishes when $r$ is either very small or very large. Integrating eq. 8.63 over all frequencies gives the time-dependent variance of the multi-degree-system response $y(t)$:

$$\sigma_y^2(t) = \int_0^\infty G_y(\omega,t)d\omega \simeq \sum_{k=1}^{n}\left(c_k^2 + \sum_{j \neq k} c_j c_k A_{kjt}\right)\sigma_{y_k}^2(t) = \sum_{k=1}^{n} \alpha_{kt} c_k^2 \sigma_{y_k}^2(t) \quad (8.65)$$

where:

$$\alpha_{kt} = 1 + \sum_{j \neq k} (c_j/c_k) A_{kjt} \quad (8.65a)$$

It is easy to see that the "cross-terms" will be insignificant, i.e., $\alpha_{kt} \simeq 1$, when: (1) the modal frequencies are well-separated; (2) the damping values are small; and (3) time $t$ is sufficiently large. Higher spectral moments have the same form as eq. 8.65; the $i$th spectral moment is:

$$\lambda_{i,y}(t) = \int_0^\infty \omega^i G_y(\omega,t)d\omega \simeq \sum_{k=1}^{n} \alpha_{kt} c_k^2 \lambda_{i,y_k}(t) \quad (8.66)$$

where $\lambda_{i,y_k}(t)$ is the $i$th spectral moment for the $k$th mode (see eqs. 8.33 and 8.42). Of course, $\lambda_{0,y_k}(t) = \sigma_y^2(t)$. Other spectral parameters and level-crossing statistics introduced in Section 8.33 can now be obtained for multi-degree system response. The peak factor $r_{s,p}$ which needs to be multiplied by $\sigma_y(s)$ to estimate the maximum-response level corresponding to a given probability $p$ and duration $s$, can be determined by the procedure described

---

* The use of eq. 8.63 to evaluate the spectral moments results in a percentage error which is of the order of the square of the damping factor.
** Also, $A_{kjt} = r(\zeta_{jt}/\zeta_{kt})A_{kjt}$, so that $A_{jkt} + A_{kjt} = 2$ if $r = 1$.

in Section 8.3.4. Briefly, the input data required are the average frequency $\Omega_y(s)$ and the shape parameter $\delta_y(s)$. These can be evaluated in terms of the corresponding *modal* quantities $\Omega_{y_k}(s)$ and $\delta_{y_k}(s)$ by means of eqs. 8.43 and 8.44. The weights $p_k$ are the fractional modal contributions to the response variance (eq. 8.65, evaluated at $t = s$), i.e., $p_k = \alpha_{ks} c_k^2 \sigma_{y_k}^2(s)/\sigma_y^2(s)$. The ratio $m = \sigma_y^2(s)/\sigma_y^2(0.5s)$, which measures the degree of nonstationarity in the response motion, is needed in eq. 8.54 to determine the equivalent stationary-response duration $s_0$. If more than one mode is significant and the modal frequencies are not very close, then the bandwidth factor $\delta_y(s)$ will be large, implying that the upper-bound curve in Fig. 8.15 can be used to obtain the peak factor $r_{s,p}$.

### 8.4.2.1 Comparison between response spectrum and random vibration approaches

Rosenblueth (1951) first suggested the widely used root-sum-square rule for combining modal maxima (eq. 8.58). This rule rests on the assumption of statistical independence among modes and applies theoretically to the *standard deviations* of the modal contributions as can be seen by putting $\alpha_{kt} = 1$ in eq. 8.65. If the differences among the peak factors of the total response and the modal responses are indeed neglected, then eq. 8.65 can be converted to an improved rule for combining modal maxima when modal interactions are important, i.e.:

$$S = \left( \sum_{k=1}^{n} \alpha_{ks} c_k^2 S_{Dk}^2 \right)^{1/2} \tag{8.67a}$$

using the notation defined in Section 8.4.1, with $\alpha_{ks} = 1 + \Sigma_{j \neq k}(c_j/c_k) A_{kjs}$ (see eq. 8.65a). Further improvement is possible, however, by accounting explicitly for the peak factors, as follows:

$$S = \bar{r} \left[ \sum_{k=1}^{n} \alpha_{ks} c^2 (S_{Dk}/\bar{r}_k)^2 \right]^{1/2} \tag{8.67b}$$

where $\bar{r}$ and $\bar{r}_k$ denote, say, the median peak factors of the total response and the response in mode $k$, respectively.

### 8.4.2.2 Absolute acceleration response

Very similar results can be obtained for the absolute acceleration response $z(t)$ at a point in an $n$-degree structure. The time-dependent spectral-density function $G_z(\omega, t)$ of the absolute acceleration response $z(t)$ has an approximate form similar to that of $G_y(\omega, t)$ in eq. 8.63:

$$G_z(\omega, t) \simeq G(\omega) | 1 + \omega^2 \sum_{k=1}^{n} c_k H_k(\omega, t)|^2 \simeq G(\omega) \left[ 1 + \omega^4 \sum_{k=1}^{n} \alpha_{kt} c_k^2 |H_k(\omega, t)|^2 \right] \tag{8.68}$$

with $\alpha_{kt}$ defined by eq. 8.65a. As before, the understanding is that the time-dependent transfer function $|H_k(\omega, t)|$ is, for practical purposes, equivalent to a complete transfer function $|H_k(\omega)|$ with an equivalent time-dependent damping parameter $\zeta_{kt}$. The approximation on the right-hand side of eq. 8.68 is suitable for determining the spectral moments of $z(t)$ and for use in random vibration analysis leading to response statistics of "secondary systems" dealt with in Section 8.5.

## 8.4.3 Time-integration analysis

The third analysis procedure requires the numerical solution of the modal equations of motion (by time- or frequency-domain methods). The output may be in the form of individual modal response functions $y_k(t)$ which can be superimposed using eq. 8.56 (or eq. 8.57). The maximum value of $y(t)$ (or $z(t)$) is then sought. This procedure can be repeated for several actual earthquakes or simulated ground motions to determine the range of possible maximum responses.

A comparison between response predictions based on random vibration analysis and repeated time integration is presented in Fig. 8.20. Three 4 degree-of-freedom models are considered which are identical except for the values of the natural periods. The pertinent dynamic properties are summarized in Fig. 8.20 which shows the cumulative probability distribution of the maximum displacement of the 4th floor relative to ground for the three

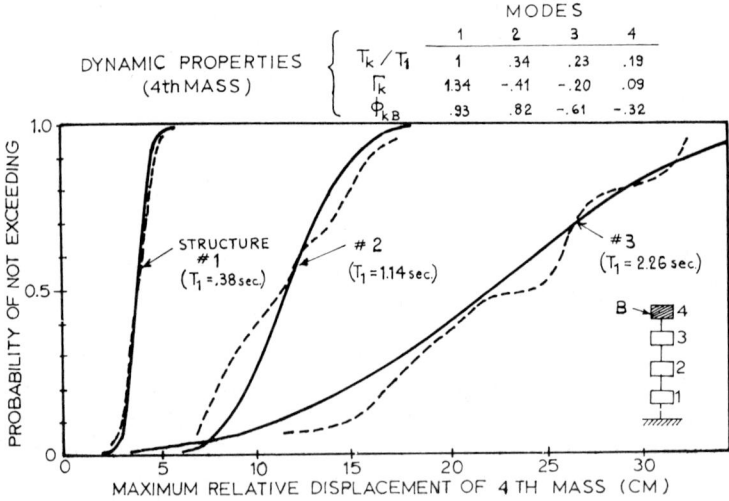

Fig. 8.20. Probability distributions of multi-degree system response obtained (1) by random vibration analysis, and (2) by time-integration analysis based on 14 simulated ground motions.

systems. A constant 2% damping value is assumed for all modes. The starting point for both methods is the smooth response spectrum for 2% damping scaled to 0.3 $g$ maximum ground acceleration (see Fig. 8.1). By assuming this spectrum to correspond to 50% exceedance probability and selecting a strong-motion duration of 17.5 seconds, the spectral-density function of the ground motion can be derived by the procedure outlined in Section 8.3.5. The random vibration predictions can then be made simply and directly, leading to the solid curves shown in Fig. 8.20. The crude frequency distributions shown as dotted lines (Fig. 8.20) are obtained by extensive numerical integration using fifteen artificial motions with common spectral content and duration.

## 8.5 LIGHT SECONDARY SYSTEMS

In a number of important problem situations arising in earthquake engineering, the dynamic model of interest consists of a primary system whose response provides the input to a secondary system. The primary system is often a linear lumped-mass system and the secondary system a simple linear oscillator. One application of this model is in the area of soil amplification where the question of concern is: "how do the soils underlying a structure modify the earthquake motion?" In this case, the soil layers constitute the primary system and the secondary system is a one-degree structure. An analysis of this soil-structure system yields the *ratio of the response spectra* at the base and at the top of the soil layer, respectively.

The second application arises in the seismic analysis of equipment in buildings. The response quantity of interest is the displacement of the equipment relative to the motion of the structure at the point of support for the equipment. The analysis of the structure-equipment system yields the so-called *floor response spectra* which are very useful in seismic design of equipment.

These combined (primary—secondary) systems can, in principle, be modeled as *one* multi-degree-of-freedom system, and the methods and results obtained in the previous section remain valid also for the case. Indeed, if the interaction effect between the soil and the structure, or between the structure and the equipment is considerable, it is necessary to combine the primary and secondary systems into a single dynamic model. It is common, however, to separate the seismic analysis of the secondary system from that of the primary system, and to view the acceleration response at the point of support (of the secondary system) as the input for the dynamic analysis of the secondary system. This uncoupled dynamic model is suitable when the inertial properties of the (massive) primary and (light) secondary system are very different, and has great practical value when an entire spectrum of secondary systems needs to be examined.

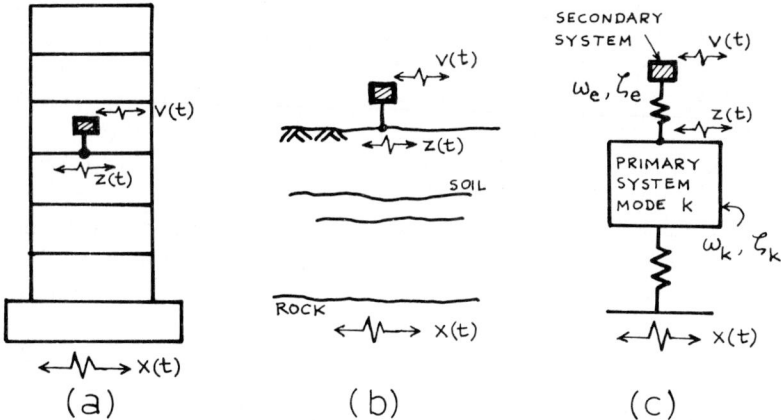

Fig. 8.21. Multi-degree-of-freedom system with attached secondary system supported at point B.

The first step in the analysis is aimed at obtaining the characteristics of the accelerogram at the support point $B$ (see Fig. 8.21). These will depend on the base acceleration and on the primary-system dynamic properties, i.e., the modal quantities $\omega_k$, $\zeta_k$, $\Gamma_k$ and $\phi_{kB}$ introduced in Section 8.4. The second step consists of the evaluation of the secondary-system response to the motion at its support. The output may be, for example, the secondary system's maximum acceleration which can be converted into equivalent static loads for stress analysis. Finally, if the secondary system is itself a multi-degree-of-freedom system, the second step in the procedure will have to be repeated for each secondary-system mode.

### 8.5.1 A response spectrum-based method

Assume that a modal analysis of the structure has been made using the response-spectrum approach described in Section 8.4.1. This analysis step yields the maximum modal accelerations at the point at which the secondary system is attached. For the second step, Biggs (1972) has suggested a procedure briefly outlined below which ensures satisfactory results when the secondary system is either very rigid or very flexible, but is frankly approximate and semi-empirical in between these limiting cases. The method predicts the maximum acceleration of a one-degree secondary system, $A_e$, as follows:

$$A_e = \left( \sum_k A_{ek}^2 \right)^{1/2} \qquad (8.69)$$

where $A_{ek}$ is the maximum secondary-system response attributable to mode

$k$ of the primary system. The form of $A_{ek}$ depends on the ratio of primary/secondary system frequencies $\omega_k/\omega_e$. At low values of $\omega_k/\omega_e$ (for relative rigid secondary system), $A_{ek}$ is viewed as an amplification of the mode-$k$ acceleration response of the primary system. In the case of a flexible secondary system ($\omega_k/\omega_e$ high), $A_{ek}$ is an amplification of the acceleration response of the secondary system as if it were founded directly on the ground. The amplification factors for both cases were developed *empirically* in graphical form as a function of $\omega_k/\omega_e$ for the damping values $\zeta_k = 0.04$ and $\zeta_e = 0.005$; they were found to be quite similar in shape to the classical amplification factor of a damped one-degree system subjected to a harmonic forcing function.

If the secondary system is itself a multi-degree system, then the above computation based on eq. 8.69 must be repeated for each secondary-system mode, and the individual modal maxima must be combined, for example, by root-sum-square procedure.

### 8.5.2 Random vibration approach

When an earthquake strikes, the secondary system will respond to the motion $z(t)$ generated at the support point $B$ on the primary system. The frequency content of the motion $z(t)$ is described by the (time-dependent) spectral-density function $G_z(\omega,t)$. If the primary system is not too lightly damped *, the variation of the frequency content of $z(t)$ *with time* will often not be important, and $G_z(\omega,t)$ will provide a reasonable representation of the input spectral-density function for the random vibration analysis of the secondary system. As in eq. 8.29, the time-dependent spectral density function $G_e(\omega,t)$ of the secondary system response can be expressed by:

$$G_e(\omega,t) = G_z(\omega,s)|H_e(\omega,t)|^2 \tag{8.70}$$

where $|H_e(\omega,t)|^2$ is the transient squared amplification function for the secondary system, which has the form of eq. 8.31 and can be approximated (as in eq. 8.34) by:

$$|H_e(\omega,t)|^2 = [(\omega_e^2 - \omega^2)^2 + 4\zeta_{et}^2 \omega_e^2 \omega^2]^{-1} \tag{8.71}$$

where $\zeta_{et} = \zeta_e[1 - \exp(-2\zeta_e\omega_e t)]^{-1}$ and $\zeta_e$ = the secondary system damping. As in earlier analyses, attention focuses on the variance of the response at its peak value, when $t = s$. Working in terms of the pseudo-acceleration

---

* For very lightly damped or very long period primary systems, the function $G_z(\omega,s)$ must be viewed as an approximate representation of the frequency content during the most intense part of the motion $z(t)$ (which will occur toward the end of the input motion and just thereafter, during the early part of the motion decay period). The analysis presented in this section should still provide a very useful approximation in this case. A more complete analysis is presented by Chakravorty and Vanmarcke (1972, 1973).

response variance $\sigma_e^2(s)$ a result analogous to eq. 8.18 is obtained:

$$\sigma_e^2(s) = \omega_e^4 \int_0^\infty G_e(\omega,s)d\omega \simeq \int_0^{\omega_e} G_z(\omega,s)d\omega + G_z(\omega_e,s) \, \omega_e \left(\frac{\pi}{4\zeta_{es}} - 1\right) \tag{8.72}$$

Inserting the expression for $G_z(\omega,s)$ given by eq. 8.68 yields:

$$\sigma_e^2(s) \simeq \sum_{k=1}^{n'} \alpha_{ks} c_k^2 \sigma_k^2(s) + \left(1 + \sum_{k=1}^{n} \alpha_{ks} c_k^2 \omega_e^4 |H_k(\omega_e,s)|^2\right)\left(\sigma_{eg}^2(s) - \sigma^2 F^*(\omega_e)\right) \tag{8.73}$$

where $\sigma_{eg}^2(s)$ and $\sigma_k^2(s)$ are the pseudo-acceleration response variances of one-degree systems supported on the ground, which correspond, respectively, to the secondary system and to mode $k$ of the primary system; also:

$$\alpha_{ks} = \left[1 + \sum_{j \neq k}^{n} (c_j/c_k) A_{kjs}\right]$$

$n$ = the number of significant primary system modes and $n'$ = the number of primary system modes for which the frequency ratio $\omega_k/\omega_e$ is less than one; also, $\sigma^2$ is the ground acceleration variance and $F^*(\omega_e)$ is the normalized cumulative spectral density of the ground acceleration which increases monotonically from 0 to 1 as $\omega_e$ increases from 0 to $\infty$ (see eqs. 8.6b and 8.21). The value of the product of $\omega_e^4$ and the $k$th modal transfer function $H_k(\omega_e,s)$ vanishes when $\omega_e \ll \omega_k$ and tends to one if $\omega_e \gg \omega_k$. If $\omega_e \ll \omega_k$, eq. 8.73 yields the desired result $\sigma_e^2(s) = \sigma_{eg}^2(s)$, and if $\omega_e \gg \omega_k$, $\sigma_e^2(s)$ approaches the variance of the primary system response acceleration (as $\sigma_{eg}^2(s) \to \sigma^2$ and $F^*(\omega_e) \to 1$).

If the primary system has only one significant mode, we have $n = 1$ and $n = 0$ or 1 depending on the frequency ratio $\omega_k/\omega_e$.

Assuming white noise input acceleration, eq. 8.73 is used to evaluate the ratio of r.m.s. values $\sigma_e(s)/\sigma_{eg}(s)$ plotted in Fig. 8.22 as a function of the frequency ratio $\omega_k/\omega_e$ for the damping values $\zeta_k = 0.04$ and $\zeta_e = 0.005$. Also shown in Fig. 8.22 is the ratio of the corresponding maximum acceleration responses $S_{Ae}/S_{Aeg}$ for four strong-motion records (Biggs, 1972). The 20% difference between the theoretical and the observed values at $\omega_k/\omega_e = 1$ is believed to be attributable to the difference between the peak factors by which $\sigma_e(s)$ and $\sigma_{eg}(s)$ must be multiplied to predict the corresponding maximum values.

The random vibration result, eq. 8.73, also provides the basis for the following new proposal to evaluate secondary-system response directly from the specified ground response spectra:

$$A_e = \left\{\sum_{k=1}^{n'} \alpha_{ks} c_k^2 S_{Ak}^2 + \left[1 + \sum_{k=1}^{n} \alpha_{ks} c_k^2 \omega_e^4 |H_k(\omega_e,s)|^2\right]\left[S_{Ae}^2 - A_g^2 F^*(\omega_e)\right]\right\}^{1/2} \tag{8.74}$$

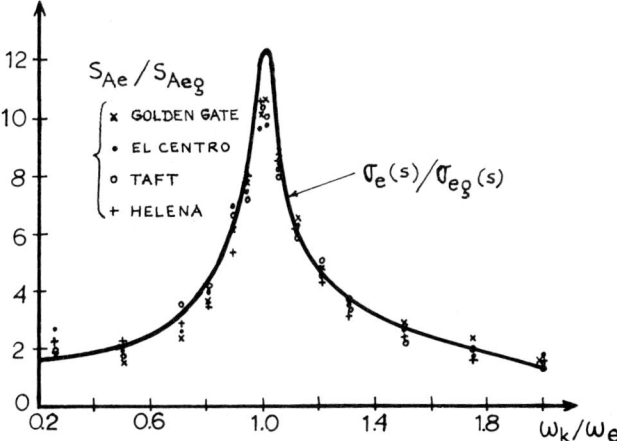

Fig. 8.22. Comparison between (1) The ratio of maximum acceleration responses $S_{Ae}/S_{Aeg}$ obtained numerically for four strong-motion records (Biggs, 1973), and (2) the ratio of standard deviations $\sigma_e(s)/\sigma_{eg}(s)$ predicted by random vibration theory based on white noise input.

in which $A_e$ = maximum acceleration of a one-degree secondary system, $S_{Ak}$ = the pseudo-acceleration (ground) response spectrum for the period and damping of mode $k$, $S_{Ae}$ = the pseudo-acceleration (ground) response spectrum for the equipment period and damping $A_g$ = maximum acceleration of the ground; $\alpha_{ks}$, $n$ and $n'$ are as defined in eq. 8.73. Amplified response spectra can be estimated directly using eq. 8.74. If the secondary system has several significant modes, eq. 8.74 can be used to estimate each modal contribution.

### 8.5.3 Time-integration method

The time-integration method of dynamic analysis to obtain floor response spectra or response spectra on top of a soil stratum is straightforward, but tends to be tedious and expensive. First, the acceleration response record must be obtained at the point(s) of interest on the primary structure. This motion then provides the input to the response analysis of the secondary system. Of course, a time-integration analysis based on a single ground-motion record will result in only a single value of maximum response. The entire procedure needs to be repeated a number of times, for different input motions, before reasonably reliable information about the range of probable responses can be obtained.

## 8.6 INELASTIC SYSTEMS

In earthquake engineering perhaps more than in any other area of applied dynamics, it is necessary to face up to important nonlinear effects in the behavior of structures and soils. The stiffness and strength values appropriate at low amplitudes of motion and the mechanisms of energy dissipation change as a function of the motion level. They also depend on past seismic action and other environmental conditions. For example, in reinforced concrete structures, cracking of concrete and slip and yielding of reinforcement tend to reduce significantly both stiffness and strength. Also, natural frequencies and damping factors of real buildings are affected by difficult-to-estimate nonlinear effects attributable to nonstructural components and foundation soil.

Many types of nonlinearities can be dealt with satisfactorily using "equivalent linear" analyses. Linear dynamic properties are chosen to be compatible with expected response levels. This method is frequently used in studies of soil amplification and liquefaction (Seed and Idriss, 1969). Shear modulus and damping values are adjusted until they are compatible with computed strains in the soil.

A survey of a variety of nonlinear stiffness and damping effects is given by Crandall (1974). It appears that most types of relatively *mild nonlinearities* can be dealt with successfully by equivalent linearization techniques (Goto and Iemura, 1973). But these techniques do not succeed in capturing the essence of *strongly nonlinear* hysteretic behavior, i.e., energy dissipation through the development of drift and permanent set.

Many structural systems are designed to undergo plastic deformation during very severe but infrequent earthquake shaking. Design level seismic analyses based on *linear* dynamic models of buildings often predict elastic responses in excess of elastic capacities (Blume, 1960). Building components which are ductile have far more resistance than the elastic analysis would indicate. What they lack in elastic strength is made up by their capacity to

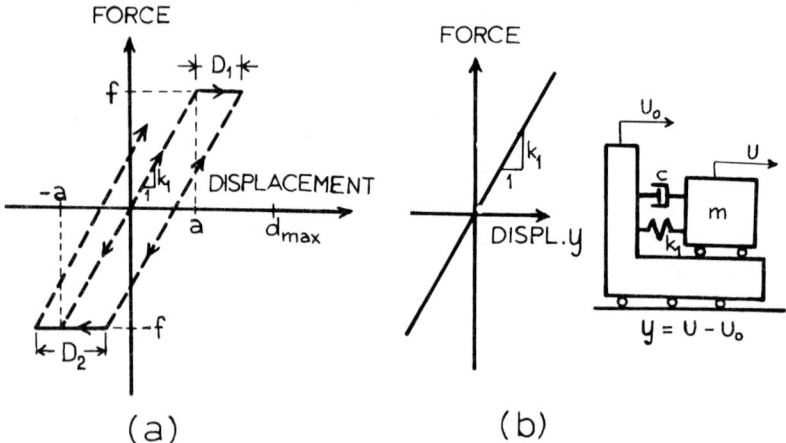

Fig. 8.23. One-degree elasto-plastic system and associated linear system.

absorb energy through inelastic action. Limits on the values of member distortions are set by available ductility or excessive deflection or overall instability.

The most common model of component nonlinear hysteretic stress—strain behavior is the elasto-plastic system shown in Fig. 8.23. It is very difficult to treat by methods other than time-integration procedures. Reliable simplified methods for predicting the response *multi-degree* inelastic systems are not yet available. Attempts at rigorous analysis have been restricted to one-degree systems (e.g., Caughey, 1960; Kobori et al., 1973; Lutes, 1967) and attest to the fact that the mathematical complexity is formidable. Two approaches to elasto-plastic one-degree system response prediction, one based on the response spectrum and the other on random vibration, are presented in the next few sections. Some results for more complex models based on step-by-step integration procedures are given in the final section.

## 8.6.1 Inelastic response spectra

Newmark and Hall (1969) have developed a simplified procedure for determining the relationship between the maximum response of an elasto-plastic system and the maximum response of the associated elastic system. The result is the inelastic response spectrum which consists of an acceleration spectrum and a (relative) displacement spectrum. Both can be plotted as a function of the (initial undamped) natural period for a specified level of the ductility ratio $\mu$ and the damping factor $\zeta$. The ductility ratio $\mu$ is the ratio of the maximum deflection to the limit of elastic deflection ($\mu = d_{max}/a$; see Fig. 8.23).

Starting from the pseudo-velocity elastic response spectrum plotted on log-log paper, the inelastic response spectrum is developed as follows. In the period range where displacement or velocity are amplified, the inelastic displacement spectrum is identical to the elastic spectrum. The inelastic acceleration spectrum in these regions is obtained by dividing the elastic spectrum by the ductility ratio $\mu$. In the amplified acceleration period range, the inelastic acceleration spectrum is obtained by dividing the elastic spectrum by $\sqrt{2\mu - 1}$, while at very high frequencies it is identical to the elastic response spectrum. The spectrum is completed by drawing a straight line from the constant acceleration line to the amplified acceleration line. Note that the inelastic displacement and acceleration spectra always differ by the factor $\mu$. Further details and background material can be found in Newmark and Rosenblueth (1971).

For example, starting from the average elastic response spectrum presented in Section 8.2.1, the inelastic response spectrum is derived for a damping of 2% and a ductility ratio of 4. The result is shown in Fig. 8.24. Response spectra for other nonlinear hysteretic systems have been obtained by Veletsos (1969).

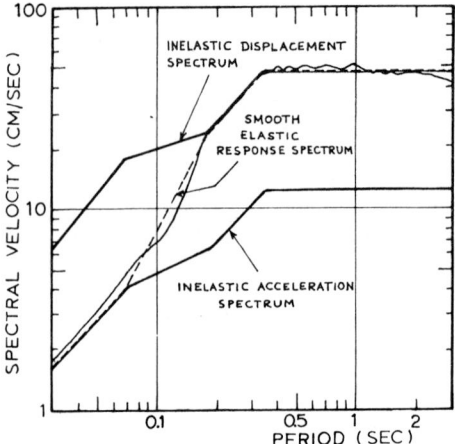

Fig. 8.24. Inelastic response spectrum for damping $\zeta = 0.02$ and ductility ratio $\mu = 4$ based on the average elastic 2% damped response spectrum (Fig. 8.1).

### 8.6.2 A probabilistic model

The stochastic model described below (Vanmarcke, 1969; Vanmarcke and Veneziano, 1973) leads to a set of approximate results for the mean, variance the probability distribution of important response measures of the elasto-plastic (E-P) system, i.e., the permanent set, the ductility factor, and the time required for yielding to progress to a given deformation level. The model is an extension of the work of Karnopp and Scharton (1966) who obtained an estimate of the average rate of energy dissipated due to yielding of E-P systems excited by stationary Gaussian white-noise excitation. The assumptions upon which the model is based have been checked by Yanev (1970) by extensive numerical simulation of E-P response using as input white noise and Tajimi-filtered white noise (eq. 8.3).

Briefly, yield-level crossings of the elasto-plastic response tend to occur in clusters. In between these bursts of inelastic action, the E-P system remains below the yield level and behaves as an elastic system. As a first approximation, clusters of yield-level crossings are treated as "points in time" at which the plastic drift $D(t)$ developed between 0 and $t$ jumps to a different value. The total plastic deformation $D(s)$ developed during $s$ seconds of E-P response motion is the sum of contributions $D_i$, each resulting from a cluster of yield-level crossings; i.e.:

$$D(s) = \sum_{i=1}^{N(s)} D_i \qquad (8.75)$$

in which $N(s)$ is the random number of contributions during the interval

$(0, s)$. $N(s)$ is assumed to have a Poisson distribution with mean $\alpha s$, where $\alpha = 2\nu_a/E[N_a]$ (see eqs. 8.37 and 8.40). The mean rate $\alpha$ depends mainly on the yield level $a$ and the r.m.s. value $\sigma_y$ of the associated linear system. *

Karnopp and Scharton (1966) derived a simple approximate expression for $\delta$, the average amount of inelastic deformation (in absolute value) resulting from a single isolated crossing of the yield level $a$, i.e., $\delta = \sigma_y^2/2a = \sigma_y/2r$. This result follows from the argument that all the kinetic energy, $m\dot{y}^2/2$ (where $m$ is the mass of the system; $\dot{y}$ is the impact velocity), will be released by yielding action. The average value of the kinetic energy at impact is approximately equal to $m\omega_n^2\sigma_y^2/2 = k_1\sigma_y^2/2$. The expected plastic deformation $\delta$ may be obtained from the energy $f\delta = k_1\sigma_y^2/2$, where $f$ denotes the yield force (see Fig. 8.23). The contributions $D_i$ are equally likely to be positive and negative with mean and variance approximately equal to 0 and $2\delta^2$, respectively. Furthermore, the distribution of $|D_i|$ is approximately exponential with mean $\delta$.

*Plastic drift or permanent set*: Under the assumptions just stated, the expected value and variance of $D(s)$ (given by eq. 8.75) can be found:

$$E[D(s)] = \alpha s E[D_i] = 0 \tag{8.76}$$

$$\text{Var}[D(s)] = \alpha s (\text{Var}[D_i] + E^2[D_i]) = 2\alpha s\delta^2 \tag{8.77}$$

where $\delta = \sigma_y^2/2a$ and $\alpha = 2\nu_a/E[N_a]$. The probability density function of $D(s)$ can also be obtained (Vanmarcke, 1969). If considerable plastic action occurs, the Central Limit Theorem can be invoked to argue that $D(s)$ has a Gaussian distribution.

*Ductility factor*: Given that plastic action occurs during the time interval $(0, s)$, the ductility factor $\mu$ can be expressed as follows:

$$\mu = \frac{1}{a}[\max_{0 \leq t \leq s} |D(t)|] + 1 = \frac{M_s}{a} + 1 \tag{8.78}$$

where $M_s$ is the peak inelastic deformation. If no plastic action occurs, it equals the ratio of the maximum elastic deformation to the yield displacement. Equation 8.78 can be used "unconditionally" when the probability of no elastic action is negligibly small.

*Peak inelastic deformation*: The peak inelastic deformation is the absolute maximum value of $D(t)$ in the time interval 0 to $s$. Its probability distribution can be approximated by viewing the crossings, at positive slope, by $|D(t)|$, of a fixed threshold $d$, as a nonstationary Poisson process with mean rate $\nu_d(t)$. $\nu_d(t)$ is proportional to $\alpha$, the mean rate of occurrence of jumps in the process $D(t)$, i.e., $\nu_d(t) = \alpha p_d(t)$, where $p_d(t)$ is the probability that a

---

* If the associated linear response does not rapidly reach a stationary condition, $\alpha$, $\sigma_y$ and $s$ may be replaced by the quantities $\alpha(s)$, $\sigma_y(s)$ and $s_0$ referred to in Section 8.3.4.

plastic set contribution at time $t$ results in an upcrossing of the level $d$. The net result is the following approximate expression for the probability distribution of the peak inelastic deformation $M_s$:

$$P[M_s \leq d] = \exp[-(e^{\alpha s} - 1)e^{-d/\delta}]; \qquad d \geq 0 \tag{8.79}$$

There is a finite probability, $P[M_s = 0]$, that no plastic action will occur. Taking $d = 0$ in eq. 8.79, this probability is found to be about $\exp(-\alpha s)$ when $\alpha s \ll 1$; it becomes negligibly small for large values of $\alpha s$. Note that for values of $d > 0$, eq. 8.79 has the form of a Type-1 Extreme Value Distribution. A characteristic value of $M_s$, found by setting $P[M_s \leq d] = e^{-1}$, equals $M_s^* = \delta \ln(e^{\alpha s} - 1)$.

*Time to first crossing of a given level of plastic deformation*: The probability distribution of the time, $T_d$, to first crossing of a given level of plastic deformation is intimately related to the distribution of $M_s$. One can write:

$$P[T_d > s] = P[M_s \leq d] \tag{8.80}$$

It suffices to view the expression for $P[M_s \leq d]$ in eq. 8.79 as a function of $s$, with $d$ as a known parameter.

*Extension to bilinear hysteretic systems*: The basic model which views the total inelastic deformation $D(t)$ as a cumulative process with increments made at random "points" in time, continues to hold when the force—deformation relationship is not of the elasto-plastic type. But the statistical properties of the sizes of the increments and of the time intervals between these "points" now vary depending upon the *state* of the system, i.e., the level of inelastic deformation at which the system operates. In particular, at any given time $t$, they depend upon the values of the positive and negative yield levels corresponding to the plastic deformation $D(t)$. (For E-P systems these yield levels remain constant regardless of the value of $D(t)$.) As an example, Fig. 8.25 shows the force—displacement diagram of a simple frame with rigid girder and columns for which gravity loads have the effect of making the second slope $k_2$ negative (Husid, 1979). During the process of drift accumulation the smallest yield level ranges from zero (at $D(t) = d_m$) to $a$ (at $D(t) = 0$), as shown in Fig. 8.25.

The particular kind of "memory" and the state-dependent nature of the cumulative damage suggest a Markov process. The Markov transition probabilities and the mean times between clumps can be evaluated using an argument similar to that which earlier led to the distribution of the plastic set increment $D$. Modifications are required, however, to account for the unequal and sometimes very low yield levels (Veneziano and Vanmarcke, 1973).

Figure 8.26 shows the results of a Markov analysis of the response to Gaussian white noise of three gravity-affected hysteretic systems. The solid lines in Fig. 8.26 give the expected number of cycles (expected time divided by natural period) to failure for the three structures, as a function of the

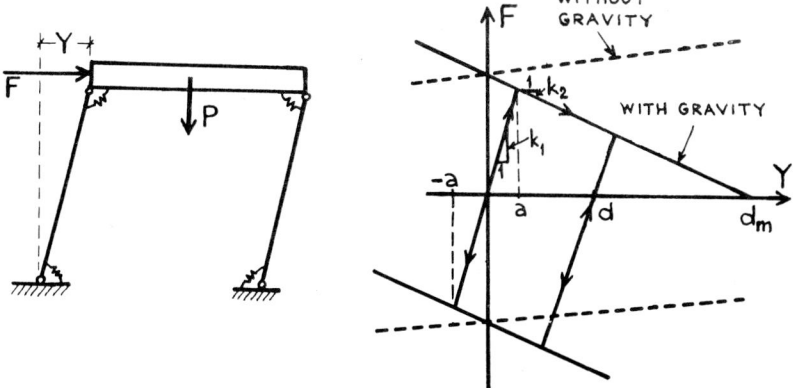

Fig. 8.25. Bilinear hysteretic system modeling the effect of gravity on a yielding structure.

ratio $r = a/\sigma_y$ (note that $\sigma_y$ is a measure of the excitation intensity). Husid (1969) used both recorded earthquakes and artificial stationary motions with Tajimi-type spectral density to develop an empirical relationship for the expected number of cycles to failure of systems of the type shown in Fig. 8.25. Husid's best estimates are represented by the dotted lines in Fig. 8.26.

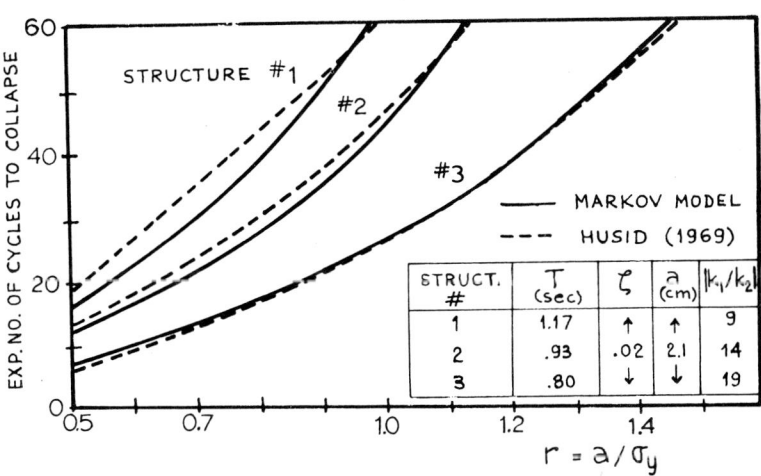

Fig. 8.26. The expected time to failure for several gravity-affected hysteretic systems. Comparison between results obtained by simulation (Husid, 1969) and by a Markov probabilistic model (Veneziano and Vanmarcke, 1973).

## 8.6.3 Time-integration analysis

Many researchers (e.g., Husid, 1969; Penzien and Liu, 1969; Veletsos, 1969; Goto and Iemura, 1973) have obtained response statistics of simple nonlinear hysteretic systems through step-by-step integration procedures using recorded and simulated ground motions. Of course, these procedures can be extended to much more complex models of the structure's behavior in the inelastic range. The choice of an appropriate analytical model poses a major problem, though. Experimental work has revealed that component behavior is quite complex, including deterioration of stiffness and strength with number of cycles and in function of the level of deformation. A true-to-life inelastic analysis of a building frame, if possible, is clearly impractical. The most common model is a "shear building" with elasto-plastic or bilinear springs in each story. The ultimate capacity of each spring may be the sum of the ultimate column end moments divided by the story height.

Time-integration analyses based on inelastic models of real buildings (Clough, 1970) have revealed that the overall deformation of the building is close to that predicted based on an elastic model with the "initial" properties of the inelastic model, but that the maximum strains in the individual components may be larger by a factor of 6. Biggs and Grace (1973) found that a crude estimate of the *average* inelastic interstory displacement can be made using elastic, first mode, interstory displacements. Dividing the latter by the average elastic limit interstory displacement gives an estimate of the average ductility ratio of the building. The pattern of inelastic action, i.e., the distribution of ductility versus height, shows large variations from earthquake to earthquake.

ACKNOWLEDGMENTS

The writer gratefully acknowledges the encouragement and assistance offered by Professors C.A. Cornell, J.M. Biggs and S.H. Crandall of M.I.T. The many comments provided by the editors and by David Elms and Peter Arnold were also very helpful. Computational assistance was offered by Bob Frank and Peter Arnold in the preparation of Figs. 1 through 3 and Fig. 20. Support for the writer's research in earthquake engineering was provided by the National Science Foundation under Grants GK-4151, GK-26296 and ATA 74-06935.

REFERENCES

Amin, M., Tsao, H.S. and Ang, A.H., 1969. Significance of nonstationarity of earthquake motions. *Proc. 4th WCEE, Santiago*, pp. 97—113.

Arias, A. and Husid, R., 1962. Influencia del amortiguamiento sobre de la respuesta de estructuras sometidas a temblor. *Rev. IDIEM*, 1 (3).
Barstein, M.F., 1960. Application of probability methods for the design against the effect of seismic forces on engineering structures. *Proc. 2nd WCEE, Tokyo.*
Biggs, J.M., 1972. Seismic response spectra for equipment design in nuclear power plants. *Proc. 1st Int. Conf. Struct. Mech. in Reactor Technology, Berlin*, 5: 329—343.
Biggs, J.M. and Grace, P.H., 1973. Seismic response of buildings designed by code for different earthquake intensities. *M.I.T. Dep. Civ. Eng., Res. Rep.*, R73-7.
Biggs., M., 1964. *Introduction to Structural Dynamics.* McGraw-Hill, New York.
Blume, J., 1960. Structural dynamics in earthquake resistant design. *Trans. ASCE*, 125: 1088—1139.
Bolotin, V.V., 1969. *Statistical Methods in Structural Mechanics.* Holden-Day, San Francisco, Calif.
Brady, A.G., 1966. *Studies of Response to Earthquake Ground Motion.* Thesis Calif. Inst. of Technol, Earthquake Eng. Res. Lab., Pasadena, Calif.
Bycroft, G.N., 1960. White noise representation of earthquakes. *J. Eng. Mech. Div., Proc. ASCE*, 86 (EM2): 1—16.
Caughey, T.K., 1960a. Random excitation of a system with bilinear hysteresis. *J. Appl. Mech.*, 27: 649—652.
Caughey, T.K., 1960b. Classical normal modes in damped linear systems. *J. Appl. Mech.*, pp. 269—271.
Caughey, T.K. and Stumpf, H.J., 1961. Transient response of a dynamic system under random excitation. *J. Appl. Mech.*, pp. 563—566.
Chakravorty, M., 1972. Transient spectral analysis of linear elastic structures and equipment under random excitation. *M.I.T. Dep. Civ. Eng., Rep.* R72-18.
Chakravorty, M.K. and Vanmarcke, E.H., 1973. Probabilistic seismic analysis of light equipment within buildings. *Proc. 5th WCEE, Rome.*
Clough, R.W., 1970. Earthquake response of structures. In: R.L. Wiegel (Editor), *Earthquake Engineering.* Prentice-Hall, Englewood Cliffs, N.J., Chapter 12.
Clough, R.W. and Penzien, J., 1975. *Dynamics of Structures.* McGraw-Hill, New York.
Corotis, R. and Vanmarcke, E.H., 1975. On the time-dependent frequency content of oscillator frequency content. *J. Eng. Mech. Div., Proc. ASCE*, 101 (EM5).
Corotis, R., Vanmarcke, E.H. and Cornell, C.A., 1972. First passage of nonstationary random processes. *J. Eng. Mech. Div., Proc. ASCE*, 98 (EM2): 401—414.
Cramer, H., 1966. On the intersections between the trajectories of a normal stationary stochastic process and a high level, *Ark. Mat.*, 6: 337.
Cramer, H. and Leadbetter, M.R., 1967. *Stationary and Related Stochastic Processes,* Wiley, New York.
Crandall, S.H., 1970. First-crossing probabilities of the linear oscillator. *J. Sound Vib.*, 12: 285—300.
Crandall, S.H., 1974. Nonlinearities in structural dynamics. *Shock Vib. Dig.*
Crandall, S.H. and Mark, W.D., 1963. *Random Vibration in Mechanical Systems.* Academic Press, New York.
Crandall, S.H., Chandiramani, K.L. and Cook, R.G., 1966. Some first passage problems in random vibrations. *J. Appl. Mech.*, 33 (Ser. E): 532.
Davenport, A.G., 1964. Note on the distribution of the largest value of a random function with application to gust loading. *Proc. Inst. Civ. Eng.*, 28: 187—196.
Ditlevsen, O., 1971. *Extremes and First Passage Times with Applications in Civil Engineering.* Thesis, Technical University of Denmark, Copenhagen.
Franklin, J.N., 1965. Numerical simulation of stationary and nonstationary Gaussian random processes. *SIAM Rev.* 7 (1): 68—80.
Goto, H. and Iemura, H., 1973. Earthquake response of single-degree-of-freedom hysteretic systems. *Proc. 5th WCEE, Rome*, Paper 266.

Hammond, S.K., 1968. On the response of single and multidegree of freedom systems to nonstationary random excitations. *J. Sound Vib.*, 7: 393—416.

Hou, S., 1968. Earthquake simulation models and their applications. *M.I.T. Dep. Civ. Eng., Rep.* R68-17.

Housner, G.W., 1947. Properties of strong ground motion earthquakes. *BSSA*, 37: 19—31.

Housner, G.W., 1959. Behaviour of structures during earthquakes. *Proc. ASCE*, 85 (EM4).

Housner, G.W. and Jennings, P.C., 1964. Generation of artificial earthquakes. *J. Eng. Mech. Div., Proc. ASCE*, 90, (EM1): 113—150.

Hurty, W.C. and Rubinstein, M.F., 1964. *Dynamics of Structures*. Prentice-Hall, Englewood Cliffs.

Husid, R., 1969. The effect of gravity on the collapse of yielding structures with earthquake excitation. *Proc. 4th WCEE, Santiago*, Vol. II.

Jenkins, J.M., 1961. General considerations in the analysis of spectra. *Technometrics*, 3: 133—166.

Jennings. P.C., 1964. Earthquake response of a yielding structure. *J. Eng. Mech. Div., Proc. ASCE*, 91 (EM4): 41—68.

Kanai, K., 1961. An empirical formula for the spectrum of strong earthquake motions. *Bull. Earthquake Res. Inst., Univ. of Tokyo*, 39: 85—95.

Karnopp, D. and Scharton, T.D., 1966. Plastic deformation in random vibration. *J. Acoust. Soc. Am.*, 39: 1154—1161.

Kobori, T., Minai, R. and Suzuki, Y., 1973. Statistical linearization techniques of hysteretic structures to earthquake excitations. *Bull. Disaster Prev. Res. Inst., Kyoto Univ.* 23 (215).

Lin, Y.K., 1967. *Probabilistic Theory of Structural Dynamics*. McGraw-Hill, New York.

Liu, S.C., 1970. Evolutionary power spectral density of strong-motion earthquakes. *BSSA*, 60 (3): 891—900.

Liu, S.C. and Jhaveri, D.P. 1969. Spectral and correlation analysis of ground-motion accelerograms. *BSSA*, 59: 1517—1534.

Longuet-Higgins, M.S., 1952. On the statistical distribution of the height of sea waves. *J. Marine Res.* II: 245—266.

Lutes, L.D., 1967. *Stationary Random Response of Bilinear Hysteretic Systems*. Thesis, Earthquake Eng. Res. Lab., C.I.T.

Lyon, R.H., 1961. On the vibration statistics of a randomly excited hard-spring oscillator. *J. Acous. Soc. Am.*, Part I, Vol. 32: 716—719, 1960; Part II, Vol. 33: 1395—1403.

McGuire, R., 1974. Seismic structural response risk analysis, incorporating peak response regressions on earthquake magnitude and distance. *M.I.T. Dep. Civ. Eng. Res. Rep.* R74-51.

Newmark, N.M. and Hall, W.J., 1969. Seismic design criteria for nuclear reactor facilities. *Proc. 4th WCEE, Santiago*.

Newmark, N.M. and Hall, W.J., 1973. A rational approach to seismic design standards for structures. *Proc. 5th WCEE, Rome*.

Newmark, N.M. and Rosenblueth, E., 1971. *Fundamentals of Earthquake Engineering*. Prentice-Hall, Englewood Cliffs, N.J.

Newmark, N.M., Blume, J.A. and Kapur, K.K., 1973. Seismic design spectra for nuclear power plants. *J. Power Div., Proc. ASCE*, 99 (PO2): 287—303.

Penzien, J. and Liu, S.C., 1969. Nondeterministic analysis of nonlinear structures subjected to earthquake excitations. *Proc. 4th WCEE, Santiago*.

Penzien, J. and Watabe, M., 1975. Characteristics of 3-dimensional earthquake ground motions. *Earthquake Eng. Struct. Dyn.*, 3: 365—373. 1975.

Priestley, M.B., 1967. Power spectral analysis of nonstationary random processes. *J. Sound Vib.*, 6: 86—97.

Pulgrano, L.J. and Ablowitz, M., 1969. The response of mechanical systems to bands of random excitation. *Shock Vib. Bull.*

Rascón, O. and Villareal, A., 1973. Estudio estadístico de los criterios para estimar la respuesta sísmica de sistemas lineales con dos grados de libertad. *Inst. Ingeniería, UNAM, Rep.* 323.

Rice, S.O., 1945. Mathematical Analysis of Random Noise. *Bell System Tech. J.*, Part I, Vol. 23: 282—332, 1944; Part II, Vol. 24: 46—156.

Roesset, J.M., Whitman, R.V. and Dobry, R., 1973. Modal analysis with foundation interaction. *J. Struct. Div., Proc. ASCE*, 99 (ST3).

Rosenblueth, E., 1951. *A Basis for Aseismic Design*. Thesis, Univ. of Illinois, Urbana.

Rosenblueth, E., 1964. Probabilistic Design to Resist Earthquakes. *J. Appl. Mech. Div., Proc. ASCE*, 90 (EM5): 189—219.

Rosenblueth, E. and Elorduy, J., 1969. Responses of linear systems to certain transient disturbances. *Proc. 4th WCEE, Santiago*, Al-185.

Sarrazin, M., Roesset, J.M. and Whitman, R.V., 1972. Dynamic soil-structure interaction. *J. Struct. Div., Proc. ASCE*, 98 (ST7).

Seed, H.B. and Idriss, I.M., 1969. Influence of soil conditions on ground motions during earthquakes. *J. Soil Mech. Found. Div., Proc. ASCE*, 95 (SM1): 99—137.

Sixsmith, E. and Roesset, J., 1970. Statistical properties of strong motion earthquakes. *M.I.T. Dep. Civ. Eng. Res. Rep.* R70-7.

Shinozuka, M., 1973. Digital simulation of ground accelerations. *Proc. 5th WCEE, Rome.*

Tajimi, H., 1960. A statistical method of determining the maximum response of a building structure during an earthquake. *Proc. 2nd WCEE, Tokyo*, 2: 781—797, Science Council of Japan, Japan.

Trifunac, M.D., Udwadia, F.E. and Brady, A.G., 1973. An analysis of errors in digitized strong motion accelerograms. *BSSA*, 63 (1): 157—187.

Vanmarcke, E.H., 1969. First passage and other failure criteria in narrow-band random vibration: A discrete state approach. *M.I.T. Dep. Civ. Eng. Rep.* R69—68.

Vanmarcke, E.H., 1972. Properties of spectral moments with applications to random vibration. *J. Eng. Mech. Div., ASCE*, 98: 425.

Vanmarcke, E.H., 1975. On the distribution of the first-passage time for normal stationary random processes. *J. Appl. Mech.*, 42 (Ser. E): 215—220.

Vanmarcke, E.H. and Cornell, C.A., 1972. Seismic risk and design response spectra. *Proc. ASCE Specialty Conf. on Safety and Reliability of Metal*, pp. 1—25.

Vanmarcke, E.H. and Veneziano, D., 1973. Probabilistic seismic response of simple inelastic systems. *Proc. 5th WCEE, Rome.*

Veletsos, A.S., 1969. Maximum deformations of certain nonlinear systems. *Proc. 4th WCEE, Santiago.*

Veneziano, D. and Vanmarcke, E.H., 1973. Seismic damage of inelastic systems: A random vibration approach. *M.I.T. Dep. Civ. Eng. Res. Rep.* R73-5.

Yanev, P.I., 1970. *Response of Simple Inelastic Systems to Random Excitation*. Thesis, M.I.T.

*Chapter 9*

DESIGN

R.V. WHITMAN and C.A. CORNELL

*Department of Civil Engineering, Massachusetts Institute of Technology, Cambridge, Mass., U.S.A.*

Design of engineered facilities requires a synthesis of many kinds of information. Previous chapters have emphasized the definition and analysis of the seismic threat in terms of the free-field effects of earthquakes. Chapter 8 contributed procedures for analysis of the dynamic response of idealized structures to earthquake effects of given intensity. Design implies choice from among alternatives; for each alternative solution the designer should assess the associated total risk. This total risk is a composite of both the likelihoods of various levels of the seismic threat, per se, and the uncertainty in the response and behavior of the system. The behavior of interest includes any characteristic that implies economic or other loss. The selection or decision process demands a weighing of the costs, risks, and benefits of the alternatives, and ultimately the choice of a single alternative.

While necessary in principle, this entire design process can be long and costly, requiring advanced knowledge and experience in what the reader has seen to be a very specialized subject. Therefore, for most purposes general seismic design norms or codes are prepared by professional committees for broad classes of systems (the taller buildings of a specific city, the highway bridges of a state, etc.). The existence of these norms removes much of the risk assessment and the fundamental risk-cost weighing from the responsibility of the engineer-designer of each specific structure. But the general design process outlined above, complete with risk assessment and weighing, must then have been carried out, more or less formally, with respect to the code as a whole (Ravindra and Lind, 1973).

For design of special projects (such as nuclear power plants and large dams), for large-scale projects (such as electrical distribution systems), and for land-use planning, it is usual practice to carry out detailed seismic studies. The decision process itself may be either more or less formalized. There are rapidly growing trends to quantify the seismicity component of risk analysis and to make explicitly the definition and cost-benefit assessment of a discrete set of alternatives. Decision analysis provides a framework within which all such relevant facts about seismicity, system behavior, costs and benefits, etc., can be collected. Even without formal optimization procedures, this coordinated assembly of facts is a major benefit of the decision analysis methodology.

In this chapter we shall present some of the principles and implications of seismic design decision analysis. The first four sections describe the analysis of the combined or total risk when systems with uncertain capabilities are exposed to a probabilistically defined seismic environment. The fifth portion of the chapter outlines and illustrates the principles of formal decision analysis applied to design of individual systems. In the sixth section structural building codes are discussed, including the application of formal decision theory to their development. In the final section lifeline planning in a seismic region is reviewed also within the general framework provided by decision analysis.

## 9.1 ANALYSIS OF TOTAL RISK

In decision-making one is concerned with the total risk that a particular alternative design will behave in some more or less unfavorable manner during future earthquakes. This total risk is the combined effect of the different possible seismic events that might occur and the different ways in which a particular design may respond to a given event. For example, in conventional structural design of buildings the engineer focuses upon a response parameter such as the maximum inter-story displacement (because it is correlated with economic damage and with structural failure). A specific level of such response may be either the result of moderate earthquake intensity coupled with unexpectedly high structural response or the result of a severe earthquake shaking combined with better (less) than predicted structural response (perhaps because the frequency content of the random ground motion was lower than expected in the range coinciding with the important frequencies of the structure). If both of these input-response combinations have the same net result in terms of the state of the structure (i.e., the inter-story displacement), then they are equivalent within the scope of the decision analysis. In that case it is not necessary to distinguish further between the two combinations. In estimating risk that a particular level of response will be experienced, it is the total probability associated with all possible combinations of seismic inputs and deviations in structural response that must be determined.

The general expression for the determination of the total risk can be written:

$$P[R_i] = \sum_{\text{all } j} P[R_i | S_j] P[S_j] \qquad (9.1)$$

in which $P[\cdot]$ signifies probability of the event indicated within the brackets, $R_i$ denotes the event that the state of system is $i$, $S_j$ means that the seismic input experienced is 'level' $j$, and $P[R_i | S_j]$ states the probability that the behavior (state) of the system will be $R_i$ *given* that seismic input $S_j$ takes place.

In principle, $S_j$ may represent an entire history (times, locations, magnitudes, site intensities, etc.) of future seismic activity in the region of interest, in which case $R_i$ might be a unique time-sequence description of the condition(s) of the system after each event. For example, $R_i$ might be a particular history of repair costs. In practice, simple definitions of $S_j$ and $R_i$ are usually employed. In structural design, $R_i$ may be simply the *worst* post-earthquake state (or condition, or loss, or response) of the structure during its anticipated economic lifetime. In regions of moderate seismic activity, where the probability of two or more significant seismic events in this time period is negligible compared to the likelihood of just one event, $R_i$ becomes simply the response or behavior under a single seismic event, if one occurs, and the null state ($R_0$, say) if no event of engineering importance occurs (i.e., if the seismic input is $S_0$, 'no occurrence').

The form of the total probability equation (eq. 9.1) emphasizes that it couples the two sources of uncertainty and that it accumulates all combinations having the same behavior or response. In this form of the equation the separation of capacities and responsibilities of seismology and engineering are also clarified; the seismologist should determine the values of the seismic risk $P[S_j]$ for all $j$, and the engineer must estimate the likelihood of different levels of behavior $R_i$ given different levels of seismic input ($P[R_i|S_j]$). In application, however, the basic total probability equation may take on different forms, each more suitable and efficient for the needs and definitions of the particular decision analysis at hand. Commonly, for example, the seismic input $S$ is represented as a continuous, scalar variable (e.g., peak ground acceleration) in which case eq. 9.1 becomes:

$$P[R_i] = \int P[R_i|S = s] f_S(s) \, ds \tag{9.2}$$

in which $f_S(s)$ is the probability density function (PDF) of $S$. If the response state is also continuous, its cumulative distribution function (CDF) becomes:

$$P[R \leq r] = F_R(r) = \int F_{R|S}(r;s) f_S(s) \, ds \tag{9.3}$$

in which $F_{R|S}(r;s)$ is the conditional CDF of $R$ given that $S$, the seismic input, is equal to $s$.

The following examples will serve to illustrate possible forms and applications of total risk analysis, but the cases considered are by no means exhaustive. In all cases (see Chapter 6), it is assumed that major earthquakes occur following a Poisson arrival process with mean annual rate $\nu$ and that the magnitudes $M$ and focal distances $D$ of the events are independent random variables. Uncertainty in the parameters of the seismicity model is ignored in this chapter for simplicity and clarity. The degree of complexity in these analyses is, therefore, dictated primarily by the nature of the idealization of the engineering system and its behavior. In most of these examples structural systems are implied.

## 9.2 ANALYSIS OF TOTAL RISK: TWO-STATE SYSTEMS

In the simplest analysis, a building is assumed to be in one of only two states after each earthquake: either undamaged (state 0) or damaged (state 1). State 1 represents some undesirable degree of damage. Depending upon the application, state 1 might be defined as: (a) damage involving a repair cost greater than $P\%$ (say 10%) of the replacement value of the building, or (b) a level of damage at which there is a significant possibility of serious injury and loss of life.

The transition from state 0 to state 1 occurs at the damage *threshold*. Realistically, this threshold is related to some aspect of structural response, such as inter-story displacement. In many applications, however, the threshold is expressed directly in terms of some parameter describing the intensity of the ground shaking, such as peak acceleration or modified Mercalli intensity. The threshold may be deterministically related to the shaking parameter or there may be uncertainty in this relationship. Several possibilities are illustrated by the examples in this section.

The two-state representation is a severe simplification. It does, however, lead directly to a risk statement popular with many clients and managers, namely, the probability that damage will occur during the life of a building. For example, one oft-quoted design criterion is: less than 10% probability of damage within the life of a project.

*Example 9.1*

In the simplest two-state case the behavior of the system is presumed to be deterministic given the intensity of the ground motion. This behavior is depicted in Fig. 9.1a.

Consider a building presumed to be damaged precisely at acceleration level $a$. Then, given an earthquake, the probability of damage is simply $p$, the probability that the site acceleration exceeds $a$, which (Chapter 6) can often be written in the form:

$$p = ca^{-k} \tag{9.4}$$

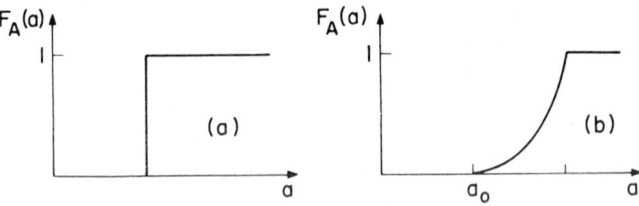

Fig. 9.1. Examples of cumulative probability distribution functions for capacity.

(See Chapter 6 for factors, such as relatively low upper-bound magnitudes, which invalidate this particular functional-form assumption for high values of $a$). If the mean rate of (significant) earthquakes is $\nu$ per year, then the annual mean rate of site accelerations greater than or equal to $a$ is (eq. 6.38):

$$\lambda_a = \nu p = \alpha a^{-k} \qquad (9.5)$$

in which $\alpha = \nu c$.

Let the state $R_i$ be that the structure is damaged $i$ times in its intended or economic life of $T$ years; then for this simple deterministic system, the system will be in state $R_i$ if and only if there are exactly $i$ events with acceleration greater than $a$ in $T$ years (i.e., only if the seismicity state is $S_i$).

From the Poisson distribution:

$$P[R_i] = P[S_i] = \exp(-\lambda_a T) (\lambda_a T)^i / i! \qquad (9.6)$$

In particular, the probability of the structure never being damaged is:

$$P[R_0] = \exp(-\lambda_a T) = \exp(-\alpha T a^{-k}) \qquad (9.7)$$

and the probability of its being damaged at least once is $1 - P[R_0]$. The implicit assumption in eq. 9.6 (but not eq. 9.7) is that after being damaged the structure is repaired, if necessary, to the point that its damage threshold is again exactly $a$.

To illustrate the use of the foregoing equations, assume:

$$\lambda_a = 0.0004 \, a^{-2} \qquad (9.8)$$

where $a$ is expressed in $g$'s. The value of the exponent is typical of a number of locations. For these parameter values, the average recurrence interval for $a \geqslant 0.1\,g$ is 25 years, which is similar to the recent history in downtown Los Angeles. Suppose that a building is designed to resist without damage a ground shaking with a peak acceleration of $0.2g$. The probability that this acceleration will be exceeded at least once in a 50-year economic life is computed as follows:

$$\lambda_a = 0.0004(0.2)^{-2} = 0.01$$

$$1 - P[R_0] = 1 - \exp(-0.01(50)) = 0.39$$

The probabilities that $0.3g$ and $0.4g$ will be exceeded during 50 years are 20% and 12%, respectively.

Such risk estimates are sometimes suggested as the basis of a design earthquake. A possible rule is: buildings should be designed for the acceleration which has a 10% probability of being exceeded during the life of the building. Using eq. 9.7, the corresponding value of $\lambda_a$ may be computed as follows:

$$P[R_0] = 0.9 = \exp(-\lambda_a T)$$

$$\lambda_a = \frac{1}{T}(-\ln 0.9) = 0.1054/T$$

Then the required $a$ may be computed from eq. 9.5:

$$a = \sqrt[k]{\alpha/\lambda_a}$$

Using the numerical values in eq. 9.8 with a 50-year life, it would be necessary to design for a peak acceleration of at least $0.44g$. If the annual mean rate is 20 times less than in eq. 9.8, then this rule leads to a design acceleration of $0.1g$. This smaller annual mean rate might, for example, be appropriate for Boston.

If ground shaking is expressed by the modified Mercalli intensity scale, then the equation for $\lambda$ takes the form:

$$\lambda_I = \mu e^{-rI} \tag{9.9}$$

Equations 9.6 and 9.7 still apply. To illustrate, assume:

$$\lambda_I = 2000 e^{-1.53 I} \tag{9.10}$$

Again the exponent value is typical for a number of locations. The average recurrence interval for $I \geq 7$ is 23 years, which again is similar to the recent history of downtown Los Angeles. The probabilities that $I = 8, 9$, and 10 will be exceeded in 50 years are 35%, 9%, and 2%, respectively.

This simple representation of structural behavior will be used again in Section 9.5 within a formal decision analysis.

*Example 9.2*

Consider next a system whose capacity (i.e., damage threshold) is uncertain owing to variation in material properties, in construction procedure, in the incomplete description of the shaking, etc. Let $F_A(a)$ be the cumulative probability distribution function of the capacity, measured, for simplicity, directly in terms of the ground acceleration. Define two states: $R_0$ = no damage of the system in $T$ years, and $R_1$ = damage of the system sometime in $T$ years. Thus $F_A(a)$ is the conditional probability of state $R_1$ given that the input state $S$ equals level $a$; that is, in eq. 9.2, $F_A(a) = P[R_1 | S = a]$. Assume that damage occurs in the most intense event, if at all.

The factor $f_S(a)da$ in eq. 9.2 is then the probability that the level $a$ is the highest level reached in $T$ years. From eq. 9.7, the CDF of $S$ (the probability of no events with acceleration greater than $a$) is $\exp(-\alpha T a^{-k})$. The desired probability density function $f_S(a)$ is formally the derivative of the distribution function with respect to its argument $a$. Then the probability of state $R_1$ is (eq. 9.2):

$$P[R_1] = \int F_A(a) \frac{d}{da} [\exp(-\alpha T a^{-k})] \, da = k\alpha T \int F_A(a) \, a^{-k-1} \exp(-\alpha T a^{-k}) \, da \tag{9.11}$$

To illustrate the use of eq. 9.11 let us assume $F_A(a)$ has the form shown in

Fig. 9.1b and given by the following equation:

$$F_A(a) = \begin{cases} 0 & a < a_0 \\ \left(\dfrac{a - a_0}{a_0}\right)^2 & a_0 \leq a \leq 2a_0 \\ 1 & a > 2a_0 \end{cases} \qquad (9.12)$$

Now eq. 9.11 becomes:

$$P[R_1] = \int_{a_0}^{2a_0} \left(\frac{a - a_0}{a_0}\right)^2 k\alpha Ta^{-k-1} \exp(-\alpha Ta^{-k}) da$$

$$+ \int_{a_0}^{2a_0} 2a_0 k\alpha Ta^{-k-1} \exp(-\alpha Ta^{-k}) da = [1 - \exp(-\alpha Ta_0^{-k})]$$

$$+ \int_{a_0}^{2a_0} \left[\left(\frac{a}{a_0}\right)^2 - 2\left(\frac{a}{a_0}\right)\right] k\alpha Ta^{-k-1} \exp(-\alpha Ta^{-k}) da \qquad (9.13)$$

The first term is the probability of exceeding $a_0$ as computed in the previous example; the integral in the second term gives the (negative) correction introduced because the building does not fail just at the design acceleration, $a_0$, but rather has an uncertain resistance in the range $a_0$ to $2a_0$. Now assume that $a_0 = 0.3g$ has been used for design, use the parameters $k$ and $\alpha$ from eq. 9.8 and take $T = 50$ years. The result is $P[R_1] = 8\%$ as compared to 20% if damage occurs with certainty right at the design acceleration. The reduction is exaggerated in this simplified illustration because the form of eq. 9.12 is such as to suggest that the resistance is likely to be much closer to $2a_0$ than $a_0$.

If the discrete modified Mercalli intensity measure is used to express the ground shaking, then the probability mass function of $S$, the maximum intensity in $T$ years, takes the form:

$$P[S = I] = P[S < I + 1] - P[S < I] = \exp(-\lambda_{I+1}T) - \exp(-\lambda_I T) \qquad (9.14)$$

where $\lambda_I$ and $\lambda_{I+1}$ are given by eq. 9.9. Hence eq. 9.11 becomes:

$$P[R_1] - \Sigma F(I)[\exp(-\lambda_{I+1}T) - \exp(-\lambda_I T)] \qquad (9.15)$$

where $F(I) = P[R_1 | S = I]$ is a discrete function expressing the probability of damage given each intensity level. For illustration, consider the data given in Table 9.I. The values of $\lambda_I$ are computed from eq. 9.10. The values of $F(I)$ come from Whitman (1973) and are thought to be valid for buildings designed according to the California Building Code provisions in force between about 1947 and 1969. These data are discussed further in Section 9.3. Here the

TABLE 9.I

Example of two-state damage probability in terms of modified Mercalli intensity

| MMI | V | VI | VII | VIII | IX | X |
|---|---|---|---|---|---|---|
| $\lambda_I$ | $9.3 \cdot 10^{-1}$ | $2 \cdot 10^{-1}$ | $4.3 \cdot 10^{-2}$ | $9.3 \cdot 10^{-3}$ | $2 \cdot 10^{-3}$ | $4.3 \cdot 10^{-4}$ |
| $\exp(-\lambda_{I+1}T)$ $-\exp(-\lambda_I T)$ | $4.5 \cdot 10^{-5}$ | 0.12 | 0.51 | 0.28 | 0.07 | 0.02 |
| $F(I) = P[R_1 \mid S = I]$ | 0 | 0 | 0 | 0.22 | 0.80 | 1.00 |
| Contribution to sum | 0 | 0 | 0 | 0.06 | 0.06 | 0.02 |

damage threshold is at 'heavy' damage, corresponding to repair costs of at least 10% of the replacement cost of the building and at least one chance in 50 of serious injury. The result is $P[R_1] = 0.14$; that is, there is a 14%-probability of damage in a 50-year period.

*Example 9.3*

In the two previous examples damage has been taken to be related to the most familiar measures of the ground motion, namely peak ground acceleration and modified Mercalli intensity. Peak ground velocity is also in common use. In many ways it would be more appropriate to relate damage to some measure of the dynamic response of a structure. One can assume that damage is a continuous function of response, in which case the state of the system might be defined directly as the dynamic response. In example 9.3 we assume a simple two-state model. Thus $R_0$ is still defined as the event that there is no damage, but now the occurrence of the event 'damage' is related to exceeding some level $y_0$ of building response $Y$ rather than just to the exceedance of some level of ground shaking. The first step in determining the probability $P[R_0]$ is to obtain the probability distribution of the response $Y$ given the level of ground shaking.

Consider a lightly damped linear, one-degree-of-freedom oscillator idealization of the structure. Then from Chapter 8, the probability distribution of the peak displacement, $Y$, given the (pseudo-stationary) r.m.s. value, $\sigma$, of the ground acceleration is approximately (eq. 8.51):

$$P[R_0|S=\sigma] = P[Y \leqslant y_0|S=\sigma] = F_{Y|S}(y_0;\sigma) = \exp\left[\frac{-\omega_n t}{\pi} \exp(-y_0^2/2\sigma^2 H^2)\right]$$
(9.16)

Note that with this approach to damage prediction the most appropriate seismic intensity variable $S$ is not the peak but the r.m.s. ground acceleration. (To emphasize our interest in $\sigma$ we have used $\sigma^2 H^2$ to represent the variance of the response of the oscillator. This product is the variance of the seismic input $\sigma^2$ times a system factor $H^2$; see Chapter 8.)

Assuming that $a$, the (expected) peak ground acceleration, and $\sigma$ are approximately proportional (see eq. 8.11), the maximum value of $\sigma$ in a period of $T$ years follows a distribution of the same general form as that of the maximum value of peak ground acceleration in $T$ years, or $\exp(-\alpha T\sigma^{-k})$.

Analogous to eq. 9.11, the probability of no damage in $T$ years takes the form:

$$P[R_0] = P[Y \leqslant y_0] = F_Y(y_0) = \int F_{Y|S}(y_0;\sigma) \frac{\mathrm{d}}{\mathrm{d}\sigma}[\exp(-\alpha T\sigma^{-k})]\mathrm{d}\sigma$$

$$= k\alpha T \int \exp\left[-\frac{\omega_n t}{\pi} \exp(-y_0^2/2\sigma^2 H^2)\right] \sigma^{-k-1} \exp(-\alpha T\sigma^{-k})\mathrm{d}\sigma \quad (9.17)$$

The state of the system response Y is now the maximum (over all the potential seismic events in T) of the peak (within each event) interstory displacements the structure experiences. It has been assumed that the maximum displacement will occur during the seismic event with the maximum intensity, $\sigma$, during the interval 0 to T. This is a reasonable assumption owing to the much broader variability in $\sigma$ compared to that in peak response (given $\sigma$). Should a case arise in which this is an inappropriate assumption, one should use in eq. 9.17 the density function of $\sigma$ during an individual event (e.g., derived by differentiation of a result like eq. 9.4) in place of that of the maximum value of $\sigma$ over T years (derived from eq. 9.7). Then eq. 9.17 would yield the system response distribution during an individual event. The probability of state $R_0$ (i.e., the maximum response in T years is less than $y_0$) would then be the probability of no events with response larger than $y_0$. In words, this probability would be calculated as e to the power minus T times one minus the revised result from eq. 9.17. It has also been assumed in eq. 9.17 either that one can ignore the possibility of two or more damaging events or that the structure is restored, if necessary, to its original condition between events.

Examination of the results of this and similar simple linear structural analyses of total seismic risk (see e.g., Borges and Castanheta, 1971) have confirmed that total variability in Y in this case is dominated by that in S, the seismic input. Roughly, the response Y can be represented as $Y = (\sigma)(F)$ in which F is the ratio $Y/\sigma$. The former term represents the (lifetime maximum) ground motion intensity and the latter is the effect of randomness of response, given the intensity $\sigma$. Assuming approximate independence of $\sigma$ and F, the coefficient of variation of Y is:

$$V_Y = \sqrt{V_\sigma^2 + V_F^2 + V_\sigma^2 V_F^2} \tag{9.18}$$

Typical values of $V_\sigma$ are in excess of 1 whereas $V_F$ may be 0.1—0.3. Therefore, ignoring the latter randomness (i.e., treating $F = Y/\sigma$ as deterministic) will not significantly underestimate the variability in Y. Design values may be underestimated by 10—30% or less. If such errors are tolerable, one need focus only on a central value (e.g., the expected value or median value) of the dynamic response given the intensity of the input. (The determination of this value still requires random vibrations analysis, however.) With this information, we are in effect saying that Y is a deterministic function of $\sigma$ (or S); call it $Y(\sigma)$. In this case eq. 9.17 simplifies to:

$$P[R_0] = F_Y(y) = P[Y(\sigma) \leq y] = P[\sigma_y \leq Y^{-1}(y)]$$
$$= F_S(\sigma_y) = \exp(-\alpha T \sigma_y^{-k}) \tag{9.19}$$

in which $\sigma_y = Y^{-1}(y)$ is the value of $\sigma$ such that $Y(\sigma) = y$. This problem is then of the same simple form as Example 9.1.

## 9.3 ANALYSIS OF TOTAL RISK: MULTIPLE DAMAGE STATES

For some applications it is desirable to predict the total loss that might occur, either as the result of a single large earthquake or as the result of a series of earthquakes that might take place over a period of years. The insurance industry and disaster planners are especially interested in estimates of the losses that might occur during a major earthquake. Friedman (1968) and O.E.P. (1972) have predicted the losses if the 1906 San Francisco event were to occur now. Estimates of losses from large underground explosions are also made (Blume and Monroe, 1971).

For either type of prediction, the two-state representation of damage used in Section 9.2 is inadequate. During a very large earthquake the total loss will be the sum of the heavy damage to the buildings in the small region of greatest ground shaking plus the less extensive damage to a greater number of buildings in the larger area of moderate ground shaking. Similarly, the losses during a series of moderate earthquakes might, in total, be as large as the greater loss from the single larger earthquake that might be expected to occur during the same period. For all such predictions it is clearly necessary to use a spectrum of damage states.

The damage state (e.g., the ratio of the cost of repairing earthquake damage to the replacement cost of a building) may be represented as continuous or discrete. The latter approach is illustrated by a recent study of the application of decision analysis to the seismic provisions of building codes (Whitman et al., 1974, 1975). Six general states of damage, $R_i$, were defined in terms of such loss-related factors as level of repair cost, degree of structural and non-structural damage, the fraction of occupants injured and killed, etc. These discrete damage states are described briefly in Table 9.II; their use will be illustrated in Example 9.4.

Because of non-uniform design and construction practice, variation in materials, properties, etc., buildings designed to a given code specification will not all experience the same level of damage during a given ground shaking. Moreover, in an individual building the degree of damage will differ in shakings having the same peak acceleration (or the same value of any single-parameter representation of the intensity of ground shaking). Thus at each intensity of shaking it is, in general, necessary to have a probability distribution function $P[R_i|S_i]$ for damage. The character of such functions at moderate and high intensities of ground shaking is illustrated in Fig. 9.2, both for a continuous damage-state variable and for a set of discrete damage states.

One use of a damage probability distribution function is to compute the probability that a building will find itself in a particular damage state $R_i$, with the associated repair cost and injuries or deaths. The first step is to determine the mean annual rate $\lambda_{R_i}$ of events in which the building experiences the damage state $R_i$. This mean rate is obtained by application of eq.

TABLE 9.II

Description of damage states

| Damage state and symbol | Description of damage | Damage ratio (%) range | Damage ratio (%) central value | Injury ratio (%) central value | Life loss ratio (%) central value |
|---|---|---|---|---|---|
| NONE – 0 | No or insignificant non-structural damage | 0–0.05 | 0 | 0 | 0 |
| LIGHT – L | Minor, localized non-structural damage | 0.05–1.25 | 0.3 | 0 | 0 |
| MODERATE – M | Widespread, extensive non-struct. damage; readily repairable struct. damage | 1.25–20 | 5 | 1 | 0 |
| HEAVY – H | Major struct. damage; possibly total non-struct. damage | 20–65 | 30 | 2 | 0.25 |
| TOTAL – T | Building condemned or replaced | 65–100 | 100 | 10 | 1 |
| COLLAPSE – C | Building partially or totally collapsed | 100 | 100 | 100 | 20 |

9.1 modified to apply to rates. Expressing the ground shaking by the modified Mercalli intensity (eq. 9.9):

$$\lambda_{R_i} = \sum_I P[R_i | S = I] (\lambda_I - \lambda_{I+1}) \tag{9.20}$$

in which $(\lambda_I - \lambda_{I+1})$ is the mean annual rate of events with $S = I$ (assuming repair, if necessary). The probability that the damage state $R_i$ will be experi-

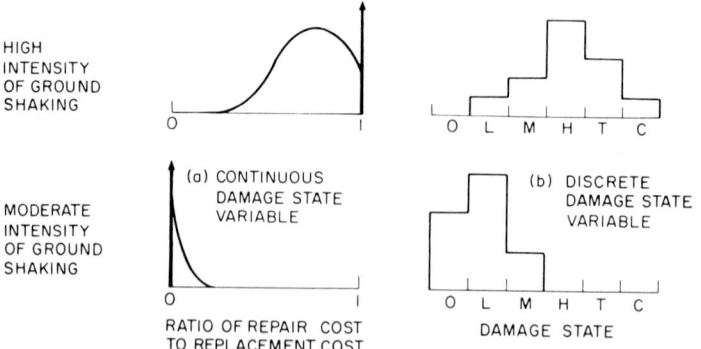

Fig. 9.2. Possible forms for damage probability distribution functions. Ordinate is probability or probability density.

enced any particular number of times during $T$ years may then be computed by using eqs. 9.6 through 9.8. Accumulating the appropriate mean annual rates of different states will yield the mean annual rate of events causing damage of any level or greater. An illustration appears in the example to follow.

For many applications it is sufficient to use simple moments derived from the damage probability distribution function. In particular, the *mean* loss ratio ($MLR$) for a given seismic intensity $S = s$ may be computed:

$$MLR(s) = \begin{cases} \int r f_{R|S}(r;s) \, dr \\ \sum LR_i P[R_i | S = s] \end{cases} \quad (9.21)$$

where $r$ is a continuous variable giving the ratio of repair to replacement cost, and $LR_i$ is the corresponding loss ratio for discrete damage states. The mean loss ratio is a non-decreasing function of the seismic intensity $S = s$. Analytical forms for such ratios have been suggested by Liu and Neghebat (1972) and Jacobson et al. (1973), and data for small residences have been collected by the USCGS (1969) and by Scholl and Farhoomank (1973). Mean life loss or mean injury ratios can be computed in similar fashion. Mean loss ratios are a primary input for expected loss computations as part of cost—risk analyses (Section 9.5). The variance of the damage probability function can also be determined and used to compute confidence limits on predictions of expected losses (Blume and Monroe, 1971).

*Example 9.4*

As part of the previously mentioned study of building code provisions, damage probability distribution functions were estimated from an empirical study of damage to multistory buildings in recent earthquakes. Observed frequencies were modified by results from linear and non-linear dynamic studies and by professional judgement. Table 9.III gives the results in the form of *damage probability matrices.* Modified Mercalli intensity was used to express the levels of ground shaking. There is a matrix for each of four levels of design, corresponding roughly to the requirements of the 1970 Uniform Building Code for Zones 0, 2, and 3 and for a 'superzone' (designated Zone S) having twice the lateral forces required for Zone 3. The matrices were intended to apply specifically to new multistory apartment buildings in Boston; because the damage potential is greatly affected by design details and by care in construction, the values in Table 9.III may not apply in other parts of the world. The matrices are subject to improvement as more knowledge and experience becomes available.

Table 9.IV gives results for mean annual rates of damage states computed using the values of $\lambda_I$ in Table 9.I. The mean annual rate of occurrences of events with damage at least as great as damage state $H$ is $\lambda_{R \geqslant H}$ or 0.0033.

TABLE 9.III

Damage probabilities (%) for pilot application of seismic design decision analysis

| Design strategy | Damage state | Modified Mercalli intensity | | | | | | |
|---|---|---|---|---|---|---|---|---|
| | | V | VI | VII | VII.5 | VIII | IX | X |
| UBC 0 | 0 | 100 | 27 | 15 | 0 | 0 | 0 | 0 |
| | L | 0 | 73 | 48 | 21 | 0 | 0 | 0 |
| | M | 0 | 0 | 33 | 45 | 20 | 0˙ | 0 |
| | H | 0 | 0 | 4 | 29 | 41 | 0 | 0 |
| | T | 0 | 0 | 0 | 5 | 34 | 75 | 25 |
| | C | 0 | 0 | 0 | 0 | 5 | 25 | 75 |
| UBC 2 | 0 | 100 | 47 | 20 | 0 | 0 | 0 | 0 |
| | L | 0 | 53 | 50 | 36 | 10 | 0 | 0 |
| | M | 0 | 0 | 29 | 52 | 53 | 0 | 0 |
| | H | 0 | 0 | 1 | 11 | 31 | 0 | 0 |
| | T | 0 | 0 | 0 | 1 | 5 | 80 | 60 |
| | C | 0 | 0 | 0 | 0 | 1 | 20 | 40 |
| UBC 3 | 0 | 100 | 57 | 25 | 5 | 0 | 0 | 0 |
| | L | 0 | 43 | 50 | 48 | 25 | 0 | 0 |
| | M | 0 | 0 | 25 | 41 | 53 | 20 | 0 |
| | H | 0 | 0 | 0 | 6 | 21 | 52 | 0 |
| | T | 0 | 0 | 0 | 0 | 1 | 23 | 80 |
| | C | 0 | 0 | 0 | 0 | 0 | 5 | 20 |
| Zone S | 0 | 100 | 67 | 30 | 10 | 0 | 0 | 0 |
| | L | 0 | 33 | 49 | 58 | 40 | 10 | 0 |
| | M | 0 | 0 | 21 | 29 | 52 | 30 | 0 |
| | H | 0 | 0 | 0 | 3 | 8 | 58 | 0 |
| | T | 0 | 0 | 0 | 0 | 0 | 2 | 90 |
| | C | 0 | 0 | 0 | 0 | 0 | 0 | 10 |

Using eq. 9.7, the probability that such damage will occur at least once in 50 years is:

$1 - \exp[-0{,}0033(50)] = 0.15$

(Compare this with the value of 0.14 in Example 9.2 where slightly different assumptions were made.) The use of such estimates of the risk of different damage states will be discussed further in Section 9.5.

Table 9.V gives the mean damage ratios and mean life loss ratios computed by applying eq. 9.21 to the damage probabilities in Table 9.III. Use of these mean loss ratios will also be discussed in Section 9.5. By combining the mean loss ratios with the mean annual rates of different intensities, an an-

TABLE 9.IV

Example of mean annual rates of damage states (based on UBC 3 in Table 9.III)

| MMI | V | VI | VII | VIII | IX | X | $\lambda_{R_i}$ | $\lambda_{R \geq i}$ |
|---|---|---|---|---|---|---|---|---|
| $\lambda_I$ | $9.3 \cdot 10^{-1}$ | $2.0 \cdot 10^{-1}$ | $4.3 \cdot 10^{-2}$ | $9.3 \cdot 10^{-3}$ | $2.0 \cdot 10^{-3}$ | $4.3 \cdot 10^{-4}$ | | |
| $\lambda_I - \lambda_{I+1}$ | $7.3 \cdot 10^{-1}$ | $1.6 \cdot 10^{-1}$ | $3.3 \cdot 10^{-2}$ | $7.3 \cdot 10^{-3}$ | $1.6 \cdot 10^{-3}$ | $4.3 \cdot 10^{-4}$ | | |
| $R_O$ | $7.3 \cdot 10^{-1}$ | $8.9 \cdot 10^{-2}$ | $0.8 \cdot 10^{-2}$ | 0 | 0 | 0 | $8.3 \cdot 10^{-1}$ * | $9.3 \cdot 10^{-1}$ ** |
| $R_L$ | 0 | $6.7 \cdot 10^{-2}$ | $1.6 \cdot 10^{-2}$ | $1.8 \cdot 10^{-3}$ | 0 | 0 | $8.6 \cdot 10^{-2}$ | $1.0 \cdot 10^{-1}$ |
| $R_M$ | 0 | 0 | $0.8 \cdot 10^{-2}$ | $3.9 \cdot 10^{-3}$ | $3.1 \cdot 10^{-4}$ | 0 | $1.3 \cdot 10^{-2}$ | $1.6 \cdot 10^{-2}$ |
| $R_H$ | 0 | 0 | 0 | $1.5 \cdot 10^{-3}$ | $8.2 \cdot 10^{-4}$ | 0 | $2.3 \cdot 10^{-3}$ | $3.3 \cdot 10^{-3}$ |
| $R_T$ | 0 | 0 | 0 | $0.1 \cdot 10^{-3}$ | $3.6 \cdot 10^{-4}$ | $3.4 \cdot 10^{-4}$ | $7.8 \cdot 10^{-4}$ | $9.4 \cdot 10^{-4}$ |
| $R_C$ | 0 | 0 | 0 | 0 | $0.7 \cdot 10^{-4}$ | $0.9 \cdot 10^{-4}$ | $1.6 \cdot 10^{-4}$ | $1.6 \cdot 10^{-4}$ |

* This is the mean annual rate of experiencing $R_O$ as the result of an earthquake with $I \geq V$. $R_O$ is also experienced if no earthquake occurs.

** This is just the mean annual rate of experiencing an earthquake with $I \geq V$ (i.e., $\lambda_{I \geq V}$).

TABLE 9.V

Mean damage and lifeloss ratios for an application of seismic design decision analysis

| Design strategy | Modified Mercalli intensity | | | | | |
|---|---|---|---|---|---|---|
| | V | VI | VII | VIII | IX | X |
| *Mean damage ratio* (% of replacement cost) | | | | | | |
| UBC 0 | 0 | 0.22 | 3.0 | 52 | 100 | 100 |
| UBC 2 | 0 | 0.16 | 1.9 | 18 | 100 | 100 |
| UBC 3 | 0 | 0.13 | 1.4 | 10 | 45 | 100 |
| Zone S | 0 | 0.10 | 1.2 | 5 | 21 | 100 |
| *Mean life loss ratio* (% of occupants) | | | | | | |
| UBC 0 | 0 | 0 | $1 \cdot 10^{-4}$ | 0.0144 | 0.058 | 0.153 |
| UBC 2 | 0 | 0 | $0.25 \cdot 10^{-4}$ | 0.0033 | 0.048 | 0.086 |
| UBC 3 | 0 | 0 | 0 | $6 \cdot 10^{-4}$ | 0.014 | 0.048 |
| Zone S | 0 | 0 | 0 | $2 \cdot 10^{-4}$ | $16 \cdot 10^{-4}$ | 0.029 |

nual expected loss ratio ($AELR$) can be computed:

$$AELR = \sum_I MLR(I) (\lambda_I - \lambda_{I+1}) \tag{9.22}$$

For the $\lambda_I$ in Tables 9.I and 9.IV, which recall maybe roughly Los Angeles seismicity levels, the annual expected damage ratio for the UBC 3 design requirement is 0.0025 (i.e., one quarter of one percent) and the annual expected life loss ratio is $5 \cdot 10^{-5}$.

## 9.4 ANALYSIS OF TOTAL RISK: DISTRIBUTED TARGETS

This section considers spatially extended systems, those whose elements are separated widely enough that different elements will experience different levels of ground shaking during a particular earthquake. The system response depends on the condition of its elements. One example is a long highway; failure of any bridge would interrupt use of the entire highway. The probability of failure of the highway system is greater than the probability that any specific bridge will fail. A second example is an electrical network with several dispersed generating facilities. The system can function effectively even if an earthquake causes the shut-down of one generating station; temporary loss of one station is not an uncommon event for a large electrical system. Simultaneous loss of several generating stations during the same earthquake would, however, constitute a 'failure' of the system. This event is less likely than the loss of any one station, provided the facilities are sufficiently widely separated. In both examples, the simple site-intensity definition of seismic input $S_j$ used in the previous sections is insufficient.

Studies of such spatially distributed 'lifelines' are very recent (see, for example, Molchan et al., 1970; Jacobson et al., 1973; Schiff et al., 1974; Whitman, 1974; Panoussis, 1974; and Taleb-Agha, 1975). Two simple examples give the character of the analyses and results.

*Example 9.5*

Consider a highway paralleling a fault, as shown in Fig. 9.3. The critical elements in the highway are the bridges, located as shown at points $u = 1, 2, \ldots, n$. Assume that a two-state definition of system behavior is sufficient. The system (highway) will be in state $R_1$ if during a given one-year period an event occurs which causes at least one of the bridges to be (seriously) damaged, and in state $R_0$ otherwise. Assume further, as in Example 9.2, that each bridge structure will be either damaged or not, and that the probability that the capacity of any particular bridge ($u = 1, 2, \ldots, n$) is less than level $a$ is $F_A(a)$ (the capacities of the $n$ bridges being mutually independent random variables).

The accelerations $a_1, a_2, \ldots, a_n$ experienced at each of the sites can be computed given that an earthquake of magnitude $M = m$ occurs at location $X = x$ (see Fig. 9.3), e.g., $a_u = g[m, d_u(x)]$. Define the seismic state, then, as the pair of values $m$ and $x$. The probability that the highway system does not fail, given this event, is the probability that *all* $n$ capacities exceed their respective experienced seismic ground accelerations, or:

$$P[R_0 | S(m,x)] = [1 - F_A(a_1)][1 - F_A(a_2)] \ldots [1 - F_A(a_n)]$$

$$= \prod_{u=1}^{n} [1 - F_A(a_u)]$$

$$= \prod_{u=1}^{n} [1 - F_A\{g(m, d_u)\}] \tag{9.23}$$

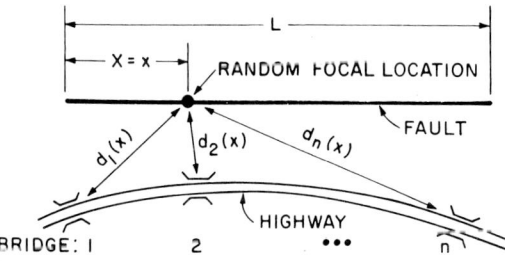

Fig. 9.3. An illustration of a fault and a highway with $n$ critical elements, showing a randomly located earthquake focus.

The probability distribution associated with the seismic state $S(m, x)$, given that an event occurs, is $f_M(m)f_X(x)$, for independent $M$ and $X$, where $f_M(m)$ might be taken as a truncated exponential distribution (Chapter 6) and $f_X(x)$ as $1/L$. (See Fig. 9.3.)

Then application of the appropriate form of the total probability equation (eq. 9.1) gives:

$$P[R_0] = P\begin{bmatrix}\text{no significant}\\ \text{event in one}\\ \text{year}\end{bmatrix} + P\begin{bmatrix}\text{an event}\\ \text{in one}\\ \text{year}\end{bmatrix} \int_m \int_x P[R_0|S(m,x)]f_M(m)f_X(x) \mathrm{d}m\, \mathrm{d}x \quad (9.24)$$

in which $P[R_0|S(m,x)]$ is given by eq. 9.23. (The probability of two or more significant events in one year is neglected.) The probability of no significant event in one year is $\exp[-(\nu)(1)]$, $\nu$ being the mean annual rate of earthquake occurrences. Figure 9.4 shows $P[R_1] = 1 - P[R_0]$ for a two-site system versus that for a one-site system, for a particular set of parameter values.

Initial studies, such as that in Fig. 9.4, suggest that, if two elements are separated from each other by more than twice the perpendicular distance to the fault, they are essentially uncoupled. The implication is that the probability of a system surviving undamaged given an event is simply the product of the corresponding marginal probabilities of the individual elements, or that the failure probability $P[R_1]$ is approximately the sum of the individual probabilities. If the elements are separated from each other by less than half the distance to the fault, they are almost totally coupled, that is, the probability of system survival is nearly equal to the probability that the one element with lowest reliability survives.

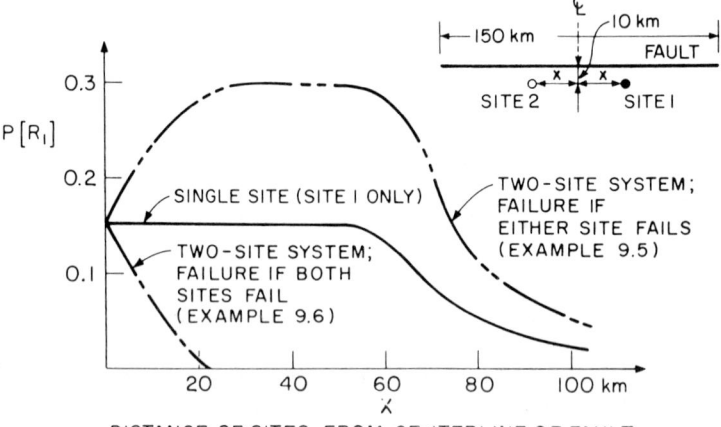

Fig. 9.4. An illustration of one- and two-site systems. (Adapted from Panoussis, 1974.)

*Example 9.6*

In this example, failure $R_1$ occurs only if both elements of a two-site system fail. This might be a case of two generating stations in a power grid or of a key bridge in each of two parallel highways.

Specifically, we consider two sites along a line parallel to a fault (Fig. 9.4) and examine the probability that both fail during the same earthquake. In this case, using the same assumptions and approximations as in Example 9.5:

$$P[R_1|S(m,x)] = P[\text{both fail}|S(m,x)] = F_A(a_1)F_A(a_2) \tag{9.25}$$

Substitution into the general form of eq. 9.2 yields the approximation (compare eq. 9.24):

$$P[R_1] = P\begin{bmatrix}\text{an event in}\\\text{one year}\end{bmatrix} \int_m \int_x F_A(a_1)F_A(a_2)f_M(m)f_X(x)\,dm\,dx \tag{9.26}$$

Sample numerical results are shown in Fig. 9.4. The probability of overall failure becomes very small when the two targets are separated by more than 20 km. In fact, if there is an upper limit to the magnitude of the earthquake that can occur along the fault, then the probability of overall failure becomes zero when the targets are so separated that the largest earthquake cannot make both of them fail.

Other types of spatially-extended systems will have other, more complex, interactions among their elements than the simple series and parallel systems of Examples 9.5 and 9.6. For example, a system may consist of elements both in series and in parallel. But as long as one can estimate the probability of a particular system state $R_i$ for a given set of the ground motion intensities at each of the sites of critical elements of the system, then the probability $P[R_i|S(m,x)]$ can be determined and substituted into eq. 9.24 or 9.26. Also the spatial location parameter $X$ can be interpreted more generally, as a two- or three-dimensional vector, permitting the use of different faults, areal 'sources', etc. (See Panoussis, 1974.)

9.5 DESIGN FOR SEISMIC RISK: OPTIMIZATION OF AN INDIVIDUAL PROJECT

Optimization of the seismic resistance balances the greater initial cost of providing increased resistance against the lesser probability of future losses. In addition to an assessment of the total risk of damage, optimization analysis requires information concerning: (1) the cost of providing earthquake resistance (referred to as the initial cost premium, or ICP), and (2) the cost of earthquake damage. The initial cost premium must be determined as a function of the design acceleration or intensity. The cost of damage should, in general, include all components of loss: human and social costs as well as repair costs. The social costs include potentially a variety of factors: injuries, death, loss of business, community impacts, etc.

If all of the initial costs and future losses are expressed in dollars, then formal optimization procedures may be utilized. Let $C_i(a)$ be the initial cost of construction, a function of the design acceleration. Let $E[C_l(a)]$ be the expected present value of the possible losses from earthquake damage; this loss is also a function of the design acceleration. The expected present value of the total cost $C_t(a)$ is:

$$C_t(a) = C_i(a) + E[C_l(a)] \tag{9.27}$$

With this criterion function the optimal decision is that design acceleration $a$ which minimizes $C_t$. Examples 9.7 and 9.8 are illustrations.

Although this economic criterion is perhaps the most popular and has been applied widely in private and public decision making, many question its application to cases involving extreme financial costs, injuries, deaths, and social impacts. Reduction of life-loss potential is often the main purpose of seismic design. Grandori and Benedetti (1973), for example, have recommended designing to such levels that the marginal number of expected lives lost per dollar invested is equilibrated in different structures and in different seismic activity regions. This criterion alone, however, provides only relative, not absolute decisions.

Modern economic analysis recommends that the objective of decision making be the maximization of expected *utility*, a more abstract value measure. In many situations dollars may be a satisfactory measure of utility, but the introduction of this more general concept of value permits formal optimization to be extended to more complex decision situations. For example, the implications of extreme dollar losses may be such that a building client is prepared to spend $100 or more to effectively eliminate a $10^{-5}$ probability of suffering a $10^6$ loss. If so, it would be said that to him the utility of spending $100 — denoted $u(-\$100)$ — is greater than $10^{-5}$ times $u(-\$10^6)$. (Because $-\$100$ is less than $(10^{-5})(-\$10^6)$, the simple dollar or economic scale does not properly reflect his preferences over this range.)

In principle an individual's or group's reaction to a spectrum of gains and losses, be they economic or other, can be assessed and assigned utility values (see, for example, De Neufville and Marks, 1974.) For example, these schemes permit the assignment of utility values that express a decision maker's relative strength of preference for, say, saving $n$ lives in a future earthquake over providing new housing to $m$ additional people this year.

Like the temperature scale, however, the utility scale is only an ordinal one. One implication is that two points on a utility versus dollar function $u(\$x)$ can be assigned arbitrary values. If this relationship happens to be linear in the range of interest, the two points can be assigned values such that the utility of any dollar loss or gain is numerically equal to the number of dollars. In this case to maximize dollar gain (or minimize dollar loss) is equivalent to maximizing utility. (See Examples 9.7 and 9.8.) Another implication is that, in general, the utilities of two quantities cannot be added. (For example, the $u(\$ x, n \text{ deaths}) \neq u(\$ x) + u(n \text{ deaths})$.) Finally these degrees of freedom in a utility scale imply that the utility values of two individuals or groups of individuals cannot be directly compared. All these restrictions have practical implications in seismic design optimization (Example 9.9).

In many problems, formal optimization may be impractical at present. Not all individuals or groups wish to base decisions upon expected (economic) value computations, and relatively few decision makers yet feel comfortable with utility notions, despite the formal proofs of axiomatic utility theory that it is irrational to do otherwise (Von Neumann and Morgenstern, 1947). In such situations the best approach is to provide the decision making individuals with the estimates of the initial cost premium, the future losses, and the probabilities of those losses for several different design accelerations in order that he may make an informed choice. (That choice will involve conscious or subconscious optimization.)

Whichever approach is adopted, it is essential to evaluate the effect of the seismic design level on the initial cost of the building. For many projects this evaluation can be done by forming and costing several alternatives, each at a

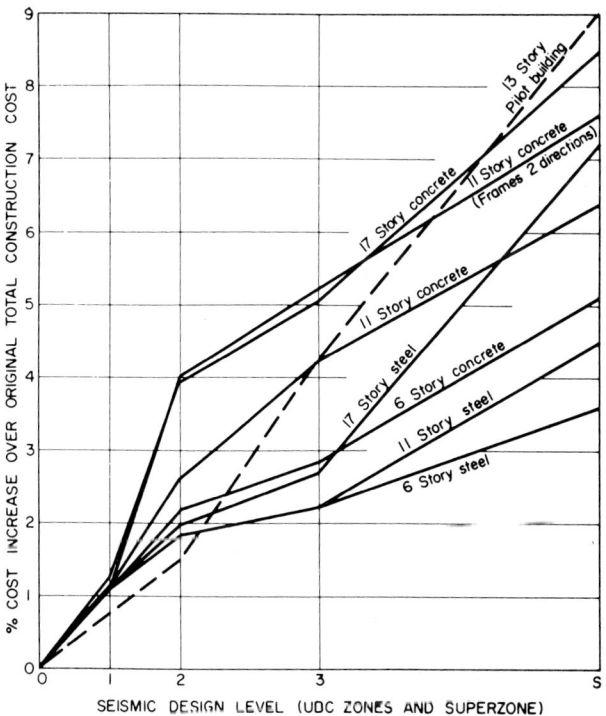

Fig. 9.5. Initial cost premiums for typical apartment buildings in Boston (Whitman et al., 1974).

different level of resistance. The engineering costs involved have largely prevented this practice, however. Few studies of such comparisons are available in the literature to provide a sound basis for estimating the initial costs. Two such studies will be described briefly. One study dealt with a hypothetical group of apartment buildings typical of those that might be built in Boston (Whitman et al., 1974). An engineering firm was asked to carry out generalized structural designs for 6, 11 and 17 story buildings with concrete moment resisting frames (CMRF), concrete shear walls (CSW), and steel moment resisting frames (SMRF). Designs were made for the requirements of the 1970 Uniform Building Code (UBC) Zones, 1, 2 and 3 and for an additional zone (a 'superzone') with required lateral forces twice those of Zone 3. The designs were costed, and the estimated increased cost of earthquake resistance was expressed as a percentage of the estimated total cost of the corresponding building designed only for wind (and of course all other normal code requirements). The cost increases are shown in Fig. 9.5. (Note that a 5% increase in total cost resulting from a strengthening of the structural system means a much larger increase, perhaps 25%, in the cost of the structural system, because the cost of the structural system in this case is typically only about 20% of the total cost of the building. This last percentage may, however, be as much as twice as high in other countries implying twice as large an increase in total cost.) Figure 9.6 gives estimated costs for upgrading the seismic resistance of an actual office building.

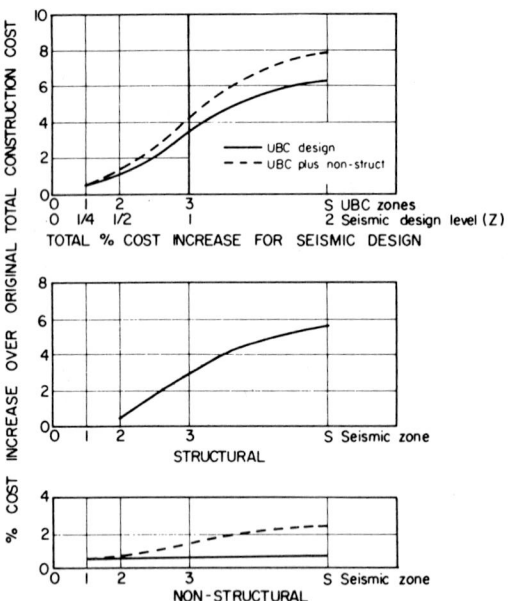

Fig. 9.6. Seismic design costs for pilot building including requirements not specified by code (Whitman et al., 1974).

TABLE 9.VI

Cost consequences of design method

| Bldg. No. | Stories: material | Construction cost | Addition to cost | Percent increase | % extra engineering cost |
|---|---|---|---|---|---|
| 2 | 19 : steel | 11,000,000 | 12,000 | 0.1% | 5% |
| 4 | 14 : concrete | 4,000,000 | 86,601 | 2.2% | 20% |
| 3 | 10 : concrete | 3,500,000 | 50,333 | 1.4% | 5% |
| 8 | 9 : concrete | 3,800,000 | 27,654 | 0.7% | 15% |
| 5 | 6 : concrete | 1,400,000 | 55,452 | 3.9% | 5% |
| 1 | 5 : concrete | 6,670,000 | 237,364 | 3.5% | 27% |
| 7 | 3 : masonry wall, metal diaph. | 534,000 | 44,371 | 8.3% | 10% |
| 6 | 2 : steel | 1,050,000 | 15,019 | 1.43% | 30% |
| 10 | 2 : steel | 2,400,000 | 9,049 | 0.38% | 15% |
| 9 | 1 : tilt-up wall, plywood diaph. | 120,200 | 11,723 | 9.7% | 30% |
| 11 | 1 : masonry wall, plywood diaph. | 1,113,000 | 11,860 | 1.0% | 15% |

(From Applied Technology Council, 1974)

The second study was recently carried out in California (ATC, 1974). It identified the costs to upgrade buildings already designed to the 1973 UBC code provisions so that the buildings would not be structurally damaged by an earthquake with a peak acceleration of about 0.25g. Table 9.VI gives the estimated increases in the total cost of the building and also the estimated increase in the engineering cost of using dynamic analysis instead of the simple static analysis usually permitted by codes in the United States.

*Example 9.7*

Consider, first, an abstract but informative illustration of optimization due to Esteva (1970). We return to Example 9.1. We assumed that the mean rate of Poisson seismic events in excess of ground acceleration $a$ is $\alpha a^{-k}$ and that a structure can be idealized to fail at exactly acceleration $a$.

Under these assumptions, $\alpha a^{-k} dt$ is the probability of the structure's failure in any time interval of length $dt$. If the structure is, as a matter of policy, re-built to the same capacity when it fails, then $E[C_f]$, the expected present value of the failure (and re-building) costs of this structure over a time horizon $T$ is:

$$E[C_f] = \int_0^T C_f \alpha a^{-k} e^{-\gamma t} \, dt = \frac{C_f \alpha a^{-k}}{\gamma} (1 - e^{-\gamma T}) \qquad (9.28)$$

in which $C_f$ is the economic cost (of repair, loss of occupancy, etc.) associated with each failure and in which $\gamma$ is the discount or interest rate, which may include an allowance for inflation. In fact, the time horizon $T$ (the intended or economic lifetime of the structure) is uncertain; in principle one should take the expectation of the last factor $(1 - e^{-\gamma T})$ with respect to a probability distribution on $T$. For realistic parameter values, however, $E[C_f]$ is very insensitive to $T$ (and even to the particular re-building policy adopted) (Esteva, 1968, and Vanmarcke, 1969); the last factor can usually simply be taken as unity, that is, as if $T$ were infinite. This conclusion of optimization theory removes an often arbitrary element, $T$, found in design strategies based on only a specified reliability level.

Let us next assume that the initial cost of construction of a structure with capacity $a$ is of the form:

$$C_i = A_0 + A_1 a^n \tag{9.29}$$

Then addition of $C_i$ and $E[C_f]$ gives the total expected economic cost of the structure. Provided economic values adequately reflect the decision maker's utility, the optimal value of the capacity, $a_0$, is found by differentiating the sum with respect to $a$ and finding the point of minimum cost. The result is:

$$a_0 = \left(\frac{\alpha k C_f}{n\gamma A_1}\right)^{1/(n+k)} \tag{9.30}$$

If we recognize that the mean return period associated with acceleration $a$ is $1/(\alpha a^{-k})$, then the solution can also be written in terms of the optimum design return period $T_0$:

$$T_0 = \frac{1}{\alpha} a_0^k = \frac{1}{\alpha}\left(\frac{\alpha k C_f}{n\gamma A_1}\right)^{k/(n+k)} \tag{9.31}$$

These two results, though idealized solutions, display a number of valuable seismic design generalizations. For example, the design return period (or the reciprocal of the design 'annual probability of failure') should be larger: if the failure costs are larger, if the marginal cost of improving the capacity is lower (i.e., if $A_1$ or $n$ is lower), if discount rate $\gamma$ is lower, if the seismicity rate ($\alpha$) is lower, or if the seismicity parameter $k$ is larger (implying that small increases in $a$ lead to large increases in return period). The numerical values in Table 9.VII indicate the sensitivity of the optimum $a_0$ and $T_0$ to such changes.

*Example 9.8*

The previous example has used only two damage states. Usually it is desirable to consider multiple damage states when making an expected loss computation.

This computation may be illustrated by using the numerical values in Table 9.IV. The mean annual rate of occurrence of damage states $\lambda_{Ri}$ for

TABLE 9.VII

Optimum seismic designs for various parameters

A. Seismicity parameters

|  | Site 1 (low seismicity) | Site 2 (high seismicity) |
|---|---|---|
| $\alpha$ | $10^{-5}$ | $10^{-3}$ |
| $k$ | 2 | 2 |
| implying for $a = 0.1$ a mean return period of | 1000 years | 10 years |

B. Structural parameters

|  | Structure 1 ("typical") | Structure 2 (greater $C_f$) | Structure 3 (lower marginal cost to strengthen) |
|---|---|---|---|
| $n$ | 2 | 2 | 2 |
| $A_1/A_0$ | 2 | 2 | 1.5 |
| $C_f/A_0$ | 2 | 20 | 20 |
| implying: |  |  |  |
| $C_f/A_1$ | 1 | 10 | 13.3 |

C. Optimum resistance ($\gamma = 10\%$)

| Site | 1 | 1 | 2 | 2 | 2 |
|---|---|---|---|---|---|
| Structure | 1 | 2 | 1 | 2 | 3 |
| Optimum resistance $a_0$ | 0.1 | 0.18 | 0.32 | 0.56 | 0.60 |
| Optimum design mean return period, $T_0$ | 1000 | 3160 | 100 | 316 | 365 |

each damage state must be multiplied by the corresponding central damage ratios (Table 9.II), giving the mean loss rate results in the left column of Table 9.VIII. Summing these cost rates over the individual damage states gives the overall mean annual rate of repair cost $C_{l_u}$ (expressed as a percentage of the replacement cost of the building). (The mean ratios in Table 9.V for design strategy UBC 3, multiplied by the corresponding $\lambda_I - \lambda_{I+1}$ and summed over $I$ would yield the same result, $C_{la}$.) Now following Example 9.7 conclusions the present value of all expected future losses may be computed from:

$$E[C_l] = \int_0^\infty C_{la}\, e^{-\gamma t}\, dt = \frac{1}{\gamma} C_{la} \quad \text{for } \gamma > 0 \tag{9.32}$$

TABLE 9.VIII

Example of mean annual loss rates by damage state (UBC 3; $\mu$ = 2000)

| Damage state | Mean annual repair cost (%) | Mean annual human cost (%) | Mean annual total loss (%) |
|---|---|---|---|
| L | 0.026 | 0 | 0.026 |
| M | 0.065 | 0.008 | 0.073 |
| H | 0.069 | 0.016 | 0.085 |
| T | 0.078 | 0.023 | 0.101 |
| C | 0.016 | 0.096 | 0.112 |
|   | 0.254 | 0.143 | 0.397 |

With a discount rate of 5%, the numerical values in Table 9.VIII lead to a present value of expected repair cost of 0.254/0.05 = 5.1%.

Next the discounted loss ratio may be added to the initial cost premium to obtain the total expected cost ratio, and the design strategy giving the minimum total expected cost may be found by inspection. The results are illustrated in Table 9.IX. In this case, which is based upon a seismic risk ($\mu$ = 2000 in eq. 9.9) that might be typical of downtown Los Angeles, use of the design requirements corresponding to Zone 3 of the Uniform Building Code gives the minimum total expected cost. Fig. 9.7 indicates the variation of total expected cost with the seismic hazard and design strategy. With a hazard ($\mu$ = 200) typical of regions of moderate seismic activity, minimum total expected cost results from using no earthquake design. The minimum expected cost design strategy may also be influenced greatly by the choice of discount rate.

In general, cost/benefit analyses which consider only repair costs indicate that strengthening a building for earthquakes is of net economic benefit only in highly seismic areas. Repair costs are, however, only a part of the problem.

A primary aim of earthquake design is to reduce the threat to human

TABLE 9.IX

Total expected losses using multiple damage states ($\mu$ = 2000; $\gamma$ = 5%)

| Design strategy | Initial cost premium (%) | Discounted repair cost (%) | Total expected cost (%) |
|---|---|---|---|
| 0 | 0 | 14.3 | 14.3 |
| 2 | 2.9 | 8.5 | 11.4 |
| 3 | 4.0 | 5.1 | 9.1 |
| S | 6.7 | 3.4 | 10.1 |

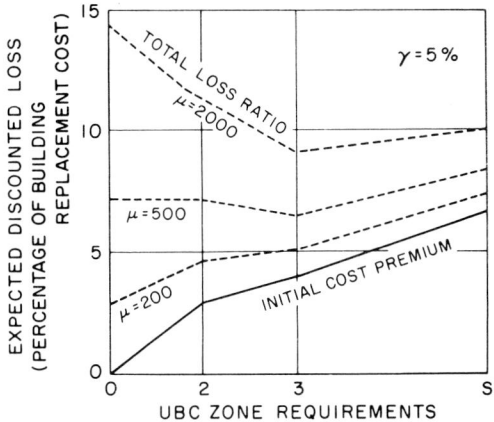
Fig. 9.7. Expected loss ratios (repair costs only) as function of seismicity.

safety. This aspect of the problem may be included directly in an economic cost/benefit analysis provided that one is prepared to assign a dollar value one is willing to pay to avoid an anonymous human injury or death. For example, suppose that a value of $ 300,000 per death and a value of $ 10,000 per injury are adopted. Introducing typical values for the number of occupants per thousand dollars of construction cost and expected fractions of occupants injured and killed, a human cost ratio may be associated with each damage state. Table 9.X gives estimated values expressed as percentages of the building replacement cost. Now, given estimates of the seismic hazard rates, the mean annual human cost rate may be determined and may be added to the mean annual repair cost rate to obtain the expected mean annual total loss rate. This is illustrated in Table 9.VIII. Finally, the present value of the expected total loss is obtained from the integrated discounted rate (eq. 9.32) and added to the initial cost premium. Figure 9.8 presents results for various assumed values of the seismic hazard parameter $\mu$.

TABLE 9.X

Human cost ratio as a function of damage state, assuming $ 300,000 per life and $ 10,000 per Injury

| Damage state | Human cost ratio (%) |
|---|---|
| 0 | 0 |
| L | 0 |
| M | 0.6 |
| H | 7 |
| T | 30 |
| C | 600 |

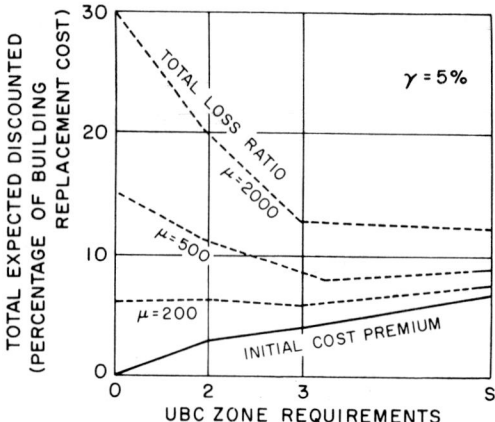

Fig. 9.8. Expected loss ratios (human and repair costs) as function of seismicity.

Provided dollar values adequately reflect the decision maker's utilities or preferences, results such as those in Fig. 9.8 can be scanned to determine the optimal (minimum expected total cost) design strategy. This simple parametric analysis suggests that earthquake design requirements aimed at increasing human safety may be justified in regions of only moderate seismic activity. Any conclusion may be influenced greatly by the particular dollar value assigned to avoiding human life and injury.

*Example 9.9*

A preliminary attempt has been made to assess the utility preference functions of several groups of people with respect to life loss and costs from earthquake damage (Whitman et al., 1974; De Neufville, 1975). Figures 9.9 and 9.10 show utility functions for two groups of people: developers and government officials. (It must be emphasized that these are very crude, preliminary estimates for the behavior of these groups. The interpretation must be considered only illustrative.) In this study, the worst situation envisioned was earthquake damage that killed 20% of the occupants of a building and made the building a total loss. This worst situation was arbitrarily assigned a utility of −100. The curves on the figures indicate the assessed relative utility of lesser losses.

Because of the necessarily arbitrary scales we must focus on relative values of increments of utility. For a given life loss level, for example, the developers' utility function is roughly linear in monetary losses; in contrast, the government officials utility functions implies 'risk aversion', a given cost increment at an already high cost level being much less desirable than the same increment at a low to moderate level. The curves in Fig. 9.10 are approximately linear in lives lost (for a given dollar loss), but the increments are

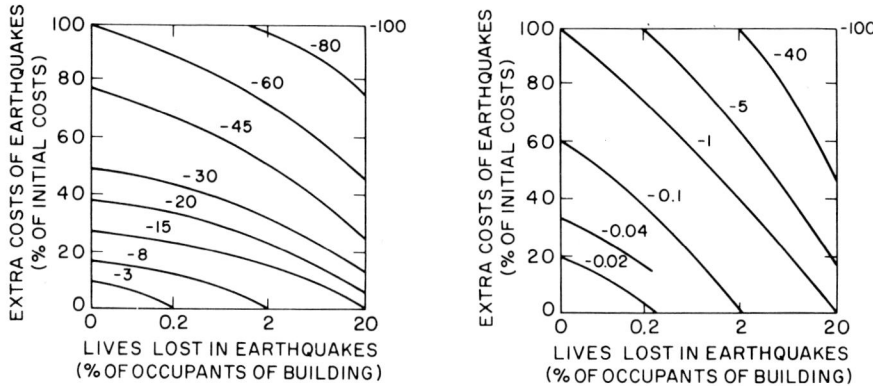

Fig. 9.9 (left). Typical utility function over costs and loss of life (developers).

Fig. 9.10 (right). Typical utility function over costs and loss of life (government officials).

larger at higher economic loss levels. In contrast in Fig. 9.9, the life loss increments are about the same at all economic levels, but on a per life basis they are much smaller at higher loss levels. The figures also suggest each group's trade-off between dollar cost and loss of life. For example, to the developer a dollar cost equal to 20% of the replacement cost is equivalent, in a decision context, to having about 5% of the occupants killed; for government officials a 20% dollar loss is equivalent to having only about 0.2% of the occupants killed.

Once utility functions are available, it is again possible to carry out a formal optimization procedure. Because the utilities discussed above were assessed as if all potential losses would occur immediately (not in the future), future losses must be discounted to present values before translating them into utilities. Thus, the spending of an initial cost premium for design strategy $k$, $ICP^{(k)}$, now and a damage loss of $CDR^{(k)}$ at time $t$ in the future (owing to an earthquake in that year causing damage state $R$) implies an economic loss of *present* economic value equal to:

$$ICP^{(k)} + CDR_R \, e^{-\gamma t} \qquad (9.33)$$

in which $\gamma$ is the economic annual discount rate. It is presumed, too, for generality, that lives lost at some future date have a 'present value' of $e^{-\beta t}$ times their actual number, $\beta$ being analogous to an annual discount rate. Then the utility of spending now for strategy $k$ and experiencing an earthquake state $R$ in year $t$ is:

$$U(CLLR_R \, e^{-\beta t}; ICP^{(k)} + CDR_R \, e^{-\gamma t}) \qquad (9.34)$$

in which $U(x;y)$ is the utility of $x$ lives lost and $y$ dollars spent or lost in year 0, and in which $CLLR_R$ is the number of lives lost if damage state $R$

occurs. (Recall both lives and dollars are expressed in fractional terms.)

These discounted losses and their utility depend directly upon the year $t$ in which the earthquake event occurs. Because, for non-linear utility functions, the discounting and adding of dollars and lives must take place *before* the utility assignment, occurrences of one or more earthquakes in future years must be considered separately. The expected utility for design strategy $k$, during economic lifetime $T$ (assuming a policy of instantaneous repairing to the initial condition), becomes:

$$E[U]^{(k)} = e^{-\lambda T} U(0, IPC^{(k)})$$

$$+ \lambda T e^{-\lambda T} \int_0^T \sum_R P_R^{(k)} U(CLLR_R \, e^{-\beta t}; ICP^{(k)} + CDR_R \, e^{-\gamma t}) \frac{1}{T} dt$$

$$+ \frac{(\lambda T)^2 e^{-\lambda T}}{2!} \int_0^T \int_0^T \sum_{R_1} \sum_{R_2} P_{R_1}^{(k)} P_{R_1}^{(k)} U(CLLR_{R_1} \, e^{-\beta t_1} +$$

$$+ CLLR_{R_2} \, e^{-\beta t_2}; ICP^{(k)} + CDR_{R_1} \, e^{-\gamma t_1} + CDR_{R_2} \, e^{-\gamma t_2}) \frac{1}{T^2} dt_1 dt_2 + \ldots$$

(9.35)

in which, for example, $R_1$ and $R_2$ (and $t_1$ and $t_2$) are the damage states experienced (and the times of occurrence) of the first and second earthquakes during time period $T$. Also, $\lambda$ is the mean rate of earthquakes and $P_R^{(k)}$ is the probability of damage state $R$ under strategy $k$ given an earthquake (Table 9.IV). Note that the leading factors in each term are the probabilities of exactly 0, 1, 2, ..., $n$ earthquakes in time interval 0 to $T$, and $1/T^n$ is the joint probability density function of the arrival times of earthquakes given the occurrence of $n$ events in the interval 0 to $T$. As discussed earlier, for sufficiently large $\gamma$, $\beta$, and $T$, the results will be insensitive to the value of $T$. In regions of low to moderate seismicity only the first two terms will be significant.

Use of this approach is illustrated by the following fictitious example. The particular seismic hazard probabilities used were selected deliberately to provide results that illustrate that the use of utilities rather than dollars may imply different optimal decisions depending upon whose utilities are employed in the analysis. For this example, the annual probabilities of occurrence of site intensities equal to or greater than those listed below are:

| MMI: | VI | VII | VII.5 | VIII | IX |
|---|---|---|---|---|---|
| | $0.25 \cdot 10^{-1}$ | $0.63 \cdot 10^{-2}$ | $0.31 \cdot 10^{-2}$ | $0.11 \cdot 10^{-2}$ | 0 |

where MMI is modified Mercalli intensity. Using the probability damage matrix in Table 9.III, the resulting annual probabilities of being in state $R$ following the occurrence of an earthquake (of site intensity VI or more)

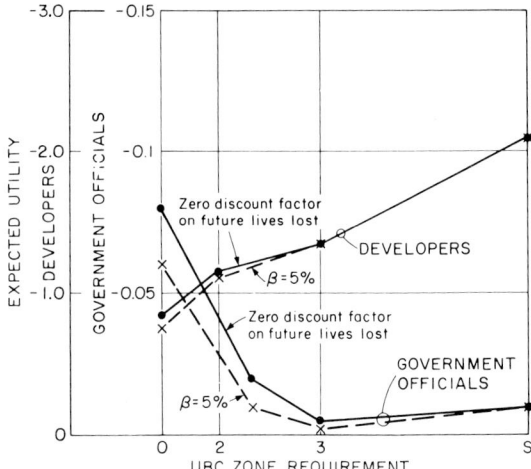

Fig. 9.11. Expected utilities versus design strategy for utility functions in Figs. 9.9 and 9.10.

were computed. Then the expected total utilities as a function of design strategy were computed; the results are shown in Fig. 9.11 for the multi-attribute utility functions corresponding to developers (Fig. 9.9) and to code-responsible officials (Fig. 9.10). A discount rate $\gamma = 5\%$ was used for economic losses, and results are shown for both $\beta = 0\%$ and $\beta = 5\%$, the discount rate on lives. A 50-year lifetime was assumed. Only the first two terms of eq. 9.35 were calculated numerically.

Observe that the best strategy consistent with the relative utilities implied by developers is UBC Zone 0 whereas UBC Zone 3 is optimal with respect to the utilities assessed from statements made by government officials. The latter conclusion reflects the relatively greater weight given to lives versus dollars by the second group. (Compare Figs. 9.9 and 9.10). Recall in interpreting Fig. 9.11 that only the relative ranking of the alternatives is important; in particular, because of the nature of the utility scale, one can give no significance to the absolute utility values nor to the ratios of utility values (see Section 9.5).

9.6 DESIGN FOR SEISMIC RISK: STRUCTURAL BUILDING CODES

The seismic design provisions in building codes typically aim to achieve several objectives. For example, the commentary for the code provisions developed by the Structural Engineers Association of California (SEAOC, 1960) states the objectives as:
— Resist minor earthquakes without damage.

— Resist moderate earthquakes without structural damage, but possibly with some non-structural damage.
— Resist major earthquakes, of the severity of the strongest experienced in California, without collapse, but possibly with some structural as well as non-structural damage.

It is thus explicitly recognized that buildings designed only for the minimum requirements of the code probably will be damaged during a major earthquake.

The codes in effect in many countries include maps or tables which subdivide the country into zones in which different design requirements apply. For example, the current zoning map for Canada (NRC, 1975), shown in Fig. 9.12, is based upon an analysis of the probability of occurrence of different levels of ground shaking (Milne and Davenport, 1969). While considerable judgment was applied in drawing this zoning map, the zones correspond roughly to regions where the mean recurrence rate is 100 years for the following accelerations:

| Zone | $a_{100}$ |
|---|---|
| 0 | $0-0.01g$ |
| 1 | $0.01-0.03g$ |
| 2 | $0.03-0.06g$ |
| 3 | $>0.06g$ |

Choice of the 100-year mean annual recurrence rate was apparently based more on judgment than on any explicit analysis of costs and losses.

Difficulties in the preparation of zoning maps have been discussed well by Housner and Jennings (1973). The principal problems, at least in the Western Hemisphere, stem from the short history of earthquake experience that is available (several centuries at most). For example, in regions of moderate or low seismicity there is no general agreement on such questions as the likelihood of very large earthquakes and the possibility that earthquakes may occur in regions of low activity in recent years. As of this writing (1975), an effort to develop a new zoning map for the United States is underway by the Applied Technology Council, based on a combination of probability analysis and judgment. The remainder of this section will illustrate and outline some of the considerations that may be used in this effort.

*Example 9.10*

An optimization approach using expected annualized losses may be made for a group of buildings grouped closely enough so that the same seismic hazard relation applies to all buildings. The expected economic loss for the group of buildings just equals the sum of the expected loss for the individual buildings. If the same mean damage ratio vs. intensity relation applies to all buildings, then this relation applies to the group as a whole.

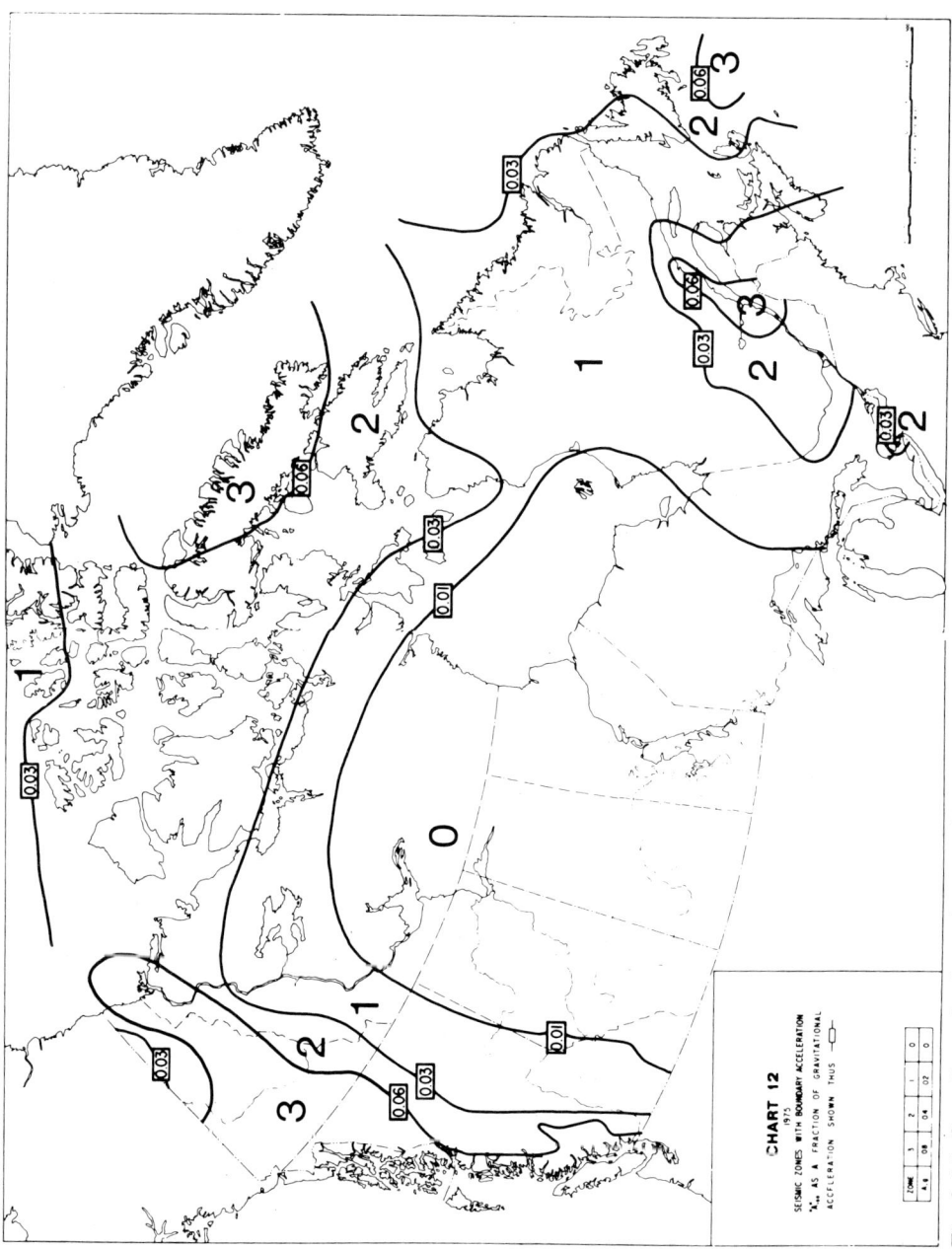

Fig. 9.12. Canadian seismic zoning map (NRC, 1975, Supplement No. 11).

It is also possible to extend this approach to computing the expected annual loss for a region or for an entire country. The region or country is subdivided into many small areas, which are then treated as points. The expected annual loss at each point is determined considering the seismic hazard relation at that point and the type and number of buildings at the point. The expected annual losses for each point are summed to give the total expected annual losses for the region or country. In this way, Wiggins et al. (1974) has estimated the annualized losses for the United States over a period of years, for several different possible policies concerning the mitigation of the earthquake hazard. The estimated reductions in losses were then compared to the expenditures required by the mitigation action.

*Example 9.11*

To many involved with the development of building codes, it is the possibility of a major catastrophe rather than the possibility of large annualized losses that is of concern. For example, insurance companies worry that it may be impossible to accumulate reserves large enough to cover losses during a truly major earthquake in a heavily populated region.

Various estimates have been made of the loss if the 1906 San Francisco or 1811 New Madrid (Missouri) earthquakes were to occur today (Friedman, 1968; Wiggins et al., 1974). These estimates have utilized data regarding damage probabilities and theoretical procedures for determining the attenuation of ground shaking with distance. These detailed studies have not, however, specifically considered the probability of occurrence of the earthquakes that might cause losses of this magnitude.

If a loss threshold defining a catastrophe can be identified, then the procedures discussed in Section 9.2 may be used to compute the probability that the threshold will be exceeded. To illustrate this approach, consider a hypothetical city with 100 buildings all having the same damage resistant characteristics and all experiencing the same ground shaking if an earthquake occurs. A 'catastrophe' might be defined as an event in which at least $n$ buildings 'fail', where failure of a building would be a damage state involving a major threat to life: e.g., collapse or even total damage. From the standpoint of government officials or the public, $n$ would be some small number (say, one to five).

If $p = F_A(a)$ is the probability that one building fails at an acceleration level $a$, then (assuming independence) the probability that $n$ buildings fail is given by the binomial formula:

$$P[N_f = n|p] = \frac{100!}{n!(100-n)!} p^n (1-p)^{100-n} \qquad (9.36)$$

The probability that $n$ or more buildings fail is then:

$$P[N_f \geqslant n|p] = 1 - P[N_f = 0|p] - \ldots - P[N_f = n-1|p] \qquad (9.37)$$

Eq. 9.37 is plotted in Fig. 9.13 for several values of $p$ and $n$. For example, if $p = 0.05$, it is virtually certain that one or more buildings will fail, and it is about 50% probable that at least five buildings will fail. With these results it is then possible to construct a function $F_{NA}(n, a)$, giving the probability that $n$ or more buildings will fail as a function of the level of ground shaking. If, for example, the dependence of $p$ on $a$ is given by $F_A(a)$ in eq. 9.12, then $F_{NA}(n, a)$ is as shown in Fig. 9.14.

The function $F_{NA}(n, a)$ can be used in eq. 9.20 (or a similar equation involving acceleration) to compute the mean annual rate of catastrophes, and to compute (using eq. 9.6) the probability of one or more catastrophes during an interval of $T$ years. Because $F_{NA}(n, a)$ rises steeply, the probability that at least a few buildings (out of many) will fail may be essentially equal to the probability that the 'design acceleration' $a_0$ is exceeded. This will be true unless the exponent $k$ in the seismic hazard equation is very large.

When developing building code requirements for a large region, a building official (such as a state official) will be influenced by the probability of a catastrophe anywhere in the region. In some cases, the population in the region may be concentrated in a few cities, each so small in area that an entire city experiences the same level of ground shaking during a given earthquake and all so widely dispersed that there is negligible risk that more than one city will be seriously shaken during any one earthquake. In such cases, the overall probability-of-catastrophe is one minus the product of the marginal probabilities that each of the particular cities has no disaster. For example, take the case of five cities all having the same number and type of buildings and the same seismic hazard, where the mean annual rate of catastrophic events is $10^{-3}$ per year for each city. The probability that any particular city will experience a catastrophe during 50 years is:

$$1 - \exp[-(50)10^{-3}] = 0.049$$

The annual probability that no city experiences a catastrophe is $(1 - 0.049)^5$

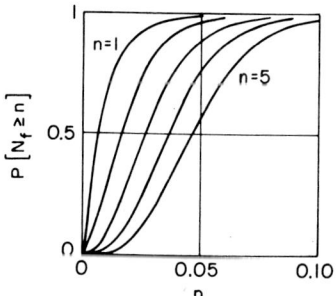

Fig. 9.13. Probability of failure of at least $n$ out of one hundred buildings, given failure probability $p$ for one building.

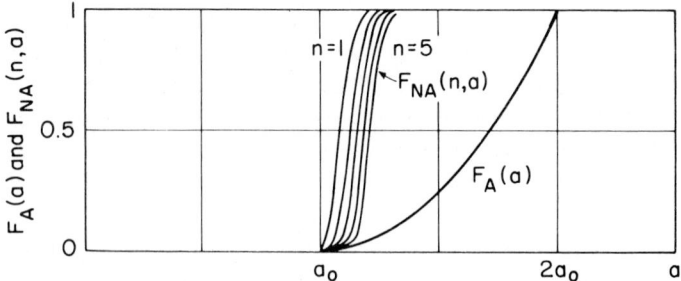

Fig. 9.14. Probability of failure of at least $n$ out of one hundred buildings, using eq. 9.12 for a single building.

= $\exp(-0.25)$. The probability that at least one city experiences a catastrophe during 50 years is $1 - \exp(-0.25) = 0.221$. This is slightly less than five times the corresponding probability for any individual city.

If the cities are located close enough together so that the very large, potentially damaging earthquake affects more than one, then the independence assumption leading to the simple product is in error and the estimate in the previous paragraph is too large. It then becomes necessary to use the principles discussed in Section 9.4.

## 9.7 DESIGN FOR SEISMIC HAZARD: LIFELINES

The word 'lifeline' refers to transportation, communication, power, fuel, water, and sanitation systems. Earthquake engineering as applied to such facilities has lagged behind earthquake engineering as applied to buildings, and during the 1971 San Fernando earthquake there were massive failures in lifeline systems even though the performance of buildings during that earthquake was generally good. Lifelines typically are geographically dispersed and often are complex systems in the sense that they possess redundancy; that is, failure of one component does not necessarily mean failure of the entire system. Another important feature of lifelines is that planning for restoration of service is an alternative strategy that must be considered when optimizing the seismic resistance of a system. While general risk-oriented guidelines have evolved for buildings (cf. Section 9.6), as yet similar, clearly-stated guidelines have not been developed for lifelines.

### 9.7.1 Possible performance criteria

One general concept adapted from design of buildings also applies to lifelines: during the largest earthquake there may be damage but there should be only a small risk of a failure that will endanger lives. This principle applies to water storage structures in developed areas, to large bridges and perhaps to gas pipelines. However, it is easy to envisage many 'failures' of lifelines that will cause great human suffering and economic loss even though the direct risk of death is very small. Perhaps the best example is loss of water for fighting fires and the attendant possibility that a rapidly spreading fire might develop. Other examples are: interruption of local traffic networks required for rescue and relief; blockage of major transportation facilities required for bringing in supplies; loss of drinking water and attendant sanitation problems; etc. It probably is not feasible to design lifeline systems to prevent all such 'failures' anywhere in a system, and guidelines are needed to indicate acceptable levels of failure.

Table 9.XI (Whitman, 1974) is a first attempt to suggest a possible set of guidelines. (See also Duke and Moran, 1975.) With reference to this table, it is difficult to provide a simple definition of 'major' and 'moderate' earthquakes, precisely because a lifeline system extends over a large area. At this stage it is not clear which type of earthquake might be more damaging to lifelines: a very large earthquake at intermediate distance which might have a significant effect on a large part of the system, or a moderate earthquake with a very intense effect upon a small part of the system.

It seems impractical to prevent all loss of service during a major earthquake, and even during a moderate earthquake. This conclusion leads to the concept of acceptable loss of service. The service lost during a major earthquake might be restored in several stages, with minimal necessary service provided quickly and full service restored over several months.

TABLE 9.XI

Possible general design guidelines for lifelines

| Category of lifeline | | Major earthquake | Moderate earthquake |
|---|---|---|---|
| | | intense ground motion or faulting in some part of system | moderate ground motion in some part of system |
| Water supply | water storage reservoir | no failure that will endanger lives | no damage |
| | local sources of water for fire-fighting | adequate supplies remain available | damage level B |
| | distribution systems | damage level A[1] | |
| Highway system | bridges, overpasses | no collapses | no structural damage |
| | roadways | damage level A | damage level B |
| Electrical system | | damage level A | damage level B |
| Gas system | | damage level A, but no contribution to fires | damage level B |

Damage level A: No more than 20% of area without service; service fully restored within one month (within one week for damage level A[1]).
Damage level B: No more than 1% of area without service; service fully restored within hours.

Example 9.13 illustrates a situation in which a simple performance criterion applies.

### 9.7.2 Modeling of lifeline systems

In general a lifeline system is a complicated, interconnected network, but for most risk analyses considerable simplification will be necessary. The way in which these simplifications should be made depends much upon the type of criteria used to judge the adequacy of network performance.

It appears useful to think of three types of simplifications: (1) treat key facilities as point 'targets' without worrying about details of connectivity; (2) approximate main features of network by simple coarse networks; and (3) treat fine networks as area 'targets'.

*Key facilities.* Many systems will involve a small number of facilities which are especially important to the overall functioning of the system as a whole. Examples would be generating stations in electric power systems, major pumping stations in water supply systems, etc. Example 9.13 treats such a case.

Generally a lifeline system can still function reasonably satisfactorily if only one of a number of key facilities is lost, especially if the facility is put out of service for only a limited period of time. For example, generating stations (or at least units in a station) are routinely shut down for maintenance. Thus, having a single key facility 'fail' during an earthquake does not necessarily mean that the entire system would fail. If, however, several such key facilities are lost unexpectedly and simultaneously (i.e., during the same earthquake), then the remaining key facilities might be unable to sustain operation of the overall system.

Once key facilities have been identified, one can perform a risk analysis to determine the annual probability that one facility will fail during some earthquake, that two targets will fail during the same earthquake, etc. Such estimates are useful input to decisions regarding the expenditure of funds to decrease system vulnerability to earthquakes.

*Simple networks.* Generally a lifeline system will contain a coarse, and in some instances reasonably simple, network of major arteries (highway system) or supply lines (high voltage lines for electrical system, major pipelines for water supply system). With simple networks, it is feasible to analyze the state of the network following all possible earthquake events and to determine whether overall system performance criteria have been fulfilled.

The simplest case of a network is a series system (Fig. 9.15b). A typical performance requirement would be: it must be possible to pass from one end of the system to the other end. Then an earthquake-induced blockage at any point of the system would mean an overall system 'failure'.

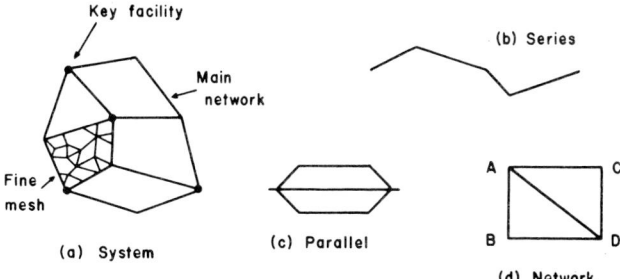

Fig. 9.15. Decomposition of system (a) into key facilities, main networks (b through d) and fine mesh.

Another simple case is a parallel system (Fig. 9.15c). If the performance criterion is being able to pass from one end to the other, then overall system failure occurs only if all branches become impassable. If the performance criterion is a minimum capacity to transmit electricity or water or vehicles, then overall system failure would occur when some fraction of the total number of branches is blocked.

Fig. 9.15d shows a simple form of a cross-linked system. Here there are a number of possible performance criteria: being able to pass from $A$ to $C$; being able to pass from $A$ to both $C$ and $D$; being able to pass from $A$ to either $C$ or $D$; being able to pass from either $A$ or $B$ to either $C$ or $D$; etc. Panoussis (1974) and Taleb-Agha (1975) have outlined how such problems may be decomposed into a series of problems involving simple series and/or parallel systems.

*Fine networks.* Many lifeline systems will involve, in some sense, a fine network of elements: the total system for moving vehicles includes city streets as well as the main arteries and a water supply system involves the street laterals in addition to the main supply lines. These fine networks typically are highly redundant, and any analysis as a network for risk purposes is computationally not feasible.

For many purposes, existence of the fine mesh may have relatively little effect upon the post-earthquake performance of the main arteries or supply lines. The system of city streets, which following an earthquake will be clogged by debris and people, would be a very poor substitute for freeways from the standpoint of moving emergency vehicles. Similarly, the street laterals in a water distribution system would be ineffective, as substitutes for main trunk supply lines, in carrying water between distant parts of a metropolitan area.

The performance of each square mile of fine mesh primarily affects the people in that square mile. Each square mile of fine mesh may therefore be regarded as an individual 'target', disconnected from all other such targets. If, however, too much of the fine mesh failed, then the ability of the overall system to accomplish the required repairs would be overtaxed. A possible measure of performance for the fine network of a lifeline system is therefore the total geographical area in which extensive damage occurs.

*Example 9.13*

In the United States, two levels of earthquake shaking are defined for the design of nuclear power plants. The smaller level of shaking is called the 'operating basis earthquake' (OBE). Regulations require that the stresses calculated using the OBE be less than normal working stresses, and that if the OBE is exceeded the plant must be shut down for inspection. There are two conflicting implications in the selection of the OBE. If the OBE is made too large, it may govern the design of some portions of the plant, with a resultant increased cost to the utility company building the plant. If the OBE is made too small, the probability of shutdown for inspection is increased, with resultant costs to the utility company and with a possible decrease in the level of service to the public.

The cost, to the utility and the public, of shutting a single plant in a power system down for post-OBE inspection is rather small. Scheduled shutdowns occur regularly in power systems, and unscheduled shutdowns are far from rare. On the other hand, if several plants experienced shaking exceeding

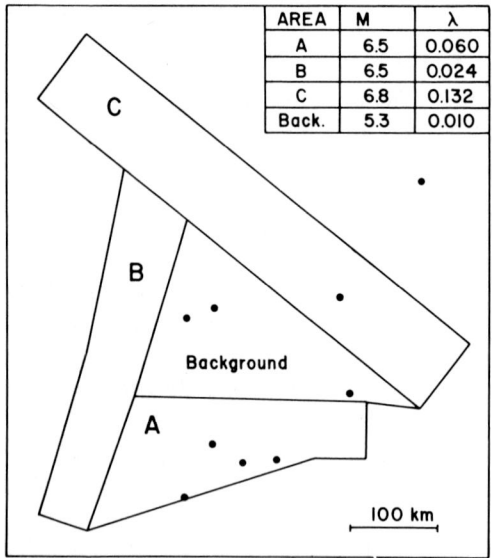

Fig. 9.16. Map of nine sites with source areas and parameters.

TABLE 9.XII

Probability that one or more sites must shut down following a single earthquake

| No. sites | $\geqslant 1$ | $\geqslant 2$ | $\geqslant 3$ | $\geqslant 4$ | $\geqslant 5$ | $\geqslant 6$ |
|---|---|---|---|---|---|---|
| Annual probability | $1.1 \cdot 10^{-2}$ | $3.0 \cdot 10^{-3}$ | $9.1 \cdot 10^{-4}$ | $1.7 \cdot 10^{-4}$ | $4.7 \cdot 10^{-7}$ | 0 |

the OBE during the same earthquake, then the necessary inspection shutdowns might mean a serious decrease in service. Therefore, when setting the OBE level, it is desirable to know the probability that the OBE might be exceeded at several plants during the same earthquake. Fig. 9.16 depicts a hypothetical situation involving nine nuclear plants distributed through a region. The earthquake source zones and maximum magnitudes ($M$) and mean seismic rates (of events of magnitude 4 or more) $\lambda$ for each source zone, are shown in the figure. Typical magnitude distributions and attenuation laws have been used. Table 9.XII gives annual probabilities that varying numbers of plants will 'fail' during the same earthquake (i.e., simultaneous shutdown for inspection will be necessary) when the OBE is set at 32 cm/sec$^2$ (0.083$g$). The analysis was made using the principles described in Section 9.4.

REFERENCES

Applied Technology Council, 1974. *An Evaluation of a Response Spectrum Approach to Seismic Design of Buildings.* Report by Applied Technology Council (San Francisco, Calif.) to Center for Building Technology, National Bureau of Standards.
Blume, J.A. and Monroe, R.E., 1971. *The Spectral Matrix Method of Predicting Damage from Ground Motion.* Report to Nevada Operations Office, US AEC, by John A. Blume and Associates Research Division.
Borges, J.F. and Castanheta, M., 1971. *Structural Safety.* Lab. Nac. de Eng. Civil, Lisbon.
De Neufville, R., 1975. How should we establish public policy on setting design codes for performance standards? *Proc. U.S. Natl. Conf. Earthquake Eng.*, Earthquake Engineering Research Institute, p. 327.
De Neufville, R. and Marks, D., 1974. *Systems Planning and Design: Case Studies in Modeling, Optimization, and Evaluation.* Prentice-Hall, Englewood Cliffs, N.J.
Duke, C.M. and Moran, D.F., 1975. Guidelines for evaluation of lifeline earthquake engineering. *Proc. U.S. Natl. Conf. Earthquake Eng.*, Earthquake Engineering Research Institute, p. 367.
Esteva, L., 1968. Bases para la formulación de decisiones de diseño sísmico. Natl. Univ. of Mexico, Inst. Eng., Rep. No. 182.
Esteva, L., 1970. Seismic risk and seismic design decisions. In: R.J. Hansen (editor), *Seismic Design for Nuclear Power Plants.* MIT Press, Cambridge, Mass.
Friedman, D.G., 1968. *Simulation of Earthquake Loss Experience on Dwelling Structures.* The Travelers Insurance Co., Hartford, Conn.

Grandori, G. and Benedetti, D., 1973. On the choice of the acceptable seismic risk. *Earthquake Eng. Struct. Dyn.*, Vol. 2.

Housner, G.W. and Jennings, P.C., 1973. Problems in seismic zoning. *Proc. 5th World Conf. Earthquake Eng.*, Rome.

Jacobson, S.E., Torabi, M. and Bonsal, P.P., 1973. *On the Consideration of Earthquake Risk in Design of Water Resource Systems*. Univ. of Calif. Water Resource Center, UCLA, Los Angeles, Calif.

Liu, S.C. and Neghebat, F., 1972. A cost optimization model for seismic design of structures. *Bell System Tech. J.*, 51(10): 2209—2225.

Milne, W.G. and Davenport, A.G., 1969. Distribution of earthquake risk in Canada. *Bull. Seismol. Soc. Am.*, 59(2): 729—754.

Molchan, G.M., Keilis-Borok, V.I. and Vilkovich, G.V., 1970. Seismicity and principal seismic effects. *Geophys. J. R. Astron. Soc.*, 21: 323—335.

N.R.C., 1975. National Building Code of Canada. National Research Council, Ottawa, Canada.

O.E.P., 1972. *A Study of Earthquake Losses in the San Francisco Bay Area*. Report prepared by Natl. Oceanic and Atmosph. Admin. for the Office of Emergency Preparedness.

Panoussis, G., 1974. Seismic reliability of lifeline networks. *Dept. Civ. Eng. Rep.*, MIT, R74—57.

Ravindra, M.K. and Lind, N.C., 1973. Theory of structural code optimization. *J. Struct. Div., Proc. ASCE*, 99(ST7): 1541—1553.

Schiff, A.J., Newsom, D., El-Abidd, A.H. and Yao, J.T.P., 1974. The use of simulation to evaluate the effects of earthquakes on lifelines with application to electric power systems. *Newslett. Earthquake Eng. Res. Inst.*, 8(2).

Scholl, R.E. and Farhoomank, I., 1973. Statistical correlation of observed ground motion with low-rise building damage. *Bull. Seismol. Soc. Am.*, 63(5).

Structural Engineers Assoc. of California, 1960. *Recommended Lateral Force Requirements and Commentary*. Subsequent versions have revised details, but the aims remain as originally stated in 1960.

Taleb-Agha, G., 1975. Seismic risk analysis of networks. Report 21, Seismic Design Decision Analysis. *MIT Dep. Civ. Eng., Publ.*, R75-43.

USCGS, 1969. *Studies in Seismicity and Earthquake Damage Statistics, 1969*. U.S. Dept. of Commerce, Environmental Service Administration, Coast and Geodetic Survey. See especially Appendix A by Steinbrugge, McClure and Snow.

Vanmarcke, E.H., 1969. First-passage and other failure criteria in narrow-band random vibration: A discrete-state approach. *Dep. Eng. Rep.*, MIT, R69-68.

Von Neumann, J. and Morgenstern, O., 1953. *Theory of Games and Economic Behavior*. Princeton University Press, Princeton, N.J.

Whitman, R.V., 1973. Damage probability matrices for prototype buildings. *Dept. Civ. Eng. Rep.*, MIT, R73-57.

Whitman, R.V., 1974. Risk-based seismic design criteria for lifelines. *ASCE Natl. Meet.*, Los Angeles, Calif. Prepr. 2148.

Whitman, R.V., Biggs, J.M., Brennan, J.III, Cornell, C.A., De Neufville, R. and Vanmarcke, E.H., 1974. Seismic design decision analysis: Methodology and pilot application. *Dep. Civ. Eng. Rep. MIT*, R74-15.

Whitman, R.V., Biggs, J.M., Brennan, J.III, Cornell, C.A., De Neufville, R. and Vanmarcke, E.H., 1975. Seismic design decision analysis. *J. Struct. Div., Proc., ASCE*, 101(ST 5).

Wiggins, J.H., Hirschberg, J.G. and Bronowicki, A.J., 1974. *Budgeting Justification for Earthquake Engineering Research*. J.H. Wiggins Co. Tech. Rep. No. 74-1201-1, Redondo Beach, Calif.

Chapter 10

SEISMOLOGICAL INSTRUMENTATION

THOMAS V. McEVILLY

*Department of Geology and Geophysics, University of California, Berkeley, Calif., U.S.A.*

10.1 INTRODUCTION

This chapter presents general information and guidelines hopefully useful in selecting instruments for seismological research and engineering applications. Two main aspects are considered: (1) defining the measurement requirements in a particular situation; (2) understanding the principles and capabilities of available sensing, signal conditioning, and recording systems and the degree to which requirements can be met by available equipment. Emphasis is on widely used conventional seismological equipment. Special devices, such as strain gages, seismoscopes, and digital acquisition systems are mentioned where appropriate, but not treated in depth.

10.2 APPLICATIONS

The following stand out among the many applications of seismological instruments in the assessment of seismic risk and in providing bases for the making of engineering decisions.
— Seismotectonic investigations: use of earthquake source parameters and spatial distribution of hypocenters to infer local or regional tectonic characteristics such as principal stress directions, extent of potential earthquake zones, deformation rates, depth range of crustal or upper-mantle earthquakes, and temporal variations of earthquake source parameters or medium properties.
— Monitoring of fault creep: detection and measurement of long-term aseismic slip on potentially active recent fault traces.
— Earthquake prediction research: monitoring temporal variations in earthquake characteristics and physical properties in the source region.
— Short-range seismicity studies: use of portable or temporary equipment in gathering spatial and temporal data on occurrence of small local earthquakes, detecting and mapping active faults, or monitoring effects of reservoir loading behind dams.
— Shallow subsurface investigations: use of a variety of active or passive seis-

mological methods in measuring soil properties, overburden thickness, lateral variation in near-surface geology, depth of water table, etc.
— Aftershock sequence studies: delineation of aftershock sequences, the nature of faulting, local grounds effects, sequence decay rates, etc.
— Strong-motion recording: detection and broadband recording of moderate to strong ground motion for specifying bases of structural design in seismic regions.
— Determination of dynamic properties of structures: measurement of structural responses to static or dynamic perturbations including, among the latter, impulsive excitation of free vibrations, forced harmonic or quasiharmonic excitation, ambient disturbances, blasts, and earthquakes.
— Model studies: monitoring of shaking-table motions and of the responses of models to dynamic excitation.

These applications require the measurement of motion or force either at a point (e.g., an accelerometer) or between two points (e.g., a strain gage). We shall concentrate on the former class of requirement and will thus treat the conventional inertial seismometer in some detail. Strain monitoring techniques for structures or across fault zones are well-covered in specialized texts. A series of papers presented at the 5th World Conference on Earthquake Engineering (1973) provide an excellent review of earthquake instrumentation for recording strong motion.

## 10.3 REQUIREMENTS: GENERAL

The nature of a particular application will usually specify the type of sensor required, the frequency range of interest, the accuracy and resolution of the record produced, and any requirements for possible interconnection of instruments, remote data telemetry, or provisions for automatic processing. Additional constraints are always present in the available budget as well as in the operational and maintenance conditions.

An important early decision involves recording requirements. At times a simple measurement of the peak value is sufficient — in other applications the time history of the parameter variation must be retained. Ordinarily direct measurement of peak values is far less expensive than time-based recording. We consider next a group of special-purpose instruments, including peak-reading devices, and follow with a more extensive discussion of widely used conventional inertial seismometers and seismographs.

## 10.4 PEAK-READING INSTRUMENTS

### 10.4.1 Peak ground motion

The information contained in the peak value of a ground-motion parameter (maximum displacement, velocity, or acceleration) is usually insufficient for much analysis beyond threshold monitoring or event counting. Instruments measuring such peak values serve frequently as complements of standard seismographs capable of supplying full time histories of ground motion. In some applications, however, peak data are of great value. A prime example is the simple low-cost shock (acceleration) indicator widely used to monitor the vibration environment during commercial transportation of delicate equipment. Another application is the peak velocity indicator used to monitor damage potential of vibrations from blasting and heavy construction. Simple automatic counting of microearthquakes larger than a given magnitude also provides a valuable activity indicator for volcanoes.

For strong-motion recording, interest is ordinarily centered on the peak horizontal ground acceleration or the peak responses of oscillators with specific periods and damping ratios. A precise measurement of peak ground acceleration is not sought. In keeping with the goals of low-cost and simplicity, and in view of the unsophisticated analysis of the resulting data, a rough estimate of peak acceleration is usually adequate. This can be obtained from several types of devices.

A falling-pin set is a primitive vibration-indicating instrument consisting of a set of slender, rigid prismatic rods of different heights, standing on a horizontal surface. Roughly, the rod slenderness determines the horizontal acceleration required to topple it. If $b$ = diameter of rod base, $h$ = rod height, and $g$ = gravity, a static horizontal acceleration of $bg/h$ is required to topple the rod. While the peak earthquake acceleration actually required to topple the rod in an earthquake may differ substantially from this value, by noting which rods have toppled over we get a gross idea of a range in which lies the maximum horizontal acceleration. The maximum and minimum values of peak ground acceleration in which we are interested determine the range of rod slenderness and the precision with which we wish to bracket the peak acceleration governs the number of rods used.

A seismoscope is an instrument which records the occurrence of an earthquake. A widely used model due to Wilmot (shown in Fig. 10.1) is a damped two-degree-of-freedom oscillator capable of recording the pendulum trajectory in the horizontal plane (see Hudson, 1958). Recording by stylus on a smoked watch glass, the instrument traces a hodogram representing the response of a lightly damped pendulum to the earthquake excitation. A reasonable model of a typical structure is embodied in a natural period of 0.75 sec and a damping factor of 0.1 critical. Having no time scale, the seismoscope record cannot be reduced to ground motion, i.e., the same record can be produced by an infinite family of widely differing ground motions. * At high-frequency excitation, the seismoscope trace deflection is proportional to

---

\* It is sometimes possible to utilize the parasitic 15—18 Hz oscillation of the suspension, excited by sharp motion and visible in the trace, as a timing reference on seismoscope records (see Scott, 1973).

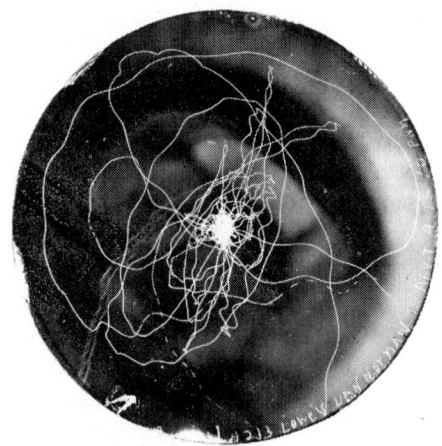

Fig. 10.1. Wilmot seismoscope (left, courtesy Kinemetrics, Inc.) and seismogram written in 1971 San Fernando earthquake (courtesy Seismological Field Survey, NOAA).

ground displacement, while at low frequencies it is proportional to ground acceleration. For any excitation, however, it fulfills the intended purpose of providing a point (0.75 sec period, 0.1 damping) on the response spectrum for the earthquake motion recorded. A series of such seismoscopes with different periods and damping can be used to obtain several points on the response spectrum directly.

Seismoscopes with natural frequencies above about 20 Hz produce, for typical earthquakes, records with amplitude directly proportional to ground acceleration and can thus be calibrated as peak-reading accelerometers. Various designs exist based on different types of oscillation, e.g., pendulums with stylus and smoked glass, cantilevers with magnetic indication of maximum deflection, elastic spheres writing ink impressions on a confining box upon flattening due to acceleration, etc.

*10.4.2 Peak structural motion*

When conducting harmonic-excitation tests of a structure, often only the motion amplitudes at different points are of interest for various excitation frequencies. The oscillations are usually of sufficiently low frequency and large amplitude; they can be measured directly at the seismometer output. When the structure is man-excited, visual observation of the seismometer output allows the operators to control their swaying. For earthquake excitation, any of the devices described in the previous section can, of course, be installed at various points within the structure.

In cases where the structure's damping ratio is to be measured, an accurate record of the decay of free oscillations must be made and peak-reading devices are inadequate. With modern instrumentation it should be easy to obtain natural frequencies and damping factors for structures to an accuracy of a few percent — quite adequate for such measurements within the linear-response range of structures.

One of the most significant structural response parameters in earthquakes is the maximum acceleration developed at selected points. Using the peak-reading accelerometers discussed in the previous section, such measurements should be possible with errors of 10% or less.

*10.4.3 Peak structural deformations*

We are often interested in measuring a building's maximum interstory relative displacements and the strains at selected sections of structural members or at joints. The former have been successfully recorded by using the device depicted in Fig. 10.2. It consists of a diagonal bar extending the whole story height, connected at the lower end to a lever that magnifies relative displacements five-fold. The lever has a pen that records on a drum possessing a clock mechanism which requires rewinding once a week. Records are obtained on paper replaced when rewinding the clock mechanism. The reason for the rotating drum in lieu of a stationary scratch device is that temperature changes and wind effects cause some story drifts which it is desirable to isolate from earthquake response. The latter is confined to such a short time interval on the record that it is impossible to derive a response-history record

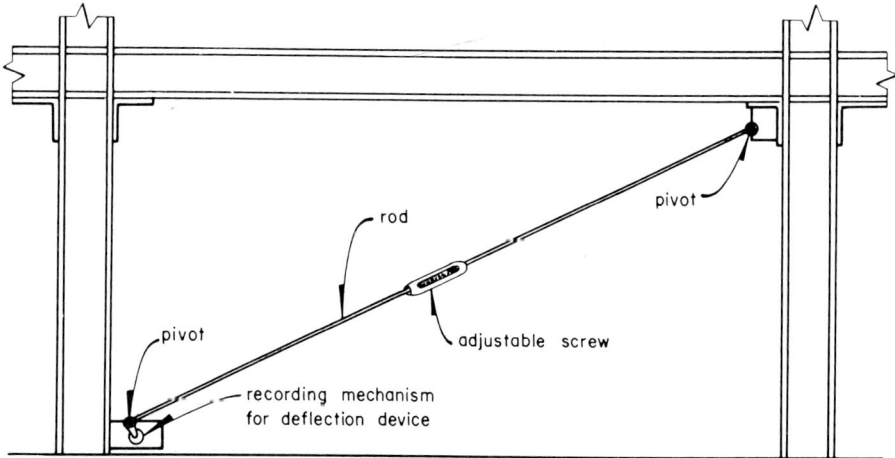

Fig. 10.2. Structural strain recording device. Diagonal element detects relative motion across entire story in building (after Zeevaert and Newmark, 1956).

Fig. 10.3. Scratch strain gage. The circular target, 2.5 cm in diameter, is rotated by the strains. End attachment plates may be separated several meters (courtesy Prewitt Associates).

but the accuracy with which maximum drifts are read is quite sufficient for practical purposes.

When strains due to phenomena other than earthquakes are negligible relative to earthquake-caused strains, a very convenient, inexpensive, and compact instrument is the scratch strain gage, shown in Fig. 10.3. This is essentially a bar connecting two fixed plates attached to members up to several meters apart. A simple magnifying device, ending in a stylus which scratches on a smoked glass or metal plate, is incorporated in the connecting rod and provides a permanent record of the strain. The record ordinarily is read through a microscope or a powerful magnifying glass.

Although the instruments described provide only peak values and their accuracy is not high, their costs are so much less than those of more conventional and complex equipment that a great many can be installed in a structure for the price of a single sophisticated instrument. It would seem that there is much to recommend the use of these simple instruments.

## 10.5 CONVENTIONAL SEISMOGRAPHIC SYSTEMS – DESIGN CONSIDERATIONS

### 10.5.1 Basic design parameters

*Bandwidth*, *sensitivity*, and *dynamic range* are the basic parameters one must define in specifying a seismographic system. A difficulty lies in their interdependence. Bandwidth is the frequency interval of greatest interest; dynamic range is the span of the largest (full scale) to the smallest (system noise) usable signal recorded and is expressed as the ratio (often in dB) of these two signals; and sensitivity refers to the absolute gain (magnification) of the seismograph within the range. Generally all three cannot be optimized simultaneously in a reasonably simple instrument. For example, the special, high-gain, long-period seismographs used in the detection of teleseismic surface waves from small nuclear explosions and earthquakes exhibit very high gain (ca. $10^5$ magnification) in a very narrow frequency band around 40 sec period (0.02—0.02 Hz). The same long-period seismograph, operating at the same site, but in the conventional WWSSN (world-wide seismographic station network) long-period mode (15—100 sec) would be capable of operating at maximum magnification of the order of $3 \cdot 10^3$, giving the same amplitude of background motion on a record as that produced by the high-gain configuration. On the other hand, if we could faithfully record the basic seismometer input with a dynamic range of $10^5$, we would capture the information in both of the foregoing configurations and we could produce either record through an appropriate playout.

A second example of parameter control can be seen in the recording methods commonly used for local earthquake motion. The bandwidth required for recording strong ground motion near the source of a major earthquake is essentially the same as that used in microearthquake recording — about 0.1—30 Hz. The actual ground motion, however, may be several centimeters for the major earthquake and several nanometers ($10^{-9}$ m) for the microearthquake, a ratio of $10^7$. A single instrument designed to do both measuring tasks simultaneously would demand a dynamic range of $10^7$, or 140 dB (20 dB is a fator of 10). This design near-impossibility gives rise to the two different system types in common use.

*10.5.2 Bandwidth*

Bandwidth can usually be specified early-on in the consideration of a seismographic system design. The required frequency range of ground motion is generally clear from the intended use of the data. Modern instruments allow great flexibility in this selection. However, if the recording medium is a visible record (as opposed to digital or analog magnetic tape to be processed later), one must take care not to attempt extreme wideband recording, as the resulting earthquake records would be of ragged appearance and difficult to interpret. Background noise amplitude also increases with bandwith, often obscuring small events that could be seen clearly on a visible record from a narrower-band instrument. An obvious example is the 6—8 sec microseism band, which is effectively avoided in the conventional short-period/long-period seismographs employed in most observatories.

If both bandwidth and visible records are required, it is often possible to split the output from a single seismometer into several desirable frequency bands, with individual gain adjustments, and to record the bands separately. A pertinent example lies in the use of the commercially available force-balance accelerometer for moderate-to-strong-motion recording. Such units exhibit flat response to ground acceleration over the frequency range 0—50 Hz with the limiting instrumental noise background smaller than $10^{-6}g$ ($g$ = gravity) for high-quality units having a full-scale capability of $1g$. Any desired frequency bands can be derived from its output on visible records at gains consistent with the inherent device output noise. Another illustration of modern wideband sensors is the new series of long-period seismometers capable of operating at natural periods of 50 sec or more in controlled environments. Equipped with appropriate transducers these instruments can produce simultaneous records of solid-earth tides (period of about 12 hours), free oscillations and mantle surface waves (100—3000 sec), conventional long-period waves (5—100 sec), conventional short-period waves (0.2—5 sec), and microearthquake signals (2—20 Hz). This exemplifies the need for separate records of the different bandwidth signals, with individual gains and recording speeds, if visible records are the prime data medium. It is clear that use of digital or analog magnetic tape for primary recording provides a means of recovering the desired bandwidth in postprocessing and thus reduces the requirements for continuous visible monitoring of all bands.

*10.5.3 Sensitivity*

Selection of the operating sensitivity for a seismograph would appear straightforward once the bandwidth is specified. In most applications the seismologist or engineer can define the maximum and minimum signals to be recorded; the interdependence of the fundamental design parameters of sensitivity and dynamic range is apparent here.

Modern high-quality inertial seismometers with moving coil transducers have inherent dynamic range capabilities of $10^5$—$10^6$ (100—120 dB) over which the output is an adequately linear and undistorted representation of the input ground motion. Only the most sophisticated and state-of-the-art recording systems (such as the digital systems in exploration seismology or in large nuclear detection arrays, with 16—19 bit word lengths) approach these values. Conventional seismographs use only a part of this dynamic range of the sensor itself.

A good quality, conventional, visible seismograph has a dynamic range of 40—50 dB (roughly, a measurement resolution in the trace position of about 0.2—0.5 mm and a full-scale swing of 50 mm). The best analog FM magnetic tape recorders exhibit roughly the same dynamic range, but generally over a wider frequency bandwidth. With digital recording the range depends on the number of bits in a data word, while the bandwidth is set by the sampling rate and seismometer response. Ten- to twelve-bit words are commonly used in seismology for dynamic range requirements of about 54 and 66 dB, repectively.

In this discussion we have assumed that the fundamental noise limitation is imposed by the recording systems, i.e., that the inherent seismometer noise (not ground noise) and the noise in the signal-processing elements (galvanometer or amplifier) are at or below the recorder noise level or resolution. This situation should be a design criterion to avoid wasting part of the usable recorder range on system noise.

Having selected the recording medium and its dynamic range, we can now specify the sensitivity required of our overall system. This is often done easily at the small-signal end of the range, where we can usually define the smallest motion we want to detect. Frequently this level is simply set at the local microseismic background level. However, in some applications it is important that earthquakes of a given size be recorded without "clipping" the record. This is particularly true for strong-motion recording where, for example, we may wish to stay on scale up to 1 or 1.5$g$. If the recorder range is not large, this requirement will result in records with no visible microseismic background. This should be considered carefully in setting the system sensitivity. Having established the operating sensitivity, its realization is a simple matter of gain between seismometer and recorder.

Finally, some scheme must exist in order that the system response can be measured in its operating configuration. This is important both in the initial set-up and in the maintenance program where periodic checks of performance are made with the system installed and operational.

*10.5.4 Response curve*

A reasonable estimate of the nature of the motion to be recorded by a proposed system thus allows specification of the fundamental instrument parameters. These characteristics are embodied in the instrument response curve, which contains all the relevant information about the seismograph's behavior under excitation. The common representation of the system response is a magnification curve $M(f)$, where $f$ is frequency, in hertz, of the excitation motion, assumed harmonic. $M$ is the ratio of record amplitude to excitation displacement; it is thus a displacement sensitivity. The record amplitude can be measured in many ways, such as millimeters of trace displacement on a record or film viewer, output voltage from an amplifier or tape recorder on playback, or the numerical value of a digital word generated by an analog-to-digital converter. The denominator in $M$ can equally well be excitation velocity or acceleration. These choices define then $V(f) = M(f)/2\pi f$ = velocity sensitivity, and $A(f) = M(f)/(2\pi f)^2$ = acceleration sensitivity.

Incomplete statements about magnification curves lend themselves to confusion. $M$, $V$, and $A$ curves can apply to the same instrument, differing solely by factors $2\pi f$. It is important that the precise meaning of the ordinates of the magnification curve be made explicit. This source of confusion is related to the practice of referring to some seismographs as "velocity meters" and to others as "accelerometers". This terminology applies to the frequency range of interest. If the $M(f)$ curve is flat in this frequency range, the instrument is termed a "displacement meter"; if $V(f)$ is flat in the range, a velocity meter; and if $A(f)$ is flat, an accelerometer. Figure 10.4 shows a set of generalized $M$, $V$, and $A$ curves for a hypothetical seismograph; this single instrument may be called a displacement meter, velocity meter, or acceler-

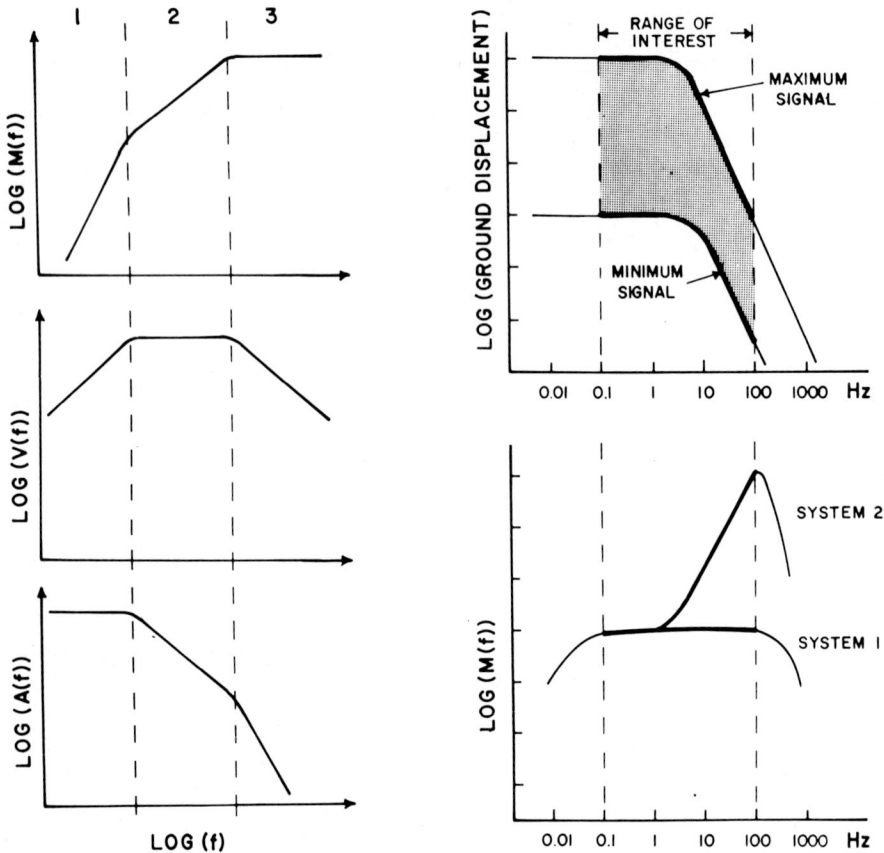

Fig. 10.4. Magnification ($M$), velocity sensitivity ($V$), and acceleration sensitivity ($A$) response curves for a seismograph, generalized to straight line segments. The seismograph may be described as an accelerometer, a velocity meter, or a displacement meter in the frequency ranges 1, 2, and 3, respectively.

Fig. 10.5. Top: Approximate spectral range, 0.1—100 Hz, of ground displacement expected from earthquakes in the magnitude range 2—5. Bottom: Alternate system magnification curves to record shaded signal range in top figure, requiring 100 dB dynamic range ('flat' response of system 1) or 60 dB ('pre-whitened' response of system 2).

ometer, depending on whether we are interested in frequency ranges 3, 2, or 1 respectively.

Analysis of the intended application of a required seismographic system should lead to the design response curve, both in shape and in absolute level. The shape reflects the bandwidth of the motion to be measured and, if possible, it should compensate for any severe frequency-dependence in the expected motion (it should, insofar as possible, "pre-whiten" the input, in the sense of making its output spectrum essentially flat in the frequency range of interest).

The latter consideration is more important the smaller the available dynamic range of the system and the wider the bandwidth desired. The response curve's absolute level is the system sensitivity and must be set recognizing the expected range of the motions to be recorded. There are upper and lower limits to the size of the motions to be recorded (their ratio being the required minimum dynamic range). Figure 10.5 illustrates these concepts with a generalized range of expected earthquake signals and two possible response curves for the bandwidth. The figure displays an earthquake-like displacement spectrum assuming that recording over three orders of earthquake magnitude is required. From the maximum signal at low frequency to the minimum at high frequency there are five orders of magnitude (100 dB). If we attempted to record in accordance with the desired bandwidth and signal range for response curve 1, we would need a resolution of $10^{-5}$ of full scale to detect the minimum signal of interest. After pre-whitening the response (curve 2), we require only a resolution of $10^{-3}$ of full scale to record the same motions. Most visible records offer little more than two orders of magnitude, and the ground-displacement spectrum is typical for earthquakes. Hence it is not surprising that difficulty is encountered in trying to record wideband earthquake ground-motion data (in the 0.05—50 Hz band, say) on systems with limited dynamic range. Even digital recording with system response curve 1 in the figure would need an 18-bit word length for the full $10^5$ range while 12 bits would suffice for the shaped response of curve 2.

Another step in defining the required response curve for a given application lies in identifying potential noise sources, in the ground or in the system, that might limit the sensitivity. The classic example of such a noise problem is the 6—8 sec microseisms so prominent in most seismograms. With amplitudes up to 10 μm during high activity, these ubiquitous surface waves have resulted in the traditional long-period/short-period division in seismometry, which effectively avoids high gain in the troublesome range.

The wide variation often resulting in response curves as we attempt to cover the useful frequency range of earthquake ground motion is illustrated in Figs. 10.6—10.8. The figures display a range of magnification curves of systems operating within the observatories of the University of California Seismographic Stations and are representative of current techniques for wide-range applications. An entire complex of instruments for recording earthquakes in the magnitude ($M$) range of $< -2$ to $> 8$ covers a phenomenal span of ground amplitudes and frequencies: $10^{-10}$ to $10^{-1}$ m or more in displacements and $3 \cdot 10^{-4}$ to $10^2$ Hz in frequencies, where we are excluding earth tides, fault creep, fault rupture, and secular strains at the low-frequency extreme, and high-frequency acoustic waves at the high end.

*10.5.5 Setting specifications*

The application for which a seismographic system is to be designed defines the key instrumental parameters required to furnish the data sought, as we

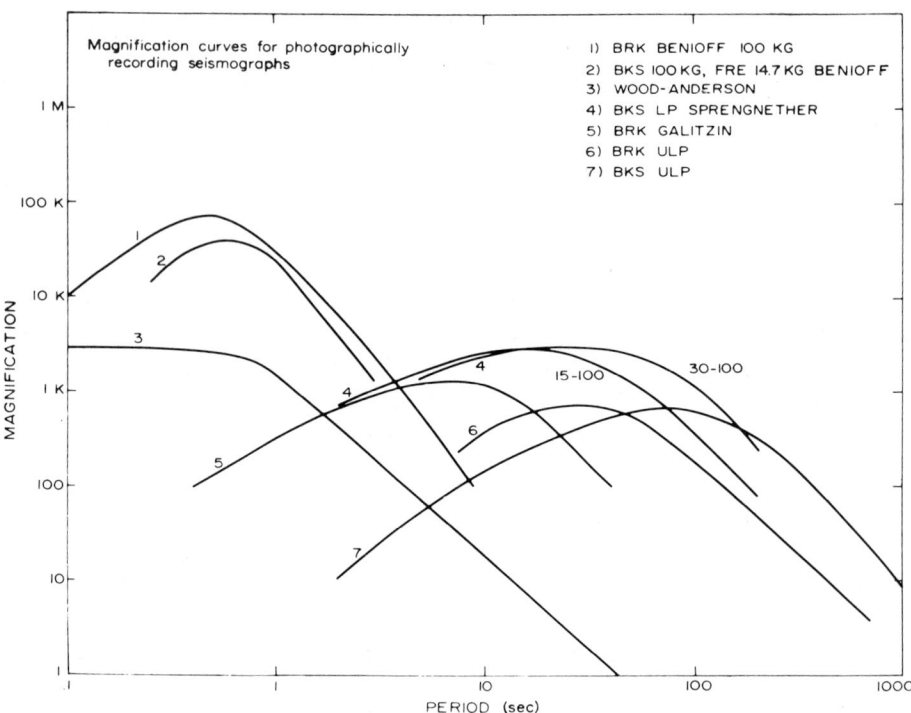

Fig. 10.6. Suite of magnification curves applicable to photographically recording seismographs in station network of University of California, Berkeley. Note wide range of maximum magnification and period of the maximum, from the conventional short-period system (curve 1) to the ultra-long period (ULP) system (curve 7).

have discussed: bandwidth, sensitivity (with possible variations over the bandwidth), and dynamic range. These result in a response curve, the absolute sensitivity levels, the required resolution at the small-signal extreme, and the full-scale capability. Bandwidth and sensitivity parameters are matters of mechanical and electrical characteristics of the seismometers and signal conditioning devices. Dynamic range, along with timing resolution and constraints on record duration format, sets additional requirements on the recording system.

The most important admonition in setting specifications for seismographic systems is to ascertain as completely as possible the nature of the ground motions we wish to record. Manufacturers of seismological equipment should not be placed in the position of interpreting requests for quotations on "short-period seismograph systems", for a "strong-motion sensor", or a broadband seismograph". The result can be a purchased system that does not do the job or that is far more elaborate and thus more expensive than needed.

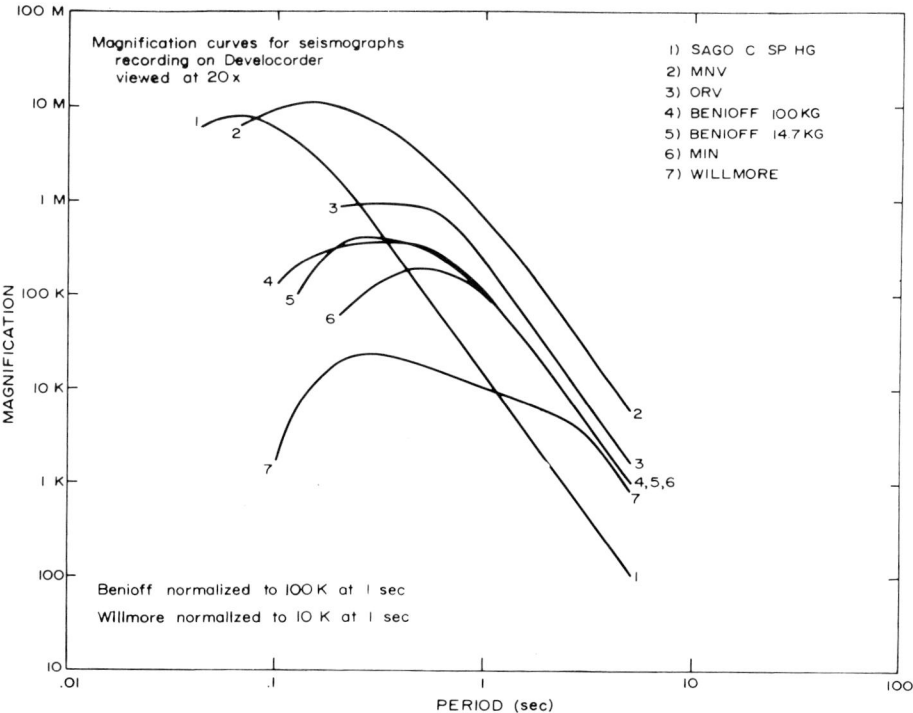

Fig. 10.7. Suite of magnification curves applicable to telemetered short-period seismographs in the University of California network which record on a single 16-mm film recorder.

A common mistake is the assumption that one should apply a "safety factor" of, say, two to the requirements in all directions to be sure of getting the system actually needed. This is seriously objectionable because seismographic design does not use a continuous spectrum of building-block components. The components are widely spaced in capabilities and in cost, and a factor of two in sensitivity and bandwidth may necessitate major steps in several components of the system and imply an order of magnitude cost increase.

When the particular application allows some flexibility in system parameters it is advisable to request quotations on packages designed for the extremes in the acceptable range. The manufacturer usually also suggests an intermediate system as optimum. On the other hand, if one knows precisely the required instrumental parameters for the application, it is well to state them exactly from the beginning. This will ordinarily insure that the most economical design will be adopted.

In this discussion of system specification we must also mention the existence of "standard" seismographs — widely used systems developed in re-

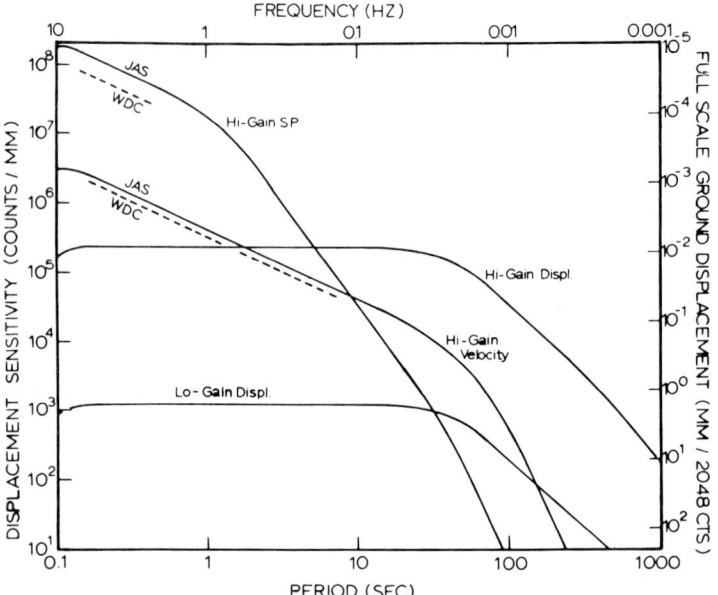

Fig. 10.8. Suite of displacement sensitivity curves applicable to telemetered broadband seismographs in the University of California network which record on FM magnetic tape at Berkeley and are digitized on playback. JAS, WDC are station names; SP is short period channel; Displ., Velocity indicate transducer used. The four response types shown are taken simultaneously from a single seismometer by appropriate selection of transducer, amplifier gain, and filter.

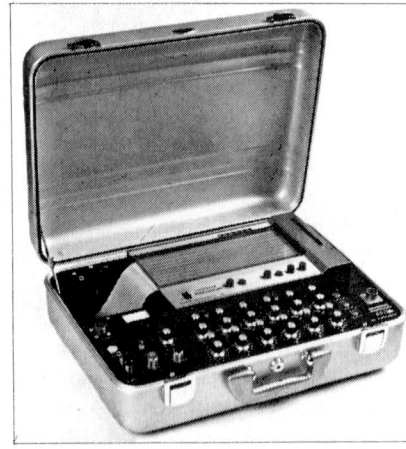

a  b

Fig. 10.9a and b. For caption see pag 396.

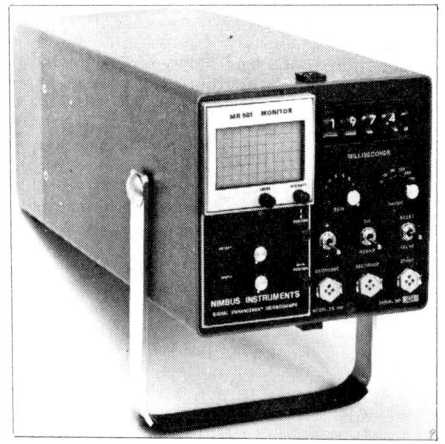

c

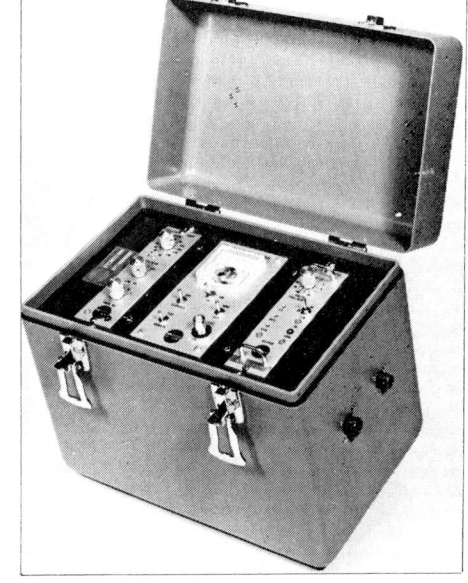

d

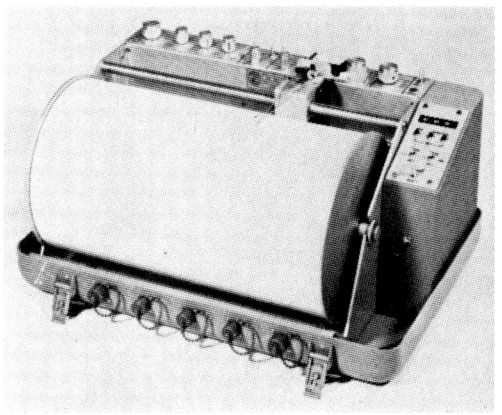

e

f

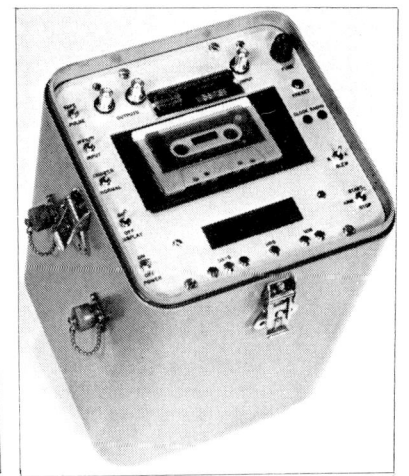

g

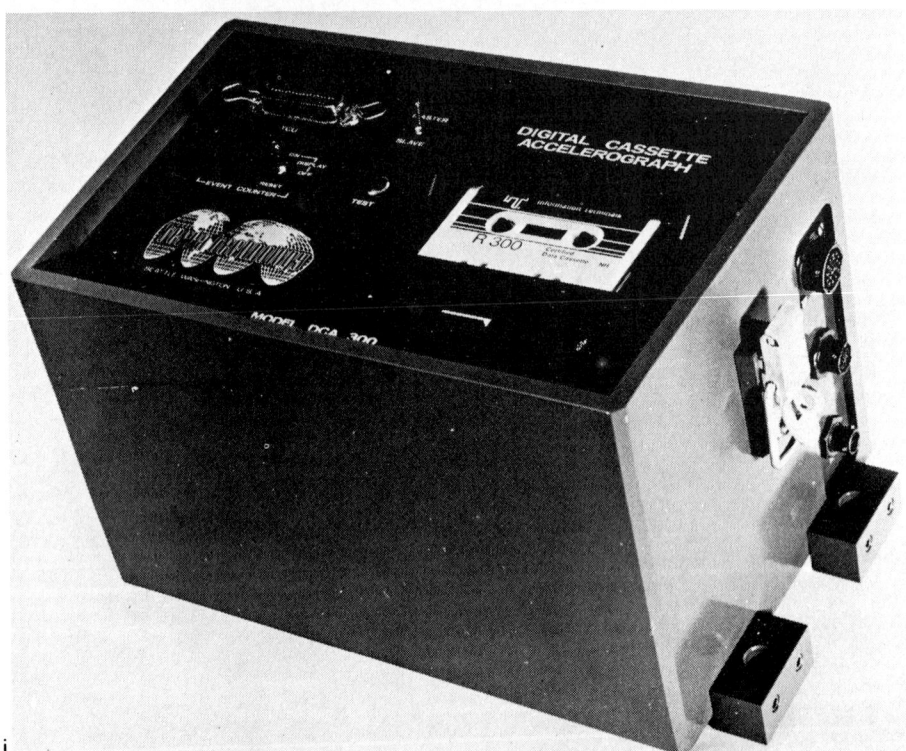

Fig. 10.9. Examples of commercially available specially packaged seismographic systems: a) strong-motion accelerometer; b) 6-channel engineering seismograph; c) signal-enhancement refraction seismograph; d) field station for FM telemetry with amplifier, voltage controlled oscillator, calibrator, and RF transmitter (inside case); e) portable microearthquake seismograph; f) digital data acquisition system with 1/2 in. magnetic tape; g) digital event recording system for high-frequency data with memory timing system and cassette magnetic tape; h) digital recorder for low-frequency data, with timing system and cassette magnetic tape; i) digital strong-motion accelerometer (courtesy Kinemetrics (a,f), Nimbus Instruments (b,c), Sprengnether Instrument (d,e,g), Memodyne (h), Terra Technology (i)).

sponse to particular applications and which consequently offer maximum performance for their cost. Several are illustrated in Fig. 10.9. It is often economically advantageous to use such packages when their specifications cover the requirements of the application in mind.

*10.5.6 Peripheral considerations*

The preceding discussion dealt with what can be called the fundamental instrumental parameters. There are usually several additional considerations in selecting a seismographic system for a given application. The more specific the application, the more significant these peripheral considerations may become. The following list provides some examples.

— Timing. One should state the required precision and resolution of the timing system. Precision involves stability of the clock, generally crystal-controlled, and its absolute time error. Resolution involves both the increment of reference pulses from the clock and the speed of the recording system.

— Record duration. The time between required changings of the paper, film, or magnetic tape.

— Physical size. The total volume and weight of the complete system are often important, especially for instruments for recording structural responses, and the matter becomes paramount in office and commercial buildings.

— Power requirements. Electrical-power consumption is crucial when one must operate the system on batteries in remote locations or with portable instruments. When maintenance costs for more conventional solutions are particularly high, use of solar cells can be appropriate.

— Telemetry. It is becoming commonplace to transmit data from a remote installation to a central or convenient recording site using land line, telephone line, or low-power radio links. There are many advantages in this practice:

(1) Reduced maintenance costs for instruments at locations of difficult accessibility.

(2) The use of a common time base for an entire network of stations.

(3) Continuous monitoring of the instruments' operation, which allows immediate detection of conditions requiring attention.

(4) Continuous recording of all data on a common medium. If magnetic tape, the uninteresting portions may be erased, retaining complete records of events only, thus avoiding the use of excessive lengths of paper, film, or tape while attaining excellent timing resolution.

(5) Reduced system costs due to elimination of local clocks, recorders, and elaborate vaults, with attendant power requirements. For station networks, these cost savings are usually not compensated by telephone-line charges, resulting in a net cost reduction for the telemetered system.

— Recording system. There is much variety in this system component. One may select smoked glass or paper; photographic paper or film; pen and helical recording on drum; strip chart; cassette, 1/4", 1/2", or 1" reel-to-reel magnetic tape, analog or digital. The user generally has several constraints that incline his preference toward a particular recorder. Thus, magnetic-tape recording adds the expensive requirement of a playback system and associated hardware but in many applications it is well justified in view of the increased dynamic range, bandwidth, and timing resolution along with flexibility in processing.

— Triggering and memory. Modern sensors and circuitry allow a wide selection of triggering thresholds and background noise/signal ratios to be incorporated into systems designed to record only sporadic events of interest. Inexpensive semiconductor memories are available that permit retention of the initial portion of the signal; yet their use is by no means widespread at present. In the coming years, however, such systems promise to become standard in seismology where the information desired consists of discrete events. Many shortcomings in present triggered seismographs, due primarily to missing the onset of the event, are eliminated by the incorporation of such memory systems.

— Instrumental protection. Instruments used at construction or near blasting sites or installed in locations exposed to vandalism or mere curiosity demand special considerations concerning their protection.

The foregoing and other additional considerations are necessary ingredients in the selection of a system configuration once the fundamental parameters have been specified. Often they drastically reduce the number of alternative system types for a given application.

## 10.6 CONVENTIONAL SEISMOGRAPHIC SYSTEMS — COMPONENT ELEMENTS

### *10.6.1 General constraints*

The preceding section treated the general problem of specifying performance and physical characteristics in a seismograph. The following discussion deals with available components used in the design of specific seismographic systems. Upon cursory analysis, it would appear that virtually any desired response curve is attainable. In theory this is the case. Any seismometer, of abitrary natural frequency, damping, mass, and sensitivity, produces a non-zero output in response to ground motion of any amplitude and frequency. Conceivably one need merely select, by some signal conditioning method (amplification and selective filtering), the sensitivity and bandwidth of interest. The concept is perfectly correct. The neglected factor is noise. As long as a system so designed will produce, in response to ground motion at the desired detection threshold, an output sufficiently above system noise, the design concept is sound. Unfortunately, this condition is not easily obtained when high sensitivity and low frequencies are involved. If it were, seismologists would need only an inexpensive exploration-type geophone, an amplifier, and a variable filter to work the entire seismic band from solid-earth tides to blast vibrations. For an extensive treatment of noise and dynamic range in inertial seismographs, the reader is referred to Melton (1976).

Realizable (as opposed to ideal) designs for seismographs are based on careful consideration and definition of the system requirements, as discussed earlier, followed by a common-sense evaluation of performance capabilities in components available for use. There will always be more than one design to accomplish a given measurement task. Satisfaction comes in optimizing the interplay of capability, complexity, and cost with the requirements of the application. Variations in the resulting solutions are evident when one considers, as an example, the wide range of seismographs in routine use worldwide as "long period" instruments. Subsequent discussions outline information on system components that, hopefully, will be helpful to the seismologist, geologist, or engineer faced with the task of specifying a seismograph for a particular use.

## 10.6.2 The complete seismograph

The basic elements integral to a seismograph are three. The *sensor*, normally a seismometer of some description, produces a measurable signal in response to a motion of its base. The response of a seismometer is characterized typically in terms of a sensitivity and some frequency dependence. This fundamental behavior can be stated compactly in its transfer function, i.e., the output (in whatever units are appropriate — e.g., volts, amperes, deflection of an optical or mechanical lever, etc.) per unit excitation (again in appropriate units — e.g., ground displacement, velocity, acceleration, strain, rotation, etc.), as a function of the frequency (in Hz) of the excitation. Generally, the transfer function is presented for convenience as a complex function of angular frequency, e.g., $S(\omega)$, where $\omega$ is in radians per second ($\omega = 2\pi f$ where $f$ is in Hz or cycles per second). The complex nature of $S(\omega)$ indicates that the output of the sensor may be out of phase with the excitation, and $S(\omega)$ could be given as $|S(\omega)|e^{i\phi(\omega)}$, where $\phi(\omega)$ is the phase lag. $S(\omega)$ completely defines the response characteristics of a seismometer. For the conventional design using a spring-mass mechanical oscillator, $S(\omega)$ can be written in terms of its natural frequency, damping, and transducer sensitivity.

The second element in the seismograph acts on the output of the sensor and is loosely described as the *signal conditioning device*. It may be no more than a mechanical or optical system of levers, or, more often in modern instruments, an electronic circuit or galvanometer providing amplification and filtering of the electrical output of the sensor. In any case, the signal conditioning element can also be characterized by its transfer function, say $C(\omega)$, giving output/input as a function of frequency in appropriate units.

While we shall see that most $S(\omega)$ are very similar in form, $C(\omega)$, on the other hand, can range from the simplicity of a numerical constant to the intricacy of an involved chain of amplifiers with low- and high-pass filters, yielding a complicated, complex functional representation. Despite potential complexity in $C(\omega)$, it is usually a straightforward procedure to write the mathematical expression for it.

The third and final element in a complete seismograph records the conditioned output of the sensor. The output of the *recorder* is a record. In its simplest form it is the deflection (e.g., in mm) of a trace from some zero position on a sheet of recording paper, film or glass. It may alternatively be a digital word (in counts) an FM tone (in Hz) or voltage written on magnetic tape. The recorder also may have a characteristic frequency response. We define its transfer function as $R(\omega)$, in the same sense as for the two previous elements. The output of the recorder is the seismogram, the fundamental medium of data presentation in seismology.

Assuming the system elements are themselves linear systems, we can define the response of the complete seismograph in terms of its components. The transfer function $T(\omega)$ of the complete seismograph is:

$$T(\omega) = S(\omega) C(\omega) R(\omega)$$

where $T(\omega)$ is the output of the recorder divided by the input to the sensor, in the appropriate units for the system. Virtually always a complex quantity, $T(\omega)$ is usually presented as $|T(\omega)|$, the amplitude response, in a magnification, velocity sensitivity, or acceleration sensitivity curve (see Fig. 10.4), along with $\phi(\omega)$, the phase response.

$T(\omega)$ completely describes the performance of a seismograph, except for system noise. It is sufficient to produce synthetic seismograms from a known input, $I(\omega)$ (where $I(\omega)$ is simply the Fourier transform, or spectrum of $I(t)$, the input ground motion time function). It serves as well to recover the input motion (in the frequency range where the input ground motion exceeds the equivalent system noise) from the recorded output $O(t)$, or its spectrum $O(\omega)$. The nature of the system transfer function is thus seen clearly in its definition:

$$O(\omega) = I(\omega) T(\omega)$$

or:

$$I(\omega) = O(\omega)/T(\omega)$$

It is evident at this stage that, in designing a required system response, all three system elements, $S$, $C$, and $R$, must be considered in shaping $T(\omega)$. With regard to the system noise level, each element also contributes. Either a level of background noise (e.g., from an amplifier) or a basic resolution (e.g., the smallest measurable signal in a recording medium) will characterize each element. These limiting values, which generally are frequency-dependent, can be "taken through" the system transfer function to give the operating noise level or resolution in $O(\omega)$.

We shall discuss the system elements in turn.

## 10.6.3 Seismometers

The seismometer is the basic sensor of ground motion, say $u(x, y, z, t)$. For inertial (spring-mass) type seismometers, the measurement is made of relative distance between the seismometer frame (ground) and an inertial reference point on the suspended mass. Normally only one component of $u$, say $u_x$, is detected with a given seismometer. The seismometer output is thus proportional in some way to $u_x(t)$, and it may, depending on the design and on the type of transducer used to detect the relative motion, actually produce a signal very nearly proportional to $u_x(t)$ or the time derivatives $\dot{u}_x(t)$ or $\ddot{u}_x(t)$ in the bandwidth of interest (see Section 10.5.5).

Extensometers or strain gages, on the other hand, measure the change in one of the components of $u$, usually in the direction of the measured component, e.g., $\partial u_x(t)/\partial x$, usually denoted as the extensional strain component $e_{xx}$. Again, depending on the transducer, which is the integral part of a seismometer converting mechanical motion into a usable output signal, the actual output may be a time derivative of $e_{xx}(t)$, e.g., $\dot{e}_{xx}(t)$.

The response equations for the conventional inertial seismometer are easily developed. For convenience in analyzing data, we desire the response in a form $S(\omega)$, where $\omega$ is angular frequency of the excitation (see Section 10.6.2).

The equilibrium equation for a damped single degree of freedom seismometer, responding to ground motion $u_x(t)$, can be written:

$$m\ddot{x} + c\dot{y} + ky = 0 \qquad (10.1)$$

where, as shown in Fig. 10.10, $x$ is the absolute displacement of $m$, $y = x - u_x$, the measurable relative displacement of $m$, $m$ is mass of the system, $c$ is the damping coefficient (force/velocity), and $k$ is the spring constant (force/displacement).

Rewriting the equation in the familiar form:

$$\ddot{y} + 2\zeta\omega_n\dot{y} + \omega_n^2 y = -\ddot{u}_x \qquad (10.2)$$

where $\omega_n = \sqrt{k/m}$, the undamped natural frequency of the system in radians/sec, and $\zeta = c/2m\omega_n$, the damping factor, i.e., the fraction of critical damping for the oscillator.

The solution to the equation of motion is readily obtained by application of the Laplace transform, i.e.:

$$Y(s) = \frac{\dot{y}(0) + y(0)(s + 2\omega_n)}{s^2 + 2\zeta\omega_n s + \omega_n^2} + \frac{F(s)}{s^2 + 2\zeta\omega s + \omega_n^2} \qquad (10.3)$$

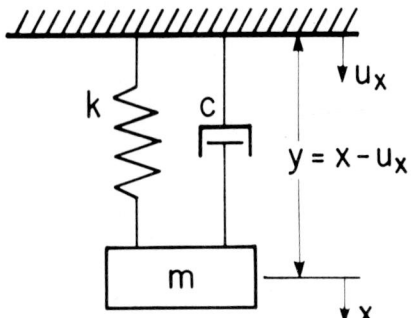

Fig. 10.10. Mechanical schematic of conventional inertial seismometer.

where $s = i\omega$, $Y(s)$ is the Laplace transform of $y(t)$, and $F(s)$ is the Laplace transform of $-\ddot{u}_x(t)$.

The first term is the complementary solution, representing transient motion due to initial conditions at $t = 0$, while the second is the particular solution for the forcing acceleration $-\ddot{u}_x(t)$.

The transient motion, or natural vibration of the oscillator can be transformed back to the time domain, giving:

$$y(t) = \exp(-\zeta\omega_n t)\left[y(0)\cos\mu\omega_n t + \left(\frac{\dot{y}(0)}{\mu\omega_n} + \frac{y(0)\zeta}{\mu}\right)\sin\mu\omega_n t\right] \tag{10.4}$$

where $\mu = (1-\zeta^2)^{1/2}$.

The physical significance of critical damping ($\zeta = 1$) is clear in this form as the value of $c$ ($= 2\zeta m\omega_n$) for which the natural vibration ceases and the motion becomes a simple exponential decay as $y(0)e^{-\omega_n t}$. This is termed critical damping. For $\zeta > 1$, the natural motion is described by exponentially decaying functions, always taking a longer time to reach a given amplitude reduction than for the critically damped case. In most seismograph designs, the natural frequency $\omega_n$ of the seismometer falls within the frequency band of the ground motion to be recorded by the system. In such cases, damping is usually set near-critical to eliminate prolonged transient oscillations which tend to mask the desired response to the ongoing forcing ground motion.

Similarly transforming the forced motion back to time yields, for a general form of $-\ddot{u}_x(t)$:

$$y(t) = \frac{-1}{\mu\omega_n}\int_0^t \ddot{u}_x(t)\exp[-\zeta\omega_n(t-\tau)]\sin[\mu\omega_n(t-\tau)]d\tau \tag{10.5}$$

This is the familiar convolution integral of excitation with impulse response seen in the analysis of linear systems response. It is clear that the output at a given time depends on the total prior time history of excitation. With damping near-critical, however, the response to a given input is reduced significantly by times of the order of several times the natural period.

Rather than a general form of $u_x$, it is convenient to consider the seismometer response to sinusoidal excitation at a particular frequency. For harmonic ground motion of the form:

$$-\ddot{u}_x(t) = A \sin \omega_e t \tag{10.6}$$

or:

$$u_x(t) = \frac{A}{\omega_e^2} \sin \omega_e t \tag{10.7}$$

where $\omega_e$ is the angular frequency of forcing motion (radians per second), the steady-state response (neglecting transient motion) is:

$$y(t) = \frac{A}{[(\omega_n^2 - \omega_e^2)^2 + (2\zeta\omega_n\omega_e)^2]^{1/2}} \sin(\omega_e t + \phi) \tag{10.8}$$

where:

$$\phi = \tan^{-1} \frac{2\zeta\omega_n\omega_e}{\omega_e^2 - \omega_n^2} \tag{10.9}$$

gives the phase shift.

This expression represents the basic measurable motion in a seismometer. Ignoring for the moment any effects of the transducer which detects this motion, we can define the basic seismometer displacement sensitivity, or magnification $M(\omega_e)$, as:

$$M(\omega_e) = \frac{|y(t)|}{|u_x(t)|} = \frac{\omega_e^2}{[(\omega_n^2 - \omega_e^2)^2 + (2\zeta\omega_n\omega_e)^2]^{1/2}} \tag{10.10}$$

Similarly, the velocity sensitivity is:

$$V(\omega_e) = \frac{M(\omega_e)}{\omega_e} = \frac{\omega_e}{[(\omega_n^2 - \omega_e^2)^2 + (2\zeta\omega_n\omega_e)^2]^{1/2}} \tag{10.11}$$

and the acceleration sensitivity is:

$$A(\omega_e) = \frac{M(\omega_e)}{\omega_e^2} = \frac{1}{[(\omega_n^2 - \omega_e^2)^2 + (2\zeta\omega_n\omega_e)^2]^{1/2}} \tag{10.12}$$

The phase lag $\phi(\omega_e)$ applies to displacement sensitivity, and is incremented by an additional $\pi/2$ and $\pi$ for velocity and acceleration sensitivities, respectively.

The equivalent expressions in the form of complex transfer functions are:

$$S_M(\omega) = \frac{-s^2}{s^2 + 2\zeta\omega_n s + \omega_n^2} \tag{10.13}$$

$$S_V(\omega) = \frac{s}{s^2 + 2\zeta\omega_n s + \omega_n^2} \tag{10.14}$$

$$S_A(\omega) = \frac{1}{s^2 + 2\zeta\omega_n s + \omega_n^2} \tag{10.15}$$

where $s = i\omega$.

These fundamental sensitivity curves are presented for various damping factors in Fig. 10.11. The corresponding phase curves are shown in Fig. 10.12. Note that an arbitrary specification of phase must be made, corresponding

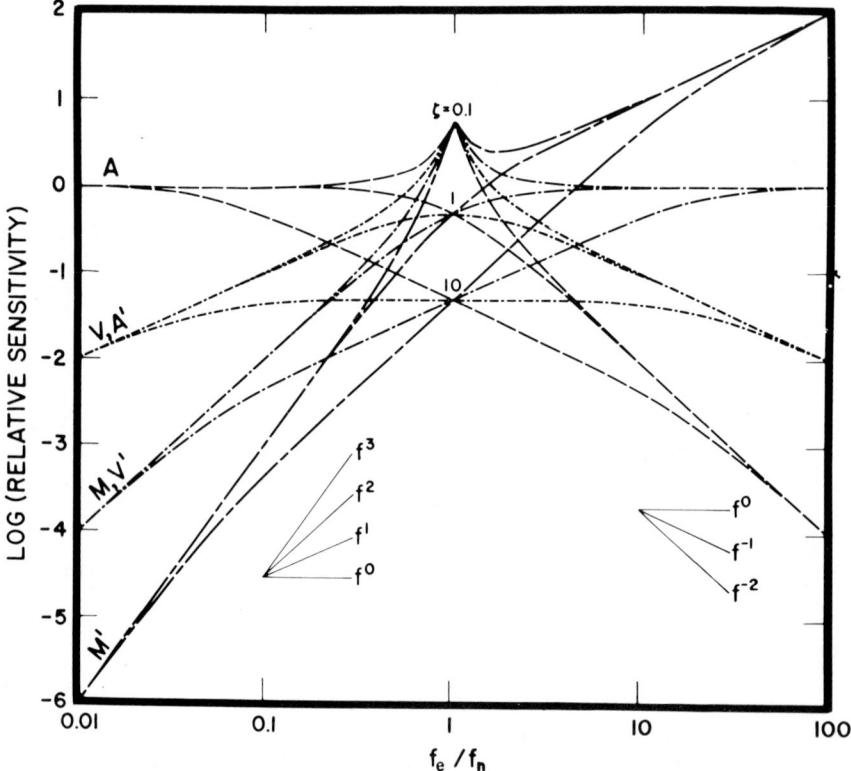

Fig. 10.11. Normalized seismometer response curves for three damping factors in terms of magnification ($M$), velocity sensitivity ($V$), and acceleration sensitivity ($A$) vs. frequency of excitation ($f_e$) relative to seismometer natural frequency ($f_n$). Unprimed symbols for seismometer with displacement transducer, primed symbols for velocity transducer, insets give asymptotic slopes.

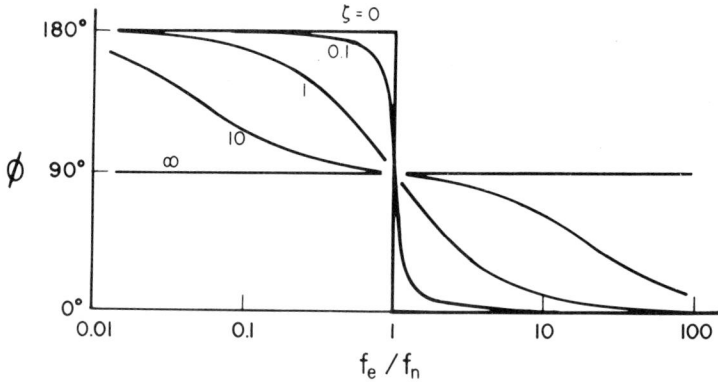

Fig. 10.12. Phase response ($\phi$) for the magnification curve $M$ in Fig. 10.11 for five damping values. High-frequency asymptotes are set by convention at $0°$.

to the sign definition on $y$ relative to $x$ and $u_x$ in Fig. 10.10. In seismometry it is conventional to specify the asymptotic phase-delay value as zero for high-frequency displacement sensitivity; i.e., we usually tap the seismometer in a given direction and identify the output polarity with the tap direction. This convention produces the phase relations shown in Fig. 10.12 for $y$ relative to $u_x$, and illustrated for high frequencies in Fig. 10.13.

The seismometer thus provides us with the measurable quantity $y$, which is related to the ground motion $u_x$ as shown in Figs. 10.11 and 10.12. Signal conditioning, as will be discussed in the next section, is normally applied to $y$ in the form of amplification and filtering. A very common form of such

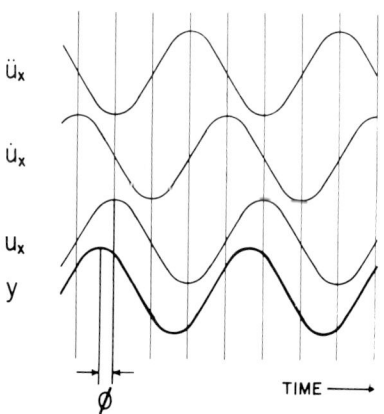

Fig. 10.13. Illustration of high-frequency ($f_e > f_n$) phase response of seismometer displacement $y$ to steady-state sinusoidal ground displacement $u_x$, velocity $\dot{u}_x$, and acceleration $\ddot{u}_x$. High-frequency limit is $0°$ by convention.

conditioning is applied within the seismometer in the process of measuring $y$. The device which performs this measurement within the seismometer is termed the transducer. Transducers in general use are designed to measure either $y$ (the displacement of the inertial reference) or $\dot{y}$. The former type, the displacement transducer, consists normally of either a passive lever system (optical or mechanical), or of an active (requiring external electrical power) electronic circuit to measure change in capacitance between a fixed and a moving plate. The velocity transducer is typically a coil of wire moving in a fixed magnetic field generated by a permanent magnet, and producing an output voltage directly proportional to the relative velocity $\dot{y}$. In this case the seismometer output is differentiated, and the response curves $M$, $V$, and $A$ in Fig. 10.11 are multiplied by $\omega_e = 2\pi f_e$, shown as $M'$, $V'$, and $A'$ in the figure. Note that it is not possible to obtain a flat magnification curve from the velocity-transducer output, unless further shaping is applied to the signal.

All the response curves in Fig. 10.11 exhibit asymptotic trends with slopes proportional to $f_e^n$ where $n$ is a positive or negative integer or zero (see insert in figure). Furthermore, these asymptotes always intersect at $f_e/f_n = 1$, i.e., at the natural frequency of the seismometer. Note also that, for heavy damping (see curves for $\zeta = 10$), a third straight segment appears around $f_e/f_n = 1$. This segment extends from $f_n/2\zeta$ to $2\zeta f_n$, and is often used in response shaping. For example, Soviet strong-motion instruments use the curve $A'$ ($\zeta \approx 10$) while U.S. designs use the curve $A$ ($\zeta \approx 1$). Willmore (1961) and Rodgers (1967) present a complete discussion of the graphical representation of response curves.

*10.6.4 Signal conditioning*

We have reviewed the nature of basic output signals presented by conventional seismometers. If the shape (frequency response) and absolute level (sensitivity) of one of these curves is essentially that required for a particular application, we need only record the signal. More often we have need for a different shape and/or sensitivity and must investigate methods of signal conditioning.

In virtually every requirement for a seismograph there exists a range of frequencies to be recorded with reasonable fidelity. This specification almost always implies injection of unwanted background noise (from either the system or the ground). In Fig. 10.14 we show a generalized background noise curve for a very quiet surface site. Superimposed on roughly a $f_e^{-1}$ or $f_e^{-2}$ wideband noise distribution are the microseism peaks, particularly the 5—8 sec band related to ocean waves near coastlines, associated with low-pressure meteorological disturbances. It is this curve which has had the greatest influence on high-gain seismograph design. The common long-period—short-period division is obviously motivated by the necessity of minimizing response around 6-sec period. The rationale for sharp peaking of the response

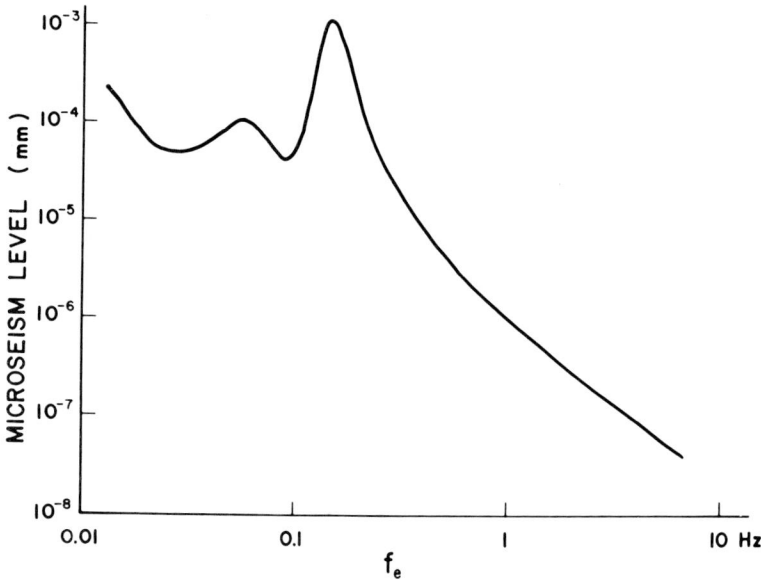

Fig. 10.14. Average background microseism level for a very quiet surface site, in mm of ground displacement, as a function of frequency.

at 35—40 sec in the modern high-gain long-period systems likewise is based on background motion considerations. Peterson and Orsini (in press) describe these considerations in the design of the modern Seismic Research Observatories (SRO) which are being installed in the world-wide network.

The degree of conditioning necessary in a system will depend on the subsequent recording of the signal. As discussed earlier in this chapter, the conventional visible seismogram used in routine seismological observations has limited dynamic range. Such records are more readable with correspondingly narrower bandwidths than are more broadband seismograms. At the other extreme, a digitally recording system with very wide dynamic range, say 100 dB, can record with a wide frequency range.

Signal conditioning in its most common implementation consists of amplification and/or filtering of the signal to produce the desired sensitivity and bandwidth. Basic parameters to be considered in conditioning are the inherent noise levels at the seismometer output and in the conditioning device.

A typical moving-coil velocity transducer with several hundred ohms resistance will have a generator constant $G$ of around 100 volts per meter per second, or 100 newtons per ampere. This quantity is a fixed property of the permanent magnet and coil, equal to the product of the flux density $B$ (webers/m$^2$) and the length of conductor $L$ (m) in the magnetic field. The quantities $Vm^{-1}s^{-1}$, N/A, or Wb/m are numerically and dimensionally identical, and all specify $G$. From Fig. 10.14, the typical background velocity at a quiet site at 1 Hz is $2\pi \cdot 10^{-6}$ mm/sec or $2\pi m\mu$/sec (or nm/sec), producing a peak voltage of 0.6 μV. The inherent Johnson noise in the coil by virtue of its resistance is smaller by more than two orders of magnitude than this voltage in a bandwidth of a few hertz, yielding a fairly clear and uncontaminated signal from the smallest ground motions we might wish to measure in the short period range. With $f_e^{-1}$ ground noise, the background particle velocity will remain essentially constant except for microseism peaks, and, so long as $y \approx u_x$ ($f_e > f_n$), the velocity-transducer output gives the most noise-free measurement of $y$ over a relatively wide frequency bandwidth.

A conventional displacement transducer using capacitance plates is inherently a device with greater noise than a coil by virtue of the fact that it is an active electronic circuit measuring small capacitance changes, thus subject to noise contributions from its components, power supply, etc. In bandwidths of a few hertz, with very careful design utilizing phase-sensitive detection, noise levels equivalent to $10^{-7}$ mm have been achieved. For the older-style design using balanced resonant circuits, a practical equivalent noise level would appear to be $10^{-5}$—$10^{-6}$ mm. See Dratler (in press) for a discussion of capacitance displacement transducers.

Modern amplifiers are capable of operating over bandwidths of several hertz with equivalent input noise levels of something less than 0.1 μV for discrete components, and less than 1 μV for the best integrated circuit amplifiers (cost about $25). Clearly, the state-of-the-art now allows substitution of solid-state amplifiers for galvanometers where high sensitivity is required and DC power (typically much less than 1 watt) is available. This development, in very recent years, has provided a new dimension in flexibility for seismograph design.

Until around the middle 1960's there was no substitute for the galvanometer as a low-noise amplifier. Equivalent input noise levels below 0.1 μV were readily available with galvanometers. Equally important, the galvanometer derives its driving power directly from the moving-coil transducer, with external power required only for the light source. It is this combination of low noise and high gain (2 to 20 mm/μV) that set the seismometer—galvanometer combination as the standard electromagnetic seismographs for many years. It also led to the phototube amplifier (PTA) when high-level electrical signals were required for subsequent visible or tape recording and other signal processing or transmission.

A conclusion apparent in the preceding discussion is important in the design of signal-conditioning electronics following a seismometer output. It is that available velocity or displacement transducers can be used with equally satisfactory results in the range of frequencies about 0.01—10 Hz, where the seismometer relative displacement $y$ can be made essentially equal to the ground displacement $u_x$, i.e., where the $M$-curve is essentially flat. In this range either transducer can be designed with inherent noise levels equivalent to ground displacements well below background in the quietest sites. At frequencies above about 10 Hz, by virtue of decreasing ground noise, the displacement-transducer noise approaches levels equivalent to background. At low frequencies, however, by virtue of the more rapid decrease in velocity ($\dot{y}$) than in displacement ($y$) ($f_e^3$ vs. $f_e^2$) dependence with longer-period ground motion (compare $M'$ with $M$ in Fig. 10.7), the displacement transducer rapidly becomes the better detector and is the logical choice for measurement of mantle surface waves, free oscillations, gravity, earth strain, and solid-earth tides. This is illustrated in a network of accelerometers for very long period seismology described by Agnew et al. (1976).

The above considerations of noise limitations in signal conditioning become more important as higher sensitivities are required. For most strong-motion and engineering-seismology applications, however, the ground motions to be measured are sufficiently large that noise levels in signal conditioning devices are not normally a serious concern.

The basic rule governing selection of signal-conditioning components is that the full dynamic range of the recording device is used for ground motion in the bandwidth of interest. An example would be in digital recording, where one would want the least significant bit in the digital data word to represent ground motion and not system noise. Similarly, the mechanical and optical design in a conventional strong-motion accelerometer should be sufficiently clean that the smallest measurable deflection of the photographic trace represents seismometer mirror deflection due to ground motion and not spurious noise in the mechanical film drive or optical systems.

*10.6.5 Recording*

The decision as to recording method for a seismograph is usually reached early in the systems design procedure. Dictated by the planned use of the data, as well as by the form and quality of the signal available for recording, the recorder selection is normally straightforward. Frequency bandwidth, dynamic range, and record length, in addition to physical and electrical constraints, are the principal parameters governing the choice.

The major division in commonly used recording systems separates visible, or hard-copy, recorders from those which record the signal on magnetic tape of some type. The former designation includes: (1) devices in which a stylus inscribes the record on smoked or otherwise coated paper, glass, or metal; (2) pen and ink recorders; (3) heated stylus on special paper; and (4) light spot on photographic film or paper. Dynamic range is limited to about 40—50 dB in the best cases. The highest frequency measurable depends on the speed of the recording medium relative to the stylus, pen, or light spot, as well as on the actual size of the trace being written. Typical recording speeds range from 0.1 mm/sec for frequencies about 0.05 Hz, up to 10 mm/sec for frequencies as high as 20—50 Hz (depending on trace width). In general, the finest trace, about 0.1 mm wide, is written by a stylus on smoked paper or glass. The problem of pen friction and associated deadband is eliminated in photographic recording where high contrast is also obtained.

The convenience of pen-ink recorders, using inexpensive paper, is often preferred in a tradeoff analysis considering resolution, contrast, and friction along with cost and convenience. A compromise with the convenience of pen and paper, but improved resolution and contrast over ink, is the heated stylus on special coated paper, which is naturally more expensive than the plain paper used with ink. A popular multi-channel recorder uses 16-mm film which is processed in the unit. Unless the seismograph is designed for a specific recording technique (e.g., the 70-mm film in many strong-motion systems), the selection of visible recorders is normally based as much on personal preference as on details in specifications.

Visible recorders, in general, are designed for low cost plus the convenience of viewing and monitoring at least one day's data on a single record. They allow measurement of time and amplitude, only estimations of frequency content, and by and large are not well suited for digitization and subsequent processing. An exception is the high-quality strong-motion accelerograph film record, which contains a single event on three well-spaced traces at relatively high recording speed.

Magnetic-tape systems constitute the other principal class of recorders commonly used in seismological measurements. Developments in low-power analog and digital circuitry during the 1970's, plus the increasing access of workers to analog playback facilities and to small computers with magnetic-tape playback capabilities, have resulted in greatly increased application of magnetic-tape recording in a wide range of seismological efforts. Recording is accomplished in three fundamentally different modes: (1) analog direct; (2) analog FM (frequency modulated); and (3) digital.

In direct recording the analog signal is converted directly into a proportional magnetization on the tape. A consequence of the dependence on tape to head velocity in the playback process is that DC signals cannot be recorded and recovered on playback, thus the method has a low-frequency cutoff of several hertz. Very low recording speeds of around 0.1 mm/sec with up to 30 dB dynamic range have been used to record data at frequencies up to 20—30 Hz. For ½-in tape with seven data tracks this is a very efficient mode of data storage. The lack of DC response and relatively low dynamic range are the major limitations of direct analog recording.

FM analog magnetic-tape recording offers the advantages of DC response, precise amplitude information, and 40—50 dB dynamic range, at a tradeoff in tape speed. 0—10 Hz bandwidth can be obtained at 1 mm/sec tape speed. Considering again seven data tracks on a ½-in tape at this slow speed, a single reel of tape can contain several weeks data from up to seven seismographs. While such recorders are used widely in seismology, they are in reality precision laboratory devices not well-suited to the rigors of field installations unless maintained constantly by qualified technicians.

Digital magnetic-tape recording suffered initially from the low information density possible relative to analog methods. For example, a 12-bit word

(66 dB dynamic range) at 25 samples per second (barely adequate for 10 Hz response, with adequate anti-alias filtering), recording on the same ½-in 7-track tape, with seven data channels multiplexed (each at 10 Hz bandwidth), would require a tape speed over 1 cm/sec — an order of magnitude less information density than for FM. Even though the lower density is balanced by an order of magnitude improvement in dynamic range, the required frequency of tape change for multi-station continuous recording poses a severe problem for seismographic data acquisition.

Two recent developments may well revolutionize the use of digital magnetic-tape recording in seismology. These are the advent of low-cost reliable digital cassette transports along with the development of integrated circuit shift register modules whereby thousands of bits of information can be held in memory on a single printed circuit board, providing a delay in the data stream while the decision is being made regarding its significance and the necessity of recording it permanently on tape. Such "self-editing" data systems, which can be used with low-cost cassette or ¼-in tape recorders for event recording, have finally brought the unquestioned superior fidelity of digital recording to routine seismographic recording where the desired data consist of discrete events, comprising but a fraction of the total monitoring time. Ambuter and Solomon (1974) and Prothero (1976) describe systems of this type.

As a final suggestion to those who would design systems based on magnetic tape as the prime recording medium — a monitor record of some type, recording continuously, is invaluable in the task of identifying, locating, and processing the events stored on the tape. Equally important is the recording on tape of a time code which can be read automatically while scanning the tape in the search and playback mode.

*10.6.6 Timing*

Timing systems commonly used in seismographs are of two types: (1) a precision oscillator driving a clock; or (2) continuous reception of a radio time signal. Even in the case of the oscillator and clock, regular radio reception is needed for time check.

Crystal oscillators with temperature compensation for time bases are available at relatively low cost and low power drain with long-term stability of $5 \cdot 10^{-8}$ or better at constant temperature. This stability makes available simple timing systems which exhibit drift rates in normal service of as little as 0.1 sec per month. The complete timing program derived from such an oscillator can be as simple as a series of minute and hour pulses for visible recorders, or as complex as a digital time code with day, hour, minute, second information repeated once per second. The basic precision is the same in either case.

Radio time signals exist in a variety of forms. Perhaps the most common is the WWV-type periodic or continuous time data transmission in the 2.5—20 MHz standard time broadcast bands. These short-wave signals can usually be received, with adequate equipment, anywhere in the world, at some time of the day. While adequate for chronometer time checks, these signals are too erratic in reception quality for use as a primary time signal. Low-frequency time broadcasts in the 15—100 kHz range, such as WWVB, can be received reliably enough for continuous recording over areas of continental dimensions. Another source of continuous time data lies in the navigational broadcasts such as the Loran system. In many areas, standard broadcast stations (500—1500 kHz) transmit daily reference signals at regular times. A micropower timer can be incorporated easily and at very low cost into field systems to turn on a pretuned receiver daily at a preset time stable to a few seconds per month.

There is the occasional system that does not require synchronization to Universal Time (UT). Special-purpose microearthquake networks, such as those used in geothermal site investigations, recording events seen only by that network, are one example. Another example is the array of deformation and pressure gages often installed in large dams or structures to record response to strong earthquake shaking. In these cases a system is adequate which provides a reference for relative timing among data channels, and sufficient precision for the analysis contemplated (e.g., high-resolution response spectra).

*10.6.7 Telemetry*

In many cases it is advantageous to emplace the seismometer some distance from the recording site. The distance may be as little as 1—2 km, where cultural or electrical noise is unacceptably high in the recorder area, or as much as hundreds of kilometers when central recording of a network of stations is desired. Modern techniques of data telemetry, FM or digital, provide a simple reliable solution to this need.

The most commonly used telemetry system for seismological data is based on frequency modulation (FM) in the audio range of 300—3000 Hz which is readily transmitted by commercial voice-grade telephone lines. Typically, seismometer signals in the 0—25 Hz range are converted by a voltage-controlled oscillator (VCO) to tones in the audio band modulated ±125 Hz by the signal. Up to nine (but preferably seven or less) independent tones (multi-components at one station, different gains, or multiple stations) can then be multiplexed onto one telephone line. At the receiver, the tones are demodulated by narrow-band discriminators which deliver the original analog seismometer signal. Dynamic range of 50—60 dB can be obtained with relatively simple systems. The signals are then available at the central recording site for recording and processing. A variation on this method uses intermediate direct recording of the multiplexed tone bundles on magnetic tape at 15/16 ips (24 mm/sec), with demodulation upon playback. This is probably the most dense packing of information available in magnetic tape recording.

The telemetry link may consist alternatively of low power (0.1—0.5 watt) radio pairs, typically transmitting in FM mode somewhere in the 50—500

MHz range, requiring line-of-sight path, and effective up to 100 km distance in favorable circumstances. With radio telemetry, as with special land lines, the 300—3000 Hz bandwidth limitation on voice-grade telephone lines can be extended, allowing more channels per link.

If the requirement exists to telemeter data with a dynamic range greater than about 60 dB, digital techniques should be investigated. While digital data telemetry is not presently in widespread use for modest seismological installations, the rapid advances in the field of digital data transmission and acquisition virtually assures that such methods are coming in seismological practice. In fact, an extensive network utilizing digital telemetry and currently being installed in Mexico is desired by Lomnitz and Gil (1976).

## 10.7 CALIBRATION

After careful consideration of the requirements and components of a proposed seismographic system has produced detailed specifications and the resulting equipment has been installed, the inevitable question is asked, "What is the system response?" This question frequently arises, sometimes as soon as the system is operational; in other cases it may be months, or even years, before data from the instruments are processed in such a manner that any more than a rough magnification value is required. Nevertheless, this question will arise sooner or later in the case of virtually any modern installation of seismographic instrumentation.

In this final section the two general and most effective approaches to answering the question of system calibration will be outlined. The two methods are: (1) theoretical, a consideration of the expected response of each element in the system and their synthesis into the overall system response; and (2) empirical, a direct measurement of the system response to an input equivalent to a known ground motion. In the former approach all the parameters which define the response of each component, must be known, i.e., seismometer natural frequency, damping, coil resistance and inductance, transducer sensitivity, amplifier and filter characteristics, and recording-system constants. Many or all of these will have to be measured in the laboratory, to a precision consistent with the desired accuracy of the theoretical response calculation. In the latter empirical method, only the means to inject a known equivalent ground-motion input, with the required precision, into the system is needed. The standard modern approach to this is the use of an independent calibration coil on the seismometer. A current passed through the coil, whose electromagnetic constant is known or can be easily measured, produces a known force on the inertial mass of the seismometer which is equivalent to an acceleration of the ground. While many apparently differing schemes for calibration are used, they are, virtually without exception, simply variations or combinations of these two methods.

With careful procedures, either approach to calibration is capable of producing response curves accurate to better than 5%. Having performed such a calibration, it is common good practice to check system response periodically with a simple weight lift or known electrical signal in the calibration coil. Variations in the system response are seen easily as changes in the response ("cal-pulse") of the system to this standard test input.

REFERENCES

Agnew, D., Berger, J., Baland, R., Farrell, W. and Gilbert, F., 1976. International deployment of accelerometers: a network for very long period seismology. EOS Trans. Am. Geophys. Union, 59: 180—188.

Ambuter, B.P. and Solomon, S.C., 1974. An event-recording system for monitoring small earthquakes. Bull. Seismol. Soc. Am., 64: 1181—1188.

Dratler, J., Jr., in press. An inexpensive linear displacement transducer using a low-power lock-in amplifier.

Hudson, D.E., 1958. The Wilmot survey type strong-motion earthquake recorder. Earthquake Eng. Res. Lab., Calif. Inst. of Technol., Pasadena.

Lomnitz, C. and Gil, J., 1976. Resmac: the new Mexican seismic array. EOS Trans. Am. Geophys. Union, 57: 68—69.

Melton, B.S., 1976. The sensitivity and dynamic range of inertial seismographs. Rev. Geophys. Space Phys., 14: 93—116.

Peterson, J. and Orsini, N.A., in press. Seismic research observatories — upgrading the world-wide network. EOS Trans. Am. Geophys. Union.

Prothero, W.A., 1976. A portable digital seismic recorder with event-recording capability. Bull. Seismol. Soc. Am., 66.

Rodgers, P.W., 1967. Overdamped second-order response. Control Eng., March 1967: 77—78.

Scott, R.F., 1973. The calculation of horizontal accelerations from seismoscope records. Bull. Seismol. Soc. Am., 63: 1637—1661.

Willmore, P.L., 1961. Some properties of heavily-damped electromagnetic seismographs. Geophys. J. R. Astron. Soc., 4: 389—404.

Zeevaert, L., and Newmark, N.M., 1956. Aseismic design of Latino Americana Tower in Mexico City. Proc. 1st World Conf. Earthquake Eng., Berkeley, Calif., 35: 1—11.

# INDEX

ABMAC technique, 249, 250, 251
Acceleration, design, 342—346, 357—379
—, effect of depth on, 168
—, El Centro earthquake, 172
—, focussing effect, 170
—, ground, 11—13, 287, 289, 290, 299, 300
—, maximum, 98, 141, 152—156, 159—162, 174, 179, 184—189, 220—221
—, modal, 318
—, recording of, 382—384, 387—390
—, source, 154—156, 159, 160, 173
— response, absolute, 317, 318, 321, 322
— sensitivity (seismographs), 389, 390, 399, 404
— spectra, 73, 97, 98, 103—105, 189
Accelerographs, 141
Accelerometers, 387, 389, 395, 408
Afghanistan, 10
African Plate, 8
Aftershocks, 13, 18, 23, 26, 158, 171, 203, 204, 206
—, recording of, 382
— related to tsunamis, 226—232
Alaska, 115, 227, 233, 254, 267, 278
Alaskan earthquake, 119, 171, 204
— — tsunami, 226, 228—232, 257, 262, 263, 270, 271
Alberta, 234
Aleutian Islands, 15, 194, 204, 255, 267, 278
— Trench, 257
Alluvial valleys, 98, 99, 170
Amchitka Islands, 257
American Plate, 8, 34, 216
Amplification factor, light secondary systems, 325
— —, single-degree systems, 303
— function, 300, 304
—, local, 87—106, 323
Amplifiers, instrumental, 388, 399, 407
Analog recording, 409
Anatolia, 15, 40, 42, 43, 44, 48, 205
Anchorage, 115; see *Alaskan earthquake*

Animal behavior, 16
Anisotropy in soils, 115
— of S-waves, 25
Annualized losses, 370, 371, 372
Antarctican Plate, 8
Apparent stress in geologic faults, 150
Arabian Plate, 8
Artificial motions, 103, 106, 288, 289, 292—295, 310, 315, 316, 322, 323, 333
Aseismic slip, 5, 16, 19, 162, 171, 210, 381
Assam, 169, 228
Asthenosphere, 7, 152
Attenuation, 103, 168, 170, 172, 173, 182—189, 219, 220, 221, 222, 316
Autocorrelation function of ground motions, 294
Autocovariance function of earthquake occurrence, 195, 203
Azores, 10

$b$-value, 13, 18, 26
Background noise, 406, 407, 408, 409
Baja California, 270, 271
Baltic Coast, 266
Band-limited noise, 294, 295, 297, 302, 315
Bandwidth, 293, 301, 305, 309, 321
— factor, 307, 310, 311, 321
—, instrumental, 386, 387, 388, 390, 391, 392, 398, 407, 408, 409
Bartlett-Lewis model of earthquake occurrence, 201
Bayes' theorem, 14, 125, 208—219, 229
Bayesian uncertainty, 215, 218, 219
Bear Valley, 168
Bentonite, 80
Bilinear hysteretic systems, 332, 333
Bingöl, 44, 47
Bismarck Sea, 8
Bligh Island, 230
Blue Mountain Lake, 20
Bore, 242, 249

Borrego Mountain, 37, 38
Boston, 344, 358, 361
Branching renewal process of earthquake occurrence, 206
Breakwaters, 266, 267
Bridges, 355, 359, 375
British Columbia, 185
Brune model of earthquake generating mechanism, 149, 158, 163
Building codes, 31, 32, 55, 66, 345, 349, 359, 369—374
Bulgaria, 49

Calibration, instrumental, 413, 414
California, 5, 6, 7, 17, 32, 34—40, 48, 51, 59, 64, 88, 185, 359, 370
— building code provisions, 345
—, tsunamis in, 225, 228, 235, 249, 255, 256, 257, 270, 271, 272, 274, 276, 277, 281, 282
Canada, 185, 195, 234, 370
Capacity, structural, probability distribution of, 342
Cape Mendocino, 249
Carbon-14 dating, 38, 39, 53
Caribbean, 8, 16
Catastrophe, 372, 373, 374
Central America, 202
Charleston, 32, 65, 66
Chatter, 49, 142
Chile, 8, 15, 18
—, tsunamis in, 225, 226, 257, 260, 264, 268, 270, 271, 273, 275, 278
China, 10, 53, 59—63
City streets, 377, 378
Clay, 76, 80, 81, 83, 85, 86, 91, 92, 103, 113, 114, 123, 124
— sensitivity, 114
Clumps, 307, 310, 330, 332
Clusters, earthquake, 201, 202, 203, 204, 205, 206, 207
—, —, space, 198
— in stochastic processes, 307, 310, 330, 332
Coarse-grained soils, 71, 72
Cocos Plate, 8, 205, 216
Codes, building, 31, 32, 55, 66, 345, 349, 359, 369—374
Coherency at geologic faults, 160, 172
Cohesionless soils, 71, 72, 74, 75, 77, 78, 82, 83, 84, 106, 112—114, 116, 117, 118, 119, 120, 121, 122, 124, 129

Cohesive soils, 76, 77, 79—81, 83, 85, 86, 87, 91, 92, 106, 112—114, 123, 124
Colombia, 269
Columbia River, 235
Communication lines, 374
Compaction, soil, 106—112, 124
Components of ground motion, 101, 289
Concrete structures, 328, 360
Consistent mass matrix, 95
Continental margins, 15
Convective circulation, 8
Coriolis force, 255
Cost, analysis, 327
—, construction, 357, 360, 361, 362, 364
—, design, 359, 360
—, human, 349, 357, 364, 365, 366, 367, 368, 369
—, instruments, 383, 392, 393, 410, 411
—, repair, 349, 363, 364, 365, 366, 367, 368, 369
—, seismic resistance, 358, 359, 360, 361, 362
Crack, circular, in geologic faults, 158, 163
—, elliptical, in geologic faults, 162, 165, 168, 170
—, semicircular, in geologic faults, 161
— stoppage in geologic faults, 162, 165
— tip in geologic faults, 159, 160
Creep slip, 5, 16, 19, 162, 171, 210, 381
Crescent City, 255, 272, 274, 275, 279, 280, 281, 282, 283
Critical deviator stress, 113, 114
— state lines in sand, 74, 75, 76, 111, 112, 113, 117, 119
— void ratio, 74, 75, 76, 111, 131
Crustal deformation, 18, 19
— tension, 169
— compression, 169
Cumulative spectral distribution, 296
Cyclic mobility, 115—118, 119—123, 132

Damage levels, 342, 349, 350
— potential of vibrations, 383
— probability, 342, 343, 344, 345, 346, 347, 348, 352, 360
— — distribution functions, 349, 350, 351
— — matrices, 351
— ratio, 350, 354
— states, 347—354, 362, 363, 364, 365, 366, 368
—threshold, 342

Damped spectra, 105, 289, 290, 291, 299, 300, 301, 302, 304, 305, 312, 313, 314
Damping, effects on peak factor, 312
—, — — spectral ordinates, 311, 312
—, equivalent linear, 98, 106, 328, 329, 333
—, hysteretic, 317
— in earthquake slip, 19
— — finite-element solutions, 101
— — ground, 173, 294
— — light secondary systems, 324, 325
— — multi-degree systems, 318, 319
— — sand, 84
— — soils, 77, 81, 82, 83, 84, 97
— — —, field determination, 127
— — seismometers, 401, 402, 403, 404, 405
— — seismoscopes, 383
— — structure, 318, 319, 384
— matrix, 96
—, modal, 96
—, radiation, 94, 95, 96
—, spurious, in numerical methods, 95
—, time dependent, 304
Dams, 31, 106, 121, 122, 192, 339
—, earth and rockfill, 106
Dardanelles, 44, 45
Dating, radiometric, 31, 38, 39, 53, 56
Dead Sea, 44
Death Valley, 36
Debris, 377
Decay period, 325
— rate, 309, 310, 312, 313
— —, time dependent, 312, 313
Decision analysis, 1, 181, 339, 357—372
Deep earthquakes, 8, 145, 168
Depth of source, effects on aftershock sequences, 204
— — —, — — ground motion, 168, 169
Design, 339—380
— acceleration, 357—379
— earthquake, 103, 343
— parameters, 221, 222
— return period, 14, 362, 363
— spectra, 103, 192, 291, 316
Deviatoric stress, 73, 85, 113, 114, 115, 123
Diffraction, seismic-wave, 158
—, tsunami, 263, 265
Digital recording, 395, 409, 410, 411
Dilatancy, 21, 22, 145
— model of earthquake generating mechanism, 21—25
— - fluid diffusion model of earthquake generating mechanism, 21, 22, 23
— - instability model of earthquake generating mechanism, 22, 23, 24
Dip-slips, 48, 169, 225
Discount rate, 362, 363, 364, 367, 368, 369
Dislocations, 19, 147, 154, 159, 161, 169, 170, 180
Displacement, fault, 4, 5, 6, 7, 18, 19, 33, 35—66, 145—147, 148, 151, 153, 154, 160, 164, 165, 171, 210, 252
—, ground, 153, 154, 164, 165, 166, 167, 188, 189, 228
—, —, near field, 158
—, —, permanent, 151
—, Holocene, 32, 35, 36, 37, 38, 47, 48, 53, 64, 65, 66
—, sensitivity of seismographs, 388, 394
—, spectral, 98
— spectral density function, 300
—, structural, exceedance probability of, 347, 348
Distributed targets, total risk, 354—357
Dominant period of ground, 103, 294
Doppler focussing, 160, 162
Double couple mechanism of earthquake generation, 143, 144, 145, 160, 294
Drainage conditions in soils, 71—73, 113, 118, 119, 121
Draw-down, tsunami, 264, 271—275
Ductility, 328, 329
— factor, 329, 330, 331, 334
Duration of ground motion, 152, 156, 159, 160, 165, 166, 167, 174, 292, 319, 331
Dynamic properties of structures, measuring, 382
— range, instrumental, 386, 388, 389, 392, 407, 409, 411, 412, 413

Earth and rockfill dams, 106, 121, 122
Earthquake energy, 4, 11, 12, 147, 148, 160, 168, 171, 172, 180
— generating mechanisms, 1, 3—11, 151—174
— — —, models of, 4—11, 141, 147, 149, 150, 151, 152, 157, 159, 161—173
— magnitude, see *Magnitude, earthquake*
— intensity, see *Intensity*
Eckernförde, 266

Eel River, 235
Effective shear stress, 153
— stress, 73, 153
El Centro, 89, 90, 169, 171, 298
Elastic energy, 142, 148
— rebound, 5, 141, 145, 147
Elasto-plastic soils, 102
— — systems, 328, 329, 330—332
Electric power stations, 359, 376
Electrical systems, design of, 339, 375
Elm slide, 235
Energy dissipation in structures, 328
—, earthquake, 4, 11, 12, 147, 148, 160, 168, 171, 172, 180, 210
—, near field, 168
— release, seismic, 10
— trapping, 168
—, tsunami, 226, 230
Ensenada, 270, 271
Envelope of stochastic process, 306, 307
Equador, 278
Equipment in structures, 287, 323—327
Equivalent amplification function in soils, 106
— damping in linear multidegree systems, 320
— — — non-linear systems, 328
— — — soils, 97, 98
— duration of ground motions, 298, 302, 313, 321, 331
— linearization of non-linear systems, 106, 328
— static forces, 318
— stationary motion, 302
— — response, 313, 321
— transfer functions in soils, 106
Erzican, 33
Essential facilities, 376
Ethics, 1
Eurasian Plate, 8
Evacuation, 2
Event counting, instrumentation, 383
Evolution of seismicity, 198
Evolutionary damping, 304, 312
— frequency content, 307
— response statistics, 295, 307, 308
— spectral-density function, 295, 307, 308, 319, 321, 325
— transfer function, 303, 319, 322
— variance of ground motion, 303
Excess life, renewal processes, 212
Exploding islands, 225

Exploration methods, 125—132
— —, geophysical, 125—127, 381, 382
— —, penetration tests, 127—132
— —, sampling, 132
Explosions, 12, 13
Exponential model of earthquake occurrence, 200, 201, 202
— — Poisson model, 14
Extensometers, 400
Extrapolation of historic records, 31

Facilities, essential, 376
Falling-pin set, 383
Far-field displacements, 144, 145
— — radiation, 144, 145
— — spectra, 145, 158, 159
Farview Peak—Dixie Valley, 169
Fatigue model of soil cyclic mobility or liquefaction, 120
— in system performance, 287
Fault breakout, 170
— creep, 5, 16, 19, 162, 171, 210, 381
— displacement, 4, 5, 6, 7, 18, 19, 33, 35—66, 145—147, 151, 153, 154, 160, 163, 164, 165, 171, 210, 252
— orientation, 169
Finite-difference techniques, 161, 170, 262, 263
— — element techniques, 95, 100, 101, 150, 170
Fire fighting, 375
First-passage problems, linear single-degree systems, 298—316
— — —, — multi-degree systems, 316—327
— — —, non-linear systems, 327—334
Floor response spectra, 323
Fluid-diffusion model of earthquake generating mechanism, 22, 23
Foam-rubber model in seismology, 161, 162—168
Focussing, 160, 162, 170, 172, 173
—, topographic, 170, 221, 222
Force-balance accelerometer, 387
Foreshocks, 16, 17, 18
Fossa Magna, 54
Fourier spectra, 89, 90, 179, 294, 297, 298
— transform, 93, 97
— —, truncated, 303, 319
— —, water-wave, 236
Frank slide, 234

Free surface amplification, 169, 170
Frequency content of ground motions, 221, 292, 294, 295—298, 302, 304, 305, 312, 316, 325
— — — — —, time dependent, 304, 312
— domain analysis, 317
Friction along fault boundaries, 142, 149
Frictional heat generation, 149
Froude number, 225, 244, 245
Fuel lines, 374

Gain, instrumental, 385, 386, 400, 403
Galvanometers, 408
Gamma process model of earthquake occurrence, 206, 207, 212
Gap, seismic, 14, 15, 16, 26, 61, 201, 202, 203, 205, 211, 212, 213, 214
Garcia wave theory, 235, 249—254, 258
Garlock Fault, 36, 44
Garm, 19
Gaussian process, 306
Gaussian white noise, 309, 310, 330, 332
Geodetic measurement of displacements, 4—7, 19
Geologic record, 31—69
Geomagnetic anomalies, 16, 22, 23
Geophysical exploration, 125—127, 381, 382
Geothermal activity, 18
— exploration, 402
Glacial till, 88
Gobi—Altai, 63
Golden Gate, 255, 257
Gorda Escarpment, 252
Graben, 60, 62
Gravel, 48, 54, 55
Gravitational sliding, 15
Gravity anomalies, 18
Greece, 49, 204
Ground acceleration, see *Acceleration, ground*
— damping 173, 294
— displacement, see *Displacement, ground*
— velocity, see *Velocity, ground*
Gulf of Alaska, 226
— — California, 8

Hard-copy recorders, 409
Hawaii, 225, 255, 256, 264, 265, 267, 268, 269, 272, 275, 279
Hazard function, 195, 202, 205
Highways, 354, 355, 359, 375

Hilo, 256, 257, 264, 265, 267, 268, 269, 272, 275, 279
Himalayas, 10, 64
Hindu-Kush, 198, 206
Hokaido, 278
Holocene displacements, 35, 36, 37, 38, 47, 48, 53, 64, 65, 66
Honshu, 49
Housner model of earthquake generating mechanism, 149
Human cost of failure, 349, 357, 364, 365, 366, 367, 368, 369
Humbolt Bay, 279
Hydrostatic stress in soils, 73
Hydrothermal, see *Geothermal*
Hysteresis, general, 328, 332, 333
—, soil, 82, 83, 96, 97, 106

Imperial Fault, 169
Impulse-response function, 303, 304, 319
Incomplete data in seismicity assessment, 218
India, 169, 228
Indian Plate, 8
Indonesia, 8
Inelastic behavior, soil, 82, 83, 96—98, 102, 106
— response spectra, 329, 330
— systems, response of, 327—334
Inflation, effect on actualization rate, 362
Initial cost of construction, 357, 360, 361, 362, 364
Instability, structural, 329, 332—334
Instantaneous stress pulse, see *Stress pulse*
Insurance, 179, 349
Integrated circuit shift, 411
Intensity, 11
— attenuation, 182—189, 222
— characteristic, 180
— correlation with magnitude and focal distance, 184
— envelope functions, 293, 294
— exceedance probability, 345, 346, 368
— — rate, 218, 219, 220, 222
—, local geology effects, 200, 221, 222
—, maximum credible, 192
—, — feasible, 192
— recurrence period, 218, 219, 220, 222
— related to damage, 351, 352, 353, 354
— scales, 13
Iran, 8, 10
Isoseismals, 182, 183, 184

Israel, 43
Itoigema—Shizuka Tectonic Line, 54, 55, 56
Izmir, 44, 46
Izu, 50, 52

Japan, 8, 15, 17, 18, 20, 49—55, 103, 104, 119, 121, 131
—, tsunamis in, 226, 228, 230, 231, 232, 255, 256, 262, 273, 278, 279
Johnson noise, 408

Kajiura wave theory, 240, 241
Kamchatka, 15, 198, 255, 256, 257, 278
Kanai-Tajimi spectra, 294, 295, 298, 302, 303, 305, 313, 314, 330, 333
Kansu, 32
Kaolinite, 80
Kelvin's principle, 238
Kenai Peninsula, 234
Keylis-Borok model of earthquake generating mechanism, 151
Kita—Izu Fault, 52
Knopoff model of earthquake generating mechanism, 151
Krakatoa, 278
Kurile Islands, 15, 256, 278
Kyoto, 49, 50

La Jolla, 271
Land use, 1, 339
Landslides, 55, 63, 65, 227
— related to tsunamis, 227, 232, 235
Lebanon, 43
Lifelines, risk of earthquake damage, 354, 356
—, design for seismic hazard, 374—379
Light secondary systems, response of, 322, 323—327
Liquefaction, 115—118, 119—123, 328
Lithosphere, 3, 7, 8, 26
Lituya Bay, 227, 233
Local amplification, 87—106, 182, 220—222, 302, 303
— —, one-dimensional models, 91—98, 102, 106, 221
— —, two-dimensional models, 98—101
— seismicity, 189—208
— — assessment, 208—218
Locking in geologic faults, 142, 149, 150, 157
Loess, 61

Long-period seismometers, 388
Los Angeles, 270, 354
— — Harbor, 270, 271
Loss of strength in soils, 106, 112—124
Love waves, 101, 142—144
Low-sun-angle photography, 54
Lumped-mass techniques, 98

Magmatism, 15, 18
Magnetic-tape recording, 395, 398, 409, 410, 411, 412
Magnification curve, instrumental, 385, 386, 389, 400, 403
Magnitude, see *Earthquake magnitude*, *Tsunami magnitude*
Markov-process analysis of bilinear systems, 332, 333
Masonry walls, 360
Matsushiro, 23
Median Tectonic Line, 51, 52, 53, 54, 55
Mediterranean, 10, 33, 42, 49
Memory, instrumental, 398, 411
Menderes Massif, 46, 47
Mendocino Escarpment, 249, 252
Metal diaphragms, 360
Metamorphism, 15
Mexico, 3, 182, 283, 195, 196, 198, 202, 205, 212, 215, 220, 270
— City, 83, 85, 91, 92
Mice Dam, 235
Microseisms counting, 383
—, local amplification, 88, 89
— networks, 412
— recording, 406, 407, 408
Microzoning, 220—222
Mid-oceanic ridges, 7, 10
Middle-American earthquakes, 202
Mino-Owari, 50, 51, 228
Missouri, 372
Modal analysis, 287, 316—319, 321
— response functions, 322
Model seismology, 26, 161, 162—176
— — test instrumentation, 382
Modified Mercalli intensity scale, 13, 182, 183, 345, 346, 351, 352, 353, 354, 368
Mongolia, 63
MSK intensity scale, 13
Multi-channel recordings, 410, 411
— - component ground motions, response to, 289
— - degree of freedom systems, 316—323
— - — — — —, light secondary, 324, 327

Naked Island, 232
Nankaido, 278
Narrow-band discriminators, instrumental, 412
— — processes, 306, 307, 309, 310
Nazca Plate, 8
Networks, design for seismic hazard, 374—379
—, risk of earthquake damage, 354—356
Neuber model of earthquake generating mechanism, 151
Nevada, 32, 65, 169
New Hebrides, 8
New Madrid, 32, 228, 372
New York State, 20
New Zealand, 20, 48, 200, 201, 202, 203
Newmark's beta method, 96
Neyman-Scott model of earthquake occurrence, 201—204
Niigata, 102, 115, 119, 120, 129, 130
Ningsia, 61
Nobi, 50, 51
Noda's wave theory, 245—248
Noise, background, 406, 407, 408, 409
—, white, 293, 294, 295, 297, 302, 303, 305, 309, 310, 315, 330
—, —, band-limited, 294, 295, 297, 302, 315
Normal faulting, 40, 47
North American Plate, 34
— Anatolian Fault, 40, 42, 44, 205
Nuclear power plants, 1, 31, 106, 124, 192, 339, 378

OBE, 378
Oceanic ridges, 10, 13
— trenches, 8
Olympia, 298
Omori's law for aftershock sequences, 203, 204
Operating basis earthquake, 378
Optimization in design, 1, 181, 339, 340, 357—369
Oregon, 257
Orogeny, 15
Orowan model of earthquake generating mechanism, 149, 150, 151, 152, 157, 159, 163
Otoigawa Shizvoka Tectonic Line, 55, 56
Overconsolidation ratio, 79
Overpusses, 375
Overshoot model of earthquake generating mechanism, 149, 150

Pacific Plate, 10, 16, 34, 215
Pacoima Dam, 157, 170, 221
Pamir—Hindu Kush, 198
Pangaea, 8, 11
Parkfield, 17, 161, 167, 169, 171
Pasadena, 88, 89
Peak inelastic deformation, single-degree systems, 331, 332
— reading instruments for ground motion, 382—384
— — — — structural motion, 384, 385
— — — — — deformation, 385, 386
— response factor, multi-degree systems, 320, 321, 326
— — —, single-degree systems, 298, 299, 308—315
Peat, 88
Peking, 61
Penetration tests, 127—132
Permanent deformation in sands, 107—112
— — — structures, 328, 331
Peru, 269
Phase shift in far-field radiation, 144, 145
— — — seismometer response, 403, 404, 405
— speed, 236
Phillipine Fault, 56, 57, 58
— Plate, 8
Phillipines, 56—59
Photographic recording, 392—409
Photographs, low-sun-angle, 54
Physics of earthquake strong motion, 141—177
Plastic drift, 331
Plasticine, 81
Plasticity index, 79
Plate boundaries, 10, 14—17, 18, 26, 34, 40, 49, 142
— convergence, 8
— divergence, 7
— subduction, 8, 10, 13, 215, 216
— tectonics, 1, 4, 7—11, 14, 142, 210
— —, modelling of, 25, 26
— transcursion, 8
Plates, major, 8, 9
Playback system, 397
Plutonism, 15
Plywood diaphragms, 360
Point source theory of earthquake generation, 142—145
Poisson-crossing, inelastic responses, 331
— —, stochastic processes, 309, 310

Poisson dispersion index, 195, 203
— distribution of number of clusters, 204
— — — — earthquakes, 13, 14, 193
— process of earthquake occurrence, 13, 14, 195, 196, 197, 198, 206, 207, 208, 211, 214
— —, non-homogeneous, 204
Poisson's ratio, dynamic, in soils, 77, 125
Pore pressure, 71, 105, 112, 113, 114, 115, 116, 121, 131
Power consumption in seismographic systems, 396
— - law model of earthquake occurrence, 200, 201, 202, 203
— lines, 374, 375, 376
— spectral density, see *Spectral density function*
—. total, of stochastic processes, 295
Predominant ground frequency, 103, 294
Present value of future losses, 357, 362, 363, 364, 367, 368, 369
Presidio, 277
Prestress in geologic faults, 148
Prewhitening, 390, 391
Prince William Sound, 230, 234
Probability distributions of structural responses, 288, 290, 291, 306
Pseudo-acceleration spectra, 289
— — —, variance of, 305
— - velocity spectra, 89, 92, 106, 188, 189, 289, 329
Puerto Montt, 115
Pumping stations, 376

Q, 172
Quaternary faulting, 35—66
Quay walls, 115
Quiescence, 16, 33, 34, 40, 43, 51, 52, 53, 54, 55

Radiation damping, 94, 95, 96
— pattern of seismic waves, 143
Radio telemetry, 412, 413
— time signals, 404
Radiocarbon dating, 38, 39, 53
Ragay Gulf, 57
Ramberg-Osgood stress—strain relation, 83, 84, 85, 92, 97, 98
Random vibration analysis, 288
— — — of light secondary systems, 325—327
— — — — multi-degree systems, 319—322

— — — — single-degree systems, 295—298
— — response spectra, 298—316
Rat Islands, 255
Rayleigh waves, 159
Recording system, 395, 396, 397, 398, 400, 408, 409—411, 412
Red Sea, 8
Regional seismicity, 218—222
Reinforced concrete structures, 328, 360
Relative density, 128, 129, 130
Renewal-process model of earthquake occurrence, 198, 205, 206, 212, 213
Repair costs, 364, 365
Resistivity, 16
Response spectra, 73, 103, 104, 105, 188, 189, 287, 289—292, 298—316
— —, inelastic, 329, 330
— — of floors, 323
— —, smooth, 288, 289, 295, 315, 316
— spectral-density function, 306
— spectrum analysis of light secondary systems, 324, 325
— — — — multi-degree systems, 317—319
— statistics of light secondary systems, 322
— — — nonlinear systems, 334
— — — single-degree systems, 322
Richter magnitude, see *Earthquake magnitude*
Ridge crests, 8
Risk aversion, 366
—, total, 339, 342, 349, 354
Roadways, 375
Rock slicdes, 225, 232
Rupture propagation velocity, 164, 165, 173

S-wave anisotropy, 25
— splitting, 25
Samoa Islands, 278
San Andreas Fault, 4, 5, 7, 15, 18, 35, 40, 44, 169, 249, 278
San Diego, 255
San Fernando, 17, 20, 40, 64, 89, 96, 110, 115, 156, 157, 158, 161, 169, 170, 221, 374, 384
San Francisco, 5, 6, 141, 152, 372
— —, tsunamis in, 228, 255, 256, 270, 272, 277, 278, 279
San Salvador, 18
Sand, 74, 103, 109, 114, 116, 117, 121,

122, 129, 131
Sanitation systems, 374
Sanriku, 256, 257, 278
Santa Monica, 271
Scarps, fault, 43, 54, 55, 58, 61
Scattering by crust inhomogeneities, 172
— — irregular interfaces, 91
Scratch strain gage, 386
Sea-floor spreading, 3
Secondary systems, response of, 322, 323—327
Sedimentary basins, 13
Seismic efficiency, 149, 150
— gaps, 14, 15, 16, 26, 61, 201, 202, 203, 205, 211, 212, 213, 214
— moment, 25, 145—147, 150
— probability maps, 220, 221
— radiation, 143, 149
— slip, 19
Seismicity, 3, 179—222
— assessment, 31—66, 179—222
—, local, 189—208
—, — assessment, 208—218
— maps, 48, 220, 221
— models, 179—181
—, regional, 218—222
Seismogram, 399
Seismographic systems, component elements, 398—414
— —, design considerations, 386—398
— —, — parameters, 386, 387
— —, general constraints, 398, 399
Seismographs, 382, 399—401
—, standard, 393—396
Seismology, 3
Seismometers, 382, 401, 406
Seismoscopes, 383, 384
Seismotectonic investigations, instrumentation, 381
Sensitivity of clays, 114
— — seismographic systems, 386, 388, 389
Sensor in seismographs, 399, 400
Seward, 228
Shaking-table tests, 101, 109, 111, 112, 117, 118, 382
Shallow earthquakes, 8, 16, 18, 168, 169, 200, 201, 202, 205
— subsurface investigation, 381
Shanghai, 61
Shansi, 60, 61
Shift register modulus, 411
Shikoku, 53, 54

Shock indicator, 383
Shutdown of power plants, 378
Sian, 60
Sigma spectrum, 314, 315
Silt, 87
Simulation, 103, 106, 288, 289, 292—295, 309, 310, 316, 322, 323, 330, 333
Sky luminescence, 17
Smith River, 235
Smooth response spectra, 288, 289, 295, 315, 316
Soil-structure interaction, 317, 323
Solid-earth tides, recording, 388, 409
— — state amplifiers, 408
Solomon Islands, 8
Somalian Plate, 8
Source complexity, 171
— dimension, 151, 152, 156, 157
— mechanism effects, 89, 90, 93, 141—172
— time function, 142
Spectral-density, cumulative, 296
— — function, 120, 179, 288, 289, 292, 293, 294, 295—298, 308, 315, 319
— — —, time dependent, 302, 303, 304
— mass, 296
— moments, 311, 319, 320, 322
— shape function, 298
— — parameters, 308
Stable soil conditions, 73—87
Stationary response of structures, 299—302, 306, 307, 309—312
Steel structures, 360
Stick-slip motions, 142, 162, 173, 174
Stochastic models of earthquake occurrence, 11—14, 195—206
— — — — —, influence on seismic risk, 206—208
Strain gages, 386, 401
— monitoring, instrumentation, 382
—, tectonic, 16, 18
Stress concentrations in lithosphere, 25
— drop, 13, 142, 148, 149, 150, 151, 152, 153, 157, 158, 159, 160, 163, 164, 173, 180, 181
— pulse model of earthquake generating mechanism, 152—159
— release, 4, 33
Strike-slip faults, 5, 24, 25, 40, 46, 48, 62, 225, 249
Structural-strain recording device, 385, 386
— synthesis, 339

Subaqueous landslides, 227
Subduction, 8, 10, 13, 26, 216
Supersonic rupture propagation, 155
Surface faulting, 31, 40, 46, 48, 49, 64, 210
— waves, 168, 169
Swarms, 18, 199
Switzerland, 235
Sympathetic activity, 15
Syria, 43
Szechwan, 62

Tadjikistan, 19
Taft, 298
Tajimi spectral-density function, 294, 295, 298, 302, 303, 305, 313, 314, 330, 333
Tango, 51, 228
Tanna Fault, 51, 52
Tectonic displacements related to tsunamis, 179, 189, 208, 211
— processes, 15
Telemetry, 298, 302, 317
Telluric currents, 32
Thixotropy, 77, 80
Thrust faults, 63, 64, 169
Tilt, crust, 16
Tilt-up walls, 360
Time integration analysis, 288
— — — for light secondary systems, 327, 328
— — — — non-linear systems, 334
— — — — single-degree systems, 316
Timing, instrumental, 397, 411
Tokachi, 279
— - Oki, 273, 283
Tokyo, 49, 50, 127
Tonga, 8
Total risk, 339, 342, 349, 354
Transducers, 406, 407, 408
Transfer function, 88, 94, 106, 304
— —, seismometer, 399, 400
Transform faults, 8, 35
Transient response of structures, 302—305, 307, 308, 312—315
Transportation line, 374
Transverse motions at source, 167, 170, 171
Trigger models of earthquake occurrence, 198—205
Triggering, seismograph, 398
Tsunami damage, related with earthquake characteristics, 226—232

— diffraction, 263, 265
—, directional characteristics, 257—260
— distribution functions, 275—280
— draw-down, 271—275
— effect along coast, 263—275
— energy, 226, 229, 230
— entrance to specific locations, 278
—, general, 225, 226
— generation, theory of, 235—244
—, hydraulic models, 244—254, 268, 269
—, large motion of boundary, 248, 249
—, Mach reflection, 264—274
— magnitude, 226, 229, 230
—, moving boundary solutions, 244—248
—, numerical solutions, 249—254, 262, 263
— refraction, 227, 264
— resonance, 270, 271
— risk, 279, 280
— run-up, 271—275
—, soil- and rock-slide generation of, 232—235
— sources, 254—257
— - tide probabilities combined, 280—283
— travel across ocean, 260—263
— warning, 279
— wave spectra, 270, 271
— — trapping, 264
Turkey, 10, 33, 40—49
Two-state systems, analysis of risk, 342—348
— — —, design, 361, 362

U.S.S.R., 18
Underground acceleration records, 87
— explosions, 349
Underthrusting, tectonic, 8, 10, 13, 26, 216
Uniform Building Code, 351
Union Bay, Seattle, 88
United States, 182, 184, 195, 359, 370, 378
Unstable soil conditions, 73—76
Upcrossings, stochastic process, 306, 332
Urayasu, 87
Ursell's wave theory, 238—241
Utility, 2, 179, 357
— networks, 106
Utsu's law for aftershock sequences, 203, 204

$v_p/v_s$ anomaly, 20, 21, 25
Vaiont, 234
Valdez, 228
Velocity anomaly, 20, 21, 25
—, ground, 103, 166, 167, 173, 174, 188, 189, 220, 221
—, —, at source, 152, 159, 160, 162, 163, 164, 165, 166, 171, 173
— response spectra, 89, 92, 106, 188, 189, 289, 329
—, rupture propagation, 153, 156, 164, 165, 176
—, seismic wave, 16, 159, 160
—, — —, field measurement, 125—127
Vere-Jones models of earthquake occurrence, 200—203
Visible recording, 410
Volcanic activity recording, 383

Volcanism, 15, 18

Water supply lines, 374, 375, 376
Waterfront bulkheads, 115
Wave trapping, 168, 264
Waves, wind generated, 266
White noise, 293, 294, 295, 297, 302, 303, 305, 309, 310, 315, 330
— —, band-limited, 294, 295, 297, 302, 315
Wide-band processes, 301, 304, 306, 309, 310
Wind-generated waves, 266

Yakutat, 228

Zoning, 35, 36, 37, 370, 371